PROBAMAT-21st Century:
Probabilities and Materials

NATO ASI Series

Advanced Science Institutes Series

A Series presenting the results of activities sponsored by the NATO Science Committee, which aims at the dissemination of advanced scientific and technological knowledge, with a view to strengthening links between scientific communities.

The Series is published by an international board of publishers in conjunction with the NATO Scientific Affairs Division

A	Life Sciences	Plenum Publishing Corporation
B	Physics	London and New York
C	Mathematical and Physical Sciences	Kluwer Academic Publishers
D	Behavioural and Social Sciences	Dordrecht, Boston and London
E	Applied Sciences	
F	Computer and Systems Sciences	Springer-Verlag
G	Ecological Sciences	Berlin, Heidelberg, New York, London,
H	Cell Biology	Paris and Tokyo
I	Global Environmental Change	

PARTNERSHIP SUB-SERIES

1.	Disarmament Technologies	Kluwer Academic Publishers
2.	Environment	Springer-Verlag / Kluwer Academic Publishers
3.	High Technology	Kluwer Academic Publishers
4.	Science and Technology Policy	Kluwer Academic Publishers
5.	Computer Networking	Kluwer Academic Publishers

The Partnership Sub-Series incorporates activities undertaken in collaboration with NATO's Cooperation Partners, the countries of the CIS and Central and Eastern Europe, in Priority Areas of concern to those countries.

NATO-PCO-DATA BASE

The electronic index to the NATO ASI Series provides full bibliographical references (with keywords and/or abstracts) to more than 50000 contributions from international scientists published in all sections of the NATO ASI Series.
Access to the NATO-PCO-DATA BASE is possible in two ways:

– via online FILE 128 (NATO-PCO-DATA BASE) hosted by ESRIN,
Via Galileo Galilei, I-00044 Frascati, Italy.

– via CD-ROM "NATO-PCO-DATA BASE" with user-friendly retrieval software in English, French and German (© WTV GmbH and DATAWARE Technologies Inc. 1989).

The CD-ROM can be ordered through any member of the Board of Publishers or through NATO-PCO, Overijse, Belgium.

3. High Technology – Vol. 46

PROBAMAT-21st Century: Probabilities and Materials

Tests, Models and Applications for the 21st Century

edited by

George N. Frantziskonis

Department of Civil Engineering
and Engineering Mechanics,
University of Arizona,
Tucson, AZ, U.S.A.

Springer-Science+Business Media, B.V.

Proceedings of the NATO Advanced Research Workshop on
PROBAMAT-21st Century: Probabilities and Materials
Tests, Models and Applications for the 21st Century
Perm, Russia
September 10–12, 1997

A C.I.P. Catalogue record for this book is available from the Library of Congress.

ISBN 978-94-010-6196-4   ISBN 978-94-011-5216-7 (eBook)
DOI 10.1007/978-94-011-5216-7

Printed on acid-free paper

TABLE OF CONTENTS

PREFACE

There are numerous technological materials – such as metals, polymers, ceramics, concrete, and many others – that vary in properties and serviceability. However, the almost universal common theme to most real materials is that their properties depend on the scale at which the analysis or observation takes place and at each scale "probabilities" play an important role. Here the word "probabilities" is used in a wider than the classical sense. In order to increase the efficiency and serviceability of these materials, researchers from NATO, CP and other countries were brought together to exchange knowledge and develop avenues for progress and applications in the 21st century.

The workshop began by reviewing progress in the subject area over the past few years and by identifying key questions that remain open. One point was how to observe/measure material properties at different scales and whether a probabilistic approach, at each scale, was always applicable and advantageous. The wide range of materials, from wood to advanced metals and from concrete to complex advanced composites, and the diversity of applications, e.g. fatigue, fracture, deformation, etc., were recognized as "obstacles" in identifying a "universal" approach.

The hierarchical nature of materials and implications to modeling, testing, and applications were discussed extensively. Modern mathematical tools such as wavelet analysis, intersection methods for random sets, truncated series expansion of random fields, showed considerable promise in "attacking" the multiscale nature of materials in modeling and in testing. Multiscaling was also discussed within the context of nonlinear dynamics. Furthermore, the concepts of scaling and criticality and their applicability in material science created interesting points of view.

Presentations addressed several materials, i.e. wood, concrete, various metals, composites, ceramics, and porous geological media. Most lectures presented both experimental and analytical/numerical results. Few were purely theoretical, e.g. on homogenization of media, and few purely numerical, e.g. modeling scale effects in concrete structures. The possibility of identifying common "points" between various approaches towards modeling different materials was discussed.

In short, the NATO ARW meeting was a great success. Discussions followed each presentation, and extensive round tables were held at the end of each day. What has been accomplished in the subject area over the past few years was clear, and hints about where we are heading, i.e. in the 21st century, were the result of the discussions. An important outcome was the general recognition that the disciplines of Physics, Materials, and Mathematics need to be bridged further. Cross disciplinary efforts were identified as important for further progress in the subject area.

September, 1997   George N. Frantziskonis

Acknowledgments

It is a pleasant duty to thank the members of the Scientific Committee and of the Local Organizing Committee for their contribution before, during, and after the Workshop. Many thanks to NATO, the Russian Fund for Basic Research, and the University of Arizona for supporting the Workshop.

Scientific Organizing Committee

Prof. George N. Frantziskonis, Department of Civil Engineering and Engineering Mechanics, University of Arizona, Tucson, AZ 85721 USA

Prof. Oleg Naimark, Institute of Continuous Media Mechanics, Urals Branch of the Russian Academy of Sciences, 1 Acad. Korolev Str., 614061 Perm, Russia

Prof. Denys Breysse, Centre de Développement de Géosciences Appliquées, Univ. Bordeaux I, Avenue des Facultés, 33405 Telence Cedex, FRANCE

Dr. Leon Mishnaevsky Jr., State Material Testing Institute (MPA), University Stuttgart, Pfaffenwaldring 32, 70569 Stuttgart, GERMANY

Prof. Fabio Casciati, Dip. di Meccanica Strutturale, Via Abbiategrasso, 211, 27110 Pavia, ITALY

Local Organizing Committee

M. Davydova
O.Plechov
L.Filimonova
S.Uvarov
D.Eremeev
D.Naldaev
V. Kolot
M.Sokovikov
V.Kudrjashov
V. Barannikov
V.Leont'ev.

Contributors to This Volume

E.C. AIFANTIS, Laboratory of Mechanics and Materials, Aristotle University, 54006 Thessaloniki, GREECE

Gabriel AUVINET, Instituto de Ingenieria, UNAM, Ciudad Universitaria, Apdo. Postal 70-472, Coyoacan 04510 MEXICO D.F.

M. AVLONITIS, Laboratory of Mechanics and Materials, Aristotle University, 54006 Thessaloniki, GREECE

J. BIAREZ, Ecole Centrale Paris, Grande Voie des Vignes, 92295 Chatenay-Malarby, FRANCE

Ludmila BOTVINA, Baikov Institute of Metallurgy, 49 Leninsky Prospekt, Moscow RUSSIA

Matthew BILY, Institute of Materials and Machine Mechanics, 75 Racianska, 83606 Bratislava SLOVAKIA

Mark BLODGETT, WL/MLLP, Wright-Patterson AFB, 2230 Tenth St. Ste 1, Bldg. 655, Area B, Ohio 45433-7817 USA

F. BONTEMPI, Dip. di Mecanica Strutturale, Via Abbiategrasso, 211, 27100 Pavia ITALY

M. BORRI-BRUNETTO, Dip. Ingenieria Strutturale, Corso Duca degli Abruzzi 24, 10129 Torino ITALY

Denys BREYSSE, Centre de Développement de Géosciences Appliquées, Univ. Bordeaux I, Avenue des Facultés, 33405 Telence Cedex, FRANCE

Fabio CASCIATI, Dip. di Mecanica Strutturale, Via Abbiategrasso, 211, 27100 Pavia ITALY

Alberto CARPINTERI, Dip. Ingenieria Strutturale, Corso Duca degli Abruzzi 24, 10129 Torino ITALY

Patrick CASTERA, LRBB, Domaine de l'Hermitage, BP 10, 33610 CESTAS Gazinet, FRANCE

T. CHELIDZE, Institute of Geophysics, 1. Alexidze str, 380093, Tbilisi. GEORGIA

Claudio CHERUBINI, Instituto di geologia applicata e geotecnica, Via Re David, 200, 70125 Bari ITALY

Bernandino CHIAIA, Dip. Ingenieria Strutturale, Corso Duca degli Abruzzi 24, 10129 Torino ITALY

E. DAMASKINSKAYA, A.F. Ioffe Physico-Technical Institute of the Russian Academy of Sciences, K-21, Polytechnicheskaja 26, S-Petersburg, 194021, RUSSIA

Marina DAVYDOVA, Inst. of Continuous Media Mechanics, Urals Branch of the Russian, Academy of Sciences, 1 Acad. Korolev Str., 614061 Perm, Russia Perm, RUSSIA

J.L. FAVRE, Ecole Centrale Paris, Grande Voie des Vignes, 92295 Chatenay-Malarby, FRANCE

V. DI FEDERICO, D.I.S.T.A.R.T., Universita di Bologna, Viale Risorgimento 2, 40136 Bologna, ITALY

Krzysztof DOLINSKI, Institute of Fund. Technological Research, 00-049 Warsaw, Swietokrzyska 21, POLAND

Lucia FARAVELLI, Dip. Ingenieria Strutturale, Corso Duca degli Abruzzi 24, 10129 Torino ITALY

George FRANTZISKONIS, Department of Civil Engineering and Engineering Mechanics, University of Arizona, Tucson, AZ 85721 USA

O. GARISHIN, Inst. of Continuous Media Mechanics, Urals Branch of the Russian, Academy of Sciences, 1 Acad. Korolev Str., 614061 Perm, Russia Perm, RUSSIA

L. GOLOTINA, Inst. of Continuous Media Mechanics, Urals Branch of the Russian, Academy of Sciences, 1 Acad. Korolev Str., 614061 Perm, Russia Perm, RUSSIA

M. GOUNELLE, Ecole Supérieure de Physique et Chimie Industrielles, 10 rue Vauquelin, 75231 Paris, FRANCE

Vanazio GRECO, University of Calabria, 87030 Roges di Rende (Cs), ITALY

Y. GUEGUEN, Ecole Normale Superieure, 24 rue Lhomond, 75231 Paris, FRANCE

Dominique JEULIN, Ecole Nationale Supérieure des Mines, 351 rue St. Honore, 77300 Paris, FRANCE

F. HACHI, Ecole Centrale Paris, Grande Voie des Vignes, 92295 Chatenay-Malarby, FRANCE

L. KOZHEVNIKOVA, Inst. of Continuous Media Mechs., Urals Branch of the Russian, Acad. of Sciences., 1 Acad. Korolev Str., 614061 Perm, Russia Perm, RUSSIA

Achintya HALDAR, Dept. Civil Engineering, University of Arizona, Tucson AZ 85721 USA

Alex HANSEN, Gruppe for Teoretisk Fysikk, University of Trondheim, 7034 Trondheim NORWAY

Fatiha HATCHI, Ecole Centrale Paris, LMSSMat/ECP, 92295 Chatenay-Malabry FRANCE

Per HEMMER, Gruppe for Teoretisk Fysikk, University of Trondheim, 7034 Trondheim NORWAY

J. HUH, Department of Civil Engineering and Engineering Mechanics, University of Arizona, Tucson, AZ 85721 USA

N. ITAGAKI, Tohuku University, Dep. of Architecture, Faculty of Engineering, Sendai 980 JAPAN

I. KELLER, Dep. Mathematical Simulation, Perm State Technical University, JSP, Komsomol av. 29/a, 614600 Perm, RUSSIA

A. KLUEV, Dep. Mathematical Simulation, Perm State Technical University, JSP, Komsomol av. 29/a, 614600 Perm, RUSSIA

V KUKSENKO, A.F. Ioffe Physico-Technical Institute of the Russian Academy of Sciences, K-21, Polytechnicheskaja 26, S-Petersburg, 194021, RUSSIA

Jacques LAMON, LCT, Domaine Universitaire, 2 allée de la Boëtie, 33600 Pessac

FRANCE

Maurice LEMAIRE, IFMA/La Rama, BP 265, 63175 Aubiere Cedex FRANCE

A.L. Maistrenko, Institute for Superhard Materials, Artozavododskaya 2, 254074 Kiev UKRAINE

M. LE RAVALEC, Geosciences Rennes, University Rennes 1, Campus Beaulieu, Bat. 15, 35042, Rennes, FRANCE

N. LISSART, LCT, Domaine Universitaire, 2 allée de la Boëtie, 33600 Pessac FRANCE

Alexander MALKIN, Institute of Applied Mechanics, 32-A Lenin Ave., 117334 Moscow RUSSIA

Hirozo MIHASHI, Tohuku University, Dep. of Architecture, Faculty of Engineering, Sendai 980 JAPAN

Leon MISHNAEVSKY Jr., State Material Testing Institute (MPA), University Stuttgart, Pfaffenwaldring 32, 70569 Stuttgart, GERMANY

Valery MOSHEV, Inst. of Continuous Media Mechanics, Urals Branch of the Russian, Academy of Sciences, 1 Acad. Korolev Str., 614061 Perm, Russia Perm, RUSSIA

Oleg NAIMARK, Inst. of Continuous Media Mechanics, Urals Branch of the Russian Acad. of Sciences, 1 Acad. Korolev Str., 614061 Perm, RUSSIA

Shlomo NEUMAN, Hydrology and Water Resources, University of Arizona, Tucson, AZ 85721 USA

L. PETRINI, Dip. di Mecanica Strutturale, Via Abbiategrasso, 211, 27100 Pavia ITALY

Ruediger RACKWITZ, Technical University Münich, Arcisstr. 21, 80290 Münich GERMANY

Stephane ROUX, Ecole Supérieure de Physique et Chimie Industrielles, 10 rue Vauquelin, 75231 Paris, FRANCE

Karam SAB, IFMA/La Rama, BP 265, 63175 Aubiere Cedex FRANCE

A. SHANYAVSKII, State Center of Flight Safety of Civil Aviation, 103340, Airport "Sheret'evo", Moscow, RUSSIA

S. SCHMAUDER, State Material Testing Institute (MPA), University Stuttgart, Pfaffenwaldring 32, 70569 Stuttgart, GERMANY

Anne TANGUY, Ecole Supérieure de Physique et Chimie Industrielles, 10 rue Vauquelin, 75231 Paris, FRANCE

N. TOMILIN, A.F. Ioffe Physico-Technical Institute of the Russian Academy of Sciences, K-21, Polytechnicheskaja 26, S-Petersburg, 194021, RUSSIA

Petr TRUSOV, Dep. Mathematical Simulation, Perm State Technical University, JSP, Komsomol av. 29/a, 614600 Perm, RUSSIA

J. VAN MIER, Delft University of Technology, Stevin Laboratory, P.O. Box 5048, 2600 GA Delft, NETHERLANDS

A. VERVUURT, Delft University of Technology, Stevin Laboratory, P.O. Box 5048, 2600 GA Delft, NETHERLANDS

M. ZAISER, Max-Planck-Institut fur Metallforschung, Institut fur Physic, P.O. Box 800 665, D-70506 Stuttgart, GERMANY

SIMULTANEOUS FAILURES IN FIBER BUNDLES

ALEX HANSEN and PER C. HEMMER
Institutt for Fysikk
Norges teknisk-naturvitenskapelige universitet
N-7034 Trondheim, Norway

Abstract

We discuss two models for the failure of bundles of parallel fibers under increasing load. By studying the burst distribution, i.e., the distribution of the number of fibers that simultaneously fail when the external load is controlled, we show analytically for one of them how the breakdown process approaches a critical point.

1. Introduction

There is in the physics community an increasing interest in fracture processes. Much of the work that has so far originated from this interest is based on ideas from statistical physics [1]. In this paper we demonstrate how certain aspects of the failure of fiber bundles under increasing load may be analysed using the language of second order phase transitions.

When a weak structural element in a material with stochastically distributed strengths fails, the increased load on the remaining elements may cause further ruptures, and thus induce a burst avalanche of a certain size Δ, i.e., one in which Δ elements fail simultaneously. When the load is further increased, new avalanches occur. The distribution of avalanche sizes, either at a fixed load, or the cumulative distribution from zero load until complete break-down of the material, depends on several factors, in particular the threshold strength distribution and the mechanism for load sharing between the elements.

We will in this paper discuss the distribution of such avalanches of simultaneous failures depending on 1) how the forces redistribute themselves in the material after the failure of a structural element, and 2) the distribution of strengths of the structural elements.

The interplay between the changing force distribution and the distribution of strengths of the structural materials generates correlations in the failure process that typically renders it inaccessible with analytical methods. Rather, one has to resort to numerical computations. However, some configurations of the structural elements are simple enough to make a complete analysis possible. We consider such configurations in this paper.

Consider N fibers of equal length l_0 and clamped at both ends (Fig. 1) All fibers have the same elastic constant, while the maximum load they can sustain, t, is picked from a cumulative probability distribution

$$\text{Prob}(t' < t) = P(t) = \int_0^t p(u)du . \tag{1}$$

1

G. N. Frantziskonis (ed.), PROBAMAT – 21st Century: Probabilities and Materials, 1–10.

2

This is a quite realistic model for long flexible cables or low-twist yarn: The assumption that all the disorder in the model appears in the strength distribution rather than in the elastic constants, may be argued by noting that the effective elastic constant of a single fiber is essentially the average of the local elastic constant along the fiber, while its strength is determined by its weakest point [2,3]. A generalization of this model in terms of series-parallel systems may be found in e.g. [4,5].

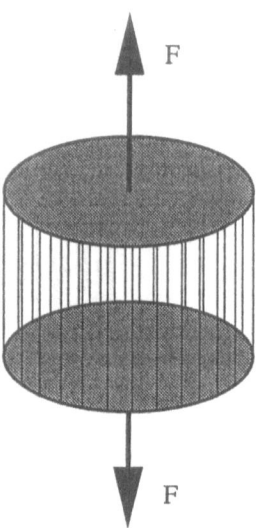

Figure 1. A fiber bundle with periodic boundary conditions. The externally applied force F is the control parameter.

For simplicity, we set the elastic constants equal to unity. If we then stretch fiber i to a length $l_0 + x$, it responds with a force f_i given by

$$f_i = \begin{cases} x & \text{, if } x < t_i; \\ 0 & \text{, if } x \geq t_i. \end{cases} \tag{2}$$

Let us define x_K as the *ordered* sequence of the failure thresholds t_i: $x_1 \leq x_2 \leq x_3 \leq \ldots \leq x_{N-1} \leq x_N$. Since we are assuming equal load sharing — that is, when a fiber breaks, the load it carried is equally distributed among the surviving fibers

— the total load on the fiber bundle when the Kth fiber is about to fail is

$$F_K = (N + 1 - K)x_K . \tag{3}$$

The average load-elongation characteristics, averaging over an ensemble of fiber bundles, is

$$\langle F \rangle / N = \langle f \rangle = [1 - P(x)]x , \tag{4}$$

where we have used that $P(x_K) \simeq K/(N + 1)$.

This model has been much studied [6-12] since the early result of Daniels [3], who showed that for large N the distribution of the *maximum* strength of fiber bundles, $S = \max_K F_K$, is gaussian around the value

$$\langle S \rangle / N = \langle s \rangle = \max_x [1 - P(x)]x . \tag{5}$$

The assumption of equal load-sharing among surviving fibers is often unrealistic, and it is natural to consider models in which the extra stresses by a fiber rupture are taken up by the fibers in the immediate vicinity. The extreme version is to assume that only the *nearest-neighbor* surviving fibers take part in the load-sharing. In a one-dimensional geometry, as in Fig. 1, precisely two fibers, one on each side, share the extra stress. When the strength thresholds take only two values, the bundle strength distribution has been found analytically [13-15].

In the next section we discuss the distribution of burst avalanches in the equal loading-sharing model. In Section 3, we review some of the analytical results recently obtained for the local load-sharing model. In contrast to the equal load-sharing model, the burst distribution does *not* follow a power law. We make some final remarks in Section 4, including pointing out that numerical investigations of two-dimensional models network models places them within numerical uncertainty in the universality class of the equal load-sharing model.

2. The Burst Distribution in the Equal Load-Sharing Model

The property of the fiber bundle model of interest in the present context, is the *burst distribution.* In order to define this property, note that the sequence of external loads F_K is not monotonously increasing. This may be readily seen from equation (3); the total load is the product of a monotonously increasing *fluctuating* quantity x_K and a monotonously *decreasing* quantity $(N + 1 - K)$. Suppose now that our control parameter is the total load F, and that $K - 1$ fibers have broken. In order to be in this situation, $F > F_K > F_J$ for all $J < K$. The latter inequality ensures that the situation we are studying is not unstable. We increase F until it reaches F_K, at which fiber K breaks. If now $F_{K+1} \leq F_K$, then fiber $K + 1$ will also break without the external load F being further increased. The same may be true for F_{K+2} and so on until the $(K + \Delta - 1)$th bond breaks. Thus, $F_{K+J} \leq F_K$ for $J < \Delta$. If now $F_{K+\Delta} > F_K$, the avalanche of breaking bonds then stops at this point, and we have experienced a burst event of size Δ.

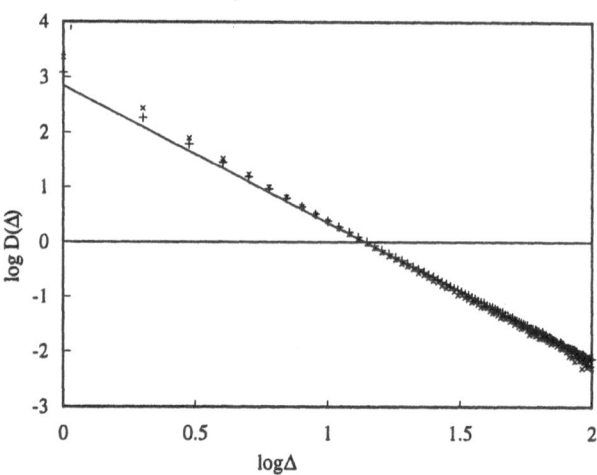

Figure 2. The burst distribution $D(\Delta)$ for the equal load-sharing model from numerical simulations: (+) 50 000 samples, each with $N = 5000$, and a threshold distribution $P(t) = t$, $0 < t < 1$; (×) 10 000 samples, each with $N = 5000$, and a threshold distribution $P(t) = t^{1/3}$, $0 < t < 1$. The log scale used is base 10.

It was shown by Hemmer and Hansen [16] that the average number of burst events of size Δ per fiber, $D(\Delta)/N$, follows a power law of the form

$$D(\Delta)/N = C\Delta^{-\tau} \tag{6}$$

in the limit $N \to \infty$. Here,

$$\tau = \tfrac{5}{2} \tag{7}$$

is the *universal* burst exponent. The value (7) is, under very mild assumptions, independent of the threshold distribution $P(t)$. We demonstrate this in Fig. 2. The prefactor C in (6) is given by

$$C = \frac{x_c p(x_c)^2}{\sqrt{2\pi}[x_c p'(x_c) + 2p(x_c)]} , \tag{8}$$

where x_c is the solution of the equation

$$x_c p(x_c) = 1 - P(x_c) , \tag{9}$$

and is the value of x for which the characteristics (4) has a maximum. Equations (6) to (9) were derived in [16] by detailed combinatorial arguments. An alternative derivation of the same result based on mappings between the fiber bundle problem and a Brownian process has been presented in Ref. [17].

The probability $\Phi(\Delta, x)$ that a burst event at elongation x will have the size Δ is

$$\Phi(\Delta, x) = \frac{\Delta^{\Delta-1}}{\Delta!} \frac{m(x)}{1 - m(x)} \left[[1 - m(x)] e^{m(x)-1} \right]^{\Delta} , \tag{10}$$

where

$$m(x) = 1 - \frac{x p(x)}{1 - P(x)} . \tag{11}$$

Note in particular that by equation (9) $m(x_c) = 0$. Let us now assume that we do not load the fiber bundle until complete collapse, i.e., at $x = x_c$, but stop at a value $x_s < x_c$. We may then ask for $D(\Delta, x_s)/N$, the expected number of burst events of size Δ that occurs between $x = 0$ and $x = x_s$ during the breakdown process. This is given by the integral

$$\begin{aligned}
\frac{D(\Delta, x_s)}{N} &= \int_0^{x_s} p(x) dx \; \Phi(\Delta, x) \\
&= \frac{\Delta^{-3/2}}{\sqrt{2\pi}} \int_0^{x_s} dx \; p(x) \frac{m(x)}{1 - m(x)} \left[[1 - m(x)] e^{m(x)} \right]^{\Delta} ,
\end{aligned} \tag{12}$$

where on the right-hand side the Stirling approximation $\Delta! \approx \sqrt{2\pi} \Delta^{\Delta+1/2} e^{-\Delta}$ for large Δ has been used. The integrand in (12) is strongly peaked near $x = x_c$. We therefore expand it to second order in $y = x_c - x$ to find

$$\frac{D(\Delta, x_s)}{N} = \frac{\Delta^{-3/2}}{\sqrt{2\pi}} p(x_c) m'(x_c) \int_{x_c - x_s}^{\infty} dy \; y e^{-m'(x_c)^2 y^2 \Delta/2} , \tag{13}$$

where we have extended the upper integration limit to ∞. We may do this integral to get

$$\frac{D(\Delta, x_s)}{N} = C \Delta^{-5/2} e^{-m'(x_c)^2 \Delta (x_c - x_s)^2/2} , \tag{14}$$

where C is defined in Eq. (8).

We may write (14) in scaling form,

$$\frac{D(\Delta, x_s)}{N} = \Delta^{-\tau} G(\Delta, x_s) = \Delta^{-\tau} G\left(\Delta^{\sigma} (x_c - x_s) \right) , \tag{15}$$

where

$$G(y) = C \, e^{-m'(x_c)^2 y^2/2} . \tag{16}$$

In particular $G(y)$ tends to the constant C for $y \to 0$. Two universal critical exponents appear, $\tau = 5/2$, Eq. (7), and

$$\sigma = \tfrac{1}{2}. \tag{17}$$

In this sense the fracture process of the fiber bundle approaches a critical point at total breakdown: The distribution of burst events follows a power law with an upper cutoff that diverges as the bundle approaches total failure.

The equal load-sharing fiber bundle model may be reinterpreted as a mean-field version of the Burridge-Knopoff model [18] for earth quakes [19]. In [19] there is an attempt to derive the burst distribution (6) from the assumption that F_K may be directly interpreted as a biased random walk. The precise nature of this random walk is elucidated in [17]. It is a peculiar asymmetric walk with variable step length. In the limit $N \to \infty$ and continuous time variable $K/N \to t$, this random walk may be mapped onto a continuous Brownian process. Such Brownian processes have been studied in Refs. [8–10] in connection with the distribution of the strength S of fiber bundles.

3. Fiber Bundles with Local Load Sharing

The values of the two critical exponents τ and σ of the scaling form (14) of the burst distribution are universal with respect to the threshold distribution $P(x)$. As has been shown by Ding and Lu [20,21], letting the elastic properties of the single fibers constituting the fiber bundle vary from fiber to fiber does not change the form (14). Nor does adding random prestresses to the fibers. Thus, the distribution (14) is quite robust.

However, what happens if we replace *equal* load sharing by *local* load sharing? This *does* change the universality class. This was demonstrated numerically in Refs. [22,23]. In Ref. [24] the burst distribution was determined analytically for a uniform threshold distribution in the interval $0 \le t \le 1$.

The simplest geometry in which to study this model is one-dimensional so that the N fibers are ordered linearly, without or with periodic boundary conditions (Fig. 1). In this case precisely two fibers, one on each side, take up, and divide equally, the extra stress. At a total force F_{tot} on the bundle the force on a fiber surrounded by n_l previously failed fibers on the left-hand side, and n_r on the right-hand side, is then

$$\frac{F_{tot}}{N}\left(1 + \tfrac{1}{2}(n_l + n_r)\right) = x(2 + n_l + n_r). \tag{18}$$

Here

$$x = \frac{F_{tot}}{2N}, \tag{19}$$

one-half the force-per-fiber, is a convenient variable to use as the driving *force parameter*. This model has been discussed previously in a different context [13–15,25–27].

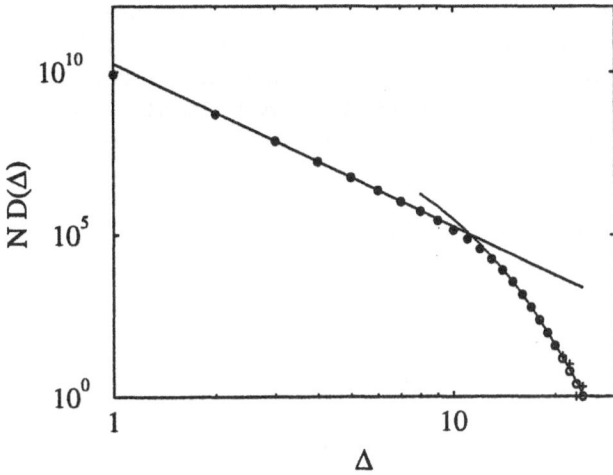

Figure 3. Burst distribution in local model as found numerically for 4 000 000 samples with $N = 20000$ fibers (+), and as calculated in [24] (o). The straight line shows the power law Δ^{-5} and the broken curve the function $\exp(-\Delta/\Delta_0)$ with $\Delta_0 = 1.1$. Note the small value of Δ_0. (From [24].)

Avalanches in the local and the equal load-sharing models have different characters. In the local model an avalanche unroll with one failure acting as the seed. If many neighboring fibers have failed, the load on the fibers on each side is high, and if they burst, the load on the new neighbors will be even higher, etc. In this way a weak region in the bundle may be responsible for the failure of the whole bundle. For a large number N of fibers the probability of a weak region somewhere is high, and this explains in a qualitative way that the maximum load the bundle are able to carry does not increase proportional to N, but slower than linear.

The load distribution rule (18) implies that an avalanche of size Δ does necessarily lead to a complete breakdown of the whole bundle if the external force is too high, i.e., if x exceeds a critical value x_{max}. Since here a fiber can at most take a load of unity, we have

$$x_{max} = \frac{1}{\Delta + 2} \, . \tag{20}$$

The burst distribution for this model (for a uniform threshold distribution) was found in Ref. [24]. We show in Table 1, $D(\Delta)/N$ for a bundle of $N = 20\,000$, together with simulation results for 4 000 000 bundles, each having 20 000 fibers. The agreement between the simulation data and the theoretical data is extremely satisfactory.

An analysis of the burst distribution obtained for this local model shows that the distribution does not follow a power law except for small values of Δ. If one nevertheless does a linear regression analysis on this part of the data set, the effective power would be of the order 5, considerable larger than the value found for the equal load-sharing model, Eq. (7). We show such a fit in Fig. 3. This effective exponent increases with increasing N [23].

TABLE 1. The burst distribution $ND(\Delta)$ for the local model with a bundle of $N = 20000$ fibers. The simulation results are based on 4 000 000 samples. The calculated values are based on the analytical results of Ref. [24]. (From [24].)

Δ	Simulation	Calculation
1	8 327 378 752	8 327 331 808
2	491 305 573	491 331 178
3	72 126 803	72 114 644
4	17 179 080	17 180 414
5	5 590 887	5 591 243
6	2 243 916	2 243 012
7	1 030 833	1 031 678
8	515 309	515 310
9	268 589	268 139
10	140 911	140 751
11	72 251	72 701
12	36 525	36 277
13	17 523	17 285
14	8 015	7 835
15	3 352	3 392
16	1 442	1 418
17	559	579
18	223	233
19	90	93.8
20	40	37.5
21	18	15.0
22	10	6.0
23	1	2.4
24	2	1.0
25	0	0.4

4. Conclusion

We have in this article discussed burst distributions in fiber bundles with two different mechanisms for load distribution when fibers rupture, viz. equal or extremely local load redistributions.

The main results are the following:

(1) For the equal model the burst distribution follows a universal power law $\Delta^{-5/2}$.

(2) For the local model the burst distribution falls off with increasing burst size much faster than for the equal model, and does not follow a power law.

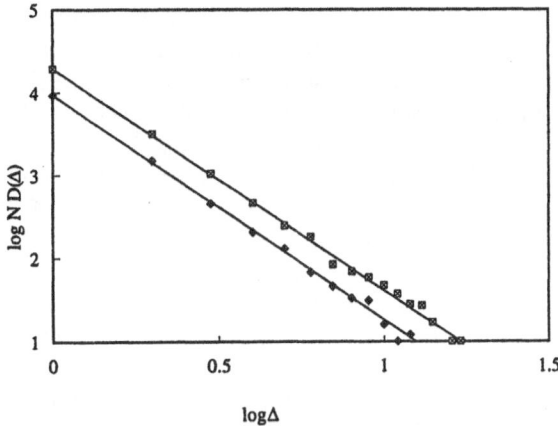

Figure 4. The distribution of bursts, $ND(\Delta)$, for square lattices of fuses, each of unit resistance, and with a current treshold picked from a treshold distribution $P(x)$ in the interval ($0 \leq x \leq 1$), where $P(x) = x$ (squares) and $P(x) = x^2$ (diamonds). The slope of the straight lines is -2.7. The simulation results are based on 100 samples of lattices of size 40×40. The log scale used is base 10. (From [22].)

The models with parallel fibers are essentially one-dimensional. What happens in higher dimensions? We have in [22] studied a square lattice of electrical fuses with stochastically distributed current tresholds. Orientation of the lattice at 45° with respect to two parallel bus bars ensures that all fuses carry the same current when all are intact. As the voltage is increased, bursts of collective burn-outs of the fuses occur. Based on simulation of several hundred lattices of size up to 50×50 we found a burst size distribution $D(\Delta) \sim \Delta^{-\tau}$ with $\tau = 2.7 \pm 0.3$ for the different threshold distributions we studied, see Fig. 4. It should be noted how close this value is to the one for the equal load-sharing model ($\tau = 5/2$). It was in Ref. [22] and in [28] argued that this might be due to the breakdown process in two dimensions being a highly correlated one [1] in which spatially close fuses tend to blow successively, especially near the end of the process. We conclude that the combined effect of dimensionality and range of load-sharing is yet to be fully understood.

References

1. H. J. Hermann and S. Roux, eds. *Statistical Models for the Fracture of Disordered Media* (North-Holland, Amsterdam, 1990).
2. F. T. Peirce, J. Text. Ind. **17**, 355 (1926).
3. H. E. Daniels, Proc. Roy. Soc. London **A183**, 405 (1945).
4. R. L. Smith and S. L. Phoenix, ASME J. Appl. Mech. **48**, 75 (1981).
5. V. V. Moshev and S. E. Evlampieva, Polym. Eng. and Sci. **37**, 1348 (1997).
6. B. D. Coleman, Trans. Soc. Rheology, **1**, 153 (1957).
7. R. L. Smith, Ph. D. Thesis, Cornell University, (1979).
8. S. L. Phoenix and H. M. Taylor, Adv. Appl. Prob. **5**, 200 (1973).
9. H. E. Daniels and T. H. R. Skyrme, Adv. Appl. Prob. **17**, 85 (1985).
10. H. E. Daniels, Adv. Appl. Prob. **21**, 315 (1989).
11. D. Sornette (1989), J. Phys. A **22**, L243.
12. D. Sornette and S. Redner, J. Phys. A **22**, L619 (1989).
13. D. G. Harlow, Proc. Roy. Soc. Lond. Ser. A **397**, 211 (1985).
14. D. G. Harlow and S. L. Phoenix, J. Mech. Phys. Solids **39**, 173 (1991).
15. P. M. Duxbury and P. M. Leath, Phys. Rev. B **49**, 12676 (1994).
16. P. C. Hemmer and A. Hansen, ASME J. Appl. Mech. **59**, 909 (1992).
17. A. Hansen and P. C. Hemmer, Trends in Statistical Physics **1**, 213 (1994).
18. R. Burridge and L. Knopoff, Bull. Seismol. Soc. Am. **57**, 341 (1967).
19. D. Sornette, J. Physique I, **2**, 2089 (1992).
20. E. J. Ding and Y. N. Lu, Phys. Rev. Lett. **70**, 3627 (1993).
21. E. J. Ding and Y. N. Lu, Commun. Theor. Phys. **19**, 283 (1993).
22. A. Hansen and P. C. Hemmer, Phys. Lett. A **184**, 394 (1994).
23. S. D. Zhang and E. J. Ding, Phys. Lett. A **193** 425 (1994).
24. M. Kloster, A. Hansen, and P. C. Hemmer, Phys. Rev. E **56**, 2615 (1997).
25. D. G. Harlow and S. L. Phoenix, Int. J. Fracture **17**, 601 (1981).
26. S. L. Phoenix and R. L. Smith, Int. J. Sol. Struct. **19**, 479 (1983).
27. C. C. Kuo and S. L. Phoenix, J. Appl. Prob. **24**, 137 (1987).
28. S. Zapperi, P. Ray, H. E. Stanley, and A. Vespignani, Phys. Rev. Lett. **78**, 1408 (1997).

SYNERGETICAL MODELS OF FATIGUE-SURFACE APPEARANCE IN METALS: THE SCALE LEVELS OF SELF-ORGANIZATION, THE ROTATION EFFECTS, AND DENSITY OF FRACTURE ENERGY

A.A. SHANYAVSKII

State Center of Flight Safety of Civil Aviation,
103340, Airport "Sheremet'evo", Moscow, Russia.

1. Introduction

In-service cyclic loading of a construction element may induce evolution of its structure on the microscopic-scale level, associated with the achievement of the critical dislocation density. Damage is accumulated in the material whose stress state, applied-loading frequency spectrum or R magnitudes, etc. may vary in a complicated way, to differ from a laboratory experiment. Storage and dissipation of energy are the two concurrent competing processes experienced by the material under loading. The material condition is evolving parallel to continued energy exchange with the environment until the crack has nucleated and grown to a critical size.

Numerous investigations of the dislocation structure of a metal indicate that the accumulating process of damage in the metal during its cyclic loading appears ordered and self-organized [1-7]. The critical condition of a material, which corresponds to the initiation of the crack and its growth during a loading cycle, may be associated with the level of damage or defect density that will be the same whatever different are the ways or conditions of the cyclic loading. Based on this idea, one can use a single, synergetical approach [8] to the analysis of the crack-growth stage .

G. N. Frantziskonis (ed.), PROBAMAT – 21st Century: Probabilities and Materials, 11–44.
© 1998 *Kluwer Academic Publishers.*

A metal with a growing crack represents an open dynamic system, which is far from equilibrium [9]; the system is exercising a series of sequential transitions from one to another stability state and the continued energy exchange with the environment. Once the system has come to a certain critical condition, the homogeneous equilibrium is not stable any more; therefore, inhomogeneities appear in the system, to be defined as dissipative structures [10]. With the dissipative structures thus formed, the inhomogeneous state of the open system acquires stability against small disturbances. In an open system, they recognize the stability state of two kinds, homogeneous or inhomogeneous. It is predominantly in the homogeneous stability state that one dissipative structure is replaced by another, which results in continuous evolution of the open system. Therefore, an open system must keep stable as long as a certain dominating mechanism of damage accumulation persists within a certain period of time, the mechanism being characteristic of a dominating type of dissipative structure.

As cyclic loading of a construction element is continuing, mechanisms of damage accumulation replace one another sequentially, each starting and keeping on for a certain time. Then, the contribution of a new mechanism may alternatively grow or cease. An open system evolves by passing through the critical states, referred to as the bifurcation points, to, alternatively, a stability or instability condition [10]. As long as the system experiences fluctuations, it cannot avoid instability immediately before a bifurcation point. The newly activated processes of damage accumulation develop or, alternatively, die out, depending on whether the system is able to the self-organized absorption of energy in the ways that shift the construction element toward a greater stability, i.e., longer life times. The general principle of self-organization is that altering the parameters of the extrinsic (applied) influence (the kind or way of cyclic loading) results not in the occurrence of the structures hierarchy but rather in switching on the fracture mechanisms operative in the system. Depending on the conditions of cyclic loading, the mechanisms may be complicated to a greater or smaller extent, and the system, change from the lower to higher level of self-organization.

In a material capable of resistance to cyclic loading the reason for the self-organized transitions through the bifurcation points is to only make operative the *least-fracture mechanism*[1]; hence, the least energy fraction is given, per loading cycle, to the formation of free surface. Such a behavior requires involvement of increasingly complicated concurrent processes.

2. The levels and hierarchy of the self-organization processes

The evolution of an open system is commonly discussed [9] in the terms of microscopic, mesoscopic, or macroscopic scale levels. The first is relevant to the effects on the scale of atomic spacing; the second, to the behavior of atomic ensembles, and the third, to the creation of bulky space structures.

At the near-threshold range of crack-growth rates (CGR) the least crack-length increment a_q per loading cycle only measures several lattice parameters [7].

Hence, we define the scale level of this stage of the crack growth as a microscopic one. At this stage, the crack-growth behavior is quite sensitive to the microstructure of the material [11], and dislocation slip is dominating. Depending on the material state, the fracture mode can be intercrystalline or transcrystalline. One can illustrate the situation by the fracture behavior of the lamellar structure of two-phase titanium alloys: the transition to a coarser scale level occurs as soon as the crack-tip plastic zone becomes larger than the average effective diameter of the subgrains [12]. This is the instant for the crack to decelerate, i.e., for the step-wise decrease of the crack-rate exponent in the Paris-type formula.

At the mesoscopic scale level, CGR is completely determined by the plastic-zone size. Here, fracture develops self-organized for creating the plastic zone in the three-dimensional stress state, whichever the way or conditions of cyclic loading. The finite-element calculations indicate [13] that, with the same crack-length, the transition from uniaxial to biaxial loading will alter the constriction extent of plastic deformation at

---

[1] The deformation mechanism that helps reducing the fracture scale to the possible lowest level.

the crack tip, the three-dimensional stress state preserved of the material.

The microscopic (Stage I) and mesoscopic (Stage II) scale levels of crack growth are in common with one another as concerns the subjects of forming shear lips at the free surface of the workpiece [14], microscopic tunneling of the crack [7], or combining shear and cleavage in the metal when subjected to uniaxial tension (Fig. 1). Isolated regions of the failed metal, formed all along the crack front; are stretched in the crack-growth direction and separated by unbroken crosspieces. Microtunnels are formed by shear during Stage I, and the crosspieces can fail, during Stage I, by the type-III shear or by growing a crack from one to other tunnel just like growing the tunnels themselves. In fact, clusters of the failed metal are formed at the fracture surface, arranged into a chain of the joined-to-one-another frustrons (Fig. 2). Together, they form the pattern of *Kelly tree* to prove the synergistic nature of the growth of fatigue cracks in metals.

However, rotational instability of deformation and fracture of the intertunnel crosspieces may become the case at Stage II [15]. Changing from the shear- to rotation-type instability in the crosspiece is associated with further complication of the way in that energy is being absorbed in the material before fracture.

Figure 3 illustrates the appearance of rotational instability in a crosspiece between microtunnels. One can easily see how a cylindrical element is being rotated around the longitudinal axis of the crosspiece or how the same element, being already a separate particle, is rotated due to the rolling effect of the stress component τ_{III}. Changing for the rotational fracture mode of the crosspieces results in the formation of spherical particles under the whatever, uniaxial-tension or biaxial regular [15] or irregular [16], loading conditions. The spherical particles are positioned in between the matrix faces formed as a result of fracture along the boundaries of the rotational-instability volumes. These particles counteract the main-crack opening to maintain the general principle of the kinetic self-organization of fatigue fracture, i.e., the least-fracture behavior.

It was fractographically confirmed that, at Stage I, the crack growth in microtunnels is associated with the development of slip: multiple-slip traces, slip steps,

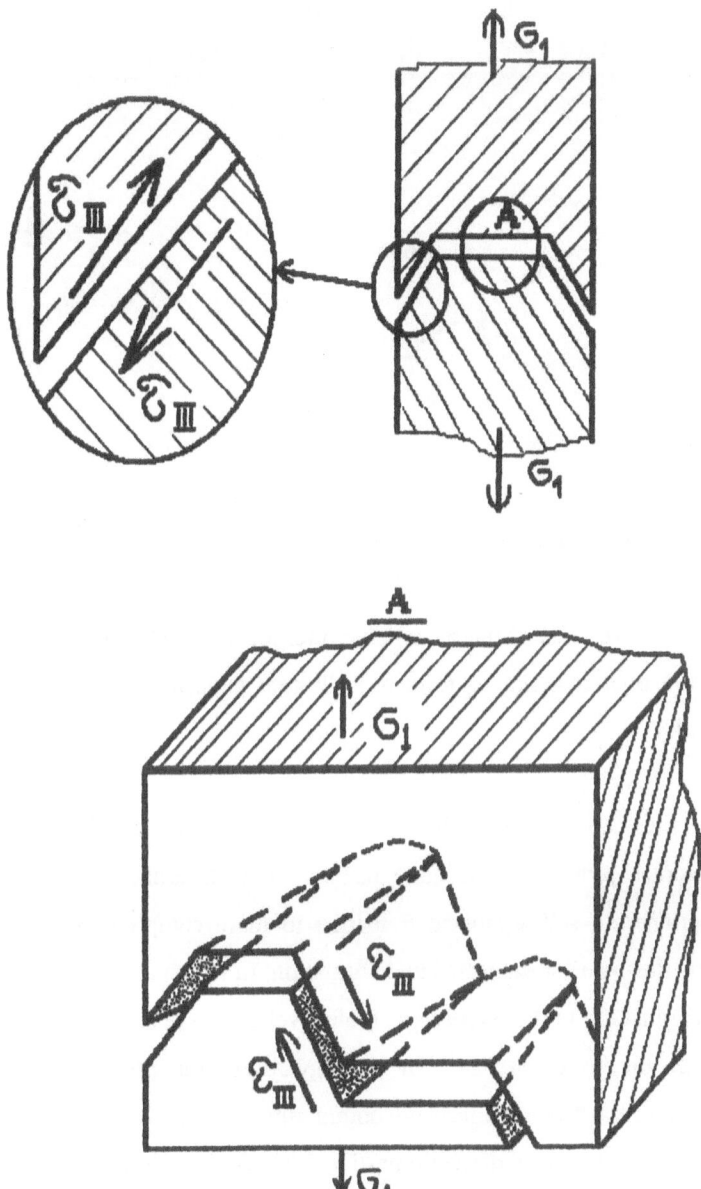

Figure 1. Combined mode-I and III opening of a fatigue crack under uniaxial-stretch conditions.

16

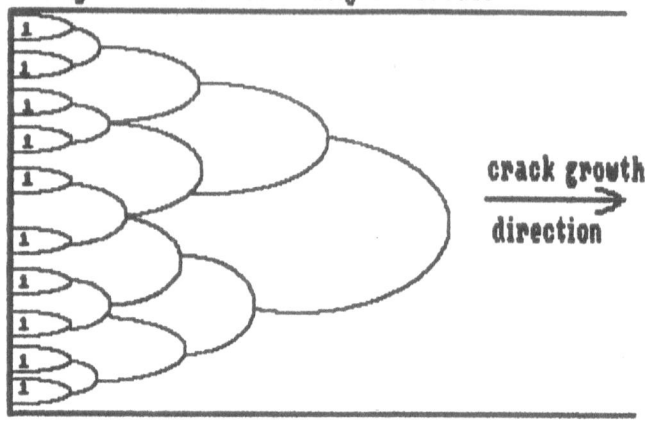

1 - places of microtunnels
along the specimen notch

Figure 2. Areas (1) of the mode-I fracture, joined to one another by the mode-III fracture of a material to form a Kelly-tree pattern of the fracture surface. (Schematic).

or extrusion sites can be seen at the background of the pseudo-striations pattern [7]. The fatigue-crack propagation is quite fast in the microtunnels. Consequently, the system experiences a self-organized transition to more complicated ways of energy absorption by the material, subjected to deformation, in which new free surface is being formed; this transition to the mesoscopic scale level occurs once the critical conditions at the crack tip were created. The energy-absorption process becomes more complicated since the rotation effects are dominating in the deformation and fracture of the material. There must be a direct proportionality between the energy amounts stored

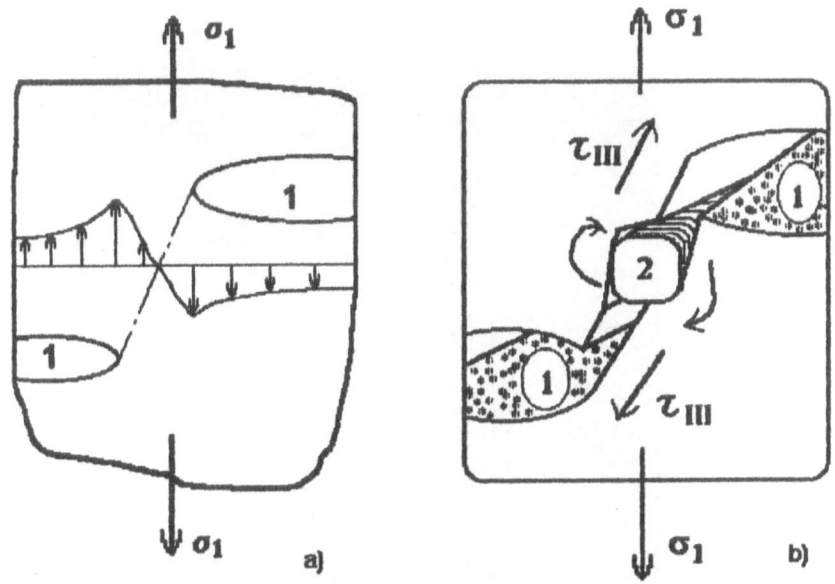

Figure 3. (*a*) and (*b*) the scheme of the occurrence of rotational instability in the intertunnel crosspieces of a material; (*c*) and (*d*) real fracture patterns of D16T Al-based alloy in the vicinity of stress-riser tip under unidirectional cyclic tension.

18

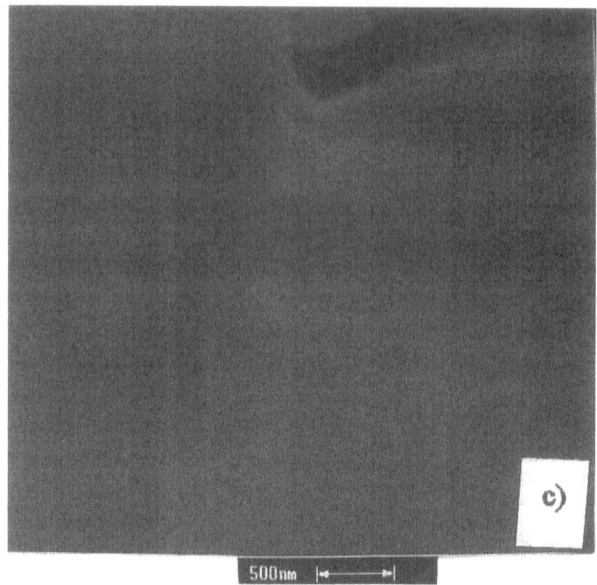

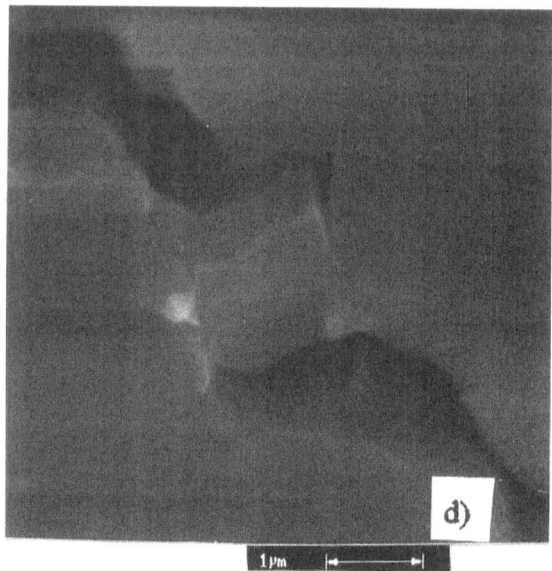

Figure 3. Continued

with plastic deformation and those dissipated at the crack tip due to the formation of free surface.

As long as the deformation mechanism is preserved, a larger plastic-zone size will correspond to a greater absorbed level of plastic-strain energy. Consequently, the crack increment will be greater within that zone. The rotational-instability phenomenon is a kind of accommodation [17] as long as it does not result in the formation of new free surface in the material; therefore, it acquires critical importance as the process of plastic deformation at the crack tip at the mesoscopic scale level. Owing to this effect, substantially greater amount of energy is absorbed at less extensive growth of the plastic zone; therefore, fracture is hampered thanks to the self-organization effects in the material.

The patterns of acoustic emission [18] were analyzed within the Elber's range of crack opening; based on those data, Stage II of crack-growth was described [19], at which the dominating fracture mechanism shows itself through the formation of fatigue striations. According to the description, plastic deformation develops, with growing applied load, within the shear lips or intertunnel crosspieces of the material. The cyclic crack opening appears *pseudo-elastic* (completely reversible) at the tunnel tips, whereas, in front of the tunnels, the material experiences rotational plastic deformation. A fresh shear-type fracture surface may be simultaneously formed, caused by the applied τ_{III} stress component, in the individual crosspieces. During the unloading half-cycle, the crack appears instantly added in length thanks to backward cracking that starts from the dislocation crack formed, by the peak loading, ahead of the main-crack tip.

In Fig. 4, this instant-cracking case is illustrated as controlled by rotational deformation. As the striation spacing is increasing with the total crack length, another complication occurs of the fracture process. Consequently, the striations become composite, concomitant with the increased crack-rate exponent in the Paris' formula. New details (local-cracking sites) are added to the fracture-surface morphology,

20

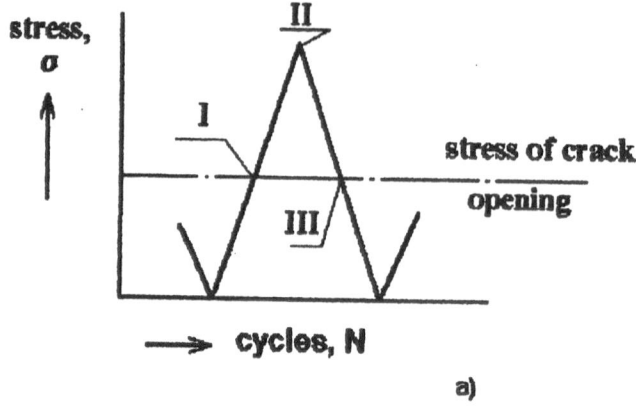

a)

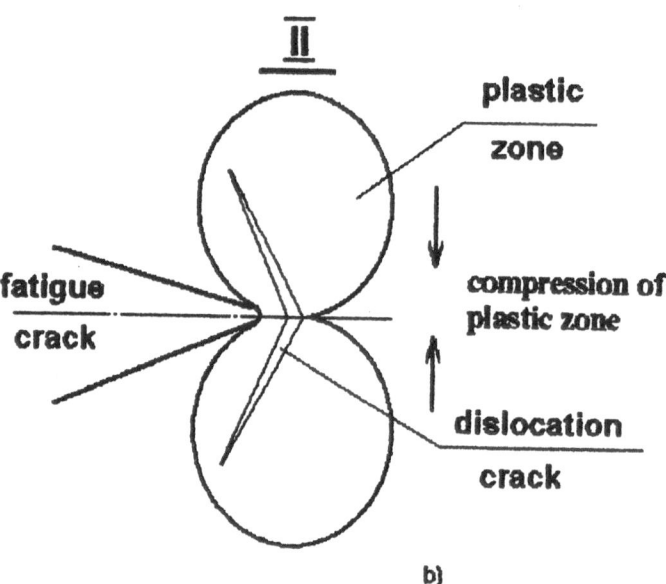

b)

Figure 4. The scheme (a-b) of the formation of fatigue striations on the stage of

(c) pseudo-elastic (V- K_I^2) or (d) pseudo-elastic-plastic (V- K_I^4) failure.

21

fatigue striation formation before spacing $\delta_i < 2\times10^{-7}$ m

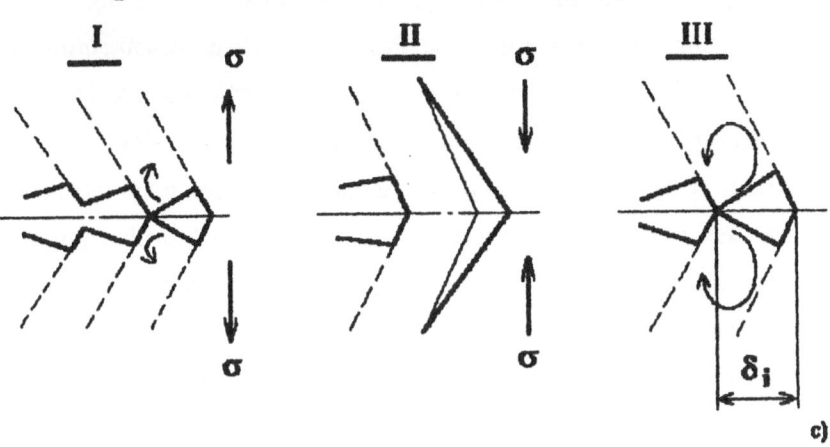

c)

fatigue striation formation after spacing $\delta_i > 2\times10^{-7}$ m

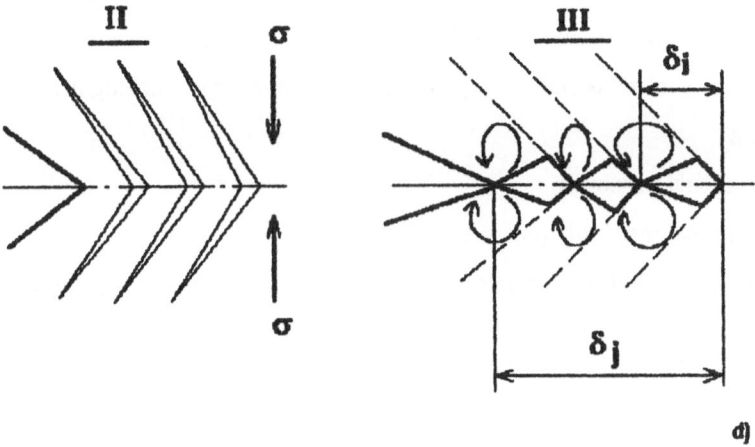

d)

Figure 4. Continued

attesting to the crack-branching effects provoked by additional slip. Patterns of ductile or brittle fracture become also visible, typical of unidirectional loading. Altogether, this is manifesting the transition from the linear (region IIa of *pseudo-elastic* fracture behavior) to nonlinear (region IIb of *pseudo-elastic-plastic* fracture behavior) [7] relationship between the crack-length and crack-increment magnitudes.

The reality of the change for the rotational effects at a crack tip at the mesoscopic scale level was confirmed through *in-situ* watching the deformation process at a fatigue-crack tip in a being stretched plate [20]. As follows from the micrographs, two slip systems became operative at the borders of the stretched element at the crack tip (Fig. 5). The instability element was revealed through its rotation. The rotation angle of 15° was measured from the micrographs published elsewhere [20]. The crack-tip propagation caused by unidirectional stretching manifested the fracture that had nothing in common with the fatigue-striation effects.

Dimple formation becomes more extensive by the termination of the stable-growth stage of the crack, indicating to the onset of a new instability; this brings the system to a next bifurcation and gives rise to the accelerated-fracture stage. One can ensure slow crack growth by keeping low the strain rate. However, the crack-length increment will substantially increase with the each new loading cycle. Thus, the hierarchy of the mechanisms involved in the self-organized kinetics of fatigue cracking is as follows,

- slip
- rotational instability
- rotational instability plus slip, and
- plastic instability associated with the crack-tip blunting, slip, and rotational instability.

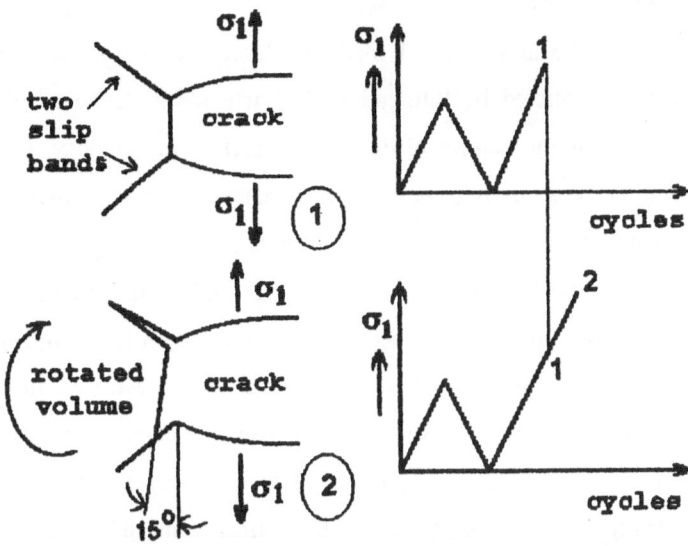

Figure 5. The scheme of the sequence of plastic shear and rotations in front of a crack tip in a thin plate subjected to unidirectional steady-state tension.

Simultaneously, there exists the hierarchy of the mechanisms involved in the failure of the intertunnel crosspieces of a material, *viz.*

♦ slip

♦ rotational instability, and

♦ steady-state (unidirectional) fracture.

The neighboring volumes of the material along the tunneling-crack front exchange information with one another to control the self-organized transition to the next fracture mechanism. Similarly, the sequence is controlled of the events of the formation of microtunnels and the failure of the intertunnel crosspieces by one of the above-mentioned fracture mechanisms.

There may be various causes, say, change of the of loading system (e.g., by transition from uniaxial to biaxial loading), level, frequency, *etc.*, for a fracture mechanism to get altered by bifurcation. Yet whichever deviation of the loading conditions, the resultant features of the fracture-surface morphology, reflective of the self-organization steps, are fully *intrinsic* to a material, i.e., not extrinsically formed. Non of the material volumes is originally informed on the conditions to be experienced next and, therefore on the kind or sequence of the forthcoming fracture mechanisms. Yet it strictly follows the intrinsic mechanism of crack-growth, pertinent to a certain range of the crack-growth rates. Hence, a material may repeatedly return to the same fracture mechanism as long as the crack is growing under nonstationary loading conditions [16]. This concept offers the methodological ground for ascertaining fractographically the level of the equivalent stresses in the elements of aviation constructions [21].

It may happen that the in-service loading conditions prevent passing from Stage I to Stage II development of fracture. This situation was found typical of the fracture behavior of the compressor discs made of titanium alloys [22, 23]. The triangle-mode cycling ensures the fatigue striations on Stage II fracture, whereas no fatigue striation is formed in the fracture surface once a stress-holding portion is introduced into the cycling mode. The effect is similar in the case of high-R cycling: striations do not form with R greater than 0.8 [7], and the pseudo-striational pattern only forms indicating to the predominance of slip (Fig.6).

Environment can tell substantially on the fracture mechanisms. An example is given by the fracture behavior of the main landing-gear track beam of a Boeing 757-28A aircraft, made of high-strength MS60 steel: it failed by intergranular cracking with though no corrosion products evident [24]. The specimens cut out of this track beam were subjected[1] to high-R cycling, 10-s stress-holdings included in

---

[1] in the State Center of Flight Safety of Civil Aviation, Moscow, Russia.

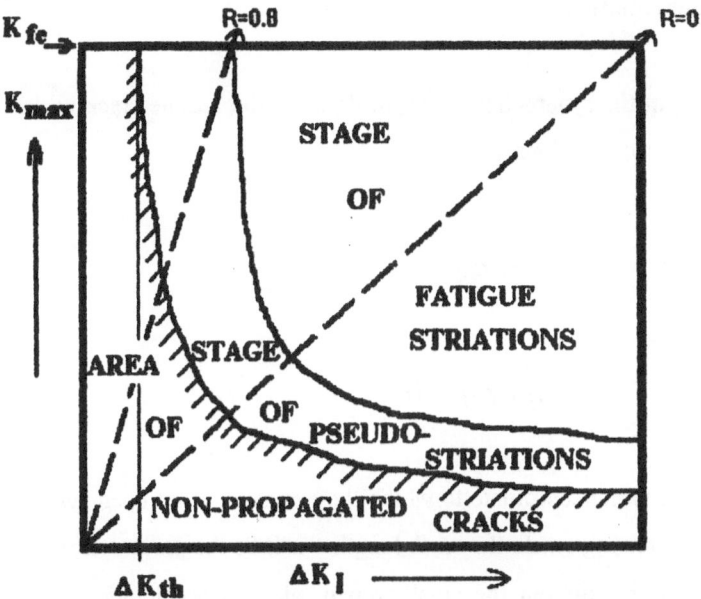

Figure 6. Regions of the fatigue striations or pseudo-striations mapped in the coordinates of stress-intensity factor K_I and stress-intensity amplitude ΔK_I.

each loading cycle, with the result of, however, transragranular fracture. The steel is well-known to have properties highly sensitive to the environment [25]; therefore, it must be the corrosion attack of the environment that ensured the intergranular crack propagation, i.e., the persistence of Stage I of the fracture process.

Based on the reality of the ordered self-organization of fatigue-crack growth one can make use of synergetics equations to simulate and control the fatigue-cracking process in the elements of aviation constructions.

3. Synergetics equations

It is common, in the synergetics terms, to describe the fatigue-cracking kinetics with the use of the basic equations as [8]

$$\dot{q} = \alpha q \tag{1}$$

$$\ddot{q} + \omega q = 0 \tag{2}$$

Parameter q is the coordinate to which the evolution of an open system is related. As concerns a growing crack, the crack length a will be the parameter, whose first and second derivatives represent the crack-growth rate and acceleration, respectively. α and ω are the ruling parameters. In the physics terms, they make possible to describe, on a full-scale, the behavior of an open system under any multiparameter conditions. The parameters retain time-independent as long as the extrinsic conditions of the system are not deviating with time. It is important that the mean properties are constant also along the crack-growth direction. Therefore, the above-mentioned parameters will only help to control the behavior of an open system if the relationship is unequivocally established between the time dependences of the ruling parameters and the parameters reflecting extrinsic service conditions of the construction element. One can use different combinations of extrinsic conditions (e.g., uniaxial tension with $R \neq 0$ or biaxial with $R = 0$) to ensure the same magnitude of the ruling parameter. Hence, the same critical state of the system in a bifurcation point can result from different combinations of the applied conditions of cyclic loading.

Equation (2) is descriptive of a bifurcation point. In this point, the second derivative is zero, which is indicative of a steady-state behavior of a system. Then, the system can be shifted to a stable or, visa versa, unstable condition. A system must be quite stable with the second derivative greater than zero. With the second derivative smaller than zero, the system is again stable, but too strong disturbance is required to reunstablize it. The latter situation can be illustrated [16] by the case of the long-term hampering of the crack, in a uniaxially or biaxially loaded material, as result of overloading. To remobilize cracking, much energy was required.

Under service conditions, the extrinsic parameters of cyclic loading are usually varying with time. As long as different combinations of those parameters can ensure the same state of the system, the ruling parameters are, in general, the multiparameter functions of the applied service conditions:

$$\alpha_i = \alpha_0[1 + f(X_1) + f(X_2) + ... + f(X_i)] \tag{3}$$

and

$$\omega_i = \omega_0[1 + p(X_1) + p(X_2) + ... + p(X_i)] \tag{4}$$

The ruling parameters α_0 and ω_0 correspond to the test conditions of the experiment, and α_i and ω_i, to the in-service loading conditions. In the latter case, the experiment conditions are not confirmed and, hence, the magnitudes or number of the parameters X_i differ from the test ones.

The crack-growth behavior of a material subjected to block-type irregular loading can be described in the terms of the general equations of synergetics. The equations will keep unchanged unless the crack-growth mechanism is altered. Yet the description of a system evolution must be modified according to the change in the behavior of the ruling parameter. As concerns the first ruling parameter, the above concept can be illustrated with the use of the following equations

$$q = \left\{ \begin{array}{c} \alpha_1 \\ \alpha_2 \\ .... \\ \alpha_i \end{array} \right\} q, \quad \text{where } \alpha_i = \left\{ \begin{array}{c} f_1[X_1, X_2, ... X_j] \\ f_2[X_1, X_2, ... X_j] \\ .... \\ f_i[X_1, X_2, ... X_j] \end{array} \right\} \quad (5)$$

In Eq. (5) the ruling parameter sequentially takes discrete magnitudes in accordance with the discrete magnitudes of a function $f_1[X_1, X_2, ... X_j]$ of the parameters $X_1, X_2 ... X_j$ of the applied conditions. Suppose that one parameter of the latter group is constant during a certain time; then the system behavior will be described by the equation that involves one of the parameters α_i.

Irregular loading is associated with the occurrence of nonlinear processes, concomitant with the intermediate loading conditions. If a sufficient amount of energy was supplied to an inert enough system its adoption may last during quite a period of time. The situation is commonly recognized as relaxation effects in a material. With the adoption period shorter than the time to start a new cycle of energy supply the same evolution equation will do. Yet it is critical, in general, to add Eq. (5) with the corrections for nonlinear process. For a multiparameter applied conditions it would run as

$$\alpha_i = \alpha_0 \{ 1 + f_j(X_1, X_2 ... X_i) + \sum_{j=1}^{k} f([X_i]_j / [X_i]_{j+1}) + \sum_{i=1}^{n} f(X_i / X_{i+1}) \} \quad (6)$$

Here, the additional terms $f([X_i]_j/[X_i]_{j+1})$ and $f(X_i/X_{i+1})$ are the corrections for the nonlinear processes provoked by altering the applied loading parameters in the level or kind, respectively.

The author used these principles of synergetics and the general most strict approaches to the description of the evolution of open systems when treating the cases of metallic materials in the constructions under various multiparameter applied conditions. The steady-state behavior of a material with a propagating fatigue-crack may last throughout the considerable part of the service life time of a construction. This primarily makes one to thoroughly analyze whether the applied conditions are safe and cannot cause the premature failure. Making use of the above synergetical relationships of the crack-growth trends, one can control the involved processes. Thereby, the stability of the material with the growing crack can be preserved at least in between the two subsequent sessions of in-service (non-breaking) inspection.

4. Description of crack growth in the synergetics terms

The first synergetical equation is applicable in the case of fatigue-crack growth as long as the strain amplitude is preserved [26, 27] throughout the metal life from the first smallest crack increment to the onset of unstable cracking. Yet with the constant stress amplitude the system behavior deviates, so that only the second synergetical equation is valid. This situation is schematically illustrated in Fig. 7 as the relationship between CGR and stress intensity factor.

Whatever the conditions or mode of cyclic loading, the achievement of definite CGR, *viz.* 5 x 10-8 m/cycle and 2.1 x 10-7 m/cycle, sequentially controlled the transitions from Stage I to Stage II and from the Substage IIa to Substage IIb, respectively. The CGR magnitudes were measured based on the Fourier analysis of the fracture-surface morphology [28].

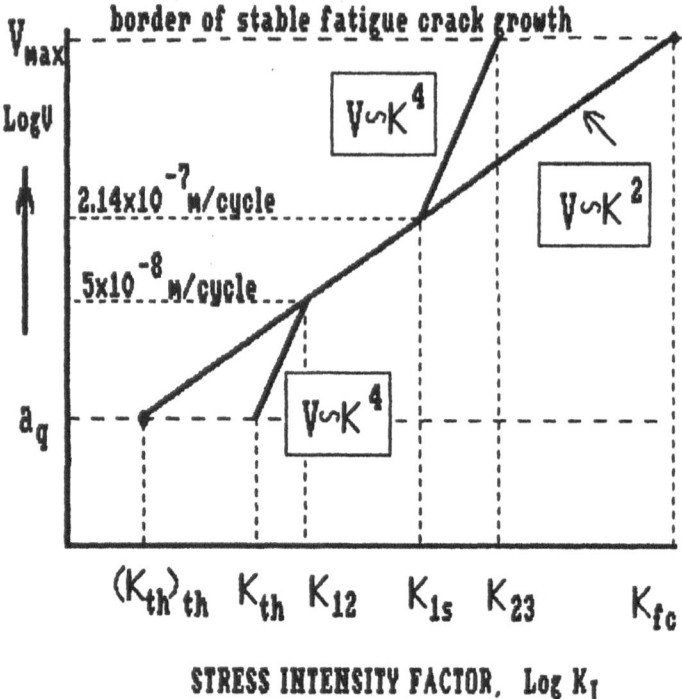

Figure 7. Growth rate of fatigue crack as a function of the stress-intensity factor for the cases of constant strain (V-K_I^2) or constant loading (V-K_I^4, V-K_I^2, and V-K_I^4) ranges. (Schematic).

Theoretical analysis of fatigue-crack growth and critical assessments of 73 published experimental Paris'-type diagrams [7, 29] provides a reason to express the ruling parameter as

$$\alpha_0 = \sigma_0^2 [(1 - v^2) / 12 E \sigma_{0.2})] \qquad (7)$$

for the case of through cracks under uniaxial-loading conditions.

The ruling parameter is reflective of the well-known relationship between the CGR and crack-tip opening. It is common to interpret this relationship in the terms of Poisson's ratio ν, Young's modulus E, and, also, off-set proof stress (usually, 0.2% YS) of the material. It was found by experiment that the ruling parameter nonlinearly depends on the applies stress σ_0 under the conditions of pulsing load. That was the uniaxial pulse-mode loading cycle with the stress range $0.3 < \sigma_0/\sigma_{0.2} < 0.4$ and loading frequency 10 to 20 Hz at 293 to 298K, 70 to 75% humidity, and 760 mm Hg pressure. $k_p = 12$ is a coefficient characteristic of the constriction extent of plastic deformation and of the crack-front profile. $k_p = 16$ for a semielliptic surface crack. k_p is varying from 12 to 24, depending on the thickness or geometry of a workpiece; this also provides corrections to take into account the effects of crack tunneling on the macroscopic as on the microscopic scale levels.

Ruling parameter α_i is involved in the case of multiparameter applied conditions. To determine α_i one needs the correction functions of the kind of $f[X_i]$. Earlier, it was done in the terms of density of fracture energy [15]. In the case of uniaxial loading, it takes a view

$$(dW_f / dV)_0 = [(1 - \nu^2) / (12E)]\sigma_0^2 \qquad (8)$$

Under the service conditions, the energy consumed in the crack growth may change, which looks as

$$(dW_f / dV)_e = (dW_f / dV)_0 + \Delta(dW_f / dV) \qquad (9)$$

Consequently,

$$\sigma_e = \sigma_0 [1 + f(X_1, X_2 \dots X_i)]^{1/2} \tag{10}$$

The density $(dW_f/dV)_e$ of fracture energy controls the crack-growth behavior for the multiparameter applied conditions however different from the reference ones. σ_e is the equivalent stress, corresponding to the uniaxial pulse-mode cycling. The stress level σ_0 may be increased or diminish, influenced by the loading conditions. It is assumed here that, as concerns the macroscopic scale level, type-I opening of the crack is always dominating. Based on the own 20-year experience of the inspection of the failed elements of aviation constructions the author is convinced that such an approach is quite realistic.

The data of 120 investigations, published between 1966 and 1989, on fatigue-crack propagation in various materials under various applied conditions were critically assessed; this made possible to classify the kinds of the $f(X_i)$ function [30]. The data were only considered if the *similarity condition* was satisfied with respect to the parameters of the applied conditions. This condition shows itself through the resultant equidistant shift of the kinetic curves, whichever crack-rate exponent in the Paris' relationship reported by the authors of the papers. It is only important that the exponent is invariant with respect to a particular parameter of applied conditions. Instead of reporting here on all the 120 correction cases, it seems worthwhile to consider some of them relevant to a case of multiaxial loading or of concurrent deviation of two parameters. Miller [13] was the first to demonstrate, on an aluminum alloy, the equidistant shift of the kinetic curves as a result of change in the conditions of biaxial loading. The correction function had been proposed as

$$\sigma_e = \sigma_0 f(\lambda) = \sigma_0 (a_0 / a_e)^{1/2} \tag{11}$$

The correction function $f(\lambda)$ for the change in the main-stresses ratio is expressed through the ratio of the crack-length values that correspond to the equal crack-growth rates at a_0 the uniaxial and a_e biaxial loading of a plate. Evidently, the correction functions (10) and (11) are similar in their structure.

☐Chan et al. [31] proposed a more complicated correction function for the tension or torsion cases of single-crystalline pipe specimens of a nickel alloy. The three components, τ_{xy}, τ_{yx}, and σ_{xx} of the stress tensor were involved, which describe the stress state of the material ahead of a crack tip. The correction function takes a form

$$\sigma_e = \sigma_0 \{1 + C_1 (\tau_{xy} / \sigma_{xx})^2 + C_2 (\tau_{yx} / \sigma_{xx})^2]^{1/2} \quad (12)$$

or

$$\sigma_e = \sigma_0 [1 + f_1(\lambda_1) + f_2(\lambda_2)]^{1/2} \quad (13)$$

Again, the structures of Equations (10) and (13) are similar. The proportionality factors C_1 and C_2 only depend on the Poisson's ratio.

The correction functions may be obtained based on the physical concepts of thermal-activation analysis. The crack-growth behavior in the high-temperature alloys was thus analyzed [32] to have revealed the effect of the concurrent deviation of the loading frequency ω and test temperature as

$$f(\omega, T) = (\omega_0 / \omega)^{p/m} \{\exp[-(Q/T)(1/T - 1/T_0)]\}^{1/m} \quad (14)$$

The cooperative effect of frequency and temperature is considered based on the concept of combined interrelated effects, each depending on the activation energy Q. p and m are the material constants, and ω_0 and T_0 are the quantities used in the test experiment.

Wei [25] took into account the cooperative influence of the aggressive environment and loading frequency, with the of correction functions formed as

$$f(\omega, pH) = [1 + (da / dN)_0 / (da / dN)]^{1/2} = [1 + (\sigma_0 / \sigma_e)^2]^{1/2} \tag{15}$$

Here, pH is the acidity level of an aggressive environment. The cooperative effect of both factors was considered in the way similar to that by Miller [13] but in the terms of the ratio of the crack-growth rates.

The forms of correction functions were also proposed [34, 35] applicable to the cases of biaxial tension, tension-compression or tension-torsion with the concurrent deviations in the respective R values. The magnitudes of the functions can be ascertained based on the concept of equivalent yield strength [35]. The characteristics of the specimens behavior can hardly be reproduced under service conditions of real construction elements and *visa versa* since the yield strength is dependent on the scale factor, kinetics of strain or relaxation processes at the crack tip, temperature, stress state, *etc.* Therefore, to authentically compare the in-tests behavior of a material in the plastic zone of a crack tip with the behavior under real-service conditions, one needs the correlation relationships, which can be established with the use of the respective correction functions.

In general, the plastic-zone volume Δv ahead of a crack tip and, hence, the CGR can be expressed through the invariants of the stress tensor as [36]

$$T_\sigma = \{1/6[(\sigma_x - \sigma_y)^2 + (\sigma_z - \sigma_y)^2) + (\sigma_x - \sigma_z)^2 + \tau_{xy}^2 + \tau_{yz}^2 + \tau_{zx}^2]\}^{1/2} \tag{16}$$

$$P_i = (1/3)(\sigma_x + \sigma_y + \sigma_z) \tag{17}$$

$$\Delta v = F\ (P_i\ /\ T_\sigma) \tag{18}$$

The greater the constriction extent of plastic deformation ahead of a crack tip the smaller the volume of the plastically deformed metal and greater yield stress of the material. The volume of plastic zone can be expressed through the zone radius using the well-known relationship

$$r = \beta(K_I\ /\ \sigma_{0.2})^2 \tag{19}$$

With the introduced dependences of yield stress $(\sigma_{0.2})_e$ on the parameters of a loading cycle, different from the parameters of the in-service conditions, one obtains

$$(\sigma_{0.2})_e = \sigma_{0.2}Q(X_1, X_2 ... X_i) \tag{20}$$

or

$$r = \beta[K_I\ /\ (\sigma_{0.2})_e]^2 = \beta(K_e\ /\ \sigma_{0.2})^2 \tag{21}$$

The proportionality factor β is introduced to take into account the constriction extent of plastic deformation along the crack front and determine the magnitude of a coefficient k_p in the ruling parameter α_i. K_e is the equivalent of the stress-intensity factor the pertinent to the pulsing-type uniaxial loading of the test specimens. Its magnitude can be determined with the use of the respective correction functions $F(X_1, X_2 ... X_i) = [Q(X_1, X_2 ... X_i)]^{-1}$ as

$$K_e = K_I F(X_1, X_2 \ldots X_i) \tag{22}$$

Using the synergetical principles and the concept of an equivalent stress-intensity factor made possible the single description [7], with respect to different-base alloys, of fatigue-crack rates at the whatever stage or step of steady-state growth of crack (Fig. 7):

$$da/dN = \begin{cases} [C_{IIs}/(K_e)_{12}]K_e^4, (K_e)_{th}/(K_e)_{12} = \Delta^{1/4}, & \text{stage I,} \\ [C_{IIs}]K_e^2, (K_e)_{Is} = \Delta^{1/4}, & \text{stage IIa} \\ [C_{IIs}/(K_e)]K_e^4, (K_e)_{Is}/(K_e)_{23} = \Delta^{1/2}, & \text{stage IIb} \end{cases} \tag{23}$$

$$C_{IIs} = (1 - v^2)/(k_p \pi E \sigma_{0.2}) \tag{24}$$

The above relationships appeared *universal* as concerned the description of crack growth behavior, for the case of K_I-mode cracking, in the elements of aviation structures; the relationships made a ground for ascertaining the levels of equivalent stress with the use of *different* parameters of the fracture-surface morphology [21].

Let us consider now the lines of further investigation into the fatigue-fracture behavior of structural materials.

5. Modeling and control of fatigue-crack growth

Simulation of fatigue-crack growth under uniaxial-loading conditions was a subject of numerous investigations (see, for instance, the review papers [16, 37]). Dealing with the multiparameter loading, one has to consider the levels and number of the loading parameters as compared with those involved in the simple test conditions; this can be done (see above) using Eqs. (1), (2), and (6). We considered the simulation of a crack-growth pattern in the case of biaxial cycling after a single overloading [38]. The obtained relationships were quite similar to the above synergetical equations and involved the correction functions. The simulation results were assessed for the case of flat cruciform specimens, and the prediction accuracy (as compared with the actual test results) appeared not worth than 80%. In the assessed case, biaxial overloads were used in the range 1,8 to 2.5 for R from 0.1 to 0.5, and main-stress ratio, -1.0 to +1.0.

Synergetical approach to the analysis of the effects of fatigue-crack growth made clear that further investigation activity must be directed to the ascertaining methodology of and criteria involved in the correction functions for Eqs. (7), (10), and (22). Unfortunately, one cannot entirely simulate the crack-growth behavior with the use of the calculations based on the concepts continuum strength and plasticity. This is immediately seen (Fig. 8) from the fatigue-fracture patterns in a plate for different crack-front profiles [15]. Traditionally, it is neglected that the relationship between the second tensile-stress component and fracture behavior of the material crosspieces between microtunnels is a function of the crack-front profile (Figs. 2 and 3).

A compression stress component is adding to the rotational-strain effects of the crosspieces in the surface-crack case. Accordingly, the crack-growth is hampered as compared with that in the specimen subjected to uniaxial loading. In the through-crack case, it is the plastic-zone size whose importance dominates. The size is grater in the biaxial than in the uniaxial-tension case. A consequence is that the transition to the biaxial tension-compression case results in an increased CGR. This is a strict demonstration of the synergistic nature of fatigue fracture: a material subjected to cyclic loading combines, by itself, the cleavage and shear fracture modes even when the load is applied uniaxially.

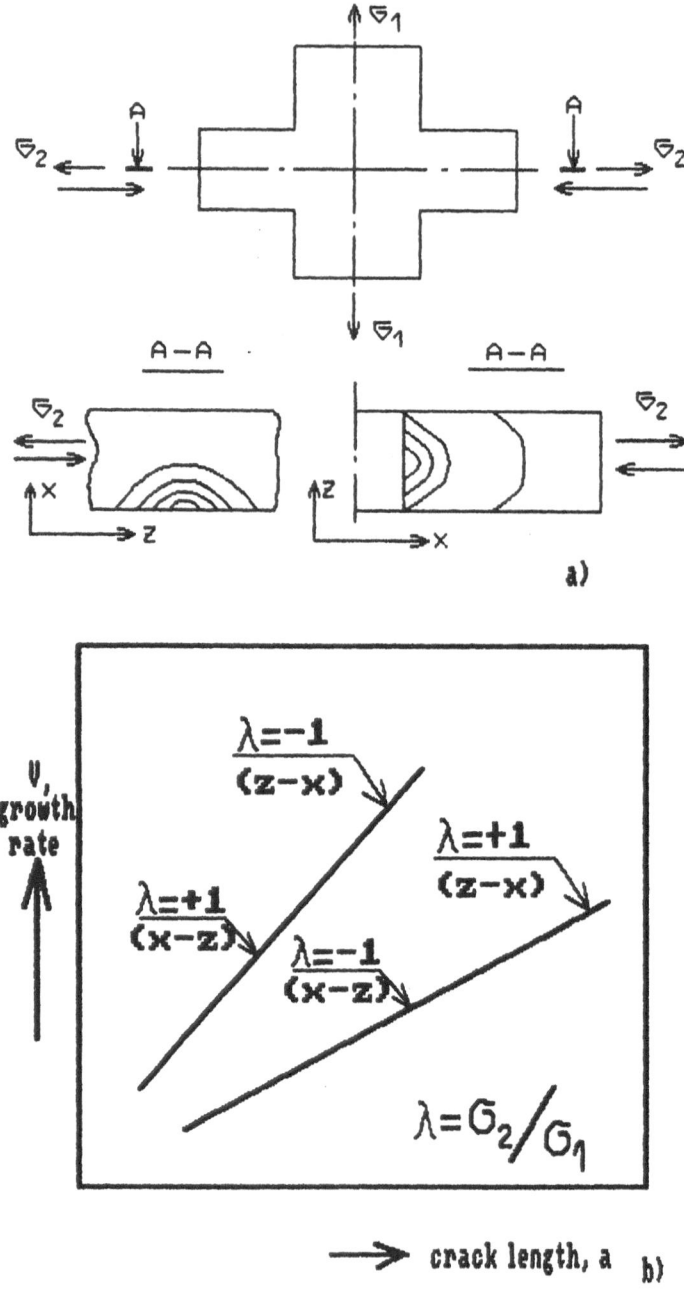

Figure 8. (*a*) position of a crack front with respect to the directions of biaxial loading in a cruciform test specimen; (*b*) CGR as a function of the stress-intensity factor [15, 34]. (Schematic).

The least addition of energy to the type III of shear will result in hampering the crack growth. This is obvious when analyzing, for various R magnitudes, the cracking kinetics as a function of the moment of torsion [33]. The crack-growth rate diminishes or increases accordingly with the twisting angle. Consequently, even under the simplest cycling conditions and, particularly, at low twisting angles the predicted crack-growth behavior may substantially deviate unless the due account is made of the nonplanar geometry of the specimen. This inspired the author to have developed a special technique, which makes possible to ascertain the effect of the torsion moment and neglect it when analyzing the crack-growth behavior under the tension-test conditions.

As a matter of fact, it is microtunneling effects of fatigue cracks that make possible, using quite simple techniques, to efficiently control the crack growth and substantially lower the increment of CGR per loading cycle [39]. The model of string can be used to illustrate the general concept of those techniques; the model simulates the sequential positions of the crack front in a construction element (Fig. 9). Additional compression applied along the crack front will add to the crack-hampering effects thanks to seizing along the general shear lips or shear surfaces formed in the material crosspieces between the microtunnels.

When designing or making new materials one has to mind the favor of splitting the crack and its subsequent propagation at the parallel levels, their distance different from a neutral reference plane. It is also favorable to have the material toughness increasing towards the material core; this would ensure, even for quite short cracks, the most efficient energy dissipation at the expense of the rotational instability.

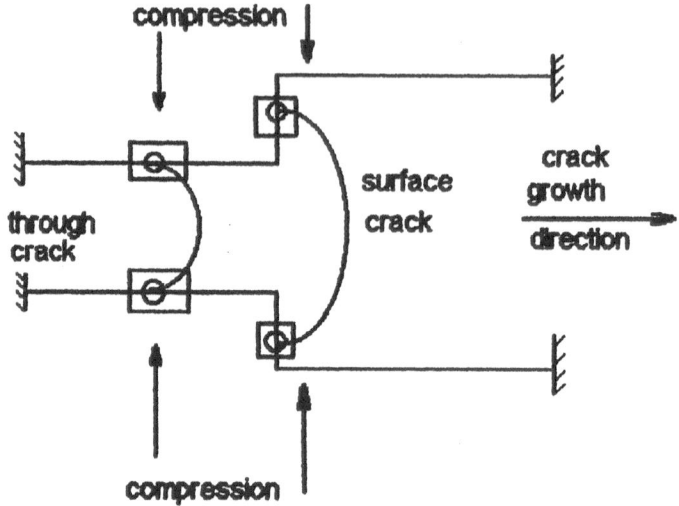

Figure 9. Crack-propagation delay, caused by compression normal to the plate faces; the effect is similar to that caused by increasing friction stress in the hinges of a string that is moving in the crack-growth direction. (Schematic).

Thus, we can outline further steps in making efficient the control of fatigue-crack growth. Primarily, one has to develop the calculation procedures of the correction functions for the parameters of applied conditions so that the ruling parameters of the process can be adequately ascertained. Thus calculated applied-condition parameters must help to take into account the synergetics of fatigue fracture and simulate the material properties that assist the effects of microtunneling of cracks. An outcome of the simulation might be the creation of the materials with the special structure (composition) formed along the expected direction of the growth of fatigue cracks.

6. References

1. Ivanova, V.S. and Terent'ev, V.F. (1978) Nature of metal fatigue (in Russian), Metallurgiya, Moscow.

2. Mughrabi, H., Ackerman, F., and Herz, K. (1979) Persistent slip bands in fatigue of face-centered and body-centered cubic metals. Fatigue Mechanisms, ASTM STP 675, ASTM, 69-105.

3. Ivanova, V.S. (1979) Fracture of metals (in Russian), Metallurgiya, Moscow.

4. Goritskii, V.M. and Terent'ev, V.F. Structure and fatigue failure of metals (in Russian), Metallurgiya, Moscow.

5. Hertzberg, R.W. (1983) Deformation and fracture mechanisms of engineering materials, John Willey and Sons, New York.

6. Ivanova V.S., Balankin, A.S., Bunin, I.Zh., and Oksogoev, A.A. (1994) Synergetics and fractals in material science (in Russian), Nauka, Moscow.

7. Ivanova, V.S. and Shanyavskii, A.A. (1988) Quantitative fractography. Fatigue fracture (in Russian), Metallurgiya, Chelyabinsk.

8. Haken, G. (1985) Synergetics. The hierarchy of instabilities in self-organized systems and devices (Russian translation), Mir, Moscow.

9. Ivanova, V.S. (1992) Synergetics. Strength and failure of metallic materials (in Russian), Nauka, Moscow.

10. Ebeling, V. (1979) Structures formed by irreversible processes. Introduction to the theory of dissipative structures. Mir, Moscow.

11. Ritchie R.O. (1977) Influence of microstructure on near threshold fatigue crack propagation in ultra-high strength steel.- Metal Science, no. 8-9, pp. 368-381.

12. Yoder G.R., Cooley L.A., and Crooker T.W. Quantitative analysis of microstructural effects on fatigue crack growth in Widmanstaetten Ti-6Al-4V and Ti-8Al-1Mo-1V. - Engng. Fract. Mech., 1979, vol. 11, pp. 805-816.

13. Miller K.J., (1977) Fatigue under complex stress.- Metal Science, no. 8-9, pp. 432-438.

14. Shanyavskii A.A. and Koronov M.Z. (1994) Shear lips on fatigue fractures of aluminum alloy sheets subjected to biaxial cyclic loads at various R-ratios. Fatigue Fract. Engng Mater. Struct., no. 9, pp. 1003-1013.

15. Shanyavskii A.A. (1996) Development of semi-elliptic fatigue cracks in AK6 aluminum alloy under biaxial loading. Fatigue Fract. Engng Mater. Struct., no. 12, pp. 1445-1458.

16. Shanyavskii A.A. and Orlov E.F. (1997) Fracture surface development in an overloaded D16T Al-alloy subjected to biaxial loading. A fractographic analysis. Fatigue Fract. Engng Mater. Struct., no. 2, pp. 151-166.

17. Panin V.A., Lihachev B.A., and Grinayev Yu.V. (1985) Sructural levels of the deformation of solids. Novosibirsk, Nauka (in Russian).

18. Elber W. (1971) The significance of fatigue crack closure.- ASTM STP 486, pp. 230-242.

19. Shanyavskii A.A. and Troyenkin D.A. (1997) Mechanisms of fatigue striations formation. - Fatigue Fract. Engng Mater. Struct. (to be published)

20. Schick E., Ude J., Michel F., and Blumenauer H. (1996) Estimation of the R-curve by in-situ measurements. - Proc. 11th Biennial Europ. Conf. Fract. - ECF11, Mechanisms and Mechanics of Damage and Failure (ed. J. Petit), Poitiers-Futuroscope, France, 3-6 September 1996, EMAS, vol. I, pp. 157-162.

21. Shanyavskii A.A. (1996) Synergetics approach to fatigue fracture analysis for fractographic stress equivalent determination in aircraft components. - Proc. 6th Intern Fatigue Congr. Berlin: Pergamon Press, vol. III, pp. 1251-1257.

22. Shanyavskii A.A. and Losev A.I. (1996) Fractographic analysis of aircraft engines compressors disks from Ti-alloys fatigued in service. Proc. 11th Biennial Europ. Conf. Fract. - ECF11, Mechanisms and Mechanics of Damage and Failure (ed. J. Petit), Poitiers-Futuroscope, France, 3-6 September 1996, EMAS, vol. II, pp. 1131-1136.

23. Shanyavskii A.A. and Stepanov N.V. (1995) Fractographic analysis of fatigue crack growth in engine compressor disks of Ti-6Al-3Mo-2Cr titanium alloy.- Fatigue Fract. Engng Mater. Struct., no. 5, pp. 539-550.

24. Fracture analysis of the L.H. and R.H. P/N 161N1611-7, main landing gear truck beam assemblies installed on Trans Aero airlines model 757-28A aircraft, registry No. EI=CLU (variable block No. NB-223). (1996) Boeing Materials Technology, Fracture Analysis Report Number MS 30512.

25. Wei R.P. (1985) Environmentally assisted fatigue crack growth.- Advances in Fatigue Science and Technology (eds. C.Moura Branco and L.Guerra Rosa), pp. 221-252.

26. Los Rios E.R., Kandil F.A., Miller K.J. and Brown M.W. (1985) Metallographic study of multiaxial creep-fatigue behavior in 316 stainless steel. In: Multiaxial Fatigue (eds. K.J.Miller and M.W.Brown), ASTM STP 853, ASTM, Philadelphia, pp. 669-687.

27. Shanyavskii A.A. (1996) The effect of pressure overload on fatigue crack growth in pressure vessels.- Fatigue Fract. Engng Mater. Struct., no. 1, pp. 1-13.

28. Sasov A.Yu. and Shanyavskii A.A. (1987) Fourier-fractography foundation of quantum mechanical nature of cracks growth.- Acta Stereol. 6(3), 925-930.

29. Shanyavskii A.A. and Grigor'ev V.M. (1989) Synergetics approach to plotting of unified kinetic diagram of fatigue crack growth in metals. In: Synergetics and fatigue fracture of metals (ed. Ivanova V.S.), Moscow, Nauka, pp. 87-98 (Russia).

30. Shanyavskii A.A. (1990) Synergetics of fatigue fracture of alloys at the second stage. In: Self-organized and fractal structures (eds. Ivanova V.S. et al.), Oil Institute in Ufa, Ufa, pp. 45-59 (in Russian).

31. Chan K.S., Hack J.E. and Leverant G.R. (1986) Fatigue crack propagation in Ni-base superalloy single crystals under multiaxial cyclic loads.- Met. Trans. A., no. 17, pp. 1739-1750.

32. Liu H.W. and Mc Gowan Y.I. (1981) A kinetic analysis of high temperature fatigue crack growth.- Scr. Met., vol. 15, pp. 507-512.

33. Shanyavskii A.A. (1996) Quantitative fractographic analyses of fatigue crack growth in longerons of in-service helicopter rotor-blades.- Fatigue Fract. Engng. Mater. Struct., no. 9, pp. 1129-1141.

34. Shanyavskii A.A., Orlov E.F., and Koronov M.Z. (1995) Fractographic analysis of fatigue crack growth in D16T alloy subjected to biaxial cyclic loads at various R-ratios.- Fatigue Fract. Engng. Mater. Struct., no. 11, pp. 1042-1148.

35. Shanyavskii A.A. (1989) Self-organization of fatigue crack kinetics. In: Synergetics and fatigue fracture of metals (ed. Ivanova V.S.), Moscow, Nauka, pp. 57-76 (in Russian).

36. Kolmogorov V.P. (1970) Strength, strains, fracture. Moscow, Metallurgy. (in Russian)

37. Shanyavskii A.A., Orlov E.F. and K.Z.Karaev (1993) Simulation of fatigue damage propagation in aircraft components under multiaxial stress-state (an Overview). Research Center of Information of Civil Aviation. Moscow (in Russian).

38. Shanyavskii A.A. and Orlov E.F. (1997) Fatigue crack growth simulation after biaxial overloads.- Fatigue Fract. Mater. Struct. (to be published).

39. Shanyavskii A.A. and Koronov M.Z. (1992) Method of specimen testing, Authorized evidence N1234567, Paper of inventions N35 (in Russian).

40. Shanyavskii A.A. (1991) Reliability of aircraft components as a result of fatigue cracks growth management (an Overview). Research Center of Information of Civil Aviation. Moscow (in Russian).

LACUNARITY OF THE CONTACT DOMAIN BETWEEN ELASTIC BODIES WITH ROUGH BOUNDARIES

Numerical analysis and scale effects

M. Borri-Brunetto, A. Carpinteri and B. Chiaia
Dipartimento di Ingegneria Strutturale, Politecnico di Torino
Corso Duca degli Abruzzi 24, 10129 Torino, Italy

1. Introduction: contact between rough surfaces

A realistic and consistent characterization of the topography of rough surfaces represents a crucial point in the modelization of interface phenomena. Many tribologic phenomena, like friction, lubrication and wear of mechanical components, strongly depend on the surface morphology [1]. At the same time, the thermo-electric conductivity between two bodies in contact is intimately related to the interface characteristics. At different scales, the shear strength of rock joints is deeply influenced by the surface morphology and shows a marked dependence on the size of the considered specimen (size-scale effect).

The earliest attempts to extend the Hertzian theory of elastic contact between smooth bodies [2] to real bodies with rough boundaries can be ascribed to Archard [3] and to Greenwood and Williamson [4]. Archard was the first to solve the discrepancy between the classical Hertzian solution and the experimentally validated Amonton's law. This law is based on the hypothesis that the frictional force T is proportional to the area of true contact A_r, which, in its turn, is proportional to the normal load F [5]. While, in fact, Hertz had shown that, in the case of smooth spheres, the real contact area A_r is related to the normal load F by a nonlinear relation ($A_r \sim F^{2/3}$), Amonton's law stated the direct proportionality between frictional force and normal load ($T = \mu F$). By means of a hierarchical model of hertzian spheres with progressively decreasing radius, Archard obtained the linear behaviour ($A_r \sim F$) requested by the experimental evidence. Although the classical Hertz solution would yield, for each sphere, a nonlinear dependence, the Archard's hierarchical assembly globally provides linearity in the limit of sphere radius tending to zero (Fig. 1).

Greenwood and Williamson [4] considered one of the bodies in contact to be rough, with contact occurring at a number of discrete points. The asperities could undergo only elastic deformation, and at the top of each peak an Hertzian micro-sphere was supposed to exist. The asperities were modelled in a way for the peaks to follow a Gaussian distribution around a mean flat surface. It was found that the Hertzian results are valid at sufficiently high loads, whereas the pressure distribution differs significantly form the Hertzian solution at lower loads. The area of true contact was determined to be

45

G. N. Frantziskonis (ed.), PROBAMAT – 21st Century: Probabilities and Materials, 45–64.
© 1998 *Kluwer Academic Publishers.*

approximately proportional to the load, so that the basic hypothesis of the theory of friction could be supported also for purely elastic contact. The authors concluded that the behaviour of rough surfaces is determined primarily by the statistical distribution of asperity heights and secondarily by their mode of deformation. Thus, although plastic deformations cannot be excluded, at least for the highest and sharpest peaks, it was argued that no new features would be introduced by plasticity.

Because the interface phenomena involve all length scales, a scale-independent description of the surfaces should be pursued. Nevertheless, classical models as the one by Greenwood and Williamson make use of statistical parameters, like the mean and variance of the surface elevations, of the slopes and of the curvatures. Unfortunately, any experimental measure is performed at a fixed resolution and hence all these quantities are strongly dependent on the resolution adopted for the description of the interface [6]. On the other hand, the theory of fractals sheds new light on the Archard's model, which is, in effect, scale-independent (Fig. 1). Archard's model can be thought of, in fact, as the first deterministic pre-fractal model applied to contact problems.

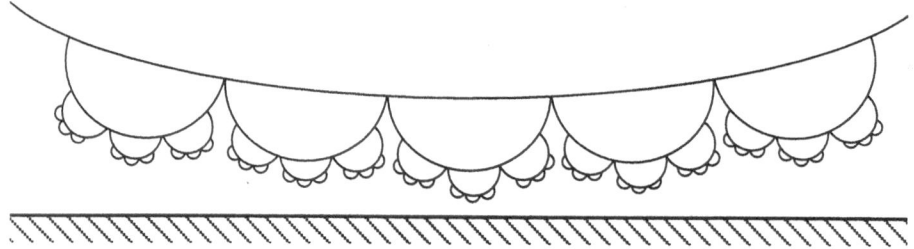

Figure 1. The Archard's hierarchical array of spheres in elastic contact with a flat plane [3].

Recently, random fractal models of rough surfaces [7] have been shown to resemble very closely the morphology of natural and man-made surfaces. The ability of fractal geometry to give a scale-independent description of reality has exerted a strong appeal for many researchers, who applied the fractal concepts to the various branches of mechanics, sometimes missing the essence of the fractal theory. The physical quantities defined on fractal sets have to change their dimensionality according to the (non integer) dimensions of the set. Thereby, scaling is provided and it is not possible to measure a scale-independent euclidean quantity, unless asymptotically [8, 9].

In the field of contact mechanics, Panagiotopoulos [10, 11] adopted an engineering-oriented fractal methodology to solve problems of unilateral and adhesive contact and problems of friction. His numerical procedure exploited the approximation of fractal boundaries through classical surfaces of integer Hausdorff dimension (Iterated Function Systems). Borodich and Mosolov [12] modelled polished fractal surfaces by means of Cantor punches. The problem of indentation was solved for this geometry, in the case of elastic as well as of plastic bodies. Using asymptotic methods, they obtained power-laws for the dependence of the load on the depth of the indentation. The exponents of the power-laws depended explicitly on the fractal dimension of the punching set. Warren and Krajcinovic [13] developed a Cantorian fractal model, similar to the one by Borodich and Mosolov, and extended the discrete model to a continuous formulation of the elastic-perfectly plastic deformation model. In this way, force-displacement relationships were derived in a closed form, allowing for simple parametric studies of the influence of surface roughness through the fractal dimension Δ_G of the interface.

A statistical study of the geometry of contact between self-affine surfaces was carried out by Plouraboué et al. [14]. By means of numerical simulations in two and three dimensions, they studied the shape of the apertures between two surfaces in contact, showing that the probability distribution of apertures displays two distinct scaling regimes for small and large apertures. They also showed that, as the surfaces are shifted with respect to each other (sliding mechanism), the contact points on one surface undergo a Lévy walk with an exponent equal to the roughness exponent of the surfaces.

Regarding recent experimental investigations, it is worth mentioning Majumdar and Bhushan [6], who adopted an optical profilometer to digitize magnetic tape surfaces. They used the Weierstrass-Mandelbrot fractal function to model the surface roughness, obtaining good correlations with experimental data. Majumdar and Tien [15] carried out tests on various machined steel surfaces and textured magnetic thin-film disks. The power-law spectral behaviour of these surfaces unequivocally proved that statistically similar images of the surfaces appear when the surface is repeatedly magnified. They also developed a fractal network model for the contact thermo-electric conductance [16]. However, although the self-similarity of the contact domain is exploited in their model, the mechanical quantities retain their usual physical dimensions.

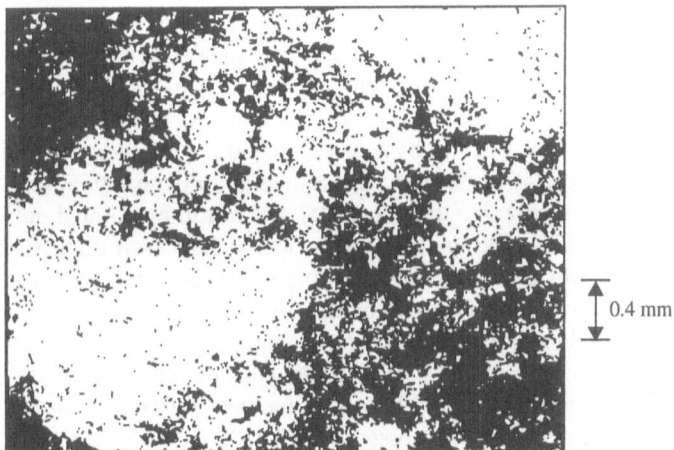

0.4 mm

Figure 2. Experimental contact domain of a natural rock joint at a mean apparent stress of 85 MPa [17].

In the field of rock mechanics, it is worth mentioning the paper by Cook [17], where most of the basic mechanical issues are addressed. An experimental technique was adopted to detect the *locus* of true contact (the black *islands* are contact points, floating in an *ocean* where contact does not occur), which, although fractals are not invoked, suggests a lacunar topology (Fig. 2). Bandis *et al.* [18], Sage *et al.* [19] and Pinto da Cunha [20] investigated the problem of scale effects on shear behaviour and closure of rock joints. This problem represents a crucial aspect for the correct characterization of natural joints and faults, starting from the data obtained on small-scale specimens.

In this paper a numerical model is presented for simulating the problem of contact between two elastic half-spaces with rough (fractal) boundaries. The results of the simulations are described, and the influence of surface fractality on the mechanical behaviour is discussed. Moreover, the topologic analysis of the contact domain (the horizontal projection of the set of points where contact truly occurs) is performed. This shows that the real contact domain $\mathcal{D}$ is a *lacunar* fractal set in the two-dimensional plane, with Hausdorff dimension Δ_σ ($\Delta_\sigma < 2.0$) progressively increasing with the load. As a major consequence, the physical quantities (mechanical stress, electricity or heat conductance) defined over this fractal set should assume noninteger (anomalous) dimensions. Therefore, scaling laws come into play. As a mechanical example, the size-effects on the shear strength of rock joints and on the interface closure deformability can be easily explained. Also, the experimentally detected departure from linearity in the classical friction laws can be interpreted in the framework of fractals.

2. Modelization of natural surfaces by self-affine fractals

Since the time of Newton, one essential hypothesis which is put forward in the description of natural physics is that of *differentiability*. Smooth euclidean shapes have been adopted in almost all modelizations of reality. This hypothesis allowed physicists to write the equations of physics in terms of differential equations. The possibility of associating gradients and curvatures to euclidean surfaces implies the smoothness (or *measurability*) of the sets and therefore their scale-independence. However, there is no *a priori* principle which imposes the fundamental laws of physics to be differentiable. Multi-scale phenomena, such as phase transitions, are nowadays successfully interpreted by means of fractal models [7]. The noninteger Hausdorff dimension of the domains on which the physical quantities are defined assumes a profound significance.

A fundamental distinction among fractal sets has to be pointed out. The so-called *invasive* fractals, that is, the spatial domains whose Hausdorff dimension $\Delta_\mathcal{G}$ is strictly larger than their initiator's topological dimension, produce *positive* scaling of the quantities defined over them (e.g. the fracture energy defined over a fracture surface [8]). The von Koch triadic curve (Fig. 3a) is an example of an invasive fractal set embedded in the bidimensional plane. This set could be considered as a profile obtained by the intersection of a fractal surface with a plane. On the contrary, *lacunar* fractals, like the middle-third Cantor set (Fig. 3b), possess Hausdorff dimension Δ_σ strictly lower than the topologic dimension of the initiator, and therefore provide *negative* scaling of the quantities defined over them (e.g. the stress defined on a contact domain like the one in Fig. 2). The damaged ligament of an heterogeneous solid subjected to tensile loading has been successfully modelled by means of lacunar fractal sets [8].

Two basic common features should be put into evidence for the domains in Fig. 3. These are the noninteger Hausdorff dimension of the sets, and their self-similarity, i.e. the symmetric dilatation property that yields similar morphologies at all the scales of observation (which can be considered as the different stages, or *pre-fractals*, in the iterative generation of the fractal sets). The fundamental difference between invasive and lacunar fractals is represented by the asymptotic behaviour of the euclidean (Lebesgue) measure of the set. While, in fact, the area of an invasive surface tends to infinity, the area of a lacunar domain tends to zero. Finite measures (Hausdorff measures) of fractal domains can be obtained only by means of noninteger dimensions ($[L]^{1.262}$ in the case of the von Koch curve and $[L]^{0.631}$ in the case of the Cantor set, respectively). Accordingly, the physical quantities defined over these sets have to assume noninteger (anomalous) dimensions.

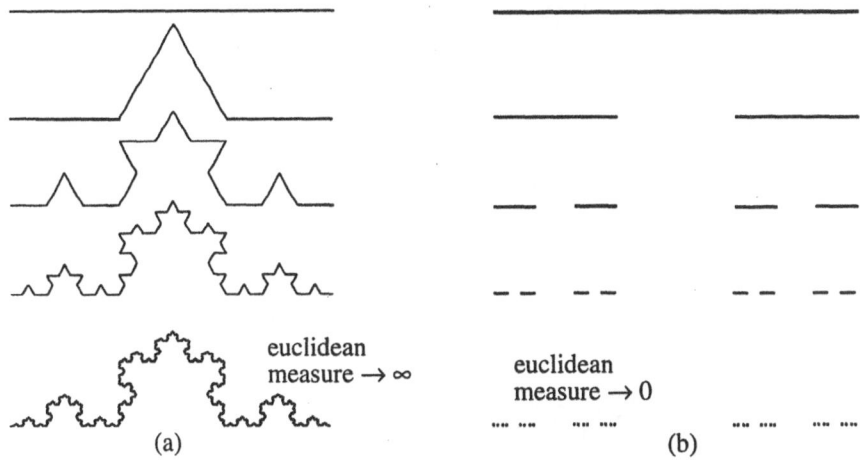

Figure 3. Deterministic models of invasive (a) and lacunar (b) fractal sets.

It is nowadays well established that many man-made surfaces, like those of magnetic tapes and of metallic components, as well as natural boundaries like fractures and rock joints, show fractal properties over a wide scale range. This implies the morphological invariance (in the statistical sense) of the surfaces under a group of affine (anisotropic) scale transformations. It is worth here to say that self-similarity is just a particular case of the more general concept of *self-affinity* [21]. Self-similarity is in fact characterized by isotropic dilatation symmetry and by the absence of an internal (cross-over) scale. The scaling relation of a self-affine surface $z(x, y)$, if r is the scaling (or magnification) factor, can be written as:

$$f(x, y, z) \rightarrow f(rx, ry, r^H z),\tag{1}$$

where $H < 1$ is called the *Hurst exponent*. Equation (1) implies that the vertical coordinates scale less than the horizontal ones. Thereby, the surface seems to become smoother as the resolution decreases. The fractal dimension is obtained as $\Delta_g = 3 - H$, and thus these "surfaces" are invasive fractals ($\Delta_g > 2.0$). According to the experimental evidence [22] and to the theory of dissipative phenomena, the Hausdorff dimension of a Brownian surface ($\Delta_g = 2.5$) often represents the upper limit of disorder. This is the case, for example, of fracture surfaces. Therefore, for most natural surfaces, $0.5 \leq H \leq 1$.

There exist several methods to generate fractal surfaces. For numerical simulation, different self-affine surfaces, with given fractal dimension Δ_g (Fig. 4), have been

50

generated by means of the *random midpoint displacement algorithm* [23]. It is important to keep in mind, however, that any algorithm can generate only an image (*pre-fractal*) of the fractal set (which is a limit concept for δ → 0). Thus, any numerical surface is characterized by a precise value of the resolution δ. This scale represents an internal length, interacting with the self-affine random field.

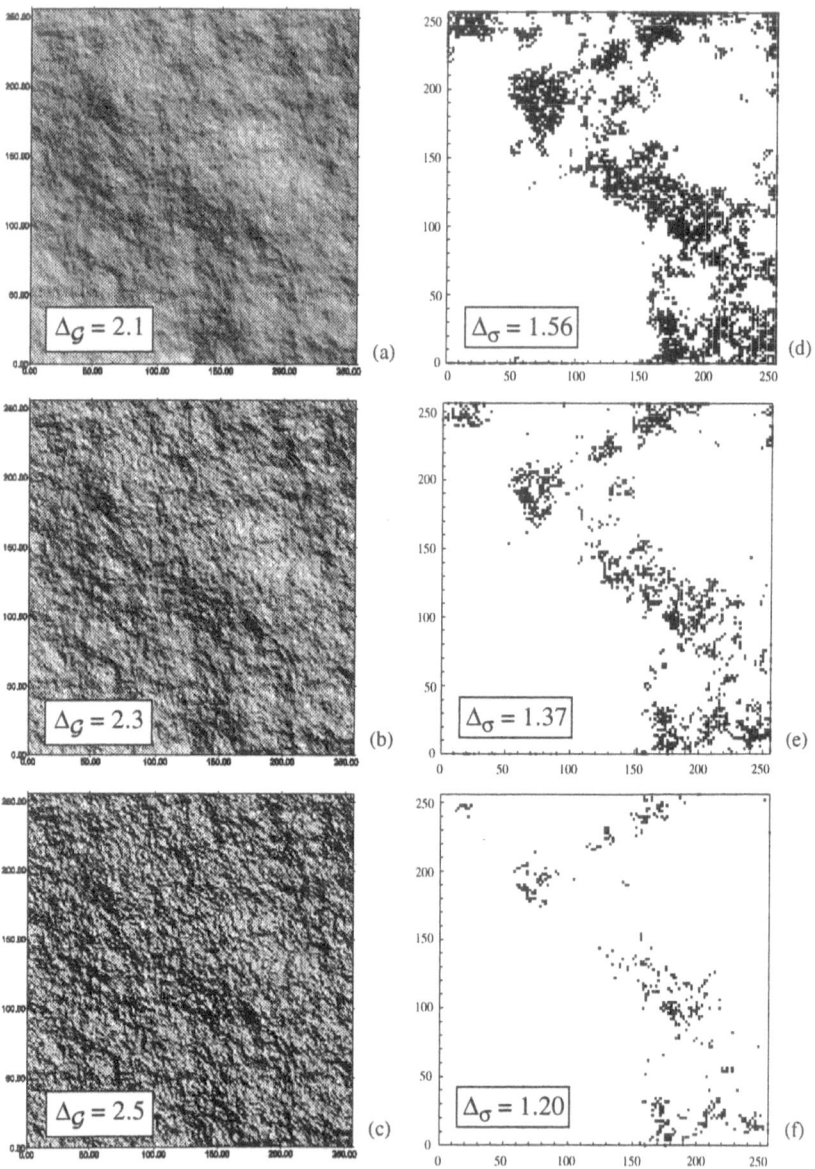

Figure 4. Fractional Brownian surfaces generated by the random midpoint displacement method (a, b, c) and corresponding contact domains (d, e, f) at the same closure displacement.

Due to the transition from a macroscopic euclidean regime to a microscopic fractal scaling, typical of self-affinity, the images in Fig. 4 can also be thought of as different magnifications of the same surface made with increasing resolution. The contact domains which result from the numerical simulations, at a fixed value of the closure displacement, are also shown in Fig. 4. Discussion on their fractal properties and on the physical implications of fractality will be made in the next sections.

For the generation of the three surfaces of Fig. 4, the same succession of random seeds was adopted, in order to highlight as much as possible the influence of the fractal dimension Δ_g on the contact mechanism. The statistical distributions of the elevations, for the three surfaces, are shown in Fig. 5. The cumulative distributions of surface heights, shown in Fig. 5d, are indeed very similar. It is also important to notice that the standard deviations of surface heights, although increasing with the fractal dimension, are pretty close to each other in the three surfaces. On the contrary, the correlation among the heights drastically decreases as the fractal dimension increases. Moreover, the skewness of the histograms is evidently larger for the smaller fractal dimensions. A quite good agreement was found by comparing these statistics with those obtained on natural rock joints [24].

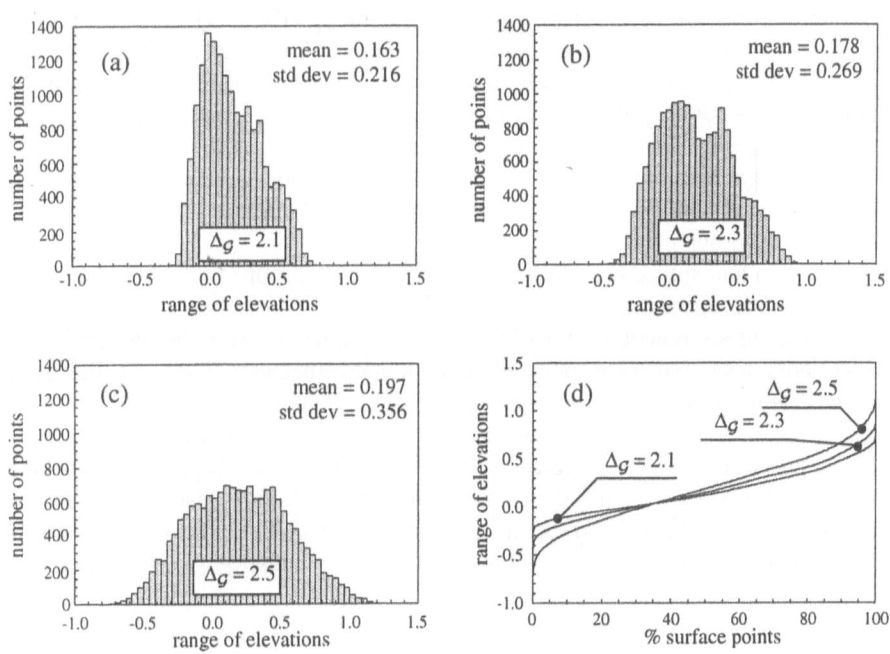

Figure 5. Histograms and cumulative distributions of surface heights for the discretization $\delta = 1/128$.

Each surface was generated, keeping constant the external size, with different discretizations, starting from a 16×16 up to a 512×512 pixels grid. Further refinements, although desirable, would imply a tremendous increase of computational time for solving the elastic problem by means of the computers at our present disposal.

3. Discretization of the elastic contact problem and numerical procedure

The goal of this work is determining the geometrical properties of the contact domain $\mathcal{D}$ between two linear-elastic half-spaces with a rough (fractal) surface. The unilateral contact problem has been extensively approached by many authors as a nonlinear optimization problem (a complete reference list can be found in [10]). In the present paper, instead, an *ad hoc* active set strategy has been preferred, in order to reduce the storage requirements for the computation. The problem we have chosen leads to a simple solution, still retaining some of the most important features of the real phenomenon. The contact zone is a part of the surfaces, where the boundary is given the requested roughness according to the self-affine hypothesis. Due to the additivity of the power spectral densities of fractal surfaces [17], the problem of the contact between two self-affine surfaces with given fractal dimension can be transformed into the problem of the contact between an equivalent self-affine surface and a plane.

A discrete version of the problem is considered, by introducing in the reference plane x-y a square grid of points with spacing δ. Of course, δ can be referred to as the adopted resolution in solving the problem and, in this case, corresponds to the resolution adopted in generating the fractal boundaries. At each node of the grid two material points can touch and transmit a force. If forces are exchanged through the interface, we assume that a small contact zone is involved around the grid point. This contact region is supposed to be an indivisible *atom* of surface, whose area is related to the spacing δ of the grid. This assumption rules out the possibility of point forces applied to the half-spaces.

A relative closure displacement (in the z direction) is imposed to the half-spaces. The solution is sought in terms of pressure and surface displacements, paying attention to the unilateral contact condition at the interface. The approximations introduced as a first approach to the problem are:
1) surface points displacements are perpendicular (z direction) to the boundary mean plane of both half-spaces;
2) displacements are functions only of the normal components (f) of the surface forces;
3) forces are related to displacements through influence functions. These coefficients are the same calculated for contact spots on smooth flat half-spaces.

The initially undeformed geometry of the solids is defined by the elevations h^A and h^B of the surface points above a reference plane $z = 0$. Let us call Ω_A (resp. Ω_B) the half-space $z \geq h^A$ (resp. $z \leq h^B$), with the assumption $h^A \geq h^B$. Without loss of generality, we assume that the body Ω_A is fixed (points far from the surface do not move), and we apply to distant points of Ω_B an upward displacement w. The position of two facing points P_r^A and P_r^B can be written as:

$$z_r^A = h_r^A + u_r^A,$$
$$z_r^B = h_r^B + u_r^B + w, \tag{2}$$

where the elastic displacements u_r^A and u_r^B are introduced to consider the deformation due to the interaction between the solids through the interface. Denoting by f_r^A (resp. f_r^B) the resultant of the tractions acting on the contact spot around P_r^A (resp. around

P_r^B), the displacements can be expressed as:

$$u_r^A = -\sum_s H_{rs}^A f_s^A,$$

$$u_r^B = \sum_s H_{rs}^B f_s^B, \tag{3}$$

where H_{rs}^A and H_{rs}^B are the influence coefficients, which can be evaluated with reference to well-known elastic solutions. For the points P_r belonging to the contact domain $\mathcal{D}$, we can write $z_r^A = z_r^B$, i.e., their positions are coincident. Thereby, from equations (2) and (3):

$$h_r^A - \sum_s H_{rs}^A f_s^A = h_r^B + w + \sum_s H_{rs}^B f_s^B, \tag{4}$$

where the equilibrium condition $f_r^A = f_r^B = f_r$ obviously holds. With the position $\Delta h_r = h_r^A - h_r^B$ (initial gap), assuming also that $H_{rs}^A = H_{rs}^B$, the final expression of the problem can be written:

$$\frac{-w + \Delta h_r}{2} = \sum_s H_{rs} f_s, \qquad \forall P_r \in \mathcal{D}. \tag{5}$$

The influence terms H_{rs} can be evaluated by considering the surface displacements u_r induced at point P_r of a half-space by a single force f_s applied at point P_s, that is:

$$u_r = H_{rs} f_s. \tag{6}$$

These displacements can be calculated, for example, by referring to the settlements induced by a unit load, applied to an elastic half space, through a rigid circular plate:

$$H_{rs} = \frac{1 - \nu^2}{\pi E a} \arcsin(a/r_{rs}), \tag{7}$$

where E, ν are respectively the Young's modulus and the Poisson ratio, a ($a = \delta/2$) is the radius of the loaded area, and r_{rs} is the distance between P_r and P_s.

By solving the system of equations (5), the contact forces between the two bodies are determined, provided the contact domain $\mathcal{D}$ is known. The solution of this problem can be conveniently achieved by means of an incremental-iterative algorithm. In matrix form, the problem, at increment i, can be written as:

$$\hat{\mathbf{u}}_i = \mathbf{H}_i \mathbf{f}_i, \tag{8}$$

where $\hat{\mathbf{u}}_i$ is the n-component vector of the displacements imposed to the points of the current contact domain $\mathcal{D}_i = \{P_1, P_2, ..., P_n\}$ for a given closure w_i; $\mathbf{f}_i$ is the vector of the unknown contact forces; $\mathbf{H}_i$ is the flexibility matrix.

The contact domain $\mathcal{D}_i$ is not known a priori. The contact condition is unilateral, i.e. only compressive forces can be transmitted through the interface points $P_r \in \mathcal{D}_i$. In general, the solution of eqs. (8) gives also tension forces at some points. An iterative procedure can be started from a tentative contact domain $\mathcal{D}_i^{(1)}$ through a sequence of sets:

$$\mathcal{D}_i^{(k+1)} \subset \mathcal{D}_i^{(k)}, \qquad k = 1, ..., m-1. \tag{9}$$

At each step k, the system (8) is solved, retaining in the contact domain $\mathcal{D}_i^{(k+1)}$ only the points where compressive forces have been found. The procedure converges to the correct domain $\mathcal{D}_i^{(m)}$ in a few steps. The initial domain $\mathcal{D}_i^{(1)}$ is chosen as the set that contains all those gridpoints which would compenetrate if contact forces were absent.

It is important to point out here that a correct solution of the elastic problem is essential to determine the set of points where contact truly occurs. Therefore, the definition of the contact domains as the horizontal cross sections of the surface at different heights, as it was argued in [6], is only a crude approximation. By neglecting elastic deformations, in fact, only the initial tentative domains $\mathcal{D}_i^{(1)}$ can be determined, which are often very different from the real contact domain.

The flexibility matrix $\mathbf{H}_i^{(1)}$ is assembled by adding, at each displacement increment i, the terms pertaining to the tentative domain $\mathcal{D}_i^{(1)}$. Thereby, the contact forces $\mathbf{f}_i^{(1)}$ are evaluated. The points where tension forces are found, are eliminated from the contact domain, obtaining the new contact domain $\mathcal{D}_i^{(2)}$ and the new matrix $\mathbf{H}_i^{(2)}$. After a few steps of the algorithm, through successive eliminations of tension points, the correct solution is attained for the displacement increment i. Then, the closure w_{i+1} is imposed, passing to a new increment.

The solution of the simultaneous equations (8) is achieved with an iterative Gauss-Seidel method, taking into account the symmetry of matrix $\mathbf{H}_i$. The contact domains obtained under the same value of the closure displacement $\overline{w}$, for three different values of the fractal dimension of the contact surfaces, are shown in Fig. 4. It can be immediately noticed that, the larger $\Delta_\mathcal{G}$, the smaller the fractal dimension Δ_σ of the contact domain (i.e., the larger its *lacunarity*).

4. Mechanical response and interfacial compliance

Numerical simulations have been carried out by means of contact surfaces with fractal dimension $\Delta_\mathcal{G}$ varying between 2.1 and 2.5. As stated before, the problem of contact between two fractal surfaces was reconducted to the contact of a fractal surface with a flat plane. All the geometrical and mechanical parameters (e.g. Young's modulus, Poisson ratio, shape of the interface) have been kept the same in all the simulations. In correspondence to any value of the imposed displacement w_i, the algorithm provides the total transmitted load, the relative displacement of all the points of the surfaces, the position of the contact points ($P_j \in \mathcal{D}$) and the pressure transmitted by each point. Note that results are presented without the units of measure, because, at this stage, only qualitative predictions and general trends are pursued from the numerical simulations.

If the area of true contact is plotted versus the applied load, at a fixed resolution δ,

(Fig. 6a), a nearly linear behaviour is observed ($A_r \sim F$) for all values of the fractal dimension. It is remarkable to notice here that disorder, which is usually associated with nonlinearity and chaos, yields in this case linear behaviour. This is in agreement with the early observations by Archard [3], and Greenwood and Williamson [4] and with many experimental results in the literature [1]. Note, however, that Majumdar and Tien [16] analytically found a nonlinear relation ($A_r \sim F^\alpha$, $1.0 \leq \alpha \leq 1.33$), based on their model of elastic-plastic contact between fractal surfaces. As it was already observed in Fig. 4, the smaller is the fractal dimension Δ_g of the punching surface, the larger is the percentage of true contact area.

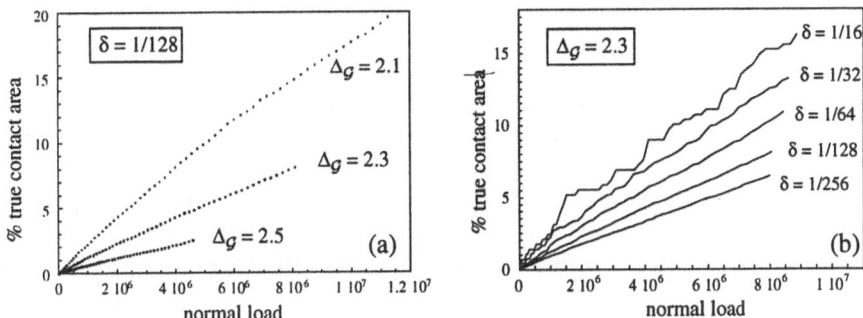

Figure 6. Diagrams of the true contact area vs. normal load for different fractal dimensions (with fixed resolution, (a)), and for different resolutions (with fixed fractal dimension, (b)).

If the real contact area A_r is plotted versus the applied load for a given surface (e.g. $\Delta_g = 2.3$), and for different discretizations of the surface, a smaller percentage of real contact area is determined as the resolution increases (Fig. 6b). This confirms the lacunarity of the contact domain through the whole loading range. Also, in Fig. 6b, one can realize that for the coarsest discretization ($\delta = 1/16$) discreteness effects come into play, yielding a staircase shape of the diagram.

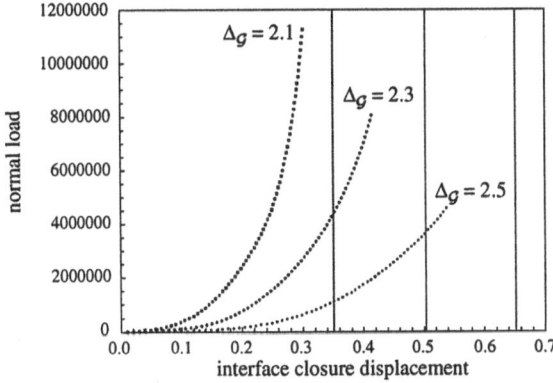

Figure 7. Computed normal load-joint closure laws and asymptotic values for different fractal dimensions.

The problem of joint closure under normal loads is very important in rock mechanics [19, 20]. The effective interface closure can be obtained (Fig. 7) by subtracting, from the total displacement w_i, the part due to the linear elasticity of the half-spaces. This bulk contribution is evaluated by considering uniform pressure on the nominal loaded area. In this way, in the load-displacement diagram (Fig. 7), the nonlinearity, due to the interface roughness, can be put into evidence. By applying hyperbolic nonlinear fitting, the asymptotes of the diagram can be estimated, which correspond to the complete closure of the joints. It can be noticed that, as the fractal dimension of the boundary increases, the extension of the closing stage increases, together with the interface compliance.

The statistical distributions of the forces exerted at the contact points, under the same external load (i.e. under the same nominal pressure) are reported, for the three surfaces, in Fig. 8. It can be concluded that the distribution of local micro-forces tends to become more uniform as the fractal dimension increases. The histograms also show that only a few points undergo very high stresses, especially for the lower fractal dimensions. This supports the early observation by Greenwood and Williamson [4] that plastic deformations should not add significant new features to the contact model, at least in a wide loading range. Moreover, it could be demonstrated that the distributions of contact micro-forces obey multifractal scaling, that is, an infinite hierarchy of exponents is necessary to describe all the moments of the distribution. This suggests the possibility of interpreting the shear collapse (and the stick-slip transition) as a critical phenomenon related to the hierarchical release of the stored elastic energy. This idea is supported by classical results by Herrmann and Roux [25] on lattice models, and by some previous attempts to model the stick-slip transition by renormalization group methods [26].

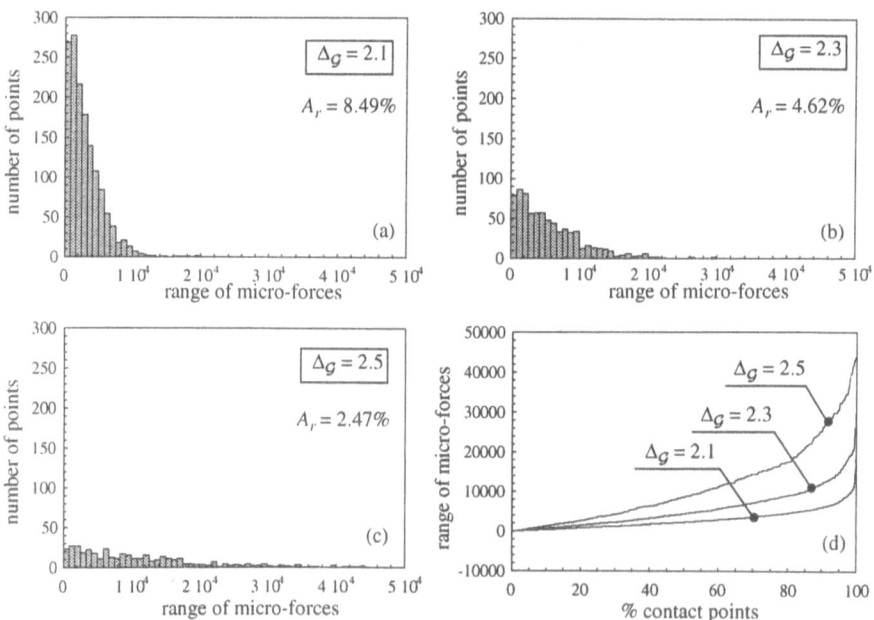

Figure 8. Histograms and cumulative distributions of the local contact forces at the same external load. Note that the total amount of contact points decreases as the fractal dimension of the surface increases.

5. Lacunarity of the contact domain and induced scale effects

When the evolutive contact problem is solved for a given surface, with given fractal dimension Δ_G but with different discretizations δ, the resulting contact domains $\mathcal{D}_\delta$ can be compared (Fig. 9). It can be easily realized that the concept of area of true contact [5], although representing a step forward with respect to the concept of apparent (nominally flat) area, is not able to describe consistently (that is, in a scale-independent manner) the interface interactions. As it clearly emerges from Fig. 9, the real contact area A_r progressively decreases with increasing the resolution of the discretization, ideally tending to zero in the theoretical limit of $\delta \to 0$.

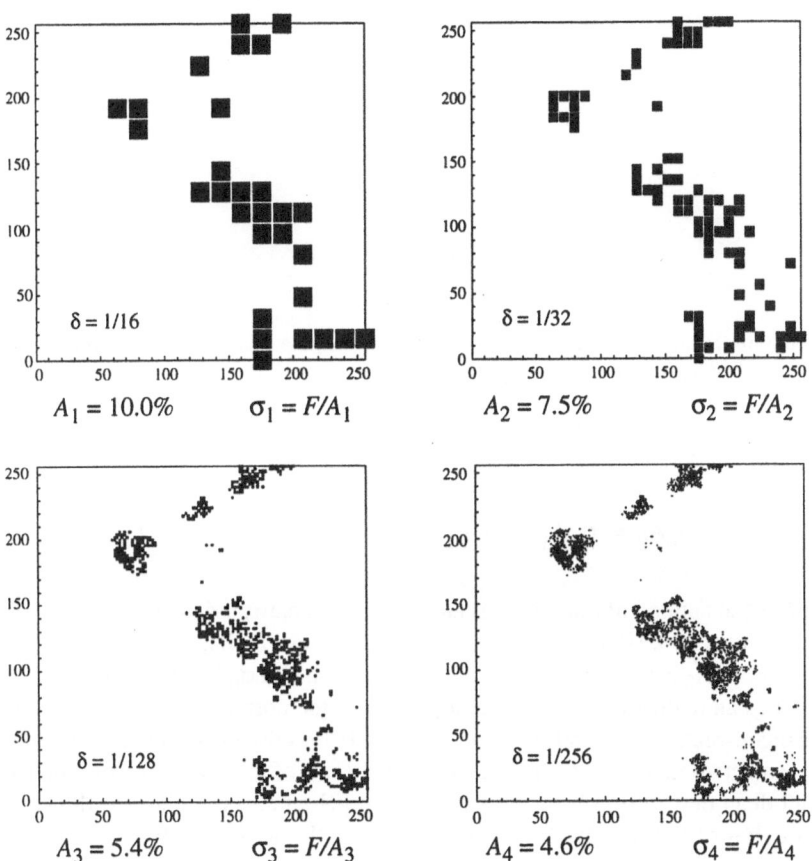

Figure 9. Decrease of the real contact area and increase of the corresponding pressure (under the same external load) with increasing resolution (surface dimension $\Delta_g = 2.1$, contact domain dimension $\Delta_\sigma = 1.39$).

This behaviour unequivocally states the lacunarity of the contact domain $\mathcal{D}$, and therefore claims the necessity of abandoning the euclidean description and moving to the fractal model, characterized by the noninteger dimension Δ_σ ($\Delta_\sigma \leq 2.0$) of the domain $\mathcal{D}$. Application of the *box-counting method* [7] to a large number of contact sets (Fig. 10)

58

allowed to compute their Hausdorff dimension Δ_σ. In this way, it was possible to compare the topology of the contact domains obtained for boundaries with different fractal dimensions Δ_g and to follow the evolution of Δ_σ, for a given fractal surface, during the loading process. It is worth to say here that the contact domains are, to a great extent, self-similar sets, that is, they show isotropic scaling. This is because they are related (but not coincident) to the so called *zero-sets* of the self-affine surfaces in contact [23], i.e., to the intersections of the surfaces with planes with equation $z = const.$

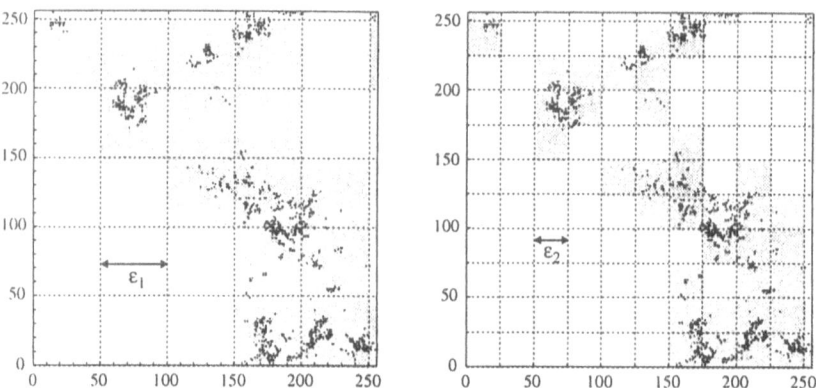

Figure 10. Box-counting method applied to the contact domain generated by a Brownian surface ($\Delta_g = 2.5$) pushing against a flat plane.

In perfect analogy with the theory of fractal scaling applied to nominal tensile strength [9, 22], let us consider the following renormalization group:

$$F = \sigma_0 A_0 = \sigma_1 A_1 = \sigma_2 A_2 = \ldots = \sigma_n A_n = \ldots = \sigma^* \mathcal{H}_{\mathcal{D}}, \tag{10}$$

where F is the applied normal load (scale-invariant quantity), A_n and σ_n are respectively the real contact area ($[L]^2$) and the real mean pressure ($[F][L]^{-2}$) measured at the pre-fractal scale n, A_0 is the apparent (nominally flat) area and σ_0 is the apparent pressure.

As one can realize from Fig. 9, at any value of the normal load F, σ_n increases with increasing resolution and tends to infinity as $\delta \to 0$, because the euclidean measure ($[L]^2$) of the contact domain $\mathcal{D}$ vanishes. Hence, in the limit of the highest resolution ($\delta \to 0$), the euclidean description loses its significance [9] and leaves place to the Hausdorff measure $\mathcal{H}_{\mathcal{D}}$ of the domain $\mathcal{D}$ (which is univocally defined only by the noninteger dimensionality $[L]^{\Delta_\sigma}$). Correspondingly, the *fractal mean pressure* σ^* can be defined as the anomalous flux of stress through the fractal interface. Owing to the dimensional homogeneity, this quantity holds the anomalous physical dimensions $[F][L]^{-\Delta\sigma}$. By equating the second and the last term in equation (10) and taking the logarithm of both sides, if b is a characteristic linear size of the interface, the following scaling law is provided:

$$\log \sigma_0 = \log \sigma^* - (2 - \Delta_\sigma) \log b. \tag{11}$$

The above relation states the dependence of the apparent normal pressure σ_0 on the size of the specimen (Fig. 14a). In close agreement with many experimental observations [17, 18], and assuming Amonton's direct proportionality between shear strength and normal pressure to hold, equation (11) affirms that the apparent shear strength $\tau_0 = T/A_0$ decreases with increasing the size of the nominal contact area, that is, the friction coefficient μ is not constant for a given interface but varies with its size. Therefore, the limit slope angle decreases with increasing the specimen size (Fig. 11). This approach may shed light on many rock slope instabilities, which are not explicable when compared to the relatively high friction values measured on smaller specimens.

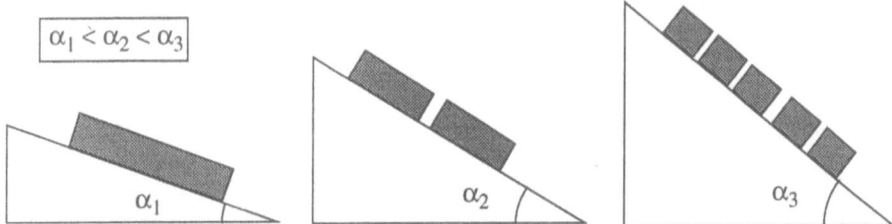

Figure 11. Schematic view of the size-scale effect on the limit slope angle (at constant apparent pressure).

Another fundamental aspect to be highlighted is the dimensional evolution of the contact domain $\mathcal{D}$, which is initially very rarefied (like a Cantor dust) and progressively increases its topological density with increasing the applied load. As the apparent pressure σ_0 between the half-spaces increases, the classical theories yield the increase of the number of contact spots, that is, the increase of the real contact area A_r. Some relations can be found in the literature [6], connecting the area of true contact A_r and the real mean pressure σ_r to the apparent pressure σ_0. However, since the euclidean measure of A_r is experimentally depending on the measurement precision (or on the numerical discretization, see equation (11)), the classical elastic contact theories do not have unique predictive capabilities.

More significantly, the continuous variation of the Hausdorff dimension Δ_σ has to be considered. As the normal load F increases, Δ_σ starts from the value 0.0 (corresponding to pointwise non-structured contact), then takes values comprised between 0.0 and 1.0 (corresponding to pointwise contact sets which are structured, or self-similar, in a well defined scale range), and subsequently takes values larger than unity as soon as linear contact structures and rarefied contact *islands* are formed (Fig. 12). The total saturation of the contact domain $\mathcal{D}$ (or, at least, of some *islands*) would imply $\Delta_\sigma = 2.0$. It may be argued that this value, in real materials, could be attained only under very high normal loads, and this would imply the extended plasticization of the material. In this limit case, the euclidean description would be consistent and the physical quantities would retain their usual integer dimensions.

Another way to investigate on the scale effects, is to perform simulations on isolated smaller parts of the original surface. For this purpose, the surface in Fig. 4b ($\Delta_g = 2.3$) was split into its four quarters and each quarter was subjected alone to the numerical simulation. Subsequently, each quarter was split into its quarters, so that sixteen surfaces with linear size equal to 1/4 of the original one were tested one by one (Fig. 13).

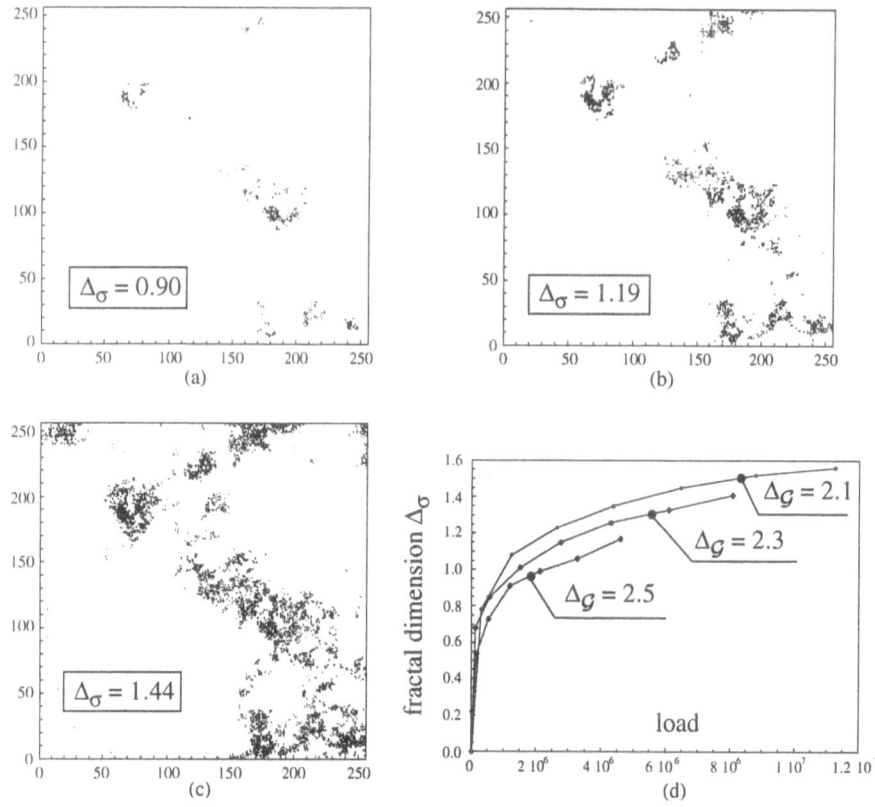

Figure 12. Three contact domains (a, b, c) at different loads for the same surface ($\Delta_g = 2.3$, $\delta = 1/256$), and evolution of the fractal dimension with load (d).

A regular trend has been observed, namely the percentage of area of true contact (A_r/A_0), at a given value of the apparent pressure σ_0, is larger for the smaller specimens. On the other hand, the Hausdorff dimension Δ_σ of the contact domains, at a fixed value of σ_0, is almost coincident regardless of the linear size of the interface. This suggests that larger contact domains (i.e. larger apparent areas A_0) are less dense in the euclidean sense, that is, the probability of the occurrence of large zones without contact (the white *oceans* in Fig. 2) increases with the size of the interface. As an example (see Fig. 13), the bottom left quarter of the original surface, due to its concavity, does not bear stresses when considered as a part of the whole, while becomes active when tested as isolated. This confirms the lacunarity of the contact domain and provides another mean to justify the size effects in the friction problem. As a 3D analogy, the euclidean density (ratio of the mass over the apparent volume) of sponge-like objects decreases with size.

Another kind of size-effect is encountered in the closure behaviour of rock joints. By comparing the numerical load-displacement diagrams obtained for the larger surfaces with the diagrams obtained for the smaller surfaces, it can be concluded that the joint closure deformability is larger for larger specimens. This important result is in agreement with the experimental data reported in [19], where a stiffer behaviour was provided by the smaller specimens.

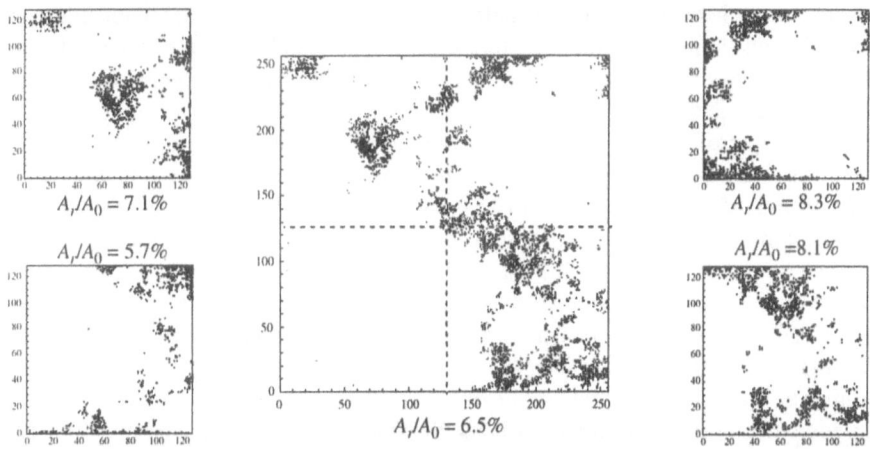

Figure 13. Contact domains, at a fixed value of nominal pressure σ_0, obtained for different parts of the original large surface (Δ_σ=2.3, δ=1/256) tested alone. Δ_σ=1.44 is approximately constant, regardless of the interface size. Averagedly, the ratio A_r/A_0 increases in the smaller surfaces.

Besides size-effect, the topological evolution of the contact domain also provides an interesting interpretation of direct shear tests on rock joints [18]. For a given specimen size, in fact, the apparent shear strength τ_0 increases with the apparent normal stress σ_0 (Fig. 14b) with progressively decreasing slope (friction coefficient) [21]. An asymptotic maximum value could be related to the (ideal) saturation of the interface contact domain. In real materials, this would imply the extended plasticization of the material. Recalling equation (11), it can also be concluded that the size-dependence of τ_0 is maximum at small external loads (small Δ_σ) and becomes progressively weaker as Δ_σ increases with the load. In the limit of $\Delta_\sigma \to 2.0$, size-effects would disappear, and a scale independent friction coefficient could be defined. Note that also in the model by Greenwood and Williamson [4], the smooth behaviour is attained for very high loads.

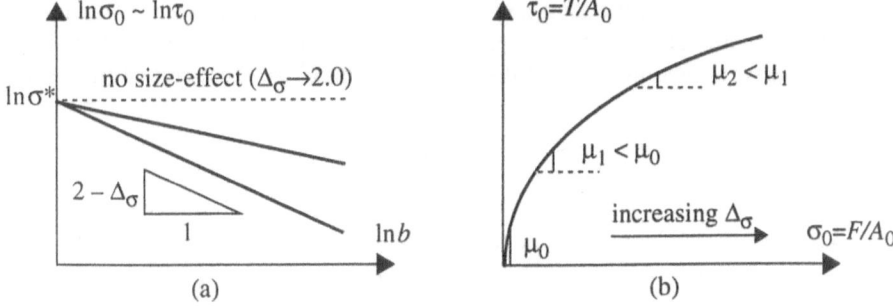

Figure 14. Size-scale effects on the nominal pressure for different lacunarities (a). Fractal interpretation of the experimentally observed decay of the apparent friction coefficient in rock joints (of fixed size) with increasing normal load (b).

6. Conclusions

A numerical model has been proposed which allows one to study the problem of unilateral contact between two elastic half-spaces with self-affine fractal boundaries. The model, although simple, permits to capture some fundamental features of the mechanical problem, which can be easily extended to a wide class of interface phenomena, such as thermal and electric conductivity.

The concepts of fractal geometry allow one to describe in a scale-independent manner the interface phenomena. The philosophical duality between *roughness hypothesis* and *adhesion hypothesis* [5] is overcome by the fractal model. The classical roughness parameters are not sufficient to explain the complex contact mechanics (smoother surfaces often yield higher friction), nor does the concept of area of true contact provide unique predictions, due to its intrinsic scale-dependence. Instead, the geometric and topologic (the fractal dimension Δ_σ and the noneuclidean measure $\mathcal{H}_\mathcal{D}$) characteristics of the contact domain $\mathcal{D}$ seem to play a preeminent role in these phenomena. The fractal model might represent the link between macroscopic and microscopic (nanotribology) models. In fact, the study of friction at the atomic scale [27] shrinks the role of roughness and indicates that the force stems from various unexpected sources, including the sound energy.

The most important conclusions can be summarized as follows:

1) the mechanical response obtained from the numerical simulations permits to affirm that, the higher the fractal dimension $\Delta_\mathcal{G}$ of the surfaces in contact, the higher the global closure deformability of the joint;

2) the horizontal projection of the surface points which undergo contact pressures is a lacunar domain (the contact domain $\mathcal{D}$), with self-similar fractal properties;

3) the higher the fractal dimension $\Delta_\mathcal{G}$ of the surfaces in contact is, the smaller the fractal dimension Δ_σ of the contact domain $\mathcal{D}$ is at the same normal load or at the same closure displacement;

4) the simulations yield a nearly linear response in the load-real area of contact diagram ($A_r \sim F$) for all the values of $\Delta_\mathcal{G}$. Thus, the local self-affine complexity of the surfaces provides macroscopic linearity;

5) if the area of true contact A_r is plotted vs. the applied load F in correspondence to different discretizations of the same surface, one notices that the finer the resolution, the smaller the real area of contact at a given value of load. This denies the validity of the concept of real contact area and confirms the noneuclidean character of the contact domain;

6) analysis of the local pressures at the contact points shows that only few points undergo very high stresses. On the other hand, the statistical distributions of the local contact forces tend to become progressively more uniform as the fractal dimension increases. Moreover, these distributions show multifractal character;

7) a major consequence of the lacunarity of the contact domain $\mathcal{D}$ is the scale-dependence of the real contact area A_r and of the real mean pressure $\sigma_r = F/A_r$. This implies the decrease of the apparent normal pressure σ_0. Assuming direct proportionality between shear strength and normal pressure, one obtains the decrease of the apparent

shear strength $\tau_0 = T/A_0$ with increasing the size of the nominal contact area A_0 (size effect). Hence, the friction coefficient μ is not constant for a given rock joint, but varies with its size;

8) numerical simulations on isolated parts of the original surfaces permit to evidence that, for a given fractal dimension, the joint closure stiffness decreases with the size of the joint. This is in agreement with experimental data. Also, the percentage of real contact area (A_r/A_0), at a given value of nominal pressure $(\sigma_0 = F/A_0)$, is larger for smaller surfaces. This comes from the lacunar fractal character of the contact domains $\mathcal{D}$;

9) for a fixed value of Δ_σ, lacunarity implies that the contact domains of larger joints are less dense in the euclidean sense. This means that the probability of presence of large zones where contact does not occur increases with the size of the interface. This represents another way to justify the size effects on shear strength;

10) the fractal dimension Δ_σ of the contact domains increases during the loading process, theoretically tending to saturate the plane ($\Delta_\sigma \rightarrow 2.0$) for very large pressures;

11) owing to the increase of Δ_σ with normal load (saturation of the contact domain), it can be argued that the size-dependence of the apparent shear strength τ_0 is larger at lower values of the normal load, and diminishes as the load increases;

12) the increase of Δ_σ with load provides an alternative mean to justify the decrease of the friction coefficient μ in the experimental τ_0-σ_0 curves obtained for a joint of fixed size.

However, many issues remain still open. First of all, a wider statistical investigation will be performed in order to draw more reliable conclusions. For example, other algorithms (FFT filtering, independent cuts, ballistic deposition) will be adopted for generating the fractal surfaces. A wider range of statistical parameters (e.g. mean and variance of the surface heights distribution) will be considered. Also, the influence of basic material parameters, e.g. Young's modulus and Poisson ratio, will be properly investigated. A more important issue, from our point of view, will be to carry out numerical simulations on real surfaces (e.g. digitization of both the rock joint surfaces) and to compare results to experimental tests. The final step will be that of simulating the mechanisms of friction, i.e., implementing the sliding displacement in the model. From an engineering viewpoint, a renormalization technique can be developed to predict the *in situ* mechanical behaviour of large rock joints, based on the combined small-scale experimental and numerical data.

7. References

1. Johnson, K.L. (1985) *Contact Mechanics*, Cambridge University Press, Cambridge.
2. Timoshenko, S. (1934) *Theory of Elasticity*, McGraw-Hill, New York.
3. Archard, J.F. (1957) Elastic deformation and the laws of friction, *Proc. Roy. Soc. London A* **243**, 190-205.
4. Greenwood, J.A. and Williamson, J.B.P. (1966) The contact of nominally flat surfaces, *Proc. Roy. Soc. London A* **295**, 300-319.
5. Bowden, F.P. and Tabor, D. (1954) *Friction and Lubrication of Solids*, Oxford University Press, Oxford.

64

6.   Majumdar, A. and Bhushan, B. (1990) Role of fractal geometry in roughness characterization and contact mechanics of surfaces, *J. Tribology (ASME)* **112**, 205-216.

7.   Mandelbrot, B.B. (1982) *The Fractal Geometry of Nature*, W.H. Freemann & Company, New York.

8.   Carpinteri, A. (1994) Fractal nature of materials microstructure and size effects on apparent mechanical properties, *Mech. Materials.* **18**, 89-101.

9.   Carpinteri, A. (1994) Scaling laws and renormalization groups for strength and toughness of disordered materials, *Int. J. Solids Struct.* **31**, 291-302.

10.   Panagiotopoulos, P.D. (1985) *Inequality Problems in Mechanics and Applications*, Birkhäuser Verlag, Basel.

11.   Panagiotopulos, P.D., Mistakidis, E.S. and Panagouli, O.K. (1992) Fractal geometry in solids and structures, *Comp. Meth. Appl. Mech. Eng.* **99**, 395-412.

12.   Borodich, F.M. and Mosolov, A.B. (1992) Fractal roughness in contact problems, *J. Appl. Math. Mech.* **56**, 681-690.

13.   Warren, T.L. and Krajcinovic, D. (1995) Fractal models of elastic-perfectly plastic contact of rough surfaces based on the Cantor set, *Int. J. Solids Struct.* **19**, 2907-2922.

14.   Plouraboué, F., Roux, S., Schmittbuhl, J. and Vilotte, J.P. (1995) Geometry of contact between self-affine surfaces, *Fractals* **3**, 113-122.

15.   Majumdar, A. and Tien, C.L. (1990) Fractal characterization and simulation of rough surfaces, *Wear* **136**, 313-327.

16.   Majumdar, A. and Tien, C.L. (1991) Fractal network model for contact conductance, *J. Heat Trans. (ASME)* **113**, 516-525.

17.   Cook, N.G.W. (1992) Natural joints in rock: mechanical, hydraulic and seismic behaviour and properties under normal stress, *Int. J. Rock Mech. Min. Sci. & Geomech. Abstr.* **29**, 198-223.

18.   Bandis, S., Lumsden, A.C. and Barton, N.R. (1981) Experimental studies of scale effects on the shear behaviour of rock joints, *Int. J. Rock Mech. Min. Sci. & Geomech. Abstr.* **18**, 1-21.

19.   Sage, J.D., Aziz, A.A. and Danek, E.R. (1990) Aspects of scale effects on rock closure, in Pinto da Cunha, A. (ed.), *Scale Effects in Rock Masses*, Balkema, Rotterdam, 175-181.

20.   Pinto da Cunha, A. (Ed.) (1993) *Scale Effects in Rock Masses*, Balkema, Rotterdam.

21.   Mandelbrot, B.B. (1985) Self-affine fractals and fractal dimension, *Phys. Scr.* **32**, 257-260.

22.   Carpinteri, A. and Chiaia, B. (1996) Power scaling laws and dimensional transitions in solid mechanics, *Chaos Solitons and Fractals* **7**, 1343-1364.

23.   Voss, R.F. (1985) Random fractal forgeries, in Earnshaw R.A. (Ed.), *Fundamental Algorithms for Computer Graphics*, NATO ASI Series F **17**, Springer-Verlag, Berlin, 805-835.

24.   Brown, S.R. and Scholz, C.H. (1985) Broad bandwidth study of the topography of natural rock surfaces, *J. Geoph. Res.* **90**, 12575-12582.

25.   Herrmann, H.J. and Roux, S. (Eds.) (1990) *Statistical Models for the Fracture of Disordered Media*, North Holland, Amsterdam.

26.   Smalley, R.F. Jr., Turcotte, D.L. and Solla, S.A. (1985) A renormalization group approach to the stick-slip behavior of faults, *J. Geoph. Res.* **90**, 1894-1900.

27.   Persson, B.N.J. and Tosatti, E. (Eds.) (1996) *Physics of Sliding Friction*, Kluwer, Dordrecht.

HOW FAR CAN WE CONTROL THE HETEROGENEITY OF WOOD AND WOOD BASED MATERIALS THROUGH DESIGN ?

P. CASTERA
LRBB (CNRS/INRA/Université Bordeaux 1)
BP 10 F-33610 Cestas
France

1. Introduction

Wood is usually considered apart from other materials by engineers and designers, related more to nature than to artifice, to tradition than to modernism. Mechanical performances of wood, especially in bending, are well known, and have been utilised in load bearing structures from historic times. To be more precise some woods are known for their mechanical performances, and actually there is a tremendous variability of the properties of commercialised woods, the origin of which lies in the diversity of species, but also in the heterogeneity of production conditions. One can take advantage of this diversity, considering that different species have different mechanical properties that can be utilised for different purposes, but this requires a specific culture of « doing with wood ».

Anticipating what will be the demand for products designed in wood is an important challenge for the wood industry. Trends at the end of this century indicate that growth in world wood demand is approximately following the growth of world population, a very significant trend since world population is expected to double within the next century.

Building with wood - or wood based materials - instead of, or in addition to other materials, will then have to take into consideration the condition that this should be an efficient use of this material. Three major approaches are followed to answer the question. The first one is the development, mainly for massive wood, of non destructive evaluation techniques (NDE) to reduce uncertainty on the mechanical behaviour. The second approach is related to the modification of wood properties through chemical or mechanical treatments : preservation, densification or stabilisation. The third issue is to make new products using wood as the major raw material, designed to fulfil the requirements of the end-user.

This paper is mainly concerned with the control of wood properties by designing new products with an increased reliability especially with respect to mechanical actions. Two illustrations of probabilistic design are presented in the second section of this paper : one is related to glulam (glued laminated timber), and the second concerns OSB (Oriented Strand Board). In the first part of this paper some consequences of heterogeneity of wood on its properties at various scales are discussed.

2. Control of properties

2.1. MICROSTRUCTURE

The physical and mechanical properties of heterogeneous materials are governed by their structure. Saying that it is important to define a relevant scale of description of the structure. For wood the unit element that is usually considered is the fibre (or tracheid in the case of conifers) and a conventional representative elementary volume would then be composed of a network of fibres arranged in the longitudinal direction and in the transverse plane of the material, as illustrated in figure 1. The characteristic length of a fibre ranges approximately from 1 to 3 mm. As we can see on this figure other elements of the structure appear at this microscale : transverse anatomic structures, such as rays, are present in most species and act as reinforcing members in the radial direction. Vessels in the earlywood of ring porous hardwoods are similar to large voids (figure 1b), and even the shape of individual fibres in the transverse plane exhibits some randomness. The actual wood structure at this scale is much more complex than this simplified representation ; however the cellular network of figure 1 is often used in micromechanical modelling of wood, and is also adequate for the description of small strands or particles.

G. N. Frantziskonis (ed.), PROBAMAT – 21st Century: Probabilities and Materials, 65–72.
© 1998 *Kluwer Academic Publishers.*

66

Various aspects of the mechanical behaviour of wood can be analysed at the scale defined in figure 1. Figure 2 presents for instance different types of buckling in the longitudinal and transverse directions, leading to different macroscopic behaviours of wood in compression [1] [2].

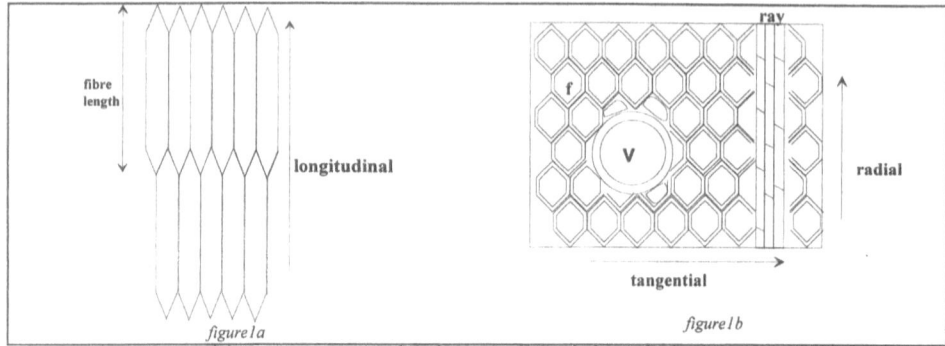

Figure 1.   Schematic representation of wood microstructure in the longitudinal and transverse directions
(v : vessel ; f : fibre or tracheid)

The mechanisms of deformation of wood in compression lead to an apparent plasticity, although damage is often observed within the cell walls. Such phenomena are strongly dependent on the moisture content in wood and on the temperature of the process, both parameters acting as softening agents. Understanding these mechanisms is of particular interest in plasticization or densification processes.

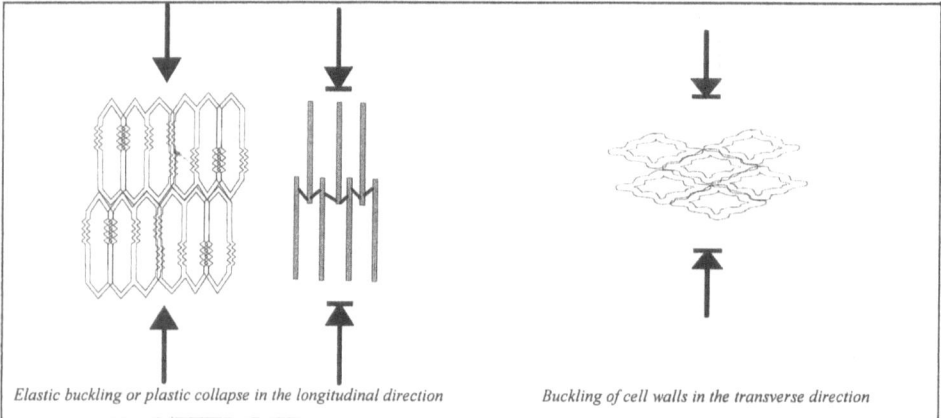

Elastic buckling or plastic collapse in the longitudinal direction   Buckling of cell walls in the transverse direction

Figure 2   Compression of wood at the microstructural scale leading to an apparent plasticity
(after Gril and Norimoto [2])

Wood in tension exhibits a brittle behaviour, with a much lower resistance in the transverse directions (radial or tangential) than in the grain direction (longitudinal). At the scale of the microstructure local discontinuities between anatomic elements, for instance at the boundary of a ray, act as weak planes for crack propagation. The middle lamellae which ensures the contact between fibres, is also a strong determinant of failure in tension.

2.2. CONTROL OF MACROSCOPIC PROPERTIES

At the mesoscale, corresponding to small clearwood specimens composed of several growth rings but excluding macrodefects such as knots, wood exhibits a pseudo-periodic structure in the transverse plane with successive layers of low density wood (earlywood) and high density wood (finalwood). This is at least true for heterogeneous woods, including most conifers. Periodicity in wood structure originates in the periodicity of wood growth in temperate regions. An important feature of the mesoscopic behaviour lies in the cylindrical orthotropic symmetry , leading to a curvature of layers when wood is close to the pith of the tree (juvenile wood).

Ring curvature may result in a heterogeneity of stresses or strains in wood submitted to simple loads (figure 3), and consequently affect failure across the grain. The example of figure 3 illustrates the case of a lamination in a GLULAM beam which undergoes transverse stresses. Depending on the boundary conditions the strains or stresses in the direction of the load vary from the sides to the centre of the lamination, leading to structural scale effects. This problem has been treated recently by [3]. The strength perpendicular to grain of glulam was shown to decrease as the width of specimen increased. Similar results have been obtained recently in our Laboratory for maritime pine glulam specimens loaded in the perpendicular-to-grain direction (figure 4). Besides the reduction of the average strength, an increase of specimen width tends to reduce the dispersion around the mean.

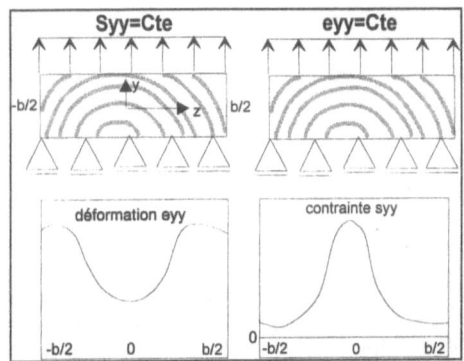

Figure 3 : heterogeneity of stresses or strains in a wood specimen submitted to radial tension

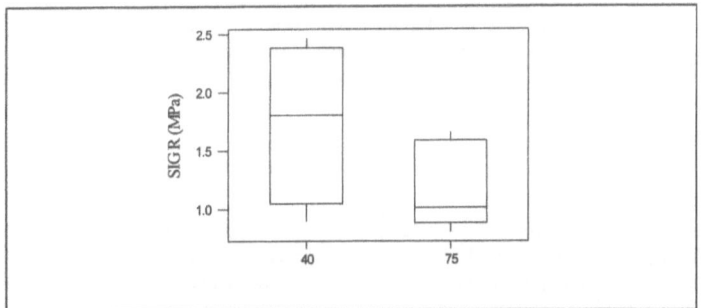

Figure 4   Width effect on the strength (SIGR) of glulam in the perpendicular-to-grain direction
(width : 40 mm on the left ; 75 mm on the right)

Compression of wood at the mesoscale often results in strain localisation effects. Macroscopic slip planes develop when compression is applied in the longitudinal direction, whereas in the tangential direction compression failure often concentrates in the earlywood zones.

The macroscale usually refers to timber, including knots and other defects (resin pockets, grain angle, drying cracks). At this scale the variability of wood properties is large, especially with regard to failure. Part of this variability can be explained by the spatial heterogeneity of the structure for large members. Visual inspection of beams is usually not efficient for the prediction of failure. At present machine strength grading is one of the most powerful methods to control wood heterogeneity and predict the strength of bending members. It is based on the correlation between a local modulus of elasticity (MOE) and a local strength. Lengthwise variations of both parameters on beams have been analysed by several authors for various visual grades of lumber [4][5].

Other methods for detecting heterogeneity in massive wood and predict strength values are currently developed. Most of them have been recently reviewed by Rouger [6]. X-ray spectral analysis seems to be one of the most promising techniques, however the problem of moisture content needs to be solved. The development of such techniques brings out a detailed information on the heterogeneity, but requires efficient statistical models for strength prediction. An illustration of the use of X-ray profiles and correlation between density and strength has been proposed recently by Renaudin and Breysse [7] to predict the failure of wood in longitudinal tension with a probabilistic approach.

3. Design of wood based materials

The objective of design here is a better control of material heterogeneity in order to reduce uncertainty on the macroscopic behaviour. This can be achieved through a reduction of the dimensions of wooden constituents within the composite. Classical types of wood elements in composite materials are shown in figure 5. One specificity of wood based materials is that wood remains the major constituent of the product. In a Medium Density Fibreboard (MDF) for instance wood fibres represents around 90% of the overall mass. The choice of a particular adhesive is mainly guided by its performances with respect to environmental conditions (moisture content, temperature...).

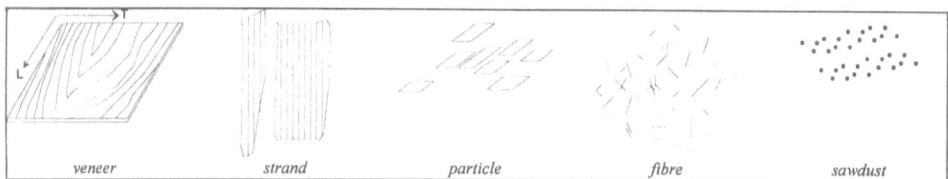

| veneer | strand | particle | fibre | sawdust |

Figure 5   different types of constitutive elements in wood based materials

On the other hand reducing the dimensions of elements has a negative effect on cohesion and creep [8]. A balance between material homogeneity and mechanical performances is then one aspect of the design of wood composites. The anisotropy of wood can be used to reinforce material properties in a particular direction. This is the case of Oriented Strand Boards (OSB) for instance. If the orientation of wooden elements in the material is completely random (particle boards), then the macroscopic behaviour becomes asymptotically isotropic.

The following examples illustrate current research at LRBB concerning the design of wood based materials. The first study is related to the elastic properties and strength of glulam members. A stochastic finite element method is used to calculate the strain and stress fields in the beam submitted to bending loads. This example shows that this type of construction of wooden members using thin laminates has a positive effect on the variability of elastic properties, but also induces a ductile behaviour due to load sharing effects. The second study concerns an elastic modelling of Oriented Strand Boards based on the description of the structure.

3.1. PROBABILISTIC DESIGN OF GLUED LAMINATED BEAMS (GLULAM)

Glulam is a multilayered beam composed of endjointed wooden laminates. The mechanical behaviour of the beam depends on the properties of laminates and of endjoints. Failure in glulam can occur either in wood or in a fingerjoint. Consequently no failure will occur in glulam if there is no failure in wood nor in endjoints, which can be written, under the assumption that the characteristics of wood and endjoints are independent :

$$1 - F_{GL} = \left(1 - F_W\right)\left(1 - F_J\right) \qquad (1)$$

Where F_{GL} : cumulative density function of glulam ; F_w : cdf of wood ; F_J : cdf of joints.
On the other hand failure in wood can occur in a clearwood zone or in a defect (knot). With a similar assumption we could write

$$1 - F_W = \left(1 - F_{CW}\right)\left(1 - F_d\right) \qquad (2)$$

where F_{CW} is the cdf of defect free wood and F_d is the cdf of weak zones. The former can be obtained from tests on small specimens, and the second is indirectly estimated from tests on lumber. The previous equations provide a basis for the statistical modelling of glulam failure, based on the heterogeneity of wood and discontinuities induced by reconstitution. However this method does not account for progressive failure, which can occur due to stress redistribution effects in the laminated beam. Moreover the variations of elastic properties between, and within laminates, affect the stress field (because of a load sharing effect) and therefore the probabilistic response of the structure.

A more complete approach has been developed by Faye *et al* [9], which accounts for the spatial variability of properties in laminates. The mechanical behaviour of fingerjoints and gluelines between two laminations is considered separately. For the prediction of elastic properties a simplified 1D beam approach with a stochastic rigidity can be used. A discretisation of the beam into elements of statistically representative length is first necessary : the representative length is related to a pre-defined <u>scale of fluctuation</u> [10][11]. For each statistical element an equivalent section modulus is calculated from the random values of local elastic properties (YOUNG and shear modulii) and a local section shear factor is computed (this factor is equal to 6/5 for a homogeneous section). This method enables to compute a deflection curve for various kinds of loads.

One interesting application of this work, within the context of this paper, was to relate the statistical properties of wood to those of glulam. It is shown for instance that glulam has a much more homogeneous behaviour than the corresponding massive wood. Figure 6 represents the coefficient of variation of the global rigidity of glulam *vs* the C.O.V. of massive wood Modulus of elasticity (MOE) : a 20% variation of MOE between laminates results in a 4% expected variation of the rigidity of glulam. This result is confirmed by published data on various woods. In figure 6 is also represented the evolution of the section shear factor *vs* the C.O.V. of elastic properties of laminates (the ratio between YOUNG and shear modulus is kept constant). This factor is slightly higher than 6/5, and increases with material heterogeneity.

NB : the section shear factor is computed from the following equations, derived from the elastic theory of composite materials :

$$k_y = \frac{1}{K_{Ui}} \sum_j E_j \int_{S_j} g(y,z) y dS$$

$$\nabla^2 g = -\frac{E_j GS_{Ui}}{G_j K_{Ui}} y \text{ in the section S}$$

$$\frac{\partial g}{\partial n} = 0 \text{ on the boundary } \partial S$$ (3)

$$G_j \frac{\partial g}{\partial y} \text{ is continuous}$$

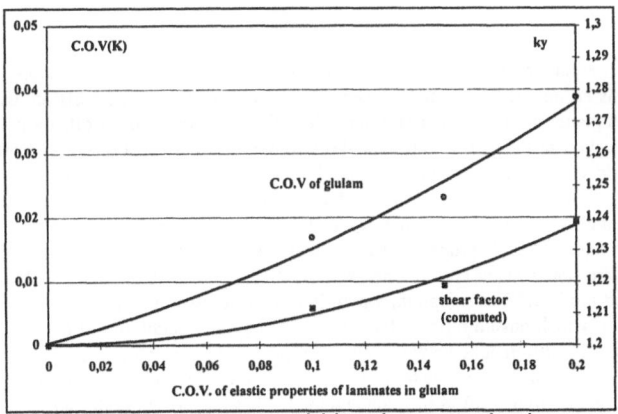

Figure 6   Heterogeneity of glulam vs heterogeneity of wood

Progressive failure : because of the variability of elastic properties within and between laminates the stress field in the beam is not uniform. Stresses are therefore computed by a finite element method. As before, a statistical mesh is defined in order to describe spatial variations of elastic constants and strength within the beam. This statistical mesh accounts for the distribution of defects and fingerjoints, so that a given element will either be a clearwood element, a defect element (weak point) or a fingerjoint element. Special elements are also introduced to compute stresses along gluelines. The finite element mesh is defined afterwards for the computation of stresses in the beam.

70

The probability of failure in one element is deduced from the equivalent Weibull stress in the element, assuming that failure in tension is brittle, whereas failure in compression is ductile (see section 1). An iterative procedure has been developed to calculate a new stress field after failure has occurred in one element and determine the next probable failure. Progressive failure can be simulated by this method, and the probabilistic behaviour of glulam is computed using Monte Carlo simulations.

Two illustrations of failure are presented in figure 7 for the same distributional characteristics of properties of laminates. Failure along gluelines induces a delamination effect, as shown in the upper example of figure 7. Failure always starts in tension at a defect or a fingerjoint. Localised failure happens when defects and/or fingerjoints are concentrated in the same area, as in the second example of figure 7.

This type of modelling becomes possible with the actual capacities of numerical computers, and should improve the possibility of designing glulam with respect to failure properties.

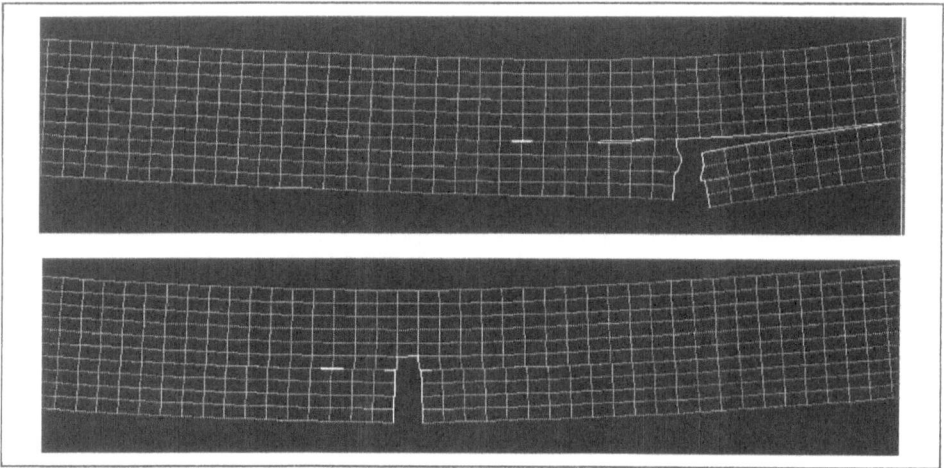

Figure 7   Computed failure in glulam using a stochastic finite element method

3.2. ELASTIC DESIGN OF OSB USING HOMOGENISATION TECHNIQUES

Panels represent a different way of using wood as a raw material for structural applications. For such products, in plane anisotropy is an important aspect of design. Structural boards such as OSB are usually composed of several mats (layers) made of strands, particles, and glue. The adhesive is most often a phenolic resin and its volume (or mass) fraction in the mat is low. The elastic properties and the directions of orthotropy can differ between layers, provided a symmetry around the central pane of the panel is respected.

Properties of OSB are governed by production variables, such as choice of raw material (strand characteristics), dimensions and orientation of strands in the mat, residual porosity after pressing.

We can define the bounds of elastic properties by classical homogenisation assumptions. On the left of figure 8 effective properties are obtained by considering that the stress field in the material is homogeneous. This is conventionally called the VOIGT assumption, which is a realistic assumption at least for small particles. For large strands however, which have a high probability to overlap the assumption of a homogeneous strain field seems more correct. The differences between effective properties predicted according to one or the other assumption are very significant. This is due to the strong anisotropy of wood (E_L/E_T=15 to 20).

A predictive model for designing OSB in bending and torsion has been recently developed at LRBB. The model will not be detailed here, the main interest for this paper being to show how far it is possible to control the properties of such products. In this model a mat is composed of 3 distinct phases : large overlapping strands, small particles, and glue. The volume fractions of large and small strands is derived from the statistical distribution of particle size in the mat, and the percentage of overlapping strands. The former variable is assessed by analysing a representative sample of strands composing a mat in production conditions. The second parameter is calculated through simulations of random mats.

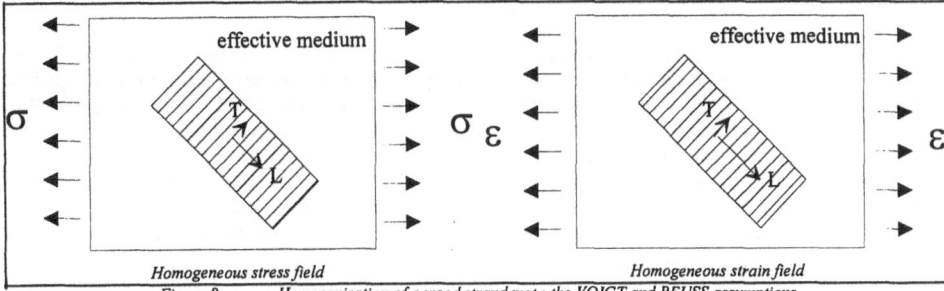

| Homogeneous stress field | Homogeneous strain field |

Figure 8   Homogenisation of a wood strand mat : the VOIGT and REUSS assumptions

Another important feature of the model concerns the orientation of strands with respect to the main axes of the mat. Orientation refers to the angle between the fibre direction in the strand and one main direction of the mat. In industrial conditions the particles are oriented in such a way that larger strands will have a higher probability to fall in the direction of the mat. The model takes into account this constraint by considering a conditional pdf of orientation given the size of strands. Practically the orientation of particles is assumed to be uniformly distributed between two size-dependent bounds. The validity of this assumption has not been verified yet, and further investigations on the structure of industrial panels are required.

As indicated before stress heterogeneity in the medium is not considered except for large strands. This assumption provides a good prediction of anisotropic elastic properties of OSB. The elastic constants calculated for various directions in the plane of the panel have been compared to off-axis results in bending of OSB specimens. The density of wood in the specimens was estimated and used as an input variable for the evaluation of strand properties, according to the following equation :

$$S_{ij}^{-1}(d,12\%) = \overline{S_{ij}^{-1}} + k_{ij}(d - 0,45)$$ (4)

where S_{ij} are the terms of the compliance tensor for wood, d is wood density at a standard moisture content, and k_{ij} is a density correction coefficient. This approach of wood property prediction has been proposed first by Guitard [12].

As shown in figure 9 the anisotropic ratio E_0/E_{90} for the panel is around 2.5, both in simulations and experiments. In the case of off-axis experiments the model tends to underestimate the experimental values. However this approach seems to provide an efficient tool for controlling the elastic properties of a wide variety of panel products.

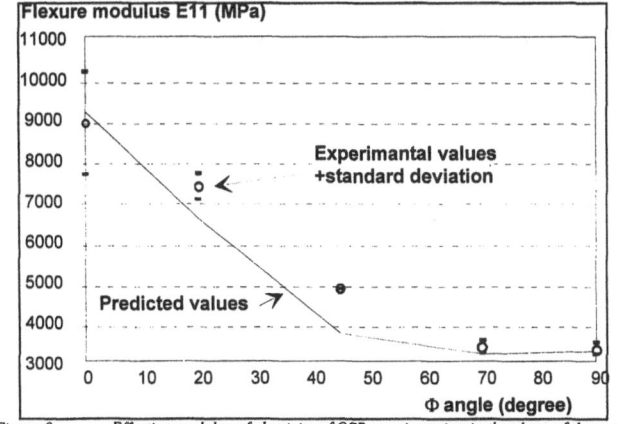

Figure 9   Effective modulus of elasticity of OSB vs orientation in the plane of the panel
Comparison between simulation and experiments

72

4. Conclusion

« Doing with wood » has now to take into consideration the diversity of products that use wood as the major raw material. Wood composites give an opportunity to optimise the use of forest resources and provide competitive products with a renewable raw material. Mechanical design of wood based materials is a developing field of research. As we have seen the mechanical behaviour of reconstituted beams or panels differ from the mechanical behaviour of wood. The result should lead to a better control of product properties but this is not always the case, because a variety of wood properties are still not controlled through design. Creep for instance is an intrinsic characteristic originating in the polymeric nature of wood fibres. Wood based panels exhibit an amplified viscoelastic behaviour as compared to massive wood. Long term performance, i.e. creep and duration of load, will therefore be the main subject of further investigations.

5. References

1. François, P. and Morlier, P. (1993) Multiaxial plasticity of wood in compression, *COST Action 508 workshop on Wood : plasticity and damage*, Limerick, April 1-2, 1993, 35-55.
2. Gril, J. and Norimoto, M. (1993) Compression of wood at high temperature, *COST Action 508 workshop on Wood : plasticity and damage*, Limerick, April 1-2, 1993, 135-144.
3. Aicher S. and Dill-Langer, G. (1996) Influence of cylindrical anisotropy of wood and loading conditions on off-axis stiffness and stresses of a board in tension perpendicular to grain, *Otto Graf Journal* 7, 216-242.
4. Czmoch, I., Thelandersson S. and Larsen, H.J. (1992) Simulation of lengthwise variability of bending strength of structural timber, *Building Aerodynamics Research Group*, Chalmers University of Technology, Göteborg, Sweden.
5. Lam, F., Wang, Y.T. and Barrett, J.D. (1994) Simulation of correlated nonstationary lumber properties, *Journal of Materials in Civil Engineering*, 6(1), 34-53.
6. Rouger, F. (1997) Classement du bois pour un emploi en structure, *International Conference Design industriel, Architecture et Rhéologie du Bois, Bordeaux, 17-21,March, 1997* (in press).
7. Renaudin, P. and Breysse D. (1996) Un modèle probabiliste pour la prédiction de la résistance de l'épicéa en traction simple, *Annales GC bois* 1, 43-52.
8. Dinwoodie, J.M. and Bonfield, P.W. (1995) Recent European research on the rheological behaviour of wood based panels, *COST Action 508 workshop on mechanical properties of panel products*, 22-23, March, 1997, Watford, 5-38.
9. Faye, C., Lac, P. and Castéra, P. (1996) Application of SFEM to the design of glued laminated timber beams, *Proceedings European Workshop on Application of Statistics and Probabilities in Wood Mechanics*, 22-23, february, 1996 (in press).
10. Castéra, P. and Faye, C. (1996) Etude de l'effet de paramètres de fabrication sur les propriétés élastiques en flexion de poutres en bois lamellé collé par une approche probabiliste, *Annales GC bois* 1,1-9.
11. Vanmarcke, E., Grigoriu, M. (1983) Stochastic finite element analysis of simple beams, *Journal of Engineering Mechanics*, vol. 109(5), 1203-1214.
12. Guitard, D. (1987) *Mécanique du matériau bois et composites*, CEPADUES Editions, Toulouse.

CRACK ROUGHNESS

A. TANGUY, M. GOUNELLE AND S. ROUX
Physique et Mécanique des Milieux Hétérogènes,
École Supérieure de Physique et Chimie Industrielles,
10 rue Vauquelin, 75231 Paris cedex 05, France.

1. Introduction

In heterogeneous materials, cracks are never flat. This observation is how-ever trivial, as the presence of different phases in the material, (voids, in-clusions, ...) produces stress intensity factor (SIF) modulation along the crack front which has to match the different phase toughness, and hence the crack is deflected, retarded or accelerated leaving in its trail a rough crack surface.

What appears as less trivial is the observation that the crack surface display modulations over scales which are *much larger* than the typical size of the heterogeneities. For instance, the power spectra of profiles $z(x)$ taken along the crack surface can generally be well fitted by a power-law

$$\langle |\tilde{z}(k)|^2 \rangle \sim k^{-1-2\zeta} \tag{1}$$

We observe that the amplitude of Fourier components increases as the wavenumber, k, decreases (or wavelength $2\pi/k$ increases).

The above power-law suggests also a surprising property: No charac-teristic length scale appears in the auto-correlation function of the height along the crack surface. This in turn implies that the surface topography is *self-affine*, (see e.g. Ref.[1] for a general introduction) i.e. it fulfills a pe-culiar scale invariance. Upon a rescaling of the distances along the mean crack plane (x, y), by a factor λ, the surface will be statistically invariant if the normal direction is scaled by factor μ which depends on λ. The scale transformation is thus

$$\begin{cases} x' &= \lambda x \\ y' &= \lambda y \\ z' &= \mu(\lambda)z \end{cases} \tag{2}$$

73

G. N. Frantziskonis (ed.), PROBAMAT – 21st Century: Probabilities and Materials, 73–91.

The new surface parametrization $z'(x',y')$ obeys the same properties as the original surface $z(x,y)$. Such transformations should form a group, and thus μ has to be a homogeneous function of λ. This implies the existence of a parameter ζ, *the roughness exponent*, such that

$$\mu = \lambda^\zeta \qquad (3)$$

where we have used the fact that $\mu(1) = 1$.

This self-affinity imposes a number of properties. A straightforward application concerns the standard deviation of the height w over a window of size δ.

$$w^2(\delta) = \left\langle \frac{1}{\delta} \int_x^{x+\delta} z(x')^2 \, dx' - \left(\frac{1}{\delta} \int_x^{x+\delta} z(x')^2 \, dx' \right)^2 \right\rangle \qquad (4)$$

The above detailed scale transformation imposes

$$w(\delta) = \delta^\zeta w(1) \qquad (5)$$

Therefore the roughness of the surface *depends on the range of scales* over which it is estimated. Another consequence is that the power spectrum of profiles taken along the surface obeys a power-law decay such as Eq.(1).

The statistical analysis of crack surfaces along these lines dates back from the pioneer work of Mandelbrot et al[2], and has been analysed and reported since then for a wide variety of materials and fracture conditions. We refer the reader to a recent review[3] for a detailed presentation of these numerous works.

Finally as a final word in the introduction, it has been noted that the experimentally determined value of the roughness exponent lies in many cases in a narrow interval around $\zeta \approx 0.8$ suggesting a *universal* value of this exponent[4]. Although this suggestion has received a large support (see e.g. [5]), in specific cases, departure from this value have been reported at extremely low velocities, where the exponent falls in the range $\zeta \approx 0.5$[6].

In spite of the large amount of experimental work devoted to this question, no satisfactory model accounts yet for the self-affine geometry of crack surfaces and *a fortiori* on the eventual universality of the roughness exponent.

2. Modelling crack front growth

One way of attacking the problem has been suggested by Bouchaud et al[7]. These authors suggested to model the motion of the crack front in the presence of noise. Let us consider a semi-infinite crack whose mean plane is the (x,y), with a crack front parallel to the y axis, and progressing along the positive x direction. At time t, the crack front can be parametrized by the two functions $x(y,t)$ and $z(y,t)$. A general equation of motion of the front may be written

$$\begin{cases} \dfrac{\partial x}{\partial t} &= v + \mathcal{L}_{//}[x,z] + \eta_{//}(x(y,t),y,z(y,t)) \\[2mm] \dfrac{\partial z}{\partial t} &= \mathcal{L}_{\perp}[x,z] + \eta_{\perp}(x(y,t),y,z(y,t)) \end{cases} \qquad (6)$$

This equation whose notations are detailed below is written in the spirit of considering a perturbation around a straight front propagation, thus retaining only the most significant terms at large distance for small deviations of z and $x - vt$, i.e. $|\partial x/\partial t - v| \ll v$, $|\partial z/\partial t| \ll v$, $|\partial x/\partial y| \ll 1$ and $|\partial z/\partial y| \ll 1$. A first order time derivative is chosen, in the spirit of a (viscous) dampened dynamics. This choice is not obvious, and in particular it neglects the eventual role of inertia. For brittle materials such as glass and ceramics, the transition from a miror crack face to the "mist" and "hackle" topography is generally believed to be due to inertia effects where the crack velocity is a fraction of the Rayleigh wave speed. A first order time derivative might however be relevant for quasistatic loading in the case of a visco-elastic material, or when the toughness exhibits a linear velocity dependence. v accounts for the mean velocity prescribed to the crack tip by the external loading in the absence of heterogeneity. The two η functions represent the material heterogeneity. The mean value of the toughness can be absorbed in the loading designed to impose a mean velocity v. Thus η is a zero-mean noise term, which does not depend on time but rather on the spatial coordinates. This term is thus characterized by its spatial correlation. A simple choice consists in assuming a white noise:

$$\begin{cases} \langle \eta(x,y,z) \rangle &= 0 \\ \langle \eta(x,y,z)\eta(x',y',z') \rangle &= D\delta(x-x')\delta(y-y')\delta(z-z') \end{cases} \qquad (7)$$

Finally, the key point lies in the expression of the two functionals $\mathcal{L}$ which give the leading order influence of the crack front geometry on the local stress intensity factor. In Ref.[7], a simple line tension term was suggested, and hence the functionals $\mathcal{L}$ contained simple linear second order differential operators. Additional quadratic terms (products of first order space derivatives) were also included on phenomenological grounds, as they are allowed by the symmetry of the problem and they affect the scaling behavior of the crack front motion. Upon the additional assumption that the noise term was time-dependent, with only short-range correlation, the problem could be mapped on the equation of motion of a directed polymer in three dimensional noisy environment, a model introduced and solved by

Ertaş and Kardar[8, 9]. In fact, as we will show below, this expression for $\mathcal{L}$ is not suited to our problem and we have to deal with a non-local kernel in space. If in addition, the crack surface behind the crack front modifies the local SIF (to leading order) then the kernel should also include a time dependence. Recently a full elastodynamic perturbation analysis of the propagation of a crack presenting weak roughness has been obtained by Movchan and Willis[10]. This opens the way to a general formulation of the propagation, a route which still remain to be explored, in spite of recent case studies in the context of elasto-dynamics[11].

3. Two-dimensional problems

In order to progress in the understanding of the crack roughness, it is natural to simplify the problem. One first step is to go to two-dimensional problems. There two problems can be considered: either the y or the z component can be eliminated in specific geometries. In the first case, we can consider fracture of a fibrous material so that the crack front is always a straight line parallel to y. The crack front is thus characterized by $x(t)$ and $z(t)$. The local SIF can be computed for different loading modes. The particular case of a self-affine geometry has been addressed for mode III cracks through a perturbation treatment performed through conformal mapping[12]. Mode I cracks can be addressed similarly[13]. Experiments have been performed in this particular case using either thin sheets of paper[14] (ignoring however buckling ahead of the crack) and wood[15]. Both studies lead to roughness exponent of the crack surface equal to $\zeta_{2D} \approx .7$. Let us note that the previous equations as introduced in Ref.[7] becomes a simple random walk, thus implying a roughness exponent $\zeta = 1/2$. From a qualitative level, this can be regarded as a success in terms of providing a theoretical framework able to account for the self-affinity of crack surfaces, with a universal roughness exponent. This suggests that this generic approach might lead to a theoretical explanation for the roughness once a more physical functional $\mathcal{L}$ is taken into account.

The second case consists in imposing $z = 0$. This case is encountered when dealing with adhesion problems, or interfacial decohesion. There obviously, no roughness is left behind the crack front. However, the crack front itself can exhibit a meandering geometry $x(y, t)$. For this geometry, Gao and Rice[16] have worked out the expression of the SIF as

$$K(y) = K_0 \left(1 + \frac{1}{2\pi} \int \frac{x(y) - x(y')}{(y - y')^2} \, dy' \right) \tag{8}$$

Let us assume that the toughness of the material is linearly velocity dependent and display a space variability, $K_c = K_{c_0}(x, y) + \kappa(x, y)v$ where v is the crack tip velocity. Equating the local SIF with K_c, and assuming a small amplitude of variation for the toughness gives a first order dynamics similar to the previously presented one, with a linear but non-local expression for the functional $\mathcal{L}_{//}$

$$\frac{\partial x(y,t)}{\partial t} = v + AG \star x + \eta(x,y) \tag{9}$$

where A is a material-dependent constant (including the velocity dependence of the toughness), $\star$ denotes a convolution product and G is the $1/y^2$ kernel of the above expression of the SIF. Let us note that the latter can also be expressed as the Hilbert transform of $\partial x/\partial y$ up to a constant multiplicative factor.

$$G \star x = \mathcal{H}\left(\frac{\partial x}{\partial y}\right) \tag{10}$$

The presence of first derivatives in the linear part of the functional $\mathcal{L}$ is forbidden by symmetry if only local terms are considered. The Hilbert transform restores the correct symmetry at the expense of non-locality. We also note that implicitly only the positive part of the r.h.s. term is considered in Eq.(9) as in the local velocity-toughness constitutive law. In the following section we will consider this interface crack model as a test problem where the basic equations are mechanically motivated.

4. A pinning-depinning transition

Let us here discuss more precisely the loading to be considered. In the above equation, the loading is hidden in the term v to be understood as the velocity of the crack in the absence of disorder $\eta = 0$ so that $\partial x/\partial y = 0$ and hence $G \star x = 0$. We note however, that the loading can be time dependent, if $v(t)$ is considered, or even more fancy control can be envisioned if v is computed from the actual crack geometry, provided v is finally expressed as a y-independent parameter.

We also note that v is not necessarily equal to the long-time average of the front position. In this sense, it is more sensible to interpret v as a *forcing term* (proportional to the large scale averaged SIF since we assumed a linear dependence of the toughness with the velocity). Actually, if v is low enough, one can consider situations where the front comes to a stop, being blocked in a front geometry where the high SIF part of the front are facing high toughness regions, and the low toughness is gathered in parts of the front where the SIF is smaller. However, if v is large enough then the front can never be trapped all along its length, and thus it advances without limit. Therefore, we may expect a critical $v = v_c$ at which the crack may actually depin and reach a non-zero average velocity. This depinning transition is of crucial interest, since apart from the meandering of the crack front which was our initial motivation, we may expect other universal features close to the depinning transition, such as a singular behavior of the large distance and time average of the crack front $V = \langle \partial x / \partial t \rangle$ with the loading above the critical value $(v - v_c)^\theta$. Using the fact that v is the large scale SIF, we recover a macroscopic velocity dependent toughness which differs from the microscopic one if $\theta \neq 1$. Moreover, the exponent θ is again expected to be universal, (independent of the microscopic details of the system). It is to be noted that a first-order transition is not excluded, in which case at the depinning forcing, the crack would jump suddenly to a finite velocity, however, it seems likely that such a discontinuous velocity-toughness relation would only occur if inertia effect are taken into account. Below the critical forcing, a straight crack will advance by a finite distance u before getting blocked in a particular configuration. This distance is expected to diverge as the critical forcing is approached as $u \propto (v_c - v)^{-\chi}$. Again the divergence of this "susceptibility" is universal.

There are two classical ways of controlling the system, one is to prescribe a constant forcing (the natural experimental procedure), while the other is to adjust the forcing instantaneously so that the macroscopic crack velocity is constant. Although much more difficult to perform experimentally, it is a simple mean to ascertain that inertia effects remain indeed negligible. In the latter case one can achieve an infinitesimal motion so that the crack will be as close as one may wish from its depinning threshold. Such a forcing is at the heart of "self-organised criticality"[17], where an infinitesimal order parameter is imposed to a system so that the system naturally evolves toward its critical point without having to fine-tune the usual control parameter to its critical value. The easiest way of achieving such a instantaneous control of the loading might be to use a fatigue test, whose amplitude is monitored by the crack motion. Otherwise, at the local level, one expects to see unstabilities as soon as an obstacle is overcomed, and then the dynamics (governed either by "viscous" or inertial effects) might play a determining role even on the macroscopic evolution. We however believe that before including such effects it is important to clarify the more academic situation of a quasistatic crack propagation.

It is important to stress the analogy of this problem with other related physical situations of pinning/depinning transitions. In this class of problems, one finds in particular the very important problem of *vortex pinning* in high-T_c superconductors[18], which maintains the superconductivity below a critical current. This field has motivated a renewal of interest in view of the inadequation of earlier descriptions of the pinning regime. At a larger scale, *solid friction* provides another field where pinning/depinning transition occurs, with few contacts between aperities between two facing surfaces, and non-local elastic interactions mediated by the bulk of the solids[19]. *Wetting* is yet another field where pinning appears naturally. The contact line between two immiscible fluid meniscus and a solid (rough ou chemically inhomogeneous) substrate can be described in similar terms as the crack front. The analogy in this case is in fact very close[20, 21], since the surface tension effects integrated along the meniscus gives rise to a non-local term which is identical to the $\mathcal{L}$ operator of our model (Hilbert transform of the contact line slope).

In these very different domains, various strategies of approach may be considered. The one we will follow belongs to a class of "extremal" models [22, 23] (extremal in the sense that only one point, the first one which will reach its "threshold", is active at a time) where sometimes a direct connection with a regular critical behavior can be established. In this field one finds "invasion percolation", or evolution models (which can be mapped to directed percolation), or "linear interface depinning" models used for dislocations, or fluid drainage/imbibition in porous media.

5. Numerical simulations

In spite of its apparent simplicity (only linear terms are retained in the evolution equation), the quenched noise makes this problem difficult to solve analytically. We have studied this model numerically.

The numerical model consists in considering a discrete crack front characterized by a series of displacements x_i at equally spaced points $i = 1, ..., N$. In order to avoid spurious edge effects, we have implemented periodic boundary conditions (identifying point $N + 1$ with 1). There are many ways of implementing the heterogeneity of the medium. We have chosen one of the simplest which limits the number of free parameters in the model. We assume that there are obstacles in the medium where the local toughness K_c assumes the same value. These obstacles are however randomly distributed in space. Let us imagine first that the loading is such that the crack is blocked at each point on one obstacle. The loading is then progressively increased up to the level where one single point is subjected to a SIF equal to the local toughness. The crack thus moves at this only point and jumps forward at a random distance (uniformly sampled between 0 and 1), where the next obstacle is supposed to lie. Simultaneously, the loading is suddenly decreased so that the crack front is again stopped. Then the loading is raised slowly up to the stage where a new point reaches its threshold, and the same process is carried over and over again. The way the SIF is computed is straightforward. The G function written above has to be modified to account for periodic boundary conditions. It can be shown that $G_{period}(n) = (\sin(\pi n/N))^{-2}$. It is convenient to define $G_{period}(0)$ as

$G_{period}(0) = -\sum_{i=1,N} G_{period}(i)$. With this convention, the local SIF is computed to be updated as

$$S_j \rightarrow S_j + G_{period}(i-j)dx_i \qquad (11)$$

where S_j is the SIF at any current site j for a reference loading, i is the site which moves by the distance dx_i. In order to compute the value of the forcing needed to have the crack advancing, we simply form the difference between the threshold 1 and the local SIF for a reference loading. The site to be moved first is then simply given by the minimum difference, and thus the maximum S.

Other implementation of the disorder can be considered such as a random distribution of toughness at regularly spaced locations, or random toughness at random places, ... We have checked that apart from short distance or small time properties (at scales which are independent of the system size) the behavior of the model was not affected.

6. Results

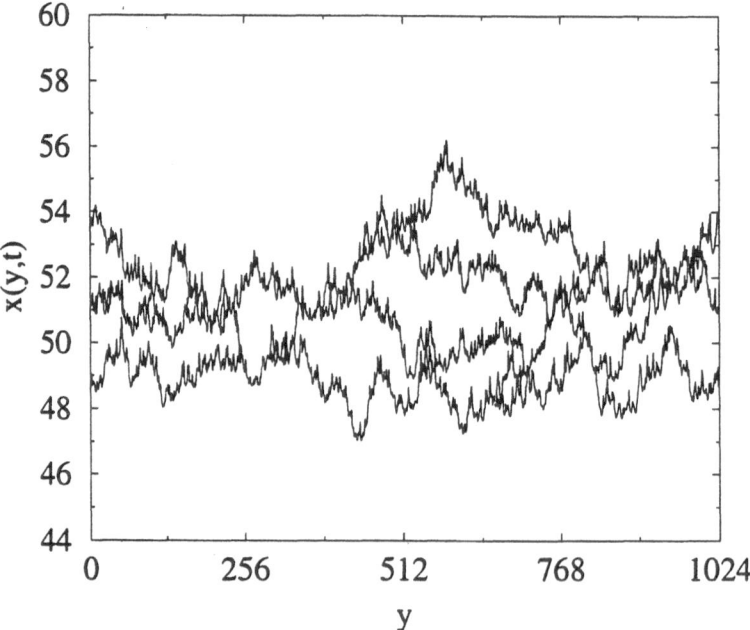

Figure 1. Crack front geometry at different time intervals. The system size is $L = 1024$.

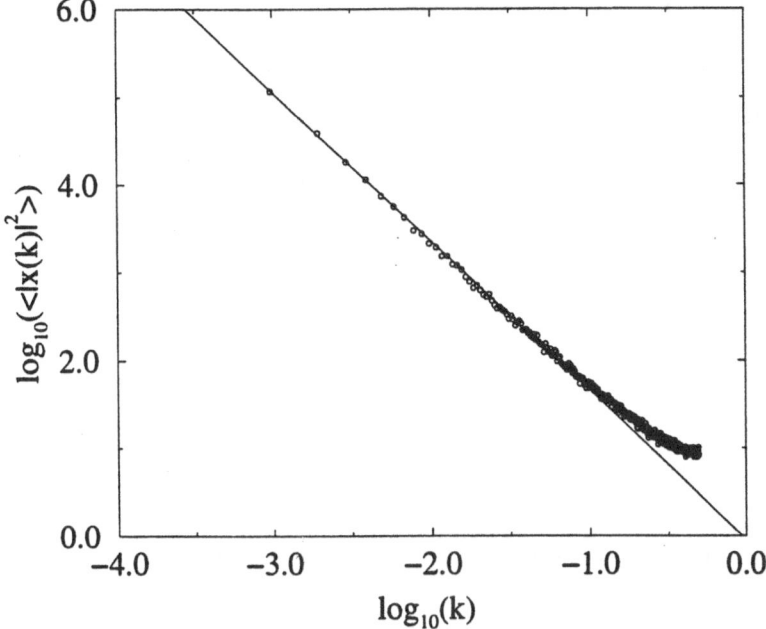

Figure 2. Average power spectrum of the crack front position in log-log scale. The dotted line shows a power-law fit which corresponds to a roughness exponent $\zeta = 0.35$. The system size is $L = 1024$.

Starting from a flat crack front, and applying the above defined rules produces a crack front which develops first a rougher and rougher geometry. This regime is however transient, and finally the roughness saturates to a steady state value after a transient time which depends strongly on the system size. In spite of the interest of this initial regime, we will in the following concentrate on the steady state.

6.1. CRACK FRONT ROUGHNESS

The first result to mention is the fact that the steady state geometry of the crack front displays a self-affine structure. Figure 1 shows the crack front geometry at different time intervals. The average power spectrum of the front position is shown in Figure 2 together with a power-law fit which allows to estimate

$$\zeta \approx 0.35 \tag{12}$$

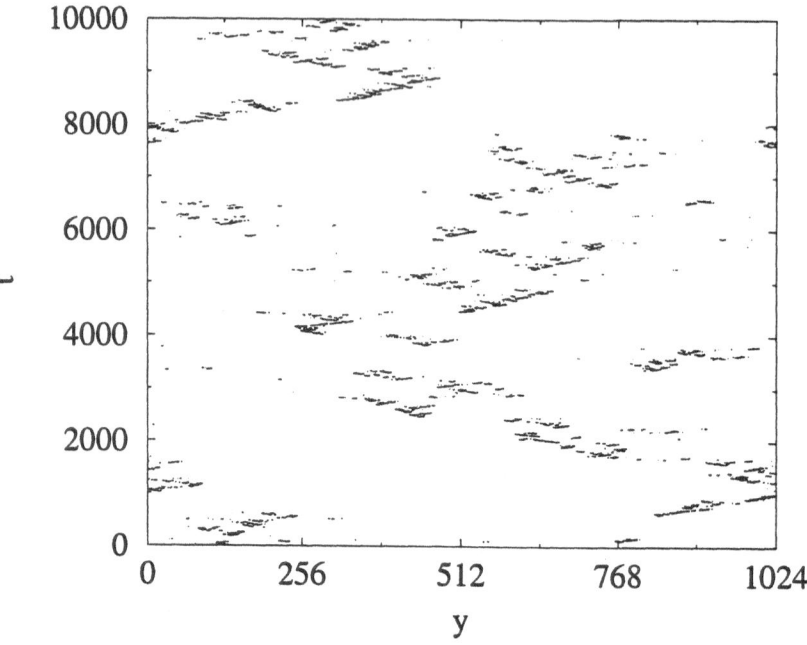

Figure 3. Activity map showing the location of the active sites as a function of time. The system size is $L = 1024$.

As earlier mentioned, the crack surface in this model is a perfect plane $z = 0$, and thus this exponent is *not* to be compared with the crack roughness exponent. It simply characterizes the front geometry.

6.2. ACTIVITY MAP

Each individual move of the crack front is taking place at a single site, and hence the long-range spatial correlation of the crack front can only result from a strongly space and time correlated location of "active sites", i.e. sites where the crack is moving. Figure 3 shows that indeed, the activity map is structured at large scales.

In order to analyse this organisation of the activity map, we propose to study the probability distribution $p(r, \Delta t)$ of distances r between active sites at a fixed time interval Δt. (Note that here time actually refers to the number of individual moves of the crack front, and thus is directly proportional to the mean distance traveled by the front.) For fixed time interval, the distribution p reveals the existence of two different regimes depending on r being smaller or larger than a characteristic scale $r^*(\Delta t)$ which does depend on Δt. At small distances $r \ll r^*$, the probability distribution is uniform, whereas at large distances, $r \gg r^*$, p decreases as a power-law of the distance $p(r) \sim r^{-b}$. Moreover r^* scales as a power-law of the time interval, $r^* \propto \Delta t^{1/z}$, with we have introduced the *dynamic exponent z*. From these two properties, we infer that p can be written in terms of a scaling function r/r^*, up to a simple normalization factor $1/r$.

$$p(r, \Delta t) = \frac{1}{r} \psi \left(\frac{r}{\Delta t^{1/z}} \right) \tag{13}$$

The scaling function ψ exhibits the following behavior

$$\psi(x) \begin{cases} \propto x & \text{for } x \ll 1 \\ \propto x^{1-b} & \text{for } x \gg 1 \end{cases} \tag{14}$$

Figure 4 shows that a good collapse of p distributions obtained at different time intervals can be obtained using a value of the dynamic exponent equal to

$$z \approx 1.30 \tag{15}$$

We also display on the same figure a dotted line which shows a power-law fit of ψ (for $r/\Delta t^{1/z}$) with an exponent $1 - b$ where

$$b \approx 2.0 \tag{16}$$

These new exponents b and z can be understood as follows. The crack front can be seen in the steady state as being pinned on a large part of its length. However, once a point fails, its influence on other points along the crack front is most important in the immediate neighborhood of this point. Therefore $p(r, \Delta t = 1)$ is peaked close to the origin. As time passes, this localized activity has spread over a distance $r^*(\Delta t) \propto \Delta t^{1/z}$, where the memory of the initiation point has been lost. If this region reaches a pinned state then the probability to jump to a remote site located at a distance $r \gg r^*$ is proportional to the influence function $G(r)$ which decays as r^{-2}, hence the value $b = 2$ of the b exponent.

Concerning the dynamic exponent, we propose an argument to relate it to the roughness exponent ζ. Let us consider a starting point (x_0, t_0) in the neighborhood of which the interface is assumed to have a self-affine geometry with the roughness exponent ζ. After a time Δt, the activity has spread over a distance $r^*(\Delta t)$ around x_0. The number of moves which have been necessary to cover the area between the crack fronts at time $t_0 + \Delta t$ and t_0 scales as $r^*.(r^*)^\zeta$, (i.e. the same scaling as $\Delta x.\Delta y$ where we simply set $\Delta y \propto \Delta x^\zeta$ as a result of self-affinity). This number of moves has to be equal to the time delay Δt times the mean crack advance during one move, i.e. a constant of order 1 in our model, hence $\Delta t \propto (r^*)^{1+\zeta}$ or

$$z = 1 + \zeta \tag{17}$$

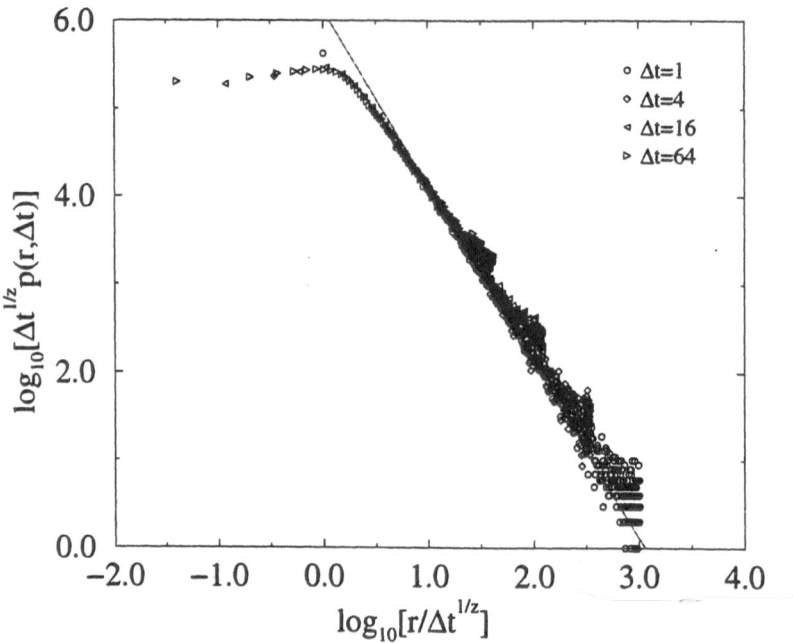

Figure 4. Check of the scaling form Eq. 13, plotting $rp(r, \Delta t)$ as a function of $r/\Delta t^{1/z}$ for different time intervals ranging from 1 to 64. The system size is here $L = 512$.

The latter result is consistent with the independent numerical determinations of $z \approx 1.30$ and $\zeta \approx 0.35$.

Another interesting property of the spatio temporal distribution of activity concern the the statistical distribution of the time interval τ during which a site at the crack front remains static. We consider the probability $p(\tau)$ that this static time is equal to τ. Figure 5 shows the log-log plot of $p(\tau)$ versus τ together with a power-law fit $\tau^{-\delta}$ with an exponent equal to $\delta = 1$.

Such a power-law is in fact a surprise in this context. Indeed an exponent δ reveals that the time series of activity at a particular site form a fractal set of dimension $\delta - 1$ for δ in the range [1,2]. Thus in our model, we reach a fractal dimension equal to zero.

In order to clarify this point, we can formally generalize our model by modifying the non-local operator G. If we choose $G(y) \propto (\sin(2\pi y/L))^{-\beta}$ the free parameter β can be tuned in order to vary the long-distance behavior of the influence function. The interfacial crack model corresponds to $\beta = 2$, whereas for $\beta = 3$ the kernel reduces to a mere second order spatial derivative, an equation known as Edwards-Wilkinson with quenched noise

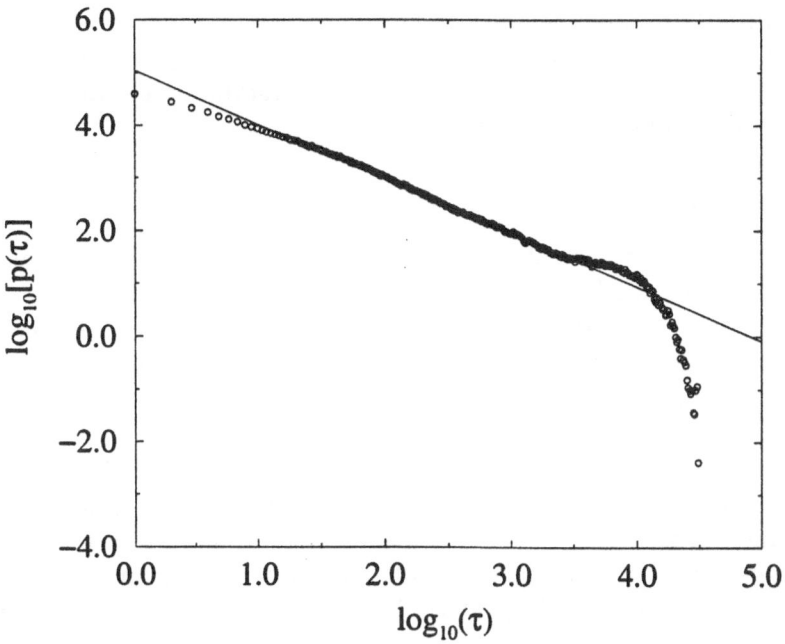

Figure 5. Log-log plot of the inactive time distribution $p(\tau)$ as a function of the time τ. The line is a power-law τ^{-1}. The system size is $L = 2048$.

(see e.g. Ref.[24]). A systematic analysis of the statistical properties of these models for different values of β shows that most exponents mentioned above and below do vary with β. In particular as β increases from 2, the δ exponent gently raises, indicating an increasing fractal dimension for the time series. For lower values of β however, the distribution p ceses to be a simple power-law but rather breaks into two distinct regimes (before falling off fast for $\tau \sim L^z$). First, for short times, p assumes a power-law form with an exponent $\delta = 1$, and then reaches a plateau where p is constant. The reason why p has to break into two regimes can be interpreted as the fact that the fractal dimension cannot assume negative values, although the It is important to note that the extent of this plateau scales as the upper cut-off L^z itself. The emergence of this plateau can then be seen on Figure 5 as a little bump at the end of the distribution. When β reaches 1 or smaller values, the activity becomes uniformly spread over the entire front, and no more space-time correlation can be observed in the map of activity. At this limiting value, one enters the domain where mean-field arguments gives a faithful description of the evolution. The surprising feature is that space and time correlation do not disappear simultaneously. At $\beta = 2$, the temporal distribution of the activity is just at the border-line of being ir-relevant, however, the spatial correlations are still very pronounced, and only disappear for $\beta = 1$. A more detailed understanding of this amazing phenomenon remains to be clarified.

6.3. AVALANCHES

We have already mentioned the fact that the interfacial crack model belongs to the world of pinning/depinning phenomena. In order to insure a constant velocity, the external loading is adjusted instantaneously so that the local SIF matches with the local toughness. Therefore it is of high interest to study the time evolution of the loading as a function of the crack advance. In the following discussion we follow closely a number of properties derived in extremal depinning models, and reviewed by Paczuski et al[23].

The maximum value of the loading defines the depinning toughness v_c. For a smaller force, the crack advances by a finite amount. Let us consider a stage where the loading takes a value v_0 close to v_c, and record the number of move $n(v)$ it could perform before encountering a configuration such that $v > v_0$, i.e. being blocked under a constant loading. This crack advance is usually termed an "avalanche" or a "burst" in the context of these statistical models. The statistical distribution, $p(n; v_0)$ of n takes the following form

$$p(n; v_0) = n^{-\tau} \phi \left(\frac{n}{n^*(v_0)} \right) \tag{18}$$

where $\phi(x)$ is a scaling function which tends to a constant for $x \ll 1$ and de-cays rapidly (faster than any power-law) for $x \gg 1$. Thus, no characteristic size of "avalanche" emerges below a maximum size $n^*(v_0)$. This maximum size is expected to diverge as v_0 approaches v_c as

$$n^*(v) \propto (v_c - v)^{-1/\sigma} \tag{19}$$

This cut-off can easily be translated into a correlation length ξ along the front which gives the spatial extent of the avalanche, through $n^* \propto \xi . \xi^\zeta$, from which we deduce the critical behavior of the correlation length at the depinning transition

$$\xi \propto (v_c - v)^{-\nu} \qquad \text{where} \qquad \nu = \frac{1}{\sigma(1 + \zeta)} \tag{20}$$

This direction which we do not persue any longer here relates the geoemtry and the dynamics of the fracture to the loading. Moreover, the connection with a critical phenomenon is made more explicit with the direct identification of v as a control parameter and v_c a critical point.

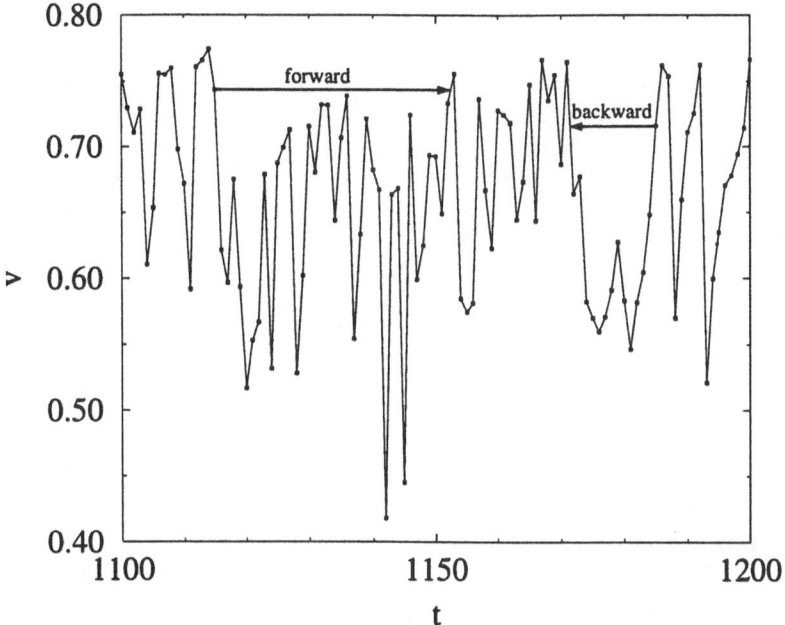

Figure 6. External loading as a function of time. From this signal one can define avalanches either by prescribing a fixed value of v, and considering the intervals where v remains below this limit, or by contructing these avalanches starting from all forcing values $v(t)$ with the statistics issued from the model, and choosing either a forward or backward time direction.

The notion of avalanches as above presented requires the introduction of a particular value of the loading. However, it is also possible to use the evolution of the model to define two distributions of avalanches which does not require the precise identification of the critical threshold. At each time step, t, one can construct the avalanches which correspond precisely to the loading $v(t)$. In fact two such avalanches can be considered, for times either larger or smaller than t, resp. the *forward* and *backward* avalanches respectively as shown in Figure 6. Integrating the avalanche distributions over all times (for all values of $v(t)$), defines the forward and backward avalanche distributions. It is convenient to introduce the cumulative distributions $N_f(t)$ and $N_b(t)$ of (resp. forward and backward) avalanches larger than t. For all models studied in Ref. [23], both of these distributions behave as power-laws (with no other cut-off than those resulting from the finite size of the system). However a striking result shown for these models is that

$$N_f(t) \propto t^{1-\tau_f} \tag{21}$$

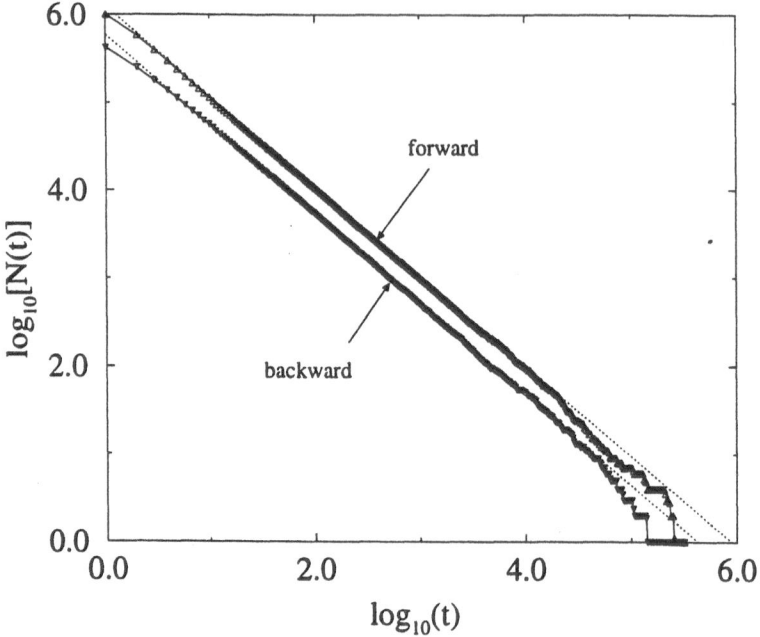

Figure 7. Log-log plot of the distribution of the forward and backward avalanches. The straight lines are power-law fits of exponent -2.

with $\tau_f = 2$ in all dimensions, a super-universal result. On the contrary, the backward avalanche distribution is characterized by an exponent τ_b which is model dependent.

Figure 7 shows both cumulative avalanche distributions for our model. Again, the super-universal result $\tau_f = 2$ is observed. The backward value is surprisingly also measured to be $\tau_b \approx 2$. Both of these fits are plotted as plain lines in this figure.

Such power-laws with the exponent $\tau = 2$ also appears in a series of un-correlated random numbers picked from the same distribution, however, we cannot infer from this result that the forces are uncorrelated. They are indeed strongly correlated. A power spectrum of the signal $v(t)$ does reveal the existence of correlations. Numerically one observes that $\langle |\tilde{v}(\omega)|^2 \rangle \propto \omega^{-0.3}$. Such an exponent reveals a strong anticorrelation. However, one deduces from this that the forward and backward avalanche distributions are indeed insensitive to those correlations and behave just as if the signal was uncorrelated. To reveal those correlations, one can integrate the $v(t)$ function after having substracted the mean value $\langle v \rangle$. The resulting function is the fluctuation part of the dissipated energy. Its power spectrum decays as

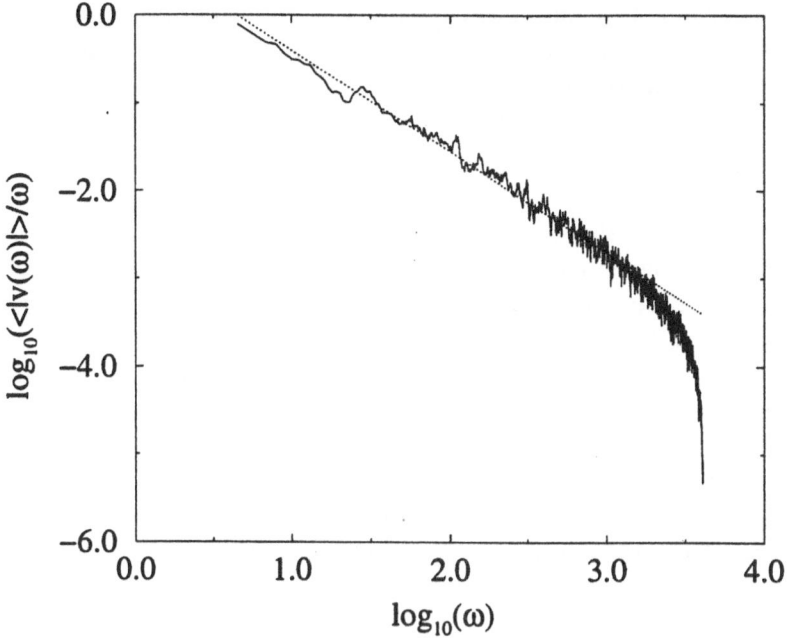

Figure 8. Spectral amplitude of the time fluctuation of dissipated energy. The straight line represents a power-law of exponent -1.15, which implies a roughness exponent of the direct space signal of 0.65.

$\omega^{-2.30}$ as shown on Figure 8, and the fluctuations of the work is a genuine self-affine signal with a roughness exponent 0.65. Working out avalanches on this integrated signal should provide a non-trivial exponent ($\tau = 1.35$ is expected). This suggests an original way of analysing the loading signal experimentally. The relation between this exponent and the previously reported ones remains to be clarified.

Finally before concluding, we note that such an experiment is being performed by Schmittbuhl and Måløy[25] by propagating an interfacial crack between two PMMA plates which have been glued together along two roughened surfaces after a few hours heating. Before decohesion the two plates are perfectly transparent, while after the crack has propagated through the material, the micro-roughness of the surfaces scatters light. A direct observation through the two plates gives access to the crack front position. The statistical analysis of the crack front geometry shows that the crack front does develop a self-affine character with a roughness exponent equal to $\zeta \approx 0.6$. Such a discrepancy between this measured value and the value extracted from the numerical model above presented $\zeta \approx 0.35$ calls for an explanation. One potential reason for this difference may be due to the slow visco-elastic relaxation in PMMA at room temperature, (well below the glass transition).

7. Conclusion

After having recalled the common observation of the self-affine nature of crack surfaces, we have focused on a simple modelling of crack propagation, specialized to the case of interfacial cracks. In spite of its simplicity, this model does reveal non trivial features which suggest that a possible mechanism for generating self-affine surface roughness in quasitatic fracture of heterogeneous materials. This model shows that in addition to the geometric features of the crack surface, the fracture process itself may reveal through intermittency and avalanches, properties which should be accessible exprimentally. This urges to perform well controlled experiments to detect those signatures (through e.g. acoustic emissions).

Acknowledgements

It is a pleasure to acknowledge a fruitful collaboration with J. Schmittbuhl, K. J. Måløy and J. P. Vilotte at an earlier stage of this work. This work has been partly supported by the Groupement de Recherche "Physique des Milieux Hétérogènes Complexes" of the CNRS.

References

1. Feder J., *Fractals*, Plenum Press, (New-York, 1989)
2. Mandelbrot B. B., Passoja D. E., and Paullay A. J., *"Fractal character of fracture surfaces of metals"*, Nature **308**, 721, (1984)
3. Bouchaud E., *"Scaling properties of cracks"*, J. Phys. Condensed Matter, **9**, 4319, (1997)
4. Bouchaud E., Lapasset G. and Planès J., *"Fractal dimension of fractured surfaces: a universal value ?"* Europhys. Lett. **13**, 73, (1990)
5. Måløy K. J., Hansen A., Hinrichsen E.L., and Roux S., *"Experimental measurements of the roughness of brittle cracks"*, Phys. Rev. Lett. **68**, 213, (1992)
6. Daguier P., Nghiem B., Bouchaud E. and Creuzet F., *"Pinning and depinning of crack fronts in heterogeneous materials"* Phys. Rev. Lett. **78**, 1062, (1997)
7. Bouchaud J. P., Bouchaud E., Lapasset G., and Planès J., *"Models of fractal cracks"*, Phys. Rev. Lett. **71**, 2240 (1993)
8. Ertaş D. and Kardar M., *"Dynamic roughening of directed lines"*, Phys. Rev. Lett. **69**, 929, (1992)
9. Ertaş D. and Kardar M., *"Dynamic relaxation of drifting polymers: a phenomenological approach"*, Phys. Rev. E **48**, 1228, (1993)
10. Movchan A. B. and Willis J. R., *"Dynamic weight functions for a moving crack. I Mode I loading"*, J. Mech. Phys. Solids **43**, 319, (1995); Movchan A. B. and Willis J. R., *"Dynamic weight functions for a moving crack. II Shear loading"*, J. Mech. Phys. Solids **43**, 319, (1995)

11. Rice J. R., Ben-Zion Y., and Kim K. S., *"Three dimensional perturbation solution for a dynamic planar crack moving unsteadily in a model elastic solid"*, J. Mech. Phys. Solids **42**, 813, (1994); Perrin G. and Rice J. R., *"Disordering of a planar crack front in a model elastic medium of randomly variable toughness"*, J. Mech. Phys. Solids **42**, 1047, (1994)

12. Vandembroucq D. and Roux S., *" Mode III stress intensity factor ahead of a rough crack"*, J. Mech. Phys. Solids **45**, 853-872, (1997)

13. Vandembroucq D. and Willis J. R., in preparation

14. Kertész J., Horváth V., and Weber F., *"Self-affine rupture lines in paper sheets"*, Fractals **1**, 67, (1993)

15. Engøy T., Måløy K. J., Hansen A. and Roux S., *"The roughness of two-dimensional cracks in wood"*, Phys. Rev. Lett. **73**, 834, (1994)

16. Gao H. and Rice J. R., *"A first order perturbation analysis of crack trapping by arrays of obstacles"*, J. Appl. Mech. **56**, 828, (1989)

17. Bak P., Tang C. and Wiesenfeld K., *"Self-organised criticality: an explanation of $1/f$ noise"*, Phys. Rev. Lett. **59**, 381, (1987)

18. Giamarchi T. and Le Doussal P., *"Statics and dynamics of disordered elastic systems"*, to be published in "Spin Glasses and random fields" ed. Young A. P., World Sci., (Singapore, 1997)

19. Caroli C. and Nozières P., *"Dry friction as a hysteretic elastic response"*, in "The physics of sliding friction", Persson B. N. ed., (1995)

20. Schmittbuhl J., Roux S., Vilotte J. P. and Måløy K. J., *"Pinning of interfacial crack: Effect of non-local interactions"*, Phys. Rev. Lett. **74**, 1787 (1995)

21. Ertaş D. and Kardar M., *"Critical dynamics of contact line depinning"*, Phys. Rev. E **49**, R2532, (1994)

22. Sneppen K., *"Self-organized pinning and interface growth in a random medium"*, Phys. Rev. Lett. **69**, 3539, (1992)

23. Paczuski M., Maslov S. and Bak P., *"Avalanche dynamics in evolution, growth and depinning models"*, Phys. Rev. E **53**, 414, (1995)

24. Roux S. and Hansen A., *"Interface roughening and pinning"*, J. Physique 4, 515, (1994)

25. Schmittbuhl J. and Måløy K. J., *"Direct observation of a self-affine crack propagation"*, Phys. Rev. Lett. **78**, 3888, (1997)

COMMON CHARACTERISTICS OF DAMAGE ACCUMULATION AND FRACTURE OF SOLIDS

L.R.BOTVINA
A.A.Baikov Institute of Metallurgy,
Russian Academy of Sciences,
49, Leninsky Prospekt, 117911, Moscow, Russia

1. Introduction

Special attention which has been payed to self-similar and self-organized phenomena during the last decade allowes to suppose that studying the general laws of kinetic processes occurring in different media will be probably one of the central scientific direction in the coming 21st century. The approaches to the analysis of these kinetic processes are to have apparently the thermodynamic basis and to be connected to each other.

Let us consider some of these approaches in detail, show the realtionships between them and their possible application to the description of the kinetic processes developing at different scale levels in various media.

2. Kinetic equations of Arrhenius - Zhurkov

It is known /1-2/ that the lifetime (t) and the fracture rate ($\dot{\varepsilon}$) of many materials with different crystal structure may be described by the exponential functions:

$$t = t_0 \exp \left[\frac{U_0 - U(\sigma)}{RT} \right] , \qquad (1)$$

$$\dot{\varepsilon} = \dot{\varepsilon}_0 \exp \left[\frac{U_{1_0} - U_1(\sigma)}{RT} \right] \qquad (2)$$

where R is the gas constant, T is the temperature, σ is the applied stress, $t_0, \dot{\varepsilon}_0$ and U_0 the constants, $U(\sigma)$ the activation energy, $\dot{\varepsilon}$ the process rate. The values of activation energy are estimated from the slope (b) of dependences

G. N. Frantziskonis (ed.), PROBAMAT – 21st Century: Probabilities and Materials, 93–107.
© 1998 *Kluwer Academic Publishers.*

lg$t$-1/T obtained at different stresses. These dependences usually form the fan of the straight lines with the pole corresponding to lgt_0=-13 and with the slope:

$$b(\sigma) = \frac{\Delta(\lg t)}{\Delta(1/T)},$$

$$U(\sigma) = U = 2,3 Rb \qquad (3)$$

Relation (1) describes the family of curves similar to those plotted in coordinates $K=f(t)$ and schematically presented in Fig.1. Here K is the measured property and P, T are the parameters characterizing the test conditions. In the case of creep, K is the deformation, t is the time, P the stress, T the temperature.

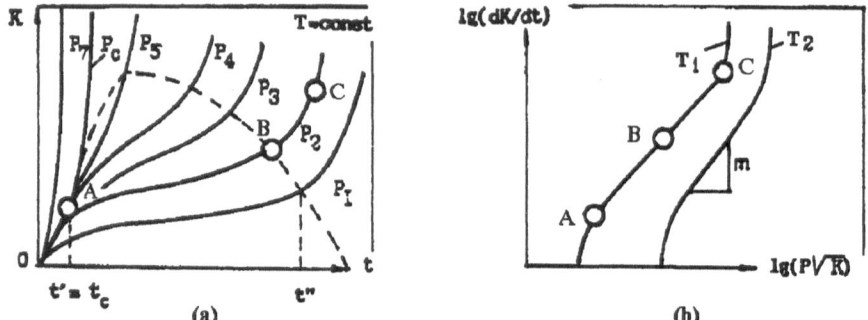

Figure 1. (a) Family of basic kinetic curves and (b) kinetic diagrams of fracture

3. Analysis of fracture processes on the base of the phase transition theory

The possibility to consider fracture processes in terms of the phase transition theory is caused by the fact that fracture kinetic curves under different loading conditions are close by their shape to the isotherms of the liquid-vapor phase transition plotted in pressure vs density coordinates. By analogy to the liquid -vapor phase transition /3-6,7/, the closeness to the critical point of the fracture process is defined by the order parameter which is a power-law function of the characteristics governing the loading conditions. An estimation of critical exponents in these power-law relations makes it possible to predict the reaching critical state of material under loading, accompanied by fracture mechanism change.

In Fig. 1 typical fracture kinetic curves $K=f(t)$ are schematically presented. They curves are obtained under a constant value of the parameter P defining the test conditions (termperature, loading rate, size of structural elements, etc.; K is a certain strain and/or fracture property; t is the time). It is clear that these curves remain similar to each other until a certain critical value of the parameter P_C is reached. For $P=P_C$, the second steady stage on the fracture curves disappears and they become closer to straight lines, as do the isotherms when the critical temperature is achieved under the liquid-gas transition conditions /7/.

Such a transition during failure is accompanied by the fracture mechanism change. According to Frenkel's idea /8/, any solid upon its loading may be considered as a two-phase material consisting of proper atoms and vacancies which are "atoms" of the second substance (voids). If the material is a two-phase one at the steady stage of fracture (i.e. consisting of damaged and undamaged volumes) we can assume that upon achievement of the critical conditions, a "phase transition" occurs with formation of a new phase: a macrocrack.

The length of the stable stage on the damage and fracture curves has been used as an order parameter for fracture process of a solid. This allowed to establish the new power-law relations and estimate the critical exponents characterizing the kinetics of pores formation during phase transition occurring in different loading conditions.

As the order parameters of fatigue and creep fracture, the lifetime on the second stage (N, t) of these processes developing under different stresses was used and the following relations were obtained:

$$\frac{N-N_C}{N_C} = A \left[\frac{\sigma - \sigma_C}{\sigma_C} \right]^{\beta} \qquad\qquad \frac{t-t_C}{t_C} = B \left[\frac{\sigma - \sigma_C}{\sigma_C} \right]^{\beta}$$

where N_C and t_C are the critical values of these order parameters at critical stress (σ_C) corresponding to disappearance of the steady stage on fracture curves. The exponents in these equations changed from 1.9 to 7 depending on material tested and loading conditions. In both cases the achievement of the critical point has led to the change of fracture mechanism. Similar analysis was fulfilled for processes of tension and impact loading /4,5/.

4. Connection of the activation energy with the critical exponent

So far as the family of the deformation and fracture curves (Fig.1,a) may be described in terms of both the phase transition theory and the Zhurkov kinetic concept, the main parameters of these approaches (the activation energy and the critical index) are to be connected.

To establish such a connection let us present the relation for the activation energy in the form:

$$U^* = 2{,}3Rb^*, \qquad \text{where} \quad U^* = U^*(\sigma),$$

$$b^* = \frac{\Delta(\lg t^*)}{\Delta T^*} \qquad t^* = \frac{t - t_c}{t_c} \qquad T^* = \frac{T_c - T}{T}$$

$$U^* = 2,3R \frac{\Delta(\lg t^*)}{\Delta T^*} \qquad (4)$$

It is supposed that the physical sense of the activation energy and the character of its dependence on stress and temperature will not change in consequence of the above-mentioned normalization.

The critical exponent of the phase transition during deformation and fracture is estimated as:

$$n = \frac{\Delta(\lg t^*)}{\Delta(\lg T^*)} \qquad (5)$$

Taking into account Equation (4), we have:

$$\frac{U^*}{n} = 2,3R \frac{\Delta(\lg T^*)}{\Delta T^*} = 2,3Rc \qquad \text{where} \qquad c = \frac{\Delta(\lg t^*)}{\Delta T^*}; \qquad n = \frac{b^*}{c}$$

Thus, the ratio of the activation energy to the critical index at the constant stress is determined by the value of the coefficient 'c'.

To verify this relation let us consider the lifetime temperature dependences of specimens from polystyrene obtained at different stresses (Fig. 2) /9/.

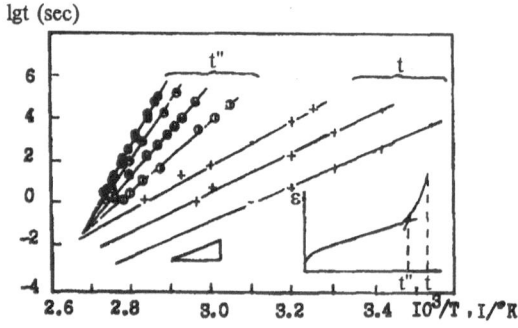

Figure 2. Temperature dependences of lifetime for specimens from polystyrene at stable (t") and final (t) stages of fracture in creep conditions under the different stresses (MPa):
10 (●), 20 (◑), 30 (◒), 40 (◍), 50,60,70 (+).

As shown in Fig.1 and Fig.2, the time t'' corresponds to the specimen lifetime at the end of the steady stage of creep before appearing the neck on the specimen. According to the scheme in Fig.1, namely this time (or the difference $t''-t'$) may be taken as the order parameter of the phase transition at creep conditions. From graphs in Fig.2, it is seen that the lifetime decreases with increasing the temperature and stress and becomes equal to one second ($\lg t = 0$). Let us choose this time as the critical value of the order parameter (t_C) and estimate the corresponding critical values of temperature ($1/T_C$) and also the normalized t^* and T^* values.

The results of estimating the activation energies (U, U^*) and the critical exponent (n) on the Equations (3-5) are presented in Fig.3.

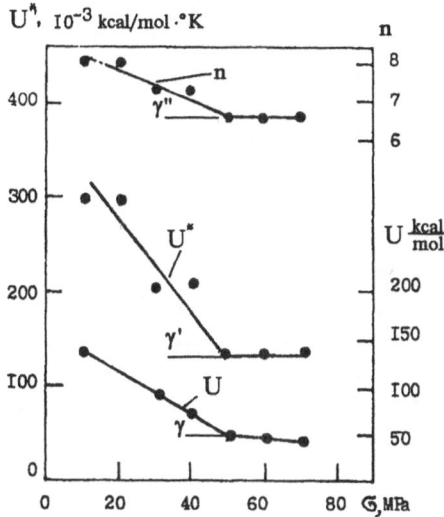

Figure 3. Dependences of the activation energy (U,U*) and the critical exponent (n) on stress for the specimens from polystyrene

It is seen that the character of the dependences of the indicated parameters on stress were found similar to each other. Accordingly, the new values of the activation energy (U^*) and the critical exponents (n) corresponding to the final fracture stage are almost independent on stress (the plateau regions on curves in Fig.3). However they display the visible sensitiveness to the stress at the steady stage of creep process. In this case the slopes of curves are characterized by the values γ, γ' and γ''.

5. Kinetic diagrams of fracture

It is known that most of strength relations are the power-type ones. Then, from the properties of the power-type functions it follows that kinetic diagrams similar to the fatigue diagrams of Paris-Erdogan may be plotted for many other deformation and fracture properties. In analogy to the kinetic diagrams of fatigue fracture, they are to be characterized by the threshold, transition and critical values of governing parameter $(P\sqrt{K})$ in points A, B, C (Fig.1). and to be described by the relation:

$$\frac{dK}{dt} = A(P\sqrt{K})^m \tag{6}$$

It seems important to distinguish these points in which the fracture mechanism changes and the transition to the another process stage occurs. For example, at the fatigue crack development in the ductile material, the accelerated crack growth initiates in the point A where fatigue striations appear. In the point B (at $\Delta K = \Delta K_S$) the first features of static fracture on the fracture surface are observed, namely, ductile dimples and shear lips. In the point C the final static rupture of specimen occurs. At plotting the kinetic damage accumulation diagram describing the crack initiation stage, the accelerated microcrack coalescence starts in the point B and as a result the macrocrack forms. Similar kinetic diagrams may also describe the processes occurring in fluids, for example, those of crysrallization of a solid phase from a metallic melt and of fluid flow /4,5/. When the crystallization occurs, the point B corresponds to the beginning of the accelerated coalescence of dendrite crystals. At fluid flow in tube the change of laminar flow by turbulent one starts in this point.

In order to obtain a single kinetic diagram for different values of the P — parameter it is necessary to introduce the limitations on the size of plastic zone (or the size of diffusion zone under crystallization process). However such a single diagram may be obtained in the normalized coordinates with using for such a normalization the coordinates of the point B (Fig.1).

The achievement of the critical values of P and t leads to coincidence of the points A and B and to immediate transition of the initial stage of kinetic process in unstable one characterized by the process mechanism corresponding to that in the region B—C of kinetic diagrams plotted under lower values of the parameter P. Thus the following relation is fulfilled:

$$(P_C\sqrt{K_A})_{t=t_C} = (P_i\sqrt{K_B})_{t=t_B} \tag{7}$$

where K_A, K_B are the values of K properties in the points A and B; t_C, t_B are the times in the critical and B point; P_C and P_i the critical and current parameters characterizing the acting factor (Fig.1). This relation was examined earlier by analysis of the family of the fatigue crack growth curves /10/. In this case reaching the critical conditions was connected with appearing the

general yielding of specimen at $K_{gy} = Y\sigma_{fy}\sqrt{\pi l_0}$ (where σ_{fy} is the cyclic yield strength, l_0 is the initial fatigue crack close by size to size of the material structural element). It was accompanied by the change of fatigue fracture mechanism and the appearance of a discontinuity on fatigue fracture kinetic diagrams. In /11/ the fatigue crack behaviour at reaching $K = K_{gy}$ was explained by appearance of self-oscillatory modes of crack development. As it was shown by Barenblatt earlier /12/, the self-oscillations also appear at the neck propagation in specimens from polymeric material tested at tension. Probably one may suppose that in these both cases the critical conditions were reached and phase transition occured.

So far as the kinetic diagrams are plotted from the same initial curves which are used for estimating the activation energy and the critical exponent, the power exponent 'm' in (6) must be connected with these parameters. Let us show the connection of the m-exponent with the activation energy. Equation (2) is obtained from relation :

$$\lg \dot{\varepsilon} = \lg \dot{\varepsilon}_0 + \frac{b_1(\sigma)}{T} \qquad b_1(\sigma) = b_1 \qquad b_1 = -\frac{\Delta(\lg \dot{\varepsilon})}{\Delta(1/T)}$$

In the case of fracture in the creep conditions Relation (6) may be presented in the form:

$$\dot{\varepsilon} = C(\sigma \sqrt{\varepsilon})^m,$$

where

$$m = \frac{\Delta(\lg \dot{\varepsilon})}{\Delta[\lg(\sigma \sqrt{\varepsilon})]} \qquad (8)$$

Then:

$$\frac{b_1}{m} = \frac{-\Delta[\lg(\sigma \sqrt{\varepsilon})]}{\Delta(1/T)}, \qquad \text{or}$$

$$\frac{U(\sigma \sqrt{\varepsilon})}{m} = \frac{-2.3R\Delta[\lg(\sigma \sqrt{\varepsilon})]}{\Delta(1/T)} = 2.3Rd, \qquad (9)$$

where

$$d = \frac{-\Delta[\lg(\sigma \sqrt{\varepsilon})]}{\Delta(1/T)}$$

It is often noted /13/ the constancy of product of the deformation rate on the steady stage of creep and the fracture time ($\dot{\varepsilon} \cdot t = $ const), i.e. the constancy of a deformation at which a fracture occurs. In this case the values of the activation energy $U(\sigma)$ and $U(\sigma \sqrt{\varepsilon})$ will be equal. Preliminary evaluation shows that the exponent "m" estimated on Equation (9) correlates with the experimental data. The more complicated relation has been suggested in /14/ for the description of the dependence of the activation energy on the exponent in the Paris-Erdogan equation.

6. Kachanov's equation of damage accumulation

According to Kachanov's approach, the damage measure (ω) is defined by the relative area of cracks or pores cumulated during loading. At the beginning of loading $\omega = 0$ (or the continuity measure $\psi = 1$) and at fracture $\omega = 1$, $\psi = 0$. The estimation of intermediate cases of the damage accumulation is difficult and Kachanov suggested to use the following kinetic equation based on the concepts of the statistical physics /15/:

$$\frac{d\omega}{dt} = F(\omega, T, \sigma) \tag{10}$$

or for the rate of damage accumulation at the constant temperature:

$$\frac{d\omega}{dt} = A \left[\frac{\sigma_{max}}{1-\omega} \right]^n$$

Recently /16/ it has been established that the damage accumulation rate is the power-law function of the product of the current damage value and the parameter defining the loading conditions (stress, temperature, structural size and so on):

$$\frac{df}{dt} = A(\sigma\sqrt{f})^n \tag{11}$$

where f is the porosity, i.e. the relative crack or pore area in the specimen, σ is the stress defining the initial conditions of loading. This equation has been obtained as a result of the analysis of experimental data under different loading conditions and, in particular, the stage sequence of the damage accumulation process. Equation (11) describes the middle region of the damage diagrams which are similar by their shape to those of Paris-Erdogan and the curves plotted in Fig.1.

7. Damage evolution in metallic specimens.

As it has been shown earlier /17/, the accumulation process for different defects in metals, polymers, and rocks develops self-similarly on various scale or structural levels. Stage sequence of the accumulation process of microdefects (slip steps, pores, microcracks) and macrodefects (macrocracks, faults) consists in initiation of defects, increase in their density till the critical constant value and jump-like transition to accumulation of the next order defects. Such a transition is caused by the growth and coalescence of defects of the previous size corresponding to the preceding scale level /16/. Any scale level of defect evolution may be characterized by the defect size distribution curve or by the dependence of the cumulative defect number (N) (Fig.4) on their sizes (l). These relationships are usually described by the power-law function /17/:

$$N = A\,l^{-\alpha} \tag{12}$$

Let us show that the power exponent (α) in this relation decreases when defects begin to coalesce. In this connection let us consider the results of studying the accumulation process of microcracks in unnotched specimens of low-carbon steel tested under cyclic loading /18/. Metallographical investigations carried out in this work have shown that at the early stages of multiple fatigue fracture the density of microcracks increases, then the growth of their number is delayed and the microcrack length starts to increase as a result of the coalescence of the adjacent microcracks. It may be seen from the curves (Fig.4) plotted with using of the above-mentioned data and connecting the lengths of microcracks with their number accumulated at the each stage of the fatigue fracture process /19/.

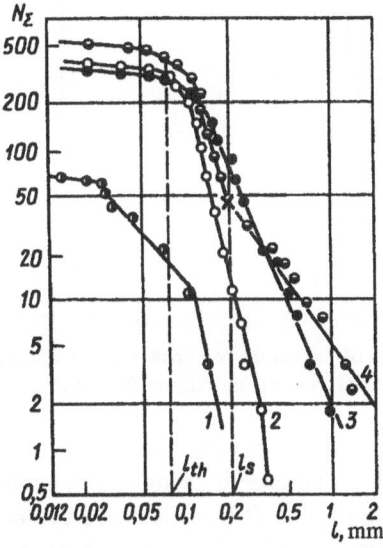

Figure 4. Cumulative frequency-length distributions of microcracks at different relative lifetime of fatigue specimens: n/n_f=0.17 (Curve 1), 0.43 (2), 0.85 (3), 0.97 (4). (Specimens of carbon steel tested at stress amplitude 333 MPa).

At the early stages of the multiple fracture, the number of defects quickly increases (upper plateau on Curve 2 in Fig.4 is raised). Then (at the relative lifetime equal to 0.85) the growth of the microcrack number actually is completed and their lengths begin to increase due to coalescence of microcracks. This process results in decrease of the slope in the size distribution curve (Curves 3 and 4 in Fig.4) which is described by the power

102

law relation (12) with the exponent decreasing almost twice for the given steel at the beginning of the main crack formation. Estimation of the concentration criterion of fracture according to relation $K = N^{-1/3}/l$ /20/ has shown that it is approximately equal to 3 at the early stages of the fatigue damage and then becomes close to 1 when the microcrack coalescence starts. In Fig.5 the curve of the main crack growth including the stage of its formation from individual microcracks and also the kinetic diagram of macrocrack formation plotted on the data of the above-mentioned study /18/ are presented.

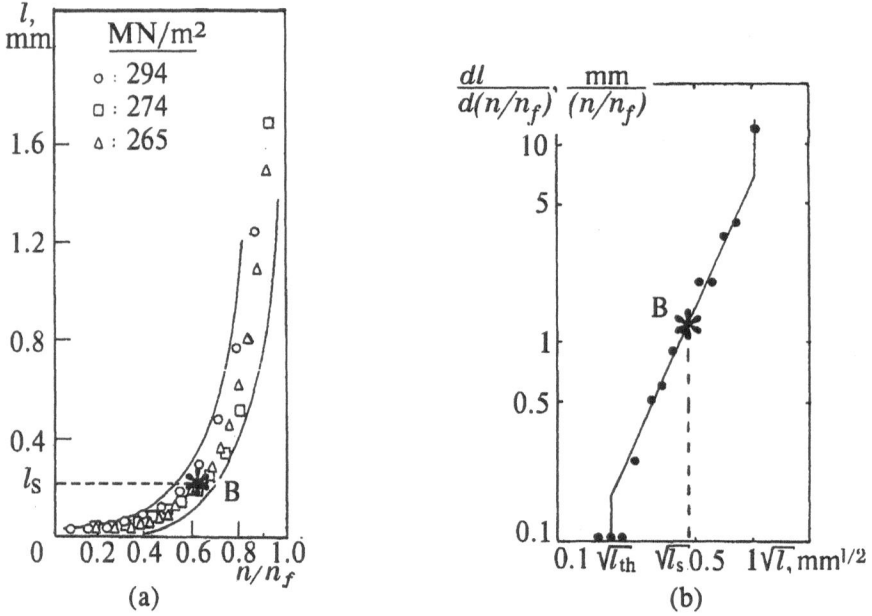

Figure 5. (a) Curve of formation and growth of the main fatigue crack /18/ and (b) the kinetic diagram of the fatigue fracture.

It is important to note the connection of some critical points observed in Fig.4 and Fig.5. So, the crack length defining the beginning the upper plateau on the cumulative damage curve (Fig.4) is close to the threshold crack length shown on the kinetic diagram (Fig.5,b). The knee point in Fig.4 where the microcrack coalescence increases (shown by the asterisk) corresponds to the completion of a stable stage and beginning the accelerated growth of the main crack (shown by the asterisk on both the crack growth curve (Fig.5,a) and the kinetic diagram (Fig.5,b)).

Thus, the threshold and transition crack length on the kinetic diagram are connected to the similar characteristics of the damage accumulation curve. The achievement of the threshold crack length (l_{th}) corresponds to the accumulation of the maximum number of the defects and the achievement of the critical crack length (l_s) before the beginning of accelerated fracture results in both the decrease of the exponent in the dependence of the cumulative defect number on the crack length (l) and the appearance of a knee point on the damage curve (Fig.4).

8. Damage evolution in the earth crust.

According to the assumption of many researchers /21,22/, the damage accumulation process occurring in the earth crust during the earthquake preparation is self-similar over a wide range of scale levels. As it has been noted above, the similar conclusion was made as a result of the analysis of the damage evolution in metallic specimens before their fracture. Moreover, it was shown that the stage sequence of the damage accumulation process on different scale levels remains also unchanged. Probably it will conserve itself on the most global level, i.e. at development of the multiple fracture in the earth crust during the earthquake preparation process. This, in turn, will mean a preservation of the above-mentioned kinetic relationships of the damage evolution observed at fracture of metallic specimens.

Let us examine this assumption by means of the analysis of the seismic activity in different world regions which characterizes (similarly to the acoustic emission) the damage accumulation processes occurring in the earth crust. The basic parameters of seismicity are the number of earthquakes (N) and their energy (E) released by an earthquake which are connected by the Gutenberg-Richter relation:

$$N = A E^{-b}, \quad \lg N = -b \lg E + C = -b M + C \tag{13}$$

where M is the magnitude or the logarithmic measure of energy, and A, b and C are numerical parameters. Usually, the dependence of the cumulated number of earthquakes on the magnitude called the frequency- magnitude relation plotted for the whole world is linear in semilogarithmic coordinates. However, in plotting this relationship for the separate regions of the world or for different time intervals before the earthquake, a bend of the lgN(M) curve upwards is noted (Fig.6) /23-25/.

As far as the energy released by an earthquake is proportional to the length of a fault arised, both the frequency-magnitude and the frequency-length relationships may be considered, probably, on the unique physical base connected with damage accumulation. So, from the noted analogy it follows that the decrease in the slope of the frequency-magnitude plot may be caused by the coalescence of the faults before the earthquake /19/. At the early stage of the earthquake preparation, the density of the intermediate faults increases, therefore the slope of the frequency-magnitude plot also increases. It is similar to the increase in the slope of the fatigue damage curve in Fig.4 at the stage of accumulation of microcracks with sizes from 0.1 till 0.4 mm. The transition from one stage of the damage process to another one may occur in the certain interval of magnitudes. In this case the plateau or discontinuity of the frequency-magnitude plot appears /25/.

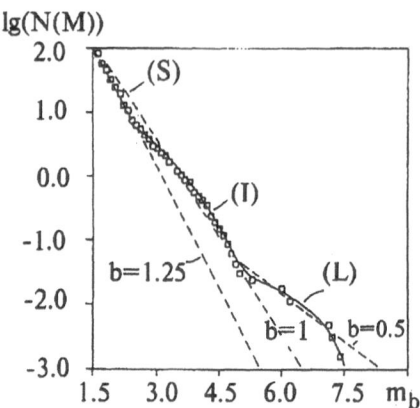

Figure 6. Cumulative frequency-magnitude distribution of the seismicity in the New Madrid zone. There appear to be different classes of event: small (S), intermediate (I) and large (L) /24/

In order to estimate the rate of the damage accumulation at the earthquake preparation, it is necessary to define the function characterizing the cumulative damage in this process which is equivalent to Kachanov's damage measure. The sum of products (EN) of numbers of seismic events and their energy /26/ or the following weighted sum of the mainshocks within (t-s,t) time and $(\underline{M}, \overline{M})$ magnitude intervals /27/ may be used as such a function:

$$S(t: \underline{M}, \overline{M}, s, \alpha, \beta) = \sum 10^{\beta(M_i - \alpha)} \qquad (14)$$

where s, α, β are the constants. For $\beta = b/3$, it is proportional to the total length of fractures, here b is the coefficient in the magnitude energy relation $\lg E = a + bM$. The average length of fracture is proportional to S/N. Use of the weighted sum of the mainshocks makes it possible to characterize the preparation process of a strong earthquake by a whole set of the preceding weak earthquakes /28/. This function seems to be the most suitable for the description of the above-mentioned process and equivalent to the damage measure introduced by Kachanov.

The analysis of the time dependence of the S — function has shown /29/ that its shape (before earthquake) is similar to the shape of the fatigue kinetic curve presented in Fig.5,a. Indeed, in this case the stable and accelerated stages of developing of the seismicity are also observed. The higher magnitude of a future earthquake, the higher the rate of increase in the cumulative energy of seismicity which is proportional to the cumulative length of fracture. Thus, kinetics of the earthquake preparation seems to be similar to the kinetics of damage accumulation before specimen fracture. Therefore, the process of development of the seismic activity as well as the process of microcrack accumulation may be described by the power-law function similar to Paris's relation. Plotting the kinetic diagrams of seismicity for several world regions confirmed this supposition. The conditions of the earthquake preparation and properties of rock will probably determine the exponent in the power-law relation describing the kinetic diagram of seismicity as well as the threshold and transition characteristics of the governing parameter. This means that metallic specimens may serve as model to study the processes occurring in the earth crust.

9. Conclusions

Thus, the conducted analysis has showed the correlation of the activation energy with the exponents of the power-law relations characterizing the kinetics of the phase transition at fracture in creep conditions. It is not surprising so far as these parameters characterize the same family of the curves. The description of the kinetic processes in terms of the phase transition theory seems more preferable in comparison with that on the base of the Arrhenius - Zhurkov equation because it allows to analyse the family of curves obtained under any acting factor distinct from temperature.

The power-type relationships between the rates of the kinetic processes and the governing parameter facilitate the estimation of the parameters which control the stages and the change of process mechanisms.

The connection of the activation energy with the critical index and the exponent in the Paris relation makes it possible to use new approaches for the description of the phenomena occurring in various media, in particular, for which the kinetic Arrhenius relation is used.

Analysis of seismic activity process in the earth crust allows to suppose that the general features of damage evolution observed in metallic specimens are preserved on the global scale level of fracture development. It shows that metallic specimens may be used for studying the general peculiarities of the earthquake preparation process.

This work was carried out with the financial support of the Russian Foundation for Basic Reseach, projects 97-01-00742 and 97-05-65952.

10. References

1. Zhurkov, S.N. (1968) Kinetic concept of solid strength, *Herald of USSR Academy of Science* 3, 46-52.
2. Regel,V.R., Slutsker, A.I., and Tomashevskij, E.E. (1974) *Kinetic nature of solid strength*, Nauka, Moscow.

3. Novikov, I.I. and Ermishkin, V.A. (1990) Description of plastic deformation of metals in the terminology of second-order phase transition theory, in *Thermophysical Properties of Substances and Materials*, Handbook (in Russian), Publishing House of Standards, Moscow, **29**, pp.56-68.

4. Botvina, L.R. (1994) Common characteristics of fracture and crystallization processes, *Metal Science and Heat Treatment* **36**, 393-403.

5. Botvina, L.R. (1997) Phase transition in fracture and crystallization processes, in J.R.Willis (ed.), *Nonlinear Analysis of Fracture*, Kluwer Academic Publishers, Dordrecht (Boston) London pp.311-320.

6. Botvina, L.R. (1996) Mechanisms and characteristics of damage and failure in terms of the theory of phase transitions, in J.Petit (ed.), *Mechanisms and Mechanics of Damage and Failure*, EMAS, Warley, West Midlands,U.K. pp.189-194.

7. Stanley, H.E. (1973) *Introduction to Phase Transitions and Critical Phenomena* (Russian translation), Mir, Moscow.

8. Frenkel, J. (1948) *Statistical Physics (in Russian)*. Publishing House of of USSR Academy of Science, Moscow-Leningrad.

9. Ratner, S.B. and Brochin, Yu.I. (1969) Temperature-time dependence of the limit stress of induced elasticity for polymers, *Reports of USSR Academy of Science* **188**, 807-810.

10. Botvina, L.R. (1981) Criterion of fatigue fracture characterizing initiation of general plastic yielding at crack tip, *in Cyclic fracture toughness of metals and alloys*, Nauka, Moscow, pp. 53-59.

11. Barenblatt, G.I. and Botvina, L.R. (1993) Self-oscillatory modes of fatigue fracture and the formation of self-similar structures at the fracture surface, *The Royal Society*, London A, 442, 489-494.

12. Barenblatt, G.I. (1974) Neck propagation in polymers, *Rheologica Acta* **13**, 924-933.

13. Petrov, V.A., Bashkarev, A.Ya. and Vettegren, V.I. (1993) Physical basis of lifetime prediction of structural materials, *Polytehnique*, Sankt-Peterburg.

14. Jeglic, F., Niessen, P., and Burns, D.J. (1973) Temperature dependence of fatigue crack propagation in an Al-2,6 Mg alloy, in Fatigue at Elevated Temperature, The *American Society for Testing and Minerals* (ASTM) STP 520, pp.139-148.

15. Kachanov, L.M. (1958) On the life-time under creep conditions, *Bulletin of USSR Academy of Sciences, Division of Technical Sciences*, **8**, 26-31.

16. Botvina, L.R. and Oparina, I.B. (1993) Regularities of damage processes at different loading conditions, *Material Science*, The International Journal, **4**, 13-23.

17. Botvina, L.R. and Barenblatt, G.I. (1985) Self-similarity of damage accumulation, *Problems of Strength*, **12**, 17-24.
18. Suh, C.M., Yuuki, R., and Kitagawa, H. (1985) Fatigue microcracks in a low carbon steel, *Fatigue & Fracture of Engineering Materials and Structures*, **8**, 193-203.
19. Botvina, L.R., Rotwain, I.M., Keilis-Borok V.I., and Oparina, I.B. (1995) On the character of the Gutenberg-Richter relation on different stages of damage accumulation and earthquake generation, *Reports of Russian Academy of Science*, **345**, 809-812.
20. Zhurkov, S.N., Kuksenko, V.S., Petrov, V.A. et al. (1978) Concentration criterion of volume-type fracture in solids, *in Physical Processes in Earthquake Sources*, Nauka, Moscow, pp. 101-116.
21. Sadovsky, M.A. (1979) On the natural piecewise structure of rocks, *Reports of USSR Academy of Sciences*, **247**, 829-840.
22. Tchalenko, J.G. (1970) Similarities between shear zones of different magnitudes, *The Geological Society of America, Bulletin*, **81**, 1625-1640.
23. Davidson, F.C.Jr. and Scholz, C.H. (1985) Frequency-moment distribution of earthquakes in the Aleutian Arc: a test of the characteristic earthquake model, *Bulletin of Seismological Society of America*, **75**, 1349 -1361.
24. Main, I.G., Peacock, S., and Meredith, P.G. (1990) Scattering attenuation and the fractal geometry of fracture systems, *PAGEOPH*, **133**, 283-304.
25. Karnik, V. and Klima, K. (1993) Magnitude-frequency distribution in the European Mediterranean earthquake regions, *Tectonophysics*, **220**, 309-323.
26. Bath, M. (1981) Earthquake recurrence of a particular type, *PAGEOPH*, **119**, 1063-1077.
27. Keilis-Borok, V.I. and Malinovskaya, L.N. (1964) One regularity in the Occurence of Strong Earthquakes, *Journal of Geophysical Research*, **69**, 3019-3025.
28. Keilis-Borok, V.I. and Kossobokov, V.G. (1990) Premonitory activation of earthquake flow: algorithm M8, *Physics of Earth Planet International*, **61**, 73-83.
29. Botvina, L.R., Oparina, I.B. and Novikova, O.V. (1997) Analysis of damage accumulation process on various scale levels, *Metal Science and Heat Treatment*, **4**, 17-22 (in Russian).

MECHANICAL RESPONSE OF DISORDERED MEDIA / PATTERNS, MEASUREMENTS AND MODELLING

D. BREYSSE
CDGA, Univ. Bordeaux I, Av. des facultés, 33405 Talence cedex, France

1. Introduction

Heterogeneity is a major characteristic of many building materials since it has a great influence on their mechanical response, particularly when failure is concerned. The failure in concrete or timber components is often regarded as a random event, difficult to predict. The wide scatter in results is the first reason for the development of semi-probabilistic approaches in european building rules (Eurocodes). However, a careful investigation of the material microstructure can provide information on what happens during failure. It can also make the engineer capable of predicting the failure response of structural elements, being assumed he has enough a priori data on the component.

The purpose of this paper is twofold :

- to show that, if one assumes that relevant local information is available, the complex process of failure of building materials can be described, from the first local failure to the ultimate load,
- to show that, more generally, if one takes this fact at the very beginning of the modelling process, it is possible to introduce in the model data that can be identified from a priori non destructive testing, and therefore, to reliably predict the component response.

To better explain the strategy used in statistical and probabilistic modelling, examples illustrate it for two materials (fiber reinforced concrete - FRC- and glulam) which have few in common, excepted the fact that they are disordered materials. In FRC, the disorder can be considered as being uniformly distributed within the volume, when glulam is more a "homogeneous" material containing a given number of defects (interfaces, joints, knots). For both, an experimental program has been developed with a double objective : (a) investigating and quantifying the microstructural disorder, (b) identifying the failure properties of specimens. Then it is shown how a material information can be used to improve the quality of the prediction. Direct applications is a better prediction of failure properties through non destructive measurements.

G. N. Frantziskonis (ed.), PROBAMAT – 21st Century: Probabilities and Materials, 109–126.
© 1998 *Kluwer Academic Publishers.*

2. Disorder in fiber reinforced concrete and consequences on its failure

2.1. EXPERIMENTAL PROGRAM

The mechanical behaviour of fiber reinforced concrete (FRC) appears to be complex, since it mixes the influence of its various components: fibers, cementitious matrix, interfaces. Therefore geometrical and material interactions at a micro-scale complicates any micro-macro model whose purpose would be to derive macro- characteristics from micro-measurements. Many models have been developed until now but they remain limited by the fact that they are often empirical or that they are limited to a very specific field of application (for instance, a given kind of material, whose characteristics have been identified through a given mechanical test).

The high number of possible parameters (called influence factors in the following) and the use of semi-empirical properties or coefficients (measuring the fiber-matrix interface quality for instance) has lead to a large amount of experimental results from which it is often difficult to draw any global and general conclusions. Thus, a specific experimental program has be designed such as to: (a) sort out the more influent parameters from the less influent ones, (b) model the experimental response. The program must also give us information about the relations existing between the micro- and macro-scale parameters.

Parameters which can influence the material response (tensile strength, fracture energy) are many [1]. A problem arises from the fact that the extensive study of all these parameters is very costful and that if one important parameter is not analyzed, the model can not be predictive. Furthermore, the possible coupling (interaction) between parameters makes things even more complex. Experimental design theory (EDT) gives a theoretical basis for the design of experimental programs in such contexts (2). The difficulty consists in choosing a well adapted *a priori* model, accounting for all possible factors and interactions. A detailed analysis of literature lead us to consider the following parameters :

- material parameters : concrete compressive strength f_c, aggregate size D_g, volumic fiber content V_f, shape of fibers F_f, length of fibers l_f, degree of anisotropy α,

- specimen and test parameters: specimen cross-section A, loading rate $\dot{\varepsilon}$.

The a priori model is, which is justified in (1) is:

$$Y = a_0 + a_1 F_f + a_2 l_f + a_3 V_f + a_4 \alpha + a_5 \frac{1}{D_g} + a_6 f_c + a_7 A + a_8 \dot{\varepsilon} + a_9 V_f^2 + a_{10} F_f f_c$$

$$+ a_{11} V_f f_c + a_{12} V_f^2 f_c + a_{13} V_f A + a_{14} V_f^2 A + a_{15} l_f \frac{1}{D_g} + a_{16} V_f \alpha + a_{17} V_f^2 \alpha$$

in which Y can be the tensile strength f_c as well as the post-cracking energy G_f. When considering the levels (values the parameters have to take - see Table 1), the EDT theory leads to a program of 48 tests (when extensive combination would have lead to 384 tests). Each test (uniaxial tension on a cylinder) is repeated several times in order to get information about the reproducibility degree.

Table 1 Parameters : number of levels and values

parameter	number	values	parameter	number	values
F_f	2	straight - with hooks	D_g	2	10 and 20 mm
l_f	2	30 and 50 mm	f_c	2	20 and 50 MPa
V_f	3	0.3, 0.5, 1%	A	2	50.26 - 95.03 cm^2
α	2	isotropic - preferential	$\dot{\varepsilon}$	2	5 and 50 µm/s

The identification of the 18 a_i coefficients after experimental results comes to the following expressions, in which all parameters have been renormalized (they vary between -1 and +1) and only the more influent parameters have been kept :

$$f_t = 3.11 + 0.63 \ f_c - 0.36 \frac{1}{D_g} + 0.33 \ \dot{\varepsilon} + 0.07 \ V_f + 0.2 \ l_f \frac{1}{D_g} + 0.07 \ A \ V_f \tag{1}$$

$$G_f = 2.99 + 0.84 \ V_f + 0.54 \ f_c + 0.26 \ l_f - 0.23 \frac{1}{D_g} + 0.2 \ V_f^2 \alpha + 0.1 \ A \ V_f \tag{2}$$

In these formula, the parameters are ranked according to their respective influence: the more influent are the compressive strength, the aggregate size and the rate of loading for the tensile strength, and the fiber volumic fraction and the compressive strength for the post-cracking energy.

2.2. UNDERSTANDING AND EVALUATING DISORDER

2.2.1. *Macroscopic variability*
The variability of macroscopic properties is a key-point if one considers structural design. Characteristic values are referred to in european building rules (Eurocode design), these values being derived from the average response and its variability. The experimental variability on strength and post-cracking energy is high, as it is shown at Table 2 for the three values of the fiber content.

Table 2. Average experimental values (E) and coefficient of variation for f_c and G_f

	all specimens	$V_f = 0.3\ \%$	$V_f = 0.5\ \%$	$V_f = 1\ \%$
E(f_t) (MPa)	3.08	3.17	2.93	3.16
c.o.v.(f_t) (%)	29	27	29	29
E(G_f) (N.mm^{-1})	2.63	1.84	2.56	3.55
c.o.v.(G_f) (%)	53	46	51	41

This variability has several origins, the first being the non-perfect reproducibility of experiments, the second being due to the fact that all (8) parameters are varying in the program. For instance, changing the fiber content V_f: from 0.3 % to 1 % will multiply by a factor 2 the average value of G_f, independently of the variability of due to other reasons (values of other controlled parameters - like the matrix strength - have changed, and it exists some "experimental noise" due to uncontrolled parameters). This is illustrated on Fig. 1 where are drawn the three cumulative density functions for the three different values of V_f.

The high variability has important implications for structural purposes. If one intends to predict the reliability level of structural components, he must model the material variability and its consequences on structural response. In the case of FRC components, models cannot be limited to the prediction of average response but they must also be able to account for the risk one has to encounter less favorable values. This point has been adressed in details in (3) and it will be discussed in the following subsection.

We will focus now on the variability on the post-cracking energy which is approximately twice larger than that exhibited on tensile strength. The reason is probably that the tensile strength is - mainly - controlled by the tensile strength of the matrix when the post-cracking energy is - also - controlled by the number of fibers which are present in the open crack.

2.2.2. *Disorder at micro-scale*

The variability of fiber content is analyzed through countings on cross-sections. On one hand, comparison between countings on various parallel cross-sections within a single specimen measures spatial variability within the volume. On the other hand, comparison between results on all failure cross-sections for all specimens provides information upon the global variability. All the results can be summarized saying that:
- for different combinations of controlled parameters (aggregate size, fiber content, fiber length, fiber shape, casting rules), the average number of fibers $E(N)$ in a given cross-section can be predicted with a limited error. The standard deviation inside the specimen corresponds to an averaged coefficient of variation (c.o.v.) of 24 %.
- the measured number of fibers in the failure cross-sections is well correlated with ($r = 0.88$) - and well below - the expected theoretical number (see fig. 2). The difference amounts to 44 % and cannot be explained by a counting bias, evaluated to be of only few percent.

2.2.3. ASSUMPTIONS AND MODEL FOR MICRO-MACRO RELATIONS

It can be assumed that the material behaviour is such that the failure cross-section is not an "average" cross-section but that failure preferentially occurs in under-reinforced sections. It is then reasonable to postulate that:
[a] the number of fibers is a random variable in the material, whose statistical properties can be characterized,
[b] the failure preferentially occurs in sections where this number is lower than average (the lowest ?).

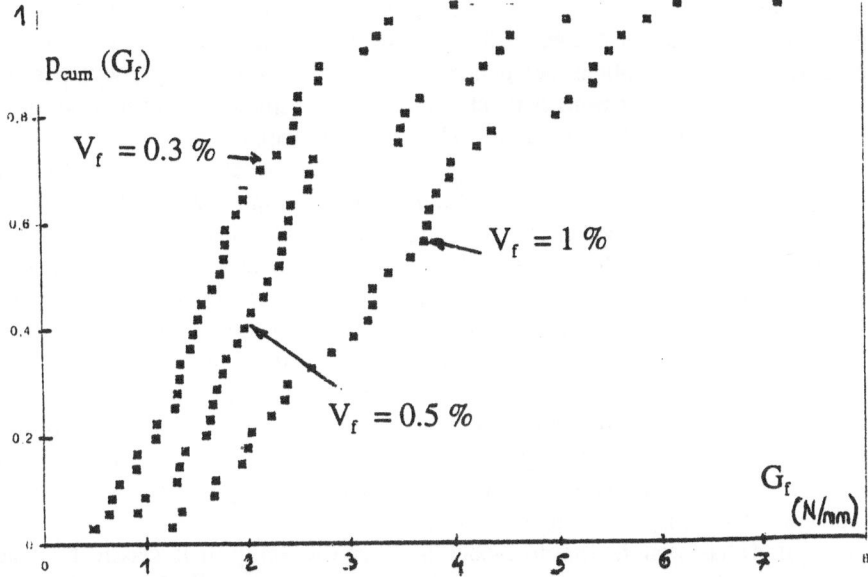

Figure 1. Cumulative density functions of G_f for the three volumic fiber contents

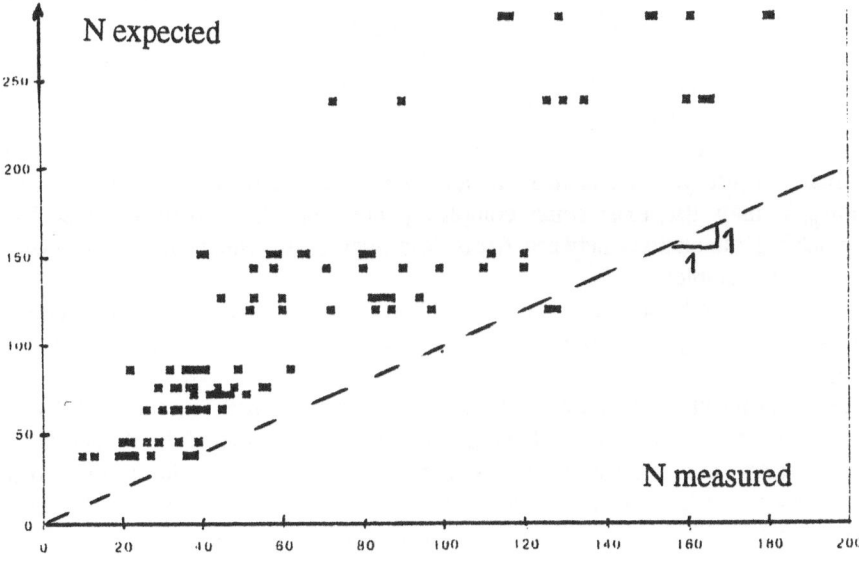

Figure 2. Measured number of fibers in the critical cross-section N and expected one N_{th}

114

In consequence, the values of mechanical properties which are influenced by the fiber content must depend (on average and regarding their variability) on the fiber content variability. These assumptions and their consequences can be verified by several ways:
- by counting fibers in various parallel cross-sections within the specimen, to quantify the spatial variability of the fiber ratio and verify if assumption [b] is valid,
- by analyzing cross-correlations between material parameters (matrix compressive strength f_c and number N of fibers in the failure cross-section) and mechanical properties of the FRC (tensile strength f_t and G_f). It has been verified that when only the matrix strength f_c influences the tensile strength f_t of the FRC ($r^2 = 0.59$), the post-cracking energy G_f is primarily correlated with the fiber content in the failure cross-section ($r^2 = 0.58$) and poorly correlated with the matrix tensile strength f_t ($r^2 = 0.15$),
- furthermore, the c.o.v. on the number of fibers N in the failure cross-section ranges between 36 and 49 % and is of the same magnitude than the c.o.v. on G_f.

All measurements confirm that the number of fibers in the section of failure is very different from (and systematically above) what would give "homogeneous material" assumptions; it is also an useful index for predicting the post-cracking energy. The challenge is now to try to predict this number.

Naaman (4) has been the first to model the fiber variability. It is classically assumed that the fiber location results from a spatial Poisson process, neglecting the fact that for larger fiber content, fibers can ball up and that these interaction can perturbate the spatial distribution (this seems to have been recently verified by Armelin (5)). A weakest link model is generally built in the longitudinal direction. Predicting N(x) and its minimum value on the specimen will require the following parameters:
- average fiber content E(N) in a cross-section, which depends on the volumic fiber content, on the cross section area, on the length and section of fibers and on the orientation degree (anisotropy),
- c.o.v.(N), which may depend on the cross-section area: a variance reduction would normally result from an increase in area, for purely statistical reasons. However, in reality, it may also exist some complex phenomena, like "balls of fibers" due to geometrical interactions between fibers (and aggregates) which make this size effect much more complex,
- correlation length for N, function of the fiber length and on the orientation degree,
- specimen size L, since, the larger the specimen, the higher the probability to find a low value.

The coefficient of variation of the fiber content has been investigated by several researchers (5, 6, 7). In the following, we will keep a c.o.v. of 25 %, similar to that identified in our measurements in parallel cross-sections. In semi-probabilistic approaches of risk, the material variability is accounted for through characteristic values X_k which quantify a given probability X has to reach the corresponding value. Similarly, one can try to estimate the number of fibers which is not obtained in a given cross-section with only a 5 percent probability. Assuming that N(x) is a random variable, with a gaussian distribution and for whose average value E(N) and standard deviation SD(N) (or c.o.v.) are known, it comes:

$$N_k = E\left(N(V_f \alpha)\right) \times \left[1 - k \ \text{c.o.v.}[N]\right] \tag{3}$$

where the scalar k depends on the chosen percentile and on the sample size (gaussian table give k = 1.64 for a 5 % risk a a sample of infinite size). Numerical application with k = 1.64, c.o.v. = 0.20 and E(N) = 0.94 N_{th} leads to:

$$N_k = 0.672 \, E \, (N(V_f, \alpha)) = 0.672 \times 0.94 \, N_{th} \, (\, V_f, \, \alpha \,) = 0.63 \, N_{th} \, (\, V_f, \, \alpha \,) \quad (4)$$

For instance, for N_{th} = 150, the characteristic number N_k comes to 95.

A specimen in uniaxial tension can be regarded as a serie of cross-sections. The minimum number of fibers is the lower value on all cross-sections. However, one cannot consider an infinite serie of independent cross-sections, since the number of fibers N(x) is a regionalized variable (in the sense of geostatistics) and since two realizations are independent only if the distance between the two sections is larger than the correlation length l_c of the regionalized variable. However, the small number of sections (due to the lower width that can be obtained by sawing) limits the quality of the estimate of the autocorrelation function. It is more efficient to consider that the correlation length is -at more- equal to the projected length of the fiber over the longitudinal axis (i.e. l = $\alpha_h l_f$ for the horizontal axis). It comes that, for a specimen of length L, only m = L/l independent values of N(x) have to be considered. For instance, if one needs to estimate the risk of having low values of N, for a specimen L = 200 mm and a correlation length l_c = 25 mm (isotropic distribution - α_h = 0.5 - and fiber length l_f = 50 mm), the minimum value of $N_{min}(x)$ will result from 8 independent realizations of N(x). The probability density function (p.d.f.) for $N_{min}(x)$ can be built from the p.d.f. for the local random variable N(x) in one cross-section. Using the independence assumption between realizations on successive cross-sections, it comes:

$$p \left(N_{min} (x) \right) < N = 1 - \left[1 - p \left(N(x) < N \right) \right]^{\pi} \quad (5)$$

where m is the number of independent cross-sections in the specimen and where the probability in the right term of the equation is directly computed from the density distribution of N(x).

This model will produce some size-effect (length-effect), as shown on fig.3. where the number of fibers in the failure cross-section (divided by the average number) varies as a function of length (divided by the material correlation length), for various levels of assumed risk. For instance, if one considers the risk p = 50% (which will give the mean value of the number of fibers in the failure cross-section measured on a large number of specimens), the number N decreases from 0.95 N_{th} when L = l_c to 0.60 N_{th} when L = 15 l_c. Length-effect is similar for other risk levels, as it can be seen on the four curves.

The model results can also be compared with experimental results : the measured number of fibers in the failure cross-section was - in average, i.e. for a risk level of 50 % - 56 % of the theoretical number N_{th}. The two figures 4 and 5 show how the prediction the number of fibers in the failure cross-section varies as a function of two parameters : specimen length (fig. 4) and c.o.v. of the number of fibers in an arbitrary cross-section (fig. 5). On Fig. 4, the cumulative experimental distribution of N/N_{th} can be compared to the predicted cumulative distribution for c.o.v.(N) = 25 % and different correlation lengths (from L/l_c = 1 to L/l_c = 15). The comparison of the curves confirm a striking similarity with the model for L/l_c of about 10, which would correspond to a

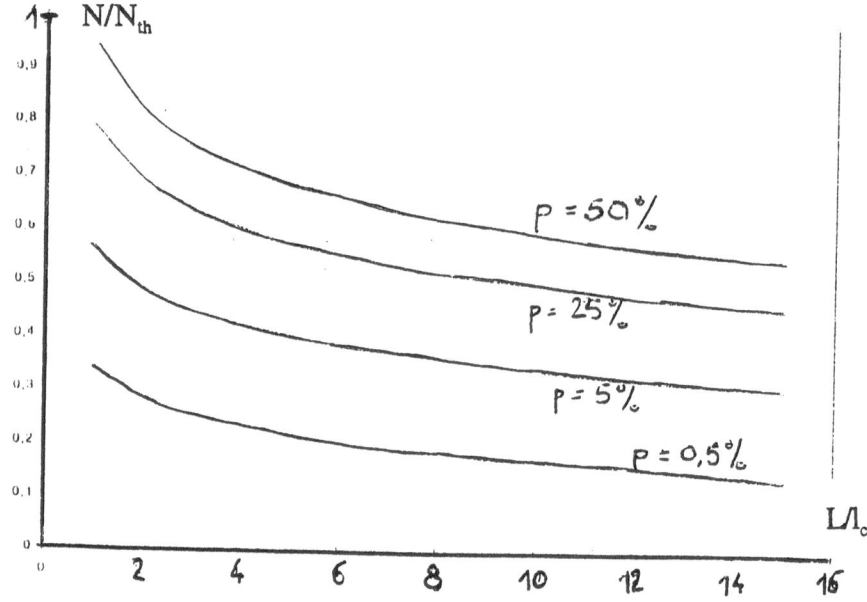

Figure 3. Size effect on the critical number of fibers for various levels of risk

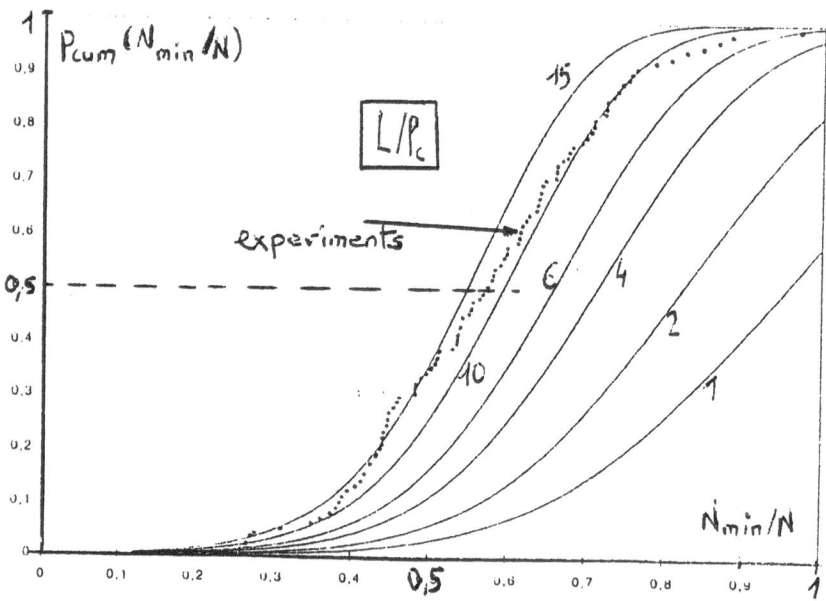

Figure 4. Risk of obtaining a given value of the critical number of fibers for different

specimen lengths. Comparison with experimental results

correlation length of 20 mm. On Fig. 5, the ratio L/l_c being taken equal to 8 (i.e. l_c = 25 mm), the average number of fibers in the critical cross-section varies from 0.88 N_{th} (c.o.v.(N) = 5 %) to 0.55 N_{th} (c.o.v.(N) = 30 %). The last result is near the experimental one.

3. Disorder in glulam and consequences on its failure

3.1. EXPERIMENTAL PROGRAM

Glulam is an heterogeneous material at many scales: the clear wood has varying characteristics within a single lam, due to natural growing process (8) and the average characteristics vary from one lam to another. Knots and, in the case of glulam finger-joints, add another kind of defects. It is important for reliability studies as well as for economical reasons (optimal grading) to predict the mechanical response of lams and glulam elements until failure. Being able to predict the response until failure means being able to quantify the influence of distributed disorder ("natural variability in clear wood") but also the influence of local accidents (joints, knots, interfaces). The question arising now is "is it possible to forecast the ultimate load of a component on which non destructive measurements have been performed ?". Basically, a positive answer needs that:
- one knows what are the relevant defects and at what scale they must be identified (see f.i. number of fibers in FRC),
- one is able to identify the characteristics of these defects (statistical distribution, correlation length, ...),
- one is able to build a model (numerical, empirical, ...) linking these data to the experimental response.

An experimental program has been designed with the aim of qualifying predictive abilities of non-destructive measurements to forecast the failure response of timber. Specimens have been loaded in tension and in bending (only results of bending tests are discussed in this paper). Before the mechanical tests, all components have been investigated by several techniques: local density, visual inspection of knotiness, measurement of knot-area-ratio (KAR = area of knot divided by the total area of the cross-section) all along the component, dynamic Young's modulus, annual ring width, slope of grain... Since the program was also devoted to the study of size effect, Table 3 summarizes the planning of 4-point bending tests. In this Table, L indicates the total length, L' the length between supports and L" the distance between the two points of loading (length in pure bending). Comparing serie A and E will give information on width effect and comparing A with B, C and D will inform us on length effect. The geometry of loading varies slightly for serie A and E for technological reasons. For each test, the MOE, the ultimate load, the location of failure and the failure mode (tensile or compressive, within a knot or not) are recorded.

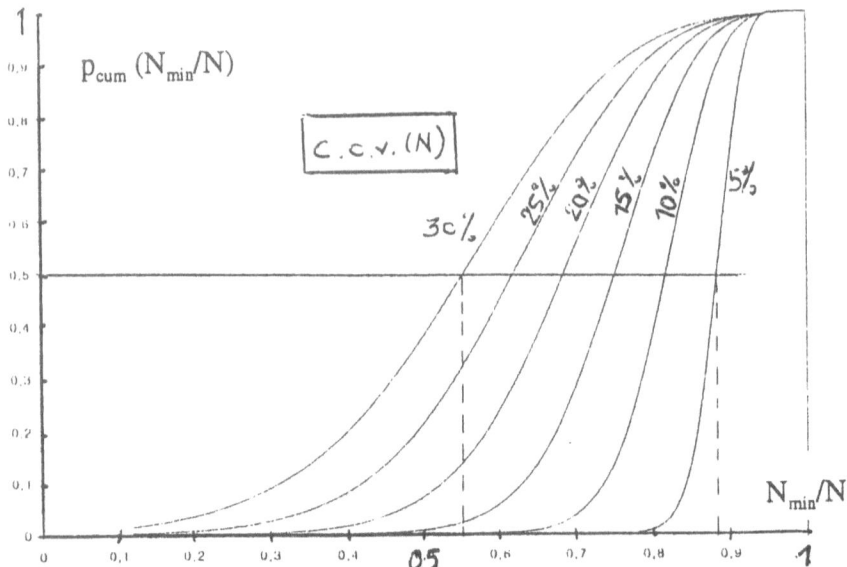

Figure 5. Risk of obtaining a given ratio N/Nth for various c.o.v.(N)

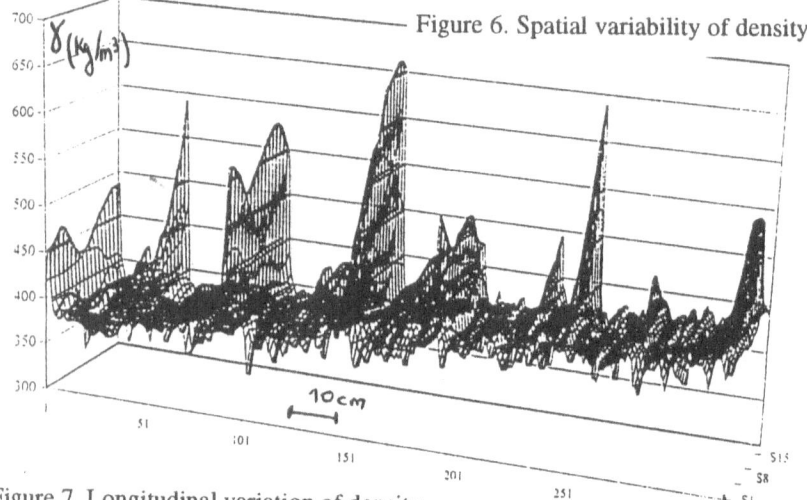

Figure 6. Spatial variability of density

Figure 7. Longitudinal variation of density

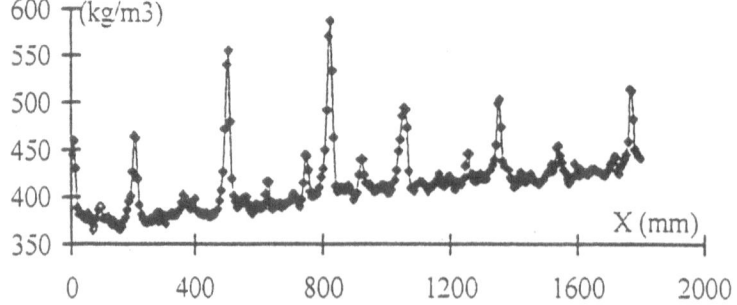

Table 3. Program of 4-point bending tests

serie	width (mm)	height (mm)	L (mm)	L' (mm)	L" (mm)	number
A	48	48	1200	1080	360	24
B	48	48	1300	1200	600	24
C	48	48	1900	1800	900	24
D	48	48	2500	2400	1200	24
E	75	48	1200	1080	360	27

3.2. UNDERSTANDING AND EVALUATING DISORDER

The failure of glulam results from defects of various nature and various scales which must all be investigated if a predictive model for failure is looked for. To simplify the problem, it can be told that defects are:
[a] defects in clear wood, linked to the variability in growing conditions: slope of grain, annual ring width, averaged elasticity modulus, local density. The density is evaluated through gamma densimetry, which, for each cross-section of any beam, gives a serie of values along the width. Its spatial variability is illustrated at Fig.6 and the longitudinal variation of values averaged on the cross-section are shown at Fig.7. The effects of the longitudinal variation of clear wood (the average value increases along the beam) and of knots (which correspond to a very high local increase in the measurement) can clearly be distinguished.
[b] localized defects, like knots and joints, which can be considered as being superimposed to the first class of defects. Knots are characterized by their size and their respective location (or distance between knots). Joints can be characterized by their strength which is difficult to identify, since it does not exist a joint without any wood around it! As an example, Fig.8 illustrates the distribution of the knot area ratio.

3.3. MODELLING DISORDER AND ITS CONSEQUENCES

Once the purpose of predicting the failure response is given, the question to know what track can be followed remains open. A general a priori rule is that, the deeper in the material we intend to go down, the higher the amount of information we will have to put in the model. In this section, we will describe two possibilities for analyzing the influence of local disorder. In the first case, a statistical model will describe how the macroscopical strength can be predicted from few material/mechanical parameters. In the second case, a micro-macro probabilistic model is built, which intends to predict the failure response of a given specimen, once its characteristics have been identified through non destructive testing. The second model will give more information but it will also, naturally, request more data. In both cases, the first task is to sort out the more influent parameters. This needs a careful analysis of the mechanical response of the material, for which the experimental program gives a lot of useful information.

3.3.1. *Statistical model for wood grading*

Here, the purpose is not to predict the response of a given specimen, but more to statistically predict what would be the response of a population containing a large number of specimens. The principles are those used in the automatical grading processes: [a] identify what material parameters X have the higher correlation degree with the failure properties, [b] build through successive regression analysis empirical models which predict the failure strength S as a function of these parameters X.

The more classical parameters measured through non destructive testing and used for visual grading are the wave velocity (linked with Young's modulus E_d), the annual ring width and the knot size. Following some criteria, the expert will grade wood, i.e. will put each specimen into a category Ci where i is the characteristic strength (in MPa). In some cases, the combination of the values is such that the specimen is rejected in a specific class R). Figure 9 illustrates the problems with such a method, since the cumulated probability for classes C22, C30 and R are poorly separated.

We can try to improve the grading process, *both* using *material information* obtained through non destructive techniques and adding a *mechanical point of of view* to the statistical analysis. Two possibilities are suggested in the following. In each case, we will build a simple statistical (empirical) model which will lead to a predicted strength S_p for each specimen. The N strength values S_p will be ranked and classes will be built, with the constraint of the same number of specimens in each class than previously obtained by visual grading in R (17 specimens), C22 (27 specimens) and C30 (54 specimens). Finally, the statistical properties (average, standard deviation, cumulated distributions) of the experimental strengths S will be computed and compared to those obtained in the usual process.

First option: for each class, an empirical correlation can be built between experimental mechanical results and non destructive parameters (density d, knot area ratio k and annual ring width a). Formulaes are obtained for S_p like (here for class C30):

$$S_p = 73.3 + 10.8 \, d - 11.1 \, a - 0.319 \, k \qquad (6)$$

Similar functions can be built for C22 and R. The regression analysis shows that k is the more influent parameter and that its influence cannot be neglected for C30 (the best specimens). These findings mean that if the knots are many and big, the specimen will fail at a lower load. If the knots are few and small, the strength is higher and will be influenced by the quality of clear wood (whose influence is very low in the other cases). Comparing predicted strengths and experimental ones, shows (Fig. 10) that the grading process is improved. On Table 4 averaged values and standard deviation for each class, obtained with usual grading and improved grading (using regression analysis) are compared. It is particularly clear that the procedure makes possible to sort out specimens which had been rejected at first and whose strength is sufficiently high to be classified in classes C22' or even in C30'. Economical consequences are straightforward.

Second option: the analysis of the failure mechanisms on all specimens has shown that all comes as if the failure was resulting from two independent mechanisms, a tensile failure in the lower part of the beam, mainly correlated to the presence of knots in this part, and a compression mechanism in the upper part, mainly correlated to the wood quality (measured through annual ring width a). A regression on all the specimens,

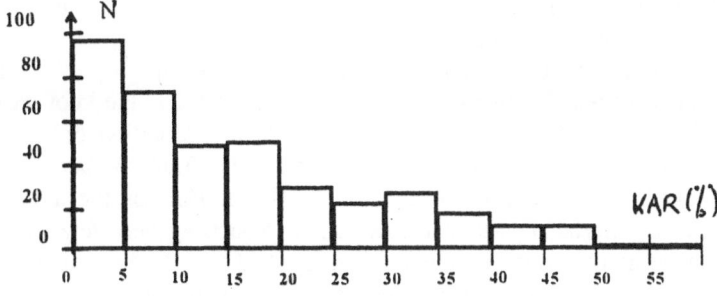

Figure 8. Histogram of the knot area ratio

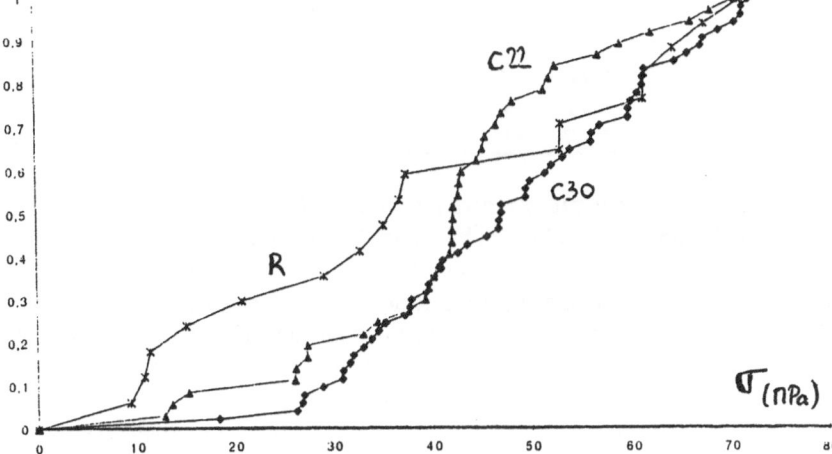

Figure 9. Cumulated density function for classes C22, C30 and R, after visual grading

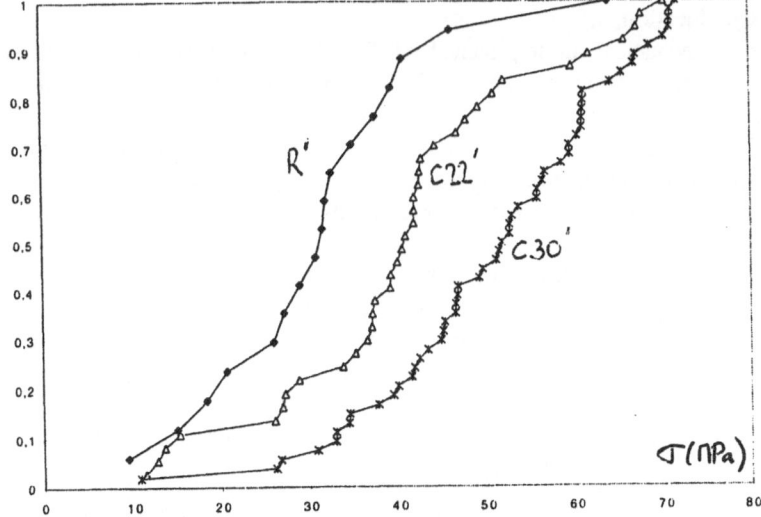

Figure 10. Quality of predictive grading using statistical correlations

122

after having divided the population in two parts according to the failure mechanisms, justifies the following empirical formula:

$$S_p = \min [70 \times (1.214 - 0.214 \, a), 70 \times (1. - 0.011 \, KAR)] \qquad (7)$$

where 70 is a reference tensile strength of clear wood and KAR is the knot area ratio. Applying this formula to all specimens makes possible the definition of classes R", C22" and C30", whose statistical properties are summarized at Table 4. Once more, the grading is improved, as it is confirmed on Fig. 11. This model has the advantage of mixing more directly mechanics and statistics and of requesting very few parameters, since all of them are in the above formula. It can be applied directly for a given species, once the reference strength value is known, and a and KAR are measured.

Table 4. Quality of predictive grading depending on the method

	visual			statistical			statistical+mechanics		
strength (MPa)	C30	C22	R	C30'	C22'	R'	C30"	C22"	R"
average	47.9	42.2	39.3	51.1	41.0	31.5	52.5	40.7	30.4
standard deviation	14.0	13.7	21.3	13.3	14.9	12.7	13.4	13.7	11.7

3.3.2. *Probabilistic numerical micro-macro model for failure*

An alternate way for predicting failure in heterogeneous materials consists in considering all material properties at micro-scale as data and in developing a deterministic mechanical model whose function is to integrate all non linearities all over the scales from micro- to macro- to obtain the component response. This requires that: (a) all relevant data are properly documented (i.e. identified in nature and value), (b) the mechanics used in the model is a good approximation of the reality. The quality of the result generally suffers from the lack of available information at the micro-scale as well as from discrepancies between the mechanical model and the real functioning. The experimental program presented above gives us a high amount of information on the material parameters at micro-scale (random fields of volumic mass f.i.) and on the defects (size, location, ...).

A simple mechanical model, named SIMWOOD, has been built to describe the mechanical response of a glulam beam. The component is discretized in elements E (of decimetric length and whose height corresponds to a lam) and in interfaces I. Each element E can be made of clear wood (Ecw), or contains a defect, being it a joint (Ej) or a knot (Ek). The mechanical description follows some simple rules (9, 10):
- any local failure can occur in tension or compression (plasticity) in an element E or by shear in interfaces I,
- the local stiffnesses and local strengths of elements are random variables which are generated from experimental information, accounting for their statistical distribution, their cross-correlations and their spatial variability (regionalized variables in the sense of geostatistics). This part has been extensively described in other papers and it is not detailed here,
- after each local failure, stresses are redistributed in the surroundings, according to rules identified from finite-element analysis.

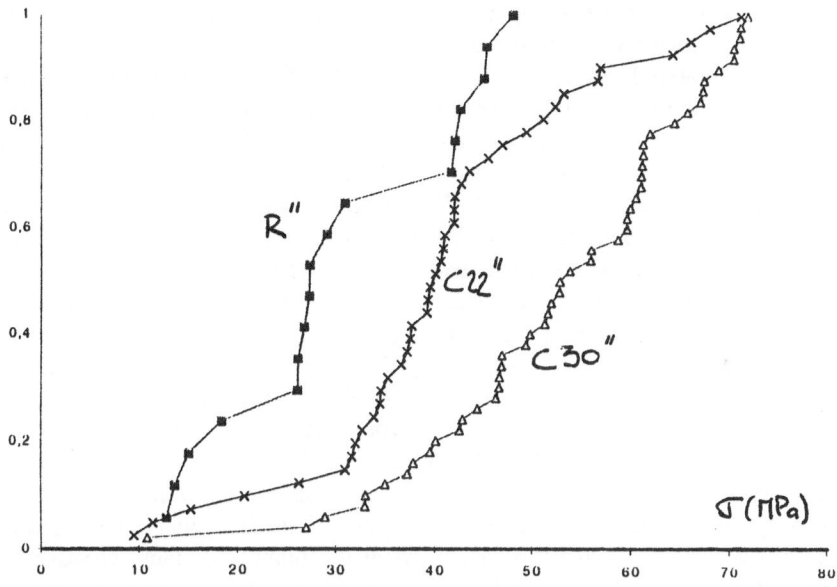

Figure 11. Quality of predictive grading using correlations and mechanical analysis

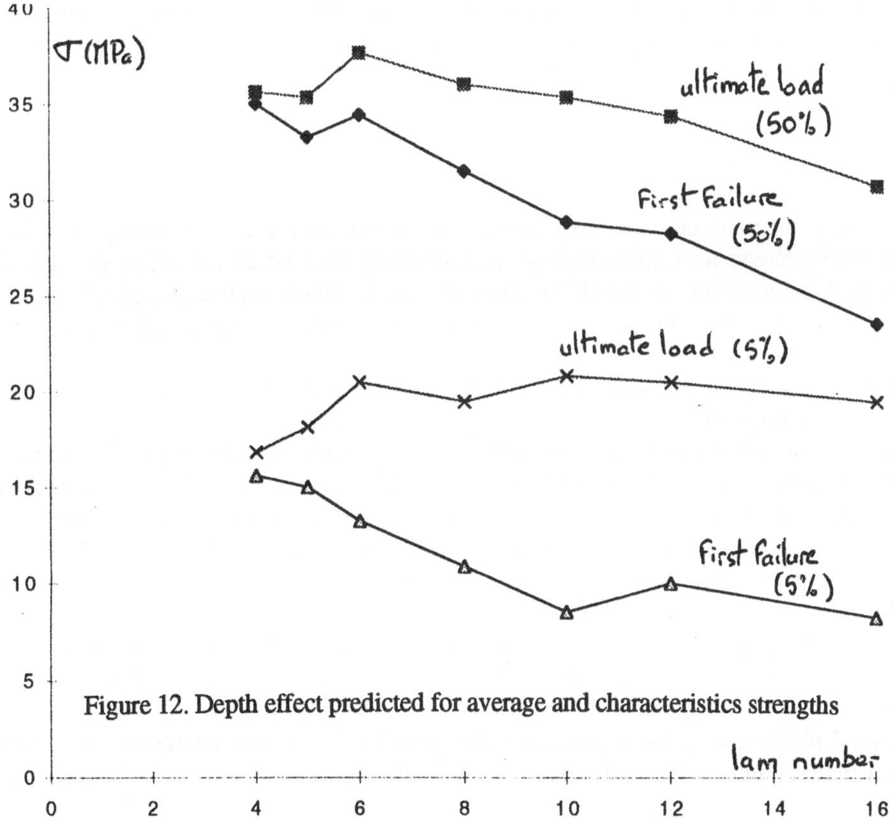

Figure 12. Depth effect predicted for average and characteristics strengths

The simulation of a given loading consists in a serie of elastic loadings, with stress redistribution after each local failure. SIMWOOD can reproduce the failure of a given specimen, whose material characteristics have been identified before loading, but it has been mainly used, up to now, to describe the response of a serie of beams with the same material statistics. Fig. 12 reproduces interesting results dealing with size-effect: the material properties remaining constant, we have studied the influence of the number of lams (beam depth) on the failure strength. We can first compare the load corresponding to the first failure with the ultimate load. For both values, some size-effect is exhibited, due to the fact that, for purely statistical reasons, the more lam we have, the higher the probability is to find a "bad lam". The size-effect intensity is larger on the first failure than on the ultimate load, since a parallel effect (a good lam "helps" a bad one) appears, which adds a (positive) size-effect to the (negative) statistical size-effect. For a beam containing 16 lams, the ultimate load and the first failure load corresponds for only 20% of the beams, when this ratio jumps to 45 % for 8 lams and 100 % for 2 lams. The other interesting point is the size-effect on characteristic strengths (corresponding to a 5% probability of failure). The difference between first failure and ultimate load is even larger, and there is no size-effect on the second one. This is a direct consequence of coupling between statistics and mechanics: when the number of lams increases, the probability of having only bad lams decreases and the positive size-effect compensates the negative one (they have here approximately the same intensity) and the result is: no apparent size-effect. Other computations have analyzed the role of defects (knots and joints) on the strength of a serie of beams, or the relative influence of distributed disorder (clear wood variability) and of local defects [10].

CONCLUSIONS

Macroscopical material variability results from microstructural variability and has important consequences on technology and on safety evaluation. Predicting the risk of failure of a component or optimizing it requires one is able to answer two questions:
- what are the more important mechanisms which play a significant role on the macroscopic response ?
- at what scale the material information must be gathered and identified to reproduce the real material response ?
We have shown, through three examples that mechanics and statistics deeply interact and that an efficient model needs both material information of quality and a detailed mechanical model. In all three cases, the failure of the component is directly induced by the material spatial variability and the question had arised of how to model this variability (or its influence on the macroscopical properties). The answer to the question of the relevant scale is not unique, as it has been shown on glulam: depending on the nature of the problem and on the amount of available information, one can choose a more global statistical approach or a more detailed micro-mechanical approach. The more important is the compatibility between all levels of the model: a detailed model is unuseful if data cannot be gathered. On the other hand, it is not interesting to collect many material information if the mechanical description is too simplified. One can add

that, obviously, as one goes down along the scales, one needs more information (which will increase the cost) and a more complex model, to go up to the macro-scale and obtain the final response (which will also increase the cost).

Then, one can be satisfied with purely empirical statistical model (for instance a formula giving the average strength of FRC and its standard deviation for various fiber contents). But one can also improve these models by adding some mechanics, like it has been done in the statistical model for glulam, where the analysis of the main mechanisms had lead to define the material parameters to include in the empirical formulae. One can finally try to describe the material microstructure with a greater accuracy, like it has been done in the last subsection. It is clear that this last approach, if the data are available and the mechanics correct, will be more fruitful. To come back to FRC, it would for instance make possible the prediction of the characteristic strength when the size changes or when bending is studied instead of tension. In wood, such a software will make possible to study the interest of using the prediction of the wood quality (by grading) to optimize the location of lams in the beam, putting the best ones on the external fibers and the bad ones near the neutral axis.

Once more, it is confirmed that the more and better information you need as output, the more and better information you have to give as input. Statistics and probabilities do not change thermodynamics and the law of increasing entropy!

ACKNOWLEDGEMENTS

I want to thank the students who have devoted part of their work on problems related to these topics, mainly A. Attar and P. Renaudin.

REFERENCES

1. Attar A. (1996), Le béton renforcé de fibres métalliques, matériau anisotrope et hétérogène, *Ph. D. Diss.*, LMT ENS Cachan, France.
2. Louvet F. (1997), Utilisation des plans d'expérience pour l'enseignement des matériaux de construction, *Coll. Enseignement des Matériaux*, ENS Cachan, 20-21/3/1997.
3. Breysse D., Attar A., Soulier B., Mesureur B. (1997), Modélisation de la réponse en traction du béton renforcé de fibres métalliques, *Mat. Str. RILEM*, to be publ. 1997.
4. Naaman A.E. et al (1974), Probabilistic analysis of FRC, *ASCE*, **100**, pp. 397-413, 4-1974.
5. Armelin H.S., Banthia N. (1997), Predicting the flexural postcracking performance of steel fiber reinforced concrete from the pullout of single fibers, *ACI Mat. J.*, 1-2/1997, pp. 18-31.
6. Bernier G., Behloul M. (1996), Effet de l'orientation des fibres sur le comportement mécanique des BPR, *Sec. Coll. Int. BRFM*, Toulouse, 7-1996.

7. Taerwe L, Van Gysel A., De Schutter G., Vyncke J. (1996), Détermination de la teneur en fibres d'acier dans du béton frais et durci, *Sec. Coll. Int. BRFM*, Toulouse, 7-1996.

8. Castéra P., Morlier P. (1994), Variability of the mechanical properties of wood: hazard and determinism, *Proceedings of PROBAMAT Conf.*, NATO 93021 Workshop, Applied Sciences Series, vol. 268, pp. 109-118, ed. Kluwer.

9. Renaudin P., Breysse D. (1996), Failure of glulam: numerical model of a random process, *Eur. Workshop on Application of Statistics and Probabilities in Wood Mechanics*, Bordeaux 22-23/2/1996.

10. Renaudin P., Breysse D. (1996), Un modèle probabiliste pour la prédiction de la résistance de l'epicea en traction simple, *Annales GC Bois*, 1, pp. 43-52.

FAILURE SCALING AS MULTISCALE INSTABILITY IN DEFECT ENSEMBLE

O.B. NAIMARK, M.M.DAVYDOVA, O.A. PLECHOV
Institute of Continuous Media Mechanics of the Russian Academy of Sciences
1, Ak. Korolev str., 614061 Perm, Russia,
e-mail: lab13 @ lab13.icmm.perm.su

1. Introduction

Fracture phenomena are some of the most intriguing processes in the nonlinear physics and material science. The nature of patterns derived from the failure processes have recently attracted a great interest. The fracture surface of quasi-brittle materials provides rich information about the structure evolution including the size, shape and localization of the critical flaw, mirror, mist, hackle and macroscopic crack branching [1]. The fracture patterns also reveal a well-defined fractal structure [2]. During the past decades a very important role of mesoscopic defects (microcracks, microshears) was established due to the study of transition from disperse accumulation of defects (damage) to fracture. Experimental data show that the defect ensemble exhibits pronounced features of the statistical multiparticle system with a strong interaction between defects and stress field, in particular, induced by macrocracks. To describe solid response to the defect growth, a statistical approach was developed in [3] using the experimental results from the direct study of microcrack evolution.The statistical approach allowed us to establish the specific features of the defect ensemble evolution depending on the characteristic size of structural heterogeneity (for instance, the size of grains in polycrystals) and, as the consequence, different modes of nonlinear solid response to the defect growth. Under loading conditions these nonlinearities are realized as specific forms of spatial-localized structures of defects. In dynamically loaded solids (shock wave loading, dynamic crack propagation) these structures have very legible structural pattern and their appearance are accompanied by a qualitative change of solid response to loading. The changes in the response occur in the form of the topological transition. The self-similarity of solid behaviour is caused by the excitation of spatial-time structures in the defect ensemble. These structures are related to the eigenfunction spectrum of the corresponding nonlinear problem which is determined by the nonlinearity (attractor) types of the equations developed in the framework of the statistical approach. The results of statistical analysis allow us to study the following phenomena:

127

G. N. Frantziskonis (ed.), PROBAMAT – 21st Century: Probabilities and Materials, 127–142.
© *1998 Kluwer Academic Publishers.*

(i) crack formation caused by generation of collective modes, which develop as instabilities with the blow-up kinetics in the defect ensemble (microcracks) localized on the spectrum of spatial scales;

(ii) self-similarity of failure kinetics under impact loading as the resonance excitation of collective modes in the defect system;

(iii) self-similarity laws of failure caused by the crack propagation (steady-state, crack branching).

2. Statistical Model

The ensemble of typical mesoscopic defects (microcracks) reveals the features of the collective behaviour [4]. The concentration of microcracks reaches the values of about 10^{12} -10^{14} cm^{-3} and the evolution of defects is close to the evolution of thermodynamic systems but with very important difference: single microcrack represents the dislocation ensemble and possesses this ensemble properties. The parameters describing the microcracks can be introduced using the results of the dislocation theory [5] and also as the variables eliminating the diffeomorphism breaking from the point of view of the gauge field theory [6]. For the penny-shape crack the microscopic tensor characterizing the volume and orientation of microcrack has the form

$$s_{ik} = s \ v_i v_k, \tag{1}$$

where s is the microcrack volume, $\bar{v}$ is the normal vector to the defect base. Microscopic kinetics of s_{ik} is determined by the Langevin equation

$$\dot{s}_{ik} = K_{ik}(s_{lm}) - F_{ik}, \tag{2}$$

where K_{ik} and F_{ik} are deterministic and fluctuating parts of interaction forces which satisfy the conditions $\langle F_{ik}(t) \rangle = 0, \langle F_{ik}(t')F_{ik}(t) \rangle = Q\delta(t-t')$, Q is the correlator of fluctuating forces (nonequilibrium potential corresponding to the microcrack nuclei). The distribution function of defects $W(s_{ik})$ can be found as the solution of the Fokker-Plank equation [7]

$$\frac{\partial}{\partial t}W = -\frac{\partial}{\partial s_{ik}}\left(K_{ik}(s_{lm})W\right) + \frac{1}{2}Q\frac{\partial}{\partial s_{ik}}\left(\frac{\partial}{\partial s_{ik}}W\right). \tag{3}$$

It was shown in [3] that the Lagrangian $E = \int 2K_{ik}ds_{ik}$ for the considered defects which represent the dislocation pile-ups can be written in a form

$$E = E_0 - H_{ik}s_{ik} + \alpha \ s_{ik}^2, \tag{4}$$

that includes the traditional "mean field" term $H_{ik}s_{ik}$ reflecting the interaction between defects $H_{ik} = \gamma\,\sigma_{ik} + \lambda\,p_{ik}$, where σ_{ik} is the stress; $p_{ik} = n\langle s_{ik}\rangle$, n is the defect concentration; λ and α are material parameters) and "self-action" term $\alpha\,s_{ik}^2$ reflecting the free energy fluctuation at the close neighbourhood of defect. The solution of the Fokker-Plank equation based on the statistical self-similarity hypothesis of the defect distribution (experimentally confirmed in [8]) allowed the presentation of the nonequilibrium distribution function as $W = Z^{-1}\exp(-E/Q)$, where Z is the generalization of the statistical integral. The form of the distribution function reflects in fact the view by Leontovich [9] on the nature of nonequilibrium processes as the succession of the equilibrium states which can be realized due the instantaneous switching-on of some effective field. The value of the latter is determined in our case by Q. Averaging s_{ik} with the distribution function W we obtain the self-consistency equation for microcrack density tensor - deformation caused by the microcracks: $p_{ik} = n\langle s_{ik}\rangle$

$$p_{ik} = n\int s_{ik}Z^{-1}\exp\!\left(\frac{E}{Q}\right)\!dsd^3\vec{v} \tag{5}$$

Equation (7) was solved in [1] for the case of uniaxial loading for various values of the dimensionless parameter $\delta = 2\alpha/(\lambda\,n)$ (Fig.1). The value of $\lambda\,n\langle s_{ik}\rangle$ determines the intensity of long-range interactions in ensemble of mesodefects, $\alpha \approx G/r^3$. determines the sensitivity of material to the defect growth, G is the elastic modulus, r is the radius of the defect nucleus.

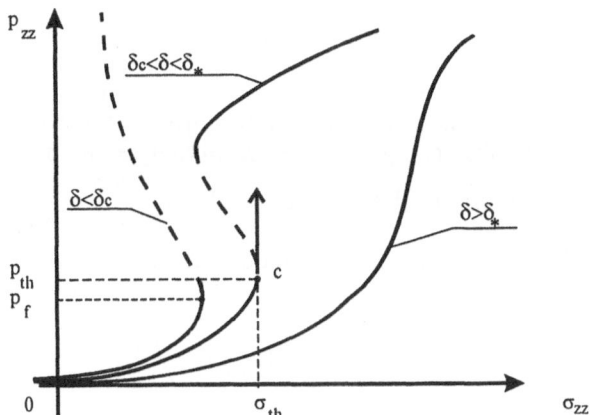

Figure 1. Nonlinear solid responses on microcrack growth.

There are three responses of material to the defect growth: monotonous $(\delta > \delta_*)$, metastable $(\delta_c < \delta < \delta_*)$ and unstable $(\delta < \delta_c)$; δ_* and δ_c being the bifurcation points correspond to the change of the asymptotes. The monotonous response

130

$\left(\delta > \delta_*\right)$ is characteristic of a weak interaction between defects. In metastable area the jump-like change of p_{ik} corresponds to the orientation ordering of the mesodefect ensemble. The pass over the δ_c-asymptotic leads to the infinite jump of p_{ik}. The passes over the asymptotics can be recognized as topological transitions that lead to the symmetry changes due to the new organization in the defect system. Mathematically speaking these transitions occur under the change of differential equation types and their group properties.

3. Constitutive Equation. Spatial-time Structures in Ensemble of Microcracks

The free energy F reflecting the spectrum of the solid responses on the defect growth (Fig.1) can be represented as the expansion

$$F = \mu\varepsilon_{ik}'^2 + k\varepsilon_{ll}^2 + \frac{1}{2}A(1-\frac{\delta}{\delta_*})p_{ik}^2 + \frac{1}{4}p_{ik}^4 + \frac{1}{6}C(1-\frac{\delta}{\delta_c})p_{ik}^6 + D\sigma_{ik}p_{ik} + \frac{1}{2}\chi(\nabla p_{ik})^2 \quad (6)$$

where A,B,C and D are the parameters of the expansion. Taking in view the polar character of the defect interaction the gradient term was introduced [10] that allowed us to describe the nonlocality effect in a "long-wave approximation", χ is the nonlocality coefficient. The forms of the coefficients upon the quadratic term and the higher term provide a qualitative changes of material responses on the defect growth in bifurcation points δ_* and δ_c.

   The thermodynamics of solid with these types of defects was developed in [3] and it was shown that for quasi-brittle materials there is exist the unique mechanism of the nonreversibility that is the generation of defects on the pre-existing nuclei. As it was established in [4] the quasi-brittle failure is characterized by the change of the mean size of the microcracks for the practically constant value of the concentration. Taking in view that the driving force of the microcrack growth kinetics is the free energy release, we obtain as the consequence of the evolution inequality (Ginsburg-Landau approach [10]) $\frac{\delta F}{\delta T} = \frac{\delta F}{\delta p_{ik}}\dot{p}_{ik} \leq 0$ $(\frac{\delta}{\delta p_{ik}}$ is the variation derivative) the kinetic equation for the microcrack density tensor p_{ik}

$$\frac{dp_{ik}}{dt} = -\Gamma\frac{\partial F}{\partial p_{ik}} + \frac{\partial}{\partial x_l}\left(\zeta\frac{\partial p_{ik}}{\partial x_l}\right), \quad (7)$$

where Γ is the kinetic coefficient, $\zeta = \Gamma\chi$. Using the potential $\Phi = F - \sigma_{ik}\varepsilon_{ik}$ for that the independent variables are σ_{ik} and p_{ik}, and the determination of deformation tensor [11] $\varepsilon_{ik} = \frac{\partial\Phi}{\partial\sigma_{ik}}$ we obtain the equation for the total deformation

$$\varepsilon_{ik} = \frac{1}{2\mu}\sigma'_{ik} + \frac{1}{2k}\sigma_{ll}\delta_{ik} + Dp_{ik}.\tag{8}$$

The equations (7), (8) represent the system of the constitutive equations of quasi-brittle solid with microcracks.

The transitions over the bifurcation points δ_c and δ_* lead to the sharp change of the symmetry of the distribution function caused by the different interaction of the scalar and the tensor modes of defects with external stress field depending on the value of δ (the size of the defect nuclei r that is determined by the characteristic size of structural heterogeneity and correlation radius of the internal stress field providing the interaction between defects). Studying the kinetic equation (7) with the free energy in the form (6) we consider the type of the solutions for the condition of the simple tension $\sigma = \sigma_{zz}$ when p_{ik} has only one component $p_{zz} = p$. The analysis of the defect evolution may be carried out due to the study of the heteroclinic solution of the equation

$$Dp + A_0(1-\delta/\delta_*)p + Bp^3 - C_0(1-\delta/\delta_*)p^5 + \frac{\partial}{\partial x}\left(\zeta\frac{\partial p}{\partial x}\right) = 0.\tag{9}$$

The behaviour of such solution can be visualized on the phase portrait (Fig.2).When $\delta > \delta_*$ solution has the form of the spatial-periodical distribution $p \to p\cdot\exp(i\phi)$. When $\delta \to \delta_*$ Eqn. (7) changes locally from elliptic to hyperbolic (separatrix S_2) and periodical solution is transformed to the solitary wave solution that corresponds to the diverge of the internal size Λ as $\Lambda \sim -\ln(\delta - \delta_*)$, (Fig.3).

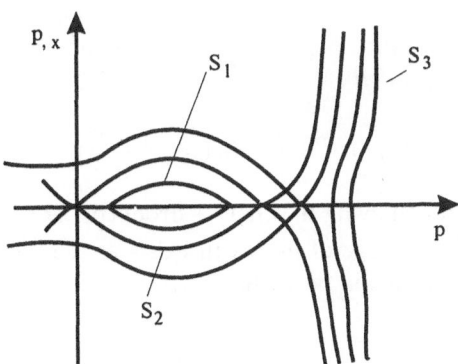

Figure 2. Heteroclinic solution.

The amplitude, rate of wave and the length of the wave front are determined by the parameters of the orientational transition and the nonlocal kinetics. Front of the solitary wave has the kink-like form

$$p(\xi) = p(x - Vt)$$

$$p = 1/2 p_a \left(1 - th\left(2\xi \, l_c^{-1}\right)\right), \quad l_c = 4/ p_a \left(2\xi \, V\right)^{1/2}. \tag{10}$$

The rate of the wave propagation V is determined by the penetration depth into the metastability area $V = \zeta^{1/2}/2L_3 \left(p_a - p_m\right)$, where $\left(p_a - p_m\right)$ is the jump of the defect density parameter due to the orientation transition.

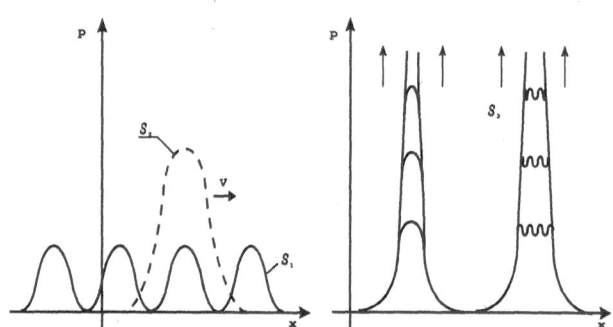

*Figure 3.*Spatial-time structures in defect ensembles.

The pass over the bifurcation point δ_c (separatrix S_3) gives the qualitative new type of spatial-temporal structures which are characterized by the explosive-like kinetics (peak regimes [12,13]) of the p-growth over some spectrum of spatial scales (Fig.3).

Let us consider the specific features of nonlinear system for $\delta \leq \delta_c$ passing the instability threshold p_c (Fig.1). In this case the p-growth is governed by the difference in the orders of higher terms of expansion (6). Assuming the power law for the nonlocality parameter $\zeta = \zeta_0(p_c)\hat{p}^\beta$ the kinetic equation for p can be written in the form [7]

$$\frac{\partial p}{\partial t} \approx S(p_c)\hat{p}^{\omega} + \frac{\partial}{\partial x}\left(\zeta_0(p_c)\hat{p}^\beta \frac{\partial \hat{p}}{\partial x}\right) \tag{11}$$

where $\hat{p} = p/p_c$ (in the following the "hat" is dropped), $\omega = 5/3$. It was shown in [12,14] that the developed stage of p-growth exists as the self-similar solution for $t \to t_c$ and can be represented in the form

$$p_A = q(t)f(\xi), \quad \xi = \frac{x}{\phi(t)}. \tag{12}$$

The substitution of (12) into (13) gives the form of functions $q(t)$ and $\phi(t)$

$$q(t) = S_0(t - t_c)^{-\frac{1}{\omega-1}}, \quad \phi(t) = S_0^{-\frac{\beta}{2(\omega-1)}}\zeta_0^{\frac{1}{2}}(t - t_c)^{\frac{\omega-\beta-1}{2(\omega-1)}} \tag{13}$$

The eigenfunction problem formulated for the function $f(\xi)$ [1] is defined by

$$-\frac{1}{\omega-1}f+\frac{\omega-\beta-1}{\omega-1}\xi f_\xi=(f^\beta f_\xi)_\xi+f^\omega,\qquad(14)$$

$$f_\xi|_{\xi=0}=0,\qquad(f^\beta f_\xi)|_{\xi=\xi_f}=f|_{\xi=\xi_f}=0,\qquad(15)$$

where ξ_f is the coordinate of the front of dissipative structure with the blow-up p-kinetics given by the self-similar solution (12), (13). Second order equation (14) must satisfy the three conditions (15), which means that the solution exists only for some $\xi_f=\hat{\xi}_f$. Thus, we arrive at the problem of the eigen values. The self-similar solution (12)-(13) describes three qualitatively different types of the spatial-temporal (time) organization in the defect media depending on the values of parameters ω and β [7]. Let us consider two important cases $\omega=\beta+1$ and $\omega>\beta+1$. As it follows from (13) the variables x and t are separated for $\omega=\beta+1$ and the self-similar solution is written as

$$p(x,t)=[S_0(t-t_c)]^{-\frac{1}{\beta}}\left(\frac{2(\beta+1)}{\beta(\beta+2)}\sin^2(\frac{\pi x}{L_T}+\pi\theta)\right)^{\frac{1}{\beta}},\qquad(16)$$

where θ is the arbitrary parameter in the range (0,1). The scale L_T is the spatial period of the solution (16) (the fundamental length [12])

$$L_T=\frac{2\pi}{\beta}(\beta+1)^{\frac{1}{2}}(\zeta_0 S_0^{-1})^{\frac{1}{2}}.\qquad(17)$$

It follows from (16) that in points $x_k=C+kL_T$ (here $C=-L_T\theta$, and k takes the values $k=\pm1,\pm2,\pm3,....$) the parameter p and the "flux" of defects $-\mu_0 p^\beta\dfrac{\partial p}{\partial x}$ are equal to zero. Therefore, equation (16) describes the process of the independent blow-up kinetics of the $p-$ growth on the fundamental scales L_T.

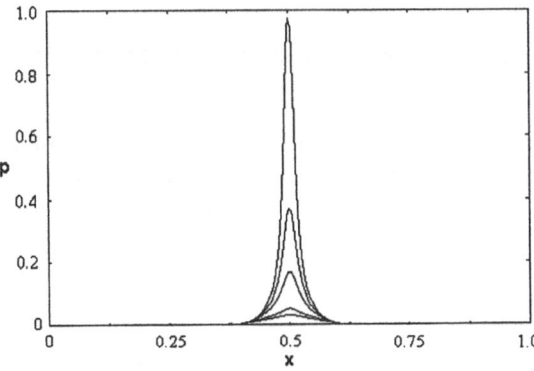

Figure 4.Spatial-temporal structure with the "peak regime" kinetics

The p-profiles are presented in Fig. 4 as a spatial distribution in time. The initial defects distribution of the triangle form is given on the scale $\Delta x_0 > L_T$ (fundamental length $L_T \approx 0.12$). The parameters in Eqn. (13) are $\zeta_0 = 1, S_0 = 1, \omega = \dfrac{5}{3}, \beta = \dfrac{2}{3}$. It is readily seen that the development of the dissipative structure in the "peak regime" is realized on the fundamental length L_T. When the scale of p-distribution coincides with the fundamental length L_T the sharp increase of p is observed and the self-similar p-profile is formed. If the initial p-distribution corresponds to the self-similar profile, the onset of self-similar regime of the dissipative structures evolution occurs immediately (resonance excitation of structures [13,15,]). For $\omega > \beta + 1$ the fundamental length L_T^* is given by the formula [12]

$$L_T^* = \pi \left(\frac{2(\omega + \beta + 1)\mu_0}{\beta(\omega - 1)S_0} \right)^{\frac{1}{2}} p_m^{\frac{1}{2}((\beta+1)-\omega)}, \tag{18}$$

where p_m is the p-maximum for the time when the self-similar profile is reached. This formula is valid when $\Delta x > L_T^*$, and the conditions of the p-kinetics in this case are close to the resonance excitation.

In a medium with microcracks passing of δ through the bifurcation points leads to the generation of spatial structures of various complexity controlled by the different attractor types. As it was shown in [3] the subjection of the continua to these attractors leads to the anomalies of the deformation behaviour. In solids they are realized as the fine grain state for $\delta > \delta_*$ [16], as the strain localization due to the shear banding for $\delta_c < \delta < \delta_*$ [17] and as the macrocrack center nucleation (mirror zones) due to the explosion-like damage localization for $\delta < \delta_c$ [18]. In the last case, the attractor type is the strange attractor and the solid behaviour has stochastic features .

4. Resonance Excitation of Multiscale Instabilities and Failure Scaling in Impacted Solid

The situation with a resonance excitation of the dissipative structures of failure was observed during the study of the transition from the quasi-static to dynamic patterns of failure in ceramic and PMMA cylindrical rods subjected to the loading of the constant stress (quasi-static loading) and impact loading when the spall failure was realized due to the propagation of the elastic wave with the stress amplitude and duration determined by the experimental set-up of the collision of the projectile with a rods.

The spall failure is characterized by small times ($10^{-7} - 10^{-6}$ sec) and large amplitude of tensile stresses, exceeding by several times the quasi-static limit of strength. Experiments were carried out on the rods ($10-12mm$ in diameter and $100-200mm$ long) of PMMA and ultraporcelain [19]. A compression impulse was initiated in the rods by impact on a light-gas cannon and due to the reflection on the free surface the rarefaction wave was formed having the triangle form. The parameters of the impulse and the spall failure kinetics were measured with a laser differential interferometer. From the results of experimental studies of the spall failure of rods, we plotted the logarithm of the fracture time τ_c versus the amplitude of the tensile stress σ_a (Fig. 5). The comparison of $\tau_c(\sigma_a)$-dependencies under quasi-static and impact loadings allowed us to establish the following regularities of the pattern formation. At some stress amplitude the sharp deviation of "fracture time - stress amplitude" dependence is observed. This phenomenon reflects the self-similarity of the failure kinetics depending on external loading conditions, so-called, "overloading effect" under spalling [20] when the failure kinetics is characterized by the weak sensitivity to the stress amplitude. The transient point from quasi-static to dynamic branch corresponds to the qualitative change of the fractographic patterns of the failure surfaces. At the quasi-static branch the typical patterns were observed when failure was initiated at the oblique surface where single mirror zones of the damage localization were observed. The pass of the transient point, that was realized dynamically, leads to the qualitative change of the fracture patterns: the multiply mirror zones were observed on the spall surfaces with a characteristic diameters that "depend" on the stress amplitude. In fact, there is exist the independent character of the fracture time on the stress amplitude, however, the stress rate is of importance and determines the "dynamic branch" slope.

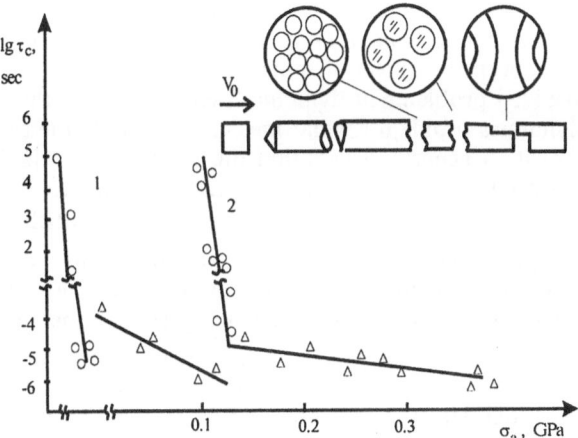

Figure 5. Time-strength dependencies of PMMA (1) and ultraporcelain (2) (○ and Δ correspond to quasi-static and dynamic branches respectively).

Fractographic pictures of fracture are of great interest in different sections of the spalling. In the section where the amplitude of loading impulse is maximal many mirror zones are seen on the spalling surface. Mirror zones appear to be zones of localized damage. In the following sections of spalling with decreased stress amplitude the patterns are similar but the scale of mirror zones increases. In the last spall section that corresponds to the transient point to the quasi-static branch only one or two mirror zones are formed at the oblique surface. The numerical simulation of the damage kinetics showed that an intensive growth of the tensile stress in the rod leads to the formation of a multicenter fracture [13]. Various damage localization scales (sizes of mirror zones) are excited in resonance regimes according to the eigenfunction spectrum of the various complexity. The stress rate is of great importance for the resonance excitation of multiply dissipative structures with the blow-up damage kinetics (mirror zones in fracture patterns). In the last case the initial statistical scattering of the defect nuclei with a large size at the oblique surface is not able to provide the leading growth of these nuclei at the surfaces. The intensive stress growth in stress wave fronts provides the resonance excitation of "peak-regime" structures due to the formation of the spatial-temporal distribution of defects corresponding to the eigen-function forms. The type of the excited eigenforms (size of the mirror zone) is determined by the stress rate at the wave front that produces the maximum p_m in (18) on the corresponding $\Delta x > L_T^*$. The resonance excitation of damage localization areas (mirror zones) according to the eigen functions seems to be as the main factor of the weak dependence of the failure time on the amplitude of the tensile stress (Fig. 5). This result explains the "overloading" effect under dynamic fracture [19].

5. Mechanism of the Instability under Crack Propagation

The dynamics of crack propagation is an interesting nonlinear pattern forming phenomenon. The dynamic behaviour of the cracks propagating through quasi-brittle material has been the object of much recent interest [20-23] There are some fundamental unsolved problems in dynamic fracture. One of the key questions in the the crack dynamics problem is what mechanism selects the crack propagation velocity and limits it. A general view is that the crack speed is limited only by the rate at which a stored elastic energy is transferred to the crack tip. The limiting speed should, therefore, be the Rayleigh wave speed V_R, which is defined by the elastic wave propagation velocity along a free fracture surface. However, as it was shown in [21], even in the brittle amorphous materials the maximum crack speed approaches the value $0.4V_R$. The enigmatic consequence of the limiting speed is the occurrence of the dynamic instability which, as it was pointed out in [21, 23], is related to the onset of new and more effective ways of energy dissipation. The instability is associated with the formation of microcraks on the side of the main crack and at higher velocities the well known phenomenon of macrobranching takes place. The branching of the main crack due to the generation of daughter cracks is the effective mechanism of the energy dissipation, which is, probably, the inherent property of nonlinear system "solid with defects".

Recent measurements [21] of crack evolution in quasi-brittle materials reveal that at some velocity of the crack propagation the system diverts this energy to not a single crack but to the formation of additional branches. It was established there exists a critical velocity V_c above which the dynamics of a crack changes drastically: the velocity of the crack exhibits strong oscillations and the roughness of the crack surface grows significantly. The rough periodic structures emerge from a smooth region and grow larger in the direction of propagation. It was found that the dynamical and statistical properties of the crack are controlled by the steady state velocity as well as the local velocity. The crack velocity exceeding the critical velocity becomes a well-defined function of the local branch length. The functional form of a given branch is independent of the mean crack velocity and can be described by a power law function of its projection in the direction of the main crack propagation. It means that the "branching angle" of a given branch is not well defined, since it is a function of the scale at which the angle is measured.

Let us apply the above developed point of view concerning the spatial-time kinetics of the microcrack ensemble evolution to study of the mechanisms that govern the dynamics of the crack propagation. It is obvious now that the interaction of the main crack and the surrounding microcrack ensemble includes two stages. The first is the formation of the defect distribution in the process zone at the crack's tip which provides the creation of the damage localization areas with a qualitative new properties. It means the formation of the new defect structure with the correlation radius of the own stresses reaching the main crack that can lead to the sensitivity of the main crack to the new structural defect at the scale level corresponding to the main crack. The kinetics of the microcrack ensemble involves generation of new scales in the process zone in the form of the dissipative structures with the peak regime kinetics on the fundamental length L_T. This scale determines the correlation radius of the interaction between the main and new cracks. As in the case of the spall failure, the generation of new structural defects (like mirror zone) is determined by the spatial microcrack distribution (including its maximum and spatial scale $\approx L_T$) that is formed when the critical stress $\hat{\sigma}_n = \max\{\sigma_c\}$(Fig.1). The value of $\hat{\sigma}_c$ is determined by the maximum value of threshold stresses σ_c given by the statistical distribution of the nuclei sizes (in fact, the initial statistical distribution of δ in the range $\delta < \delta_c$). The solution (16) contains two parameters of self-similarity: the fundamental length L_T. and the "peak time" t_c. The peak time t_c represents a sum of two times: t_1 is the period of the formation of the spatial defect distribution that is close to self-similar one, and t_2 is the so-called focusing time [12]. These two parameters determine the critical velocity of the crack propagation

$$V_c \approx \frac{L_T}{t_c}.$$

Now we consider in more detail the crack propagation under the conditions (external stress and initial crack length) , which are responsible for the steady state $(V < V_c)$ and crack branching $(V > V_c)$ scenario. The steady state propagation in the direction of the main crack is realized for external conditions providing the creation of the self-similar profile of microcrack distribution along the main crack

138

when the crack speed doesn't exceed the critical one V_c. It means that in the process zone at the crack tip the stress distribution $\sigma > \hat{\sigma}_c$ is formed on the scale L_T and there exists the time period $t = \dfrac{L_T}{V} \geq t_c$ for nucleatiom of a new daughter crack along the trace of the main crack. For the velocities $V > V_c$ this is impossible, since $\dfrac{L_T}{V} < t_c$. However, taking in view the asymptotic law of stress distribution at the crack tip

$$\sigma_{ij} \approx K_I r^{-\frac{1}{2}} f_{ij}(\theta) \tag{19}$$

(here K_I is the stress intensity factor; r, θ are the coordinates of points in the process zone) we can determine the branching angle θ_* for which the external conditions defined by the value of $K = K_I^*$ (external stress and current length of the main crack) provide the existence of L_T and $t = t_c$:

$$f_{\theta\theta}(\theta_*) = \hat{\sigma}_c K_I^{*-1} L_T^{\frac{1}{2}} \tag{20}$$

.For the given value of K_I^* the values of the angles $0 < \theta < \theta_*$ are excluded since the condition for the corresponding traces $t < t_c = \dfrac{L_T}{V_c}$ doesn't allow the formation of the self-similar profile of the microcrack distribution (daughter cracks). The next important question is what mechanism governs the increase in the main crack speed in the range $V > V_c$ when the rate of the interaction of the main crack with the closest daughter crack is limited by V_c. As it follows from the above consideration of the dynamic scaling for the spalling, the reason is the excitation of the numerous "peak regime structures" when the scale of the process zone L_{pz} with $\sigma > \hat{\sigma}_c$ expands as $L_{pz} \approx L_T k$, where $k = 1, 2 \dots$. The nucleation of these structures on the total length $L_T k$ (complex structures [7]) could considered as the subjection of the system behaviour to new attractor which is determined in the set of new independent coordinates, that is in the set of the structures of the various complexity obtained from the self-similar solution (16). It means a sharp change of the symmetry properties of the system that was predicted in [18].

6. Statistical Simulation of Crack Propagation

The regularities of transitions from damage to fracture were examined numerically. The problem of quasi-brittle fracture has been solved by the finite element method based on the equilibrium equation

$$\frac{\partial \sigma_{ij}}{\partial x_i} = 0, \tag{21}$$

the constitutive equation of elastic medium with microcracks

$$\sigma_{ij} = E_{ijkl} \left(\varepsilon_{kl} - p_{kl} \right), \tag{22}$$

and the kinetic equation of damage accumulation

$$\dot{p}_{ij} = - L_p \left(A\, p_{ij} + B\, p_{ik}\, p_{kj} + C\, p_{ik}\, p_{kl}\, p_{lj} + D\sigma_{ij} \right) + \frac{\partial}{\partial x_l} \left(\omega\, p_{ij}^{\ 2}\, \frac{\partial}{\partial x_l} p_{ij} \right). \tag{23}$$

Phenomenological parameters L_p, A, B, C, D, ω were determined as the results of approximation of typical defect density curves [24]

$$A = -3.1 \cdot 10^{11}, \quad B = 1.7 \cdot 10^{12}, \quad C = 1.0 \cdot 10^{12}, \quad \omega = 1.0 \cdot 10^{-2} \tag{24}$$

The choice of the boundary and initial conditions depends on the type of experiments which were simulated. Modeling of the deformation and fracture processes starts with random assignment of the relaxation time L_p to each element of the finite element approximation (number of the elements is 5000). The Weibull distribution function was used to obtain random value of the L_p

$$f(x) = \alpha \lambda x^{\alpha - 1} \exp\left(-\lambda x^{\alpha}\right). \tag{25}$$

At each time step we calculate a new value of the stress and strain taking into account the influence of microcrack accumulation, and than define the value of parameter p_{yy}. The element breaks down, when p_{yy} reaches the critical value p_c ($p_c = 1.0 \cdot 10^{-2}$ is an experimental estimate). The macroscopic fracture corresponds to the formation of a percolation cluster that consists of fractured elements. The final step of fracture simulation is the fractal analysis of the percolation cluster. The cluster appears to be fractal in nature and with an increase of linear dimension L of the damaged array its mass M (the number of destroyed elements) increases on the average as:

$$M(L) = AL^D, \tag{26}$$

where D is the fractal dimension, A is the effective amplitude. The mean value of A is obtained by averaging over the manifold realization of the percolation cluster. Fig. 6 compares two typical form of the percolation cluster. The process kinetic and topology of the cluster results from several reasons: (i) form of the equation (23), (ii) value of the constant in (23) and (25), (iii) loading conditions, (iv) presence and size of the initial macrodefect. The combination of the above conditions may lead to the formation of the branched cluster (Fig. 6 a) either slowly branched (Fig.6 b). This approach was used to simulate failure development in specimens with an initial macroscopic defect located in the center (the macrocrack is normal to the tension direction). The dependence $M(L)$ consists of two linear parts with the slopes determined by the fractal dimension D. Simulation of damage has demonstrated that under loading the initial stage is accompanied by preferential failure of elements located in the vicinity of the macroscopic defect ($D = 1$). The percolation cluster across the specimen results from coalescence of the cluster originating from the initial macrodefect

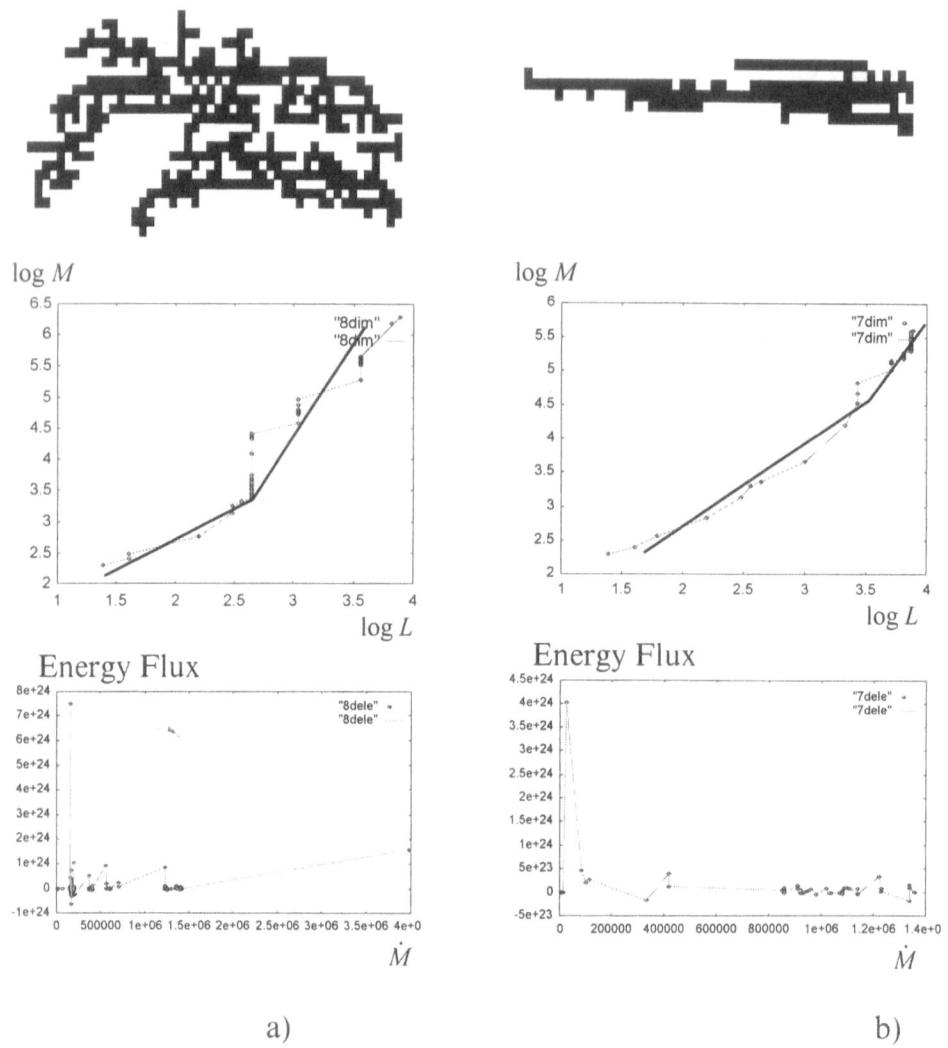

Figure 6. Typical form of the percolation cluster

with clusters in its immediate neighbourhood ($D = 1.4 - 1.7$). This is indicative of the qualitative change in the topology of damage accumulation process and the fracture mechanism replacement. The fractal dimension $D = 1$ supports the validity of approaches of fracture mechanics only at the initial stage of crack evolution. Simulation results are in agreement with the qualitative features seen in the experiments. It was shown [21] that a sharp transition from a single propagating crack to an ensembles of cracks occurs above the critical velocity of

$v_c = 0.36 V_R$ when a single crack undergoes a local change in topology and sprouts micro-branches. At that point the mean acceleration of the crack drops and oscillations of the velocity are observed whose amplitude is an increasing function of the mean velocity. Coincident with the onset of the velocity oscillations non-trivial structure is formed on the fracture surface [21]. The above results suggests that crack propagation due to interacting microcracks occurs as a randomly activated process. The present simulation also confirms the jump-like crack moving that oscillations of the energy flux illustrate (Fig. 6).

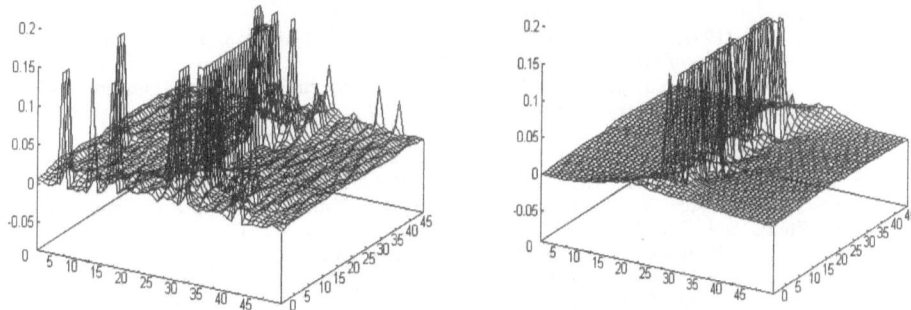

Figure 7. Spatial-time distribution of defect density

The spatial distribution of the defect density was obtained in the course of numerical simulation.

7.References

1. Mecholsky, J.J. (1985) Fracture analysis of glass surfaces. *Strength of Inorganic Glass* in C.R.Kurkjian (eds.), Plenum Press, New-York, pp.569-590.
2. Mandelbrot, B.B., Passoja, D.E. and Paullay, A.J.(1984) Fractal character of fracture surfaces in metals. *Nature*, **308**, 721-722.
3. Naimark, O.B.(1982) *On thermodynamics of deformation and fracture of solids with microcracks.* Institute of Mechanics of Continuous Media, USSR Academy of Sciences, Sverdlovsk (in Russian).
4. Betechtin, V.I., Naimark, O.B. and Silbershmidt, V.V. (1989). The fracture of solids with microcracks: experiment, statistical thermodynamics and constitutive equations. *Proceedings of Int. Conference of Fracture (ICF 7),* **6**, pp.38-45.
5. Naimark, O.B. (1995) *Structural Transitions in Solids and Mechanisms of Plasticity and Failure.* Preprint of the Institute of Continuous Media Mechanics of the Russian Academy of Sciences, Perm.
6. Kroner, E. (1986). *On the gauge theory in defect mechanics*, Lecture Notes in Physics, Heidelberg, Springer, pp.281-296
7. Naimark, O.B. (1996). Kinetic transition in ensembles of microcracks and some nonlinear aspects of fracture. In *Proceedings of the IUTAM Symposium on nonlinear analyis of fracture*, in J.R.Willis (eds.), Kluwer, Dordrecht, The Netherlands, pp. 285-298
8. Barenblatt, G.I. and Botvina, L.R. (1983). Self-similarity of fatigue fracture, *Izv.AN SSSR. Mech.Tv.Tela*, **4**, 161-165 (in Russian).
9. Leontovich, M.A. (1983). *Introduction to Thermodynamics. Statistical Physics*, Moscow, Nauka (in Russian).
10. Landau, L.D. and Lifshitz, E.M. (1980) *Course of Theoretical Physics, vol.5: Statistical Physics*, Pergamon Press, Oxford.

11. Landau, L.D. and Lifshitz, E.M. (1980). *Course of Theoretical Physics, vol.2: Mechanics of Continuous Media*, Pergamon Press, Oxford.

12. Kurdyumov, S.P. (1979) *Combustion Eigenfunctions of Nonlinear Media and Constructive Laws of Organisation.* Preprint of the Keldysh Institute of Applied Mathematics of the USSR Academy of Sciences, n.29 (in Russian).

13. Beljaev, V.V. and Naimark, O.B. (1990). Kinetics of multicenter fracture under shock wave loading. *Sov. Phys. Dokl.,* **312**, n. 2, 289-293.

14. Naimark, O.B. (1997). Structural transitions in ensembles of defects as mechanisms of failure and plastic instability under impact loading. *Proc. IX Int. Conf. Fracture.* Sydney, **6**, pp.2795-2806.

15. Naimark, O.B. (1997). Resonance excitation of multiscale instabilities and deformation anomalies in impacted solid (experimental and theoretical study), in J. Salencon (eds.), *Multiple scale analyses and coupled physical systems,* Presses Ponts et chaussees, Paris ,pp. 513-520.

16. Naimark, O.B. (1997). Nanocrystalline state as topological transitions in ensembles of grain boundary defects. *Physics of Metals.* (in press)

17. Naimark, O.B. and Ladygin, O.V.(1993) Nonequilibrium structural transitions in solids as mechanism of localization of plastic deformation. *J.of Appl. Mech. and Tech. Phys.,* **3**, 121-137 (in Russian).

18. Naimark, O.B. and Davydova, M.M. (1996) Crack initiation and crack growth as the problem of localized instability in microcrack ensemble. *J. Physique III,* **6**, 259-267.

19. Bellendir, E.N., Beljaev, V.V. and Naimark, O.B. (1989). Kinetics of multicenter fracture under spalling conditions. *Sov. Tech. Phys. Lett.,* **15 (13)**, 90-93

20. Freund, L.B. (1990) *Dynamical Fracture Mechanics*, Cambridge University Press, New York.

21. Sharon E., Gross, S.P., Fineberg J. (1995) Local crack branching as a mechanism for instability in dynamic fracture, *Phys. Rev. Lett* **74**, 5097-5099

22. Marder, M. and Gross, S. (1995) Origin of crack tip instabilities. *J. Mech. Phys. Solids.* **43**, 1, 1-48.

23. Boudet, J.F., Ciliberto, S. and Steinberg, V. (1996) Dynamics of crack propagation in brittle materials, *J. Phys. II France,* **6**, 1493-1516.

24. Betechtin V. I., Vladimirov V. J. (1979) Kinetics of microfracture of crystalline bodies, Problems of Strength and Plasticity of Solids, pp. 142-154 (in Russian).

A STATISTICAL-PROBABILISTIC APPROACH TO MICROSTRUCTURE-STRUCTURE RELATIONS IN FAILURE AND DAMAGE OF CERAMIC COMPOSITES

J. LAMON and N. LISSART
Laboratoire des Composites Thermostructuraux
UMR 47 (CNRS-SEP-UB1)
Domaine Universitaire, 3 Allée La Boétie
33600 Pessac, France

Abstract

A model is proposed for predicting the nonlinear stress-strain relations and ultimate failure of ceramic minicomposites. Description of the micro-mechanisms which dictate the mechanical response (i.e. multiple cracking of the matrix and fiber failures) is based upon fracture statistics. The minicomposite specimen consists of a bundle of parallel fibers coated with a layer of interfacial material and a matrix. It is representative of the constituents in the actual textile composites.

Predictions from characteristics of the constituents, including the fibers, the matrix and the interphase, compared satisfactorily with the tensile force-strain curves measured on practical SiC/SiC minicomposites tested under uniaxial tension. Finally the influence of various constituent characteristics on the mechanical behavior was determined.

1- Introduction

Various microstructural mechanisms influence the mechanical response of ceramic matrix composites (CMCs). Fiber/matrix interphases favor crack deflection, eventual further matrix cracking and fiber pull-out [1-4]. A dense network of matrix cracks is created in the presence of rather strong fiber/matrix interactions and limited interfacial damage [3, 5] whereas fiber pull-out requires rather weak fiber/matrix bonding [1]. The related tensile stress-strain curves exhibit typical features. In the presence of weak fiber/matrix interfacial regions, a plateau-like stress-strain behavior is obtained : the non-linear domain attributed to matrix multiple cracking extends over a narrow range of stresses and strains and the stress at matrix cracking saturation is distinct from ultimate strength. In contrast, in the presence of rather strong interphases, the stresses are higher, the non-linear domain is wider and the stress at saturation is now close to or coincides with ultimate strength [2, 3, 5].

The primary intent of the present paper is therefore to establish a model which describes the influence of constituent properties upon the stress-strain response of minicomposites

143

G. N. Frantziskonis (ed.), PROBAMAT – 21st Century: Probabilities and Materials, 143–160.

specimens. Minicomposites are representative of matrix infiltrated bundles which behave as physical entities in textile ceramic matrix composites (CMCs) reinforced with fabrics of woven fiber bundles [6]. Formation of matrix cracks and fiber failures are described using strength - probability equations derived from statistical approaches to brittle failure. The present approach to matrix cracking differs from the conventional one which considers the total volume of cracked material. In the present paper, matrix cracking is described as the brittle failure of uncracked volume elements (matrix fragments). This allows one to solve problems involving non-uniform stress-states.

Under uniform tensile loading conditions, multiple cracking of the matrix occurs first. Formation of the cracks is a random phenomenon as a result of the presence of randomly distributed fracture inducing flaws. The matrix cracks and the associated debonds affect locally the stress field operating on the fibers. The approach assimilating the bundles to a series of chain or links of constant length given by the crack spacing distance, is inappropriate here, due to random location of the matrix cracks. Instead, the gauge length is determined by the typical stress field operating on the fibers.

Individual fiber breaks within bundles have been investigated extensively in polymer matrix composites [7-9]. The minicomposites exhibit typical features due to the presence of multiple cracks across the matrix and associated interfacial debonds. When one fiber breaks, it is taken up equally by the remaining intact fibers. Furthermore, the fibers fail only once [10, 11]. The minicomposites were assimilated to bundles of fibers subject to the specific stress field induced by multiple cracking of the matrix. Ultimate fracture occurs when a critical number of fibers have failed individually.

The model was used for predicting the mechanical response of SiC/SiC minicomposites consisting of a SiC matrix reinforced with a bundle of 500 parallel SiC fibers. Then the influence of several constituent characteristics including the interfacial shear stress and statistical flaw strength parameters pertinent to the fibers and the matrix was anticipated.

2- Approaches to matrix cracking and ultimate failure

2.1 Matrix cracking

Matrix cracking in minicomposites as well as in infiltrated fiber bundles within practical CMC, involves cracks that initiate from pre-existing flaws and then propagate catastrophically through the cross section [12]. Formation of a new crack in the matrix results from the brittle failure of a uncracked matrix fragment. The fragment is thus splitted into two new fragments. In the presence of n matrix cracks, the matrix is subdivided into (n + 1) fragments. Matrix cracking can be described by applying weakest link statistics to the brittle failure of each fragment.

The probability of brittle failure of the ith matrix fragment of length $2l_i$ is given by the following equation :

$$P_{Mi} = 1 - \exp\left[-2S_M \cdot \int_0^{\ell i} \left(\frac{\sigma_M}{\sigma_{0M}}\right)^{m_M} \cdot dx\right] \qquad (1)$$

where S_M is the cross sectional area of matrix ; m_M and σ_{0M} are the statistical parameters pertinent to the matrix; σ_M is the stress operating on the matrix..

Formation of a matrix crack causes debonding of the fibers over a distance l_d:

$$2l_d = R_f \left[\frac{\sigma_{max}^f - \left(\sigma_{min}^f + \sigma_f^T \right)}{\tau} \right] \qquad (2)$$

where σ_f^T represents the axial residual stresses, σ_{max}^f is the peak stress in the fiber, τ is the interfacial shear stress along the debond (figure 1).

Debonding induces a local stress gradient in the matrix (figure 1) given by the following equation [13] :

$$\sigma_M(x) = \frac{\sigma}{1 - V_F} \cdot \frac{a}{1 + a} \cdot \frac{x - u - \ell_0}{\ell_d - \ell_0} \quad x \ge u + \ell_0 \qquad (3)$$

where σ is the stress applied to the minicomposite, and $a = \dfrac{E_M(1 - V_F)}{E_F V_F}$ (E_M and E_F are moduli of the matrix and the fibers respectively).

σ_M is uniform in the bonded portions of the minicomposites :

$$\sigma_M = \frac{\sigma a}{V_M(1 + a)} \qquad (4)$$

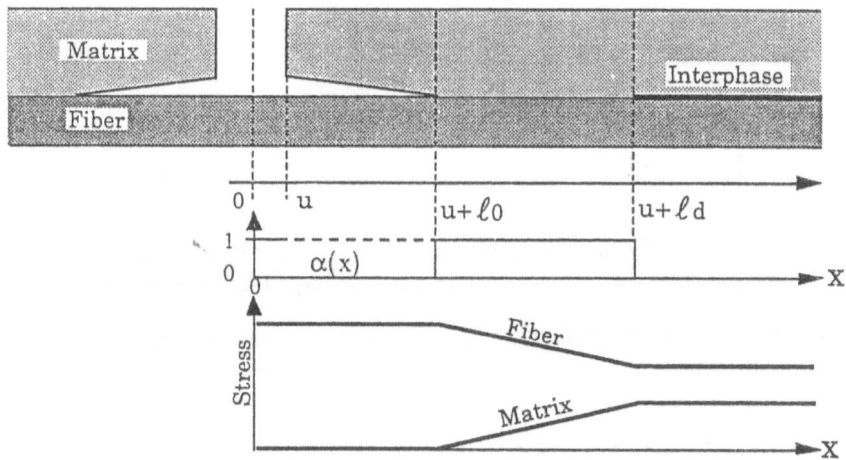

Fig.1 : Schematic diagram showing the local stress-state in the fiber and in the matrix by a matrix crack [13]

The matrix fragments delineated by the transverse cracks may be grouped into two families: (i) the type I fragments for which the effective volume of matrix is not yet equal to zero and (ii) the type II elements for which the effective volume of matrix became equal to zero. The effective volume refers to the volume of matrix located between the debonded portions in a matrix fragment. The stress field includes uniform stresses in the type I elements (eq. (3) and (4)) and only stress gradients in the type II elements (eq. (3)) (figure 2). The matrix cracks were assumed to form in the effective volumes (figure 2). Saturation was assumed to occur when the effective volume of the

146

largest matrix element becomes equal to zero. At this stage the interfacial cracks extend over the entire length of the minicomposite.

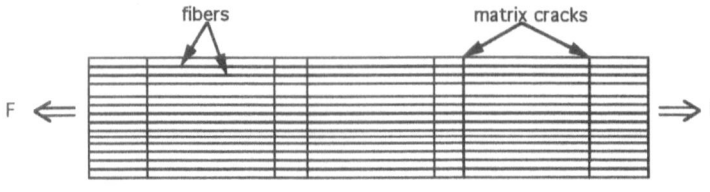

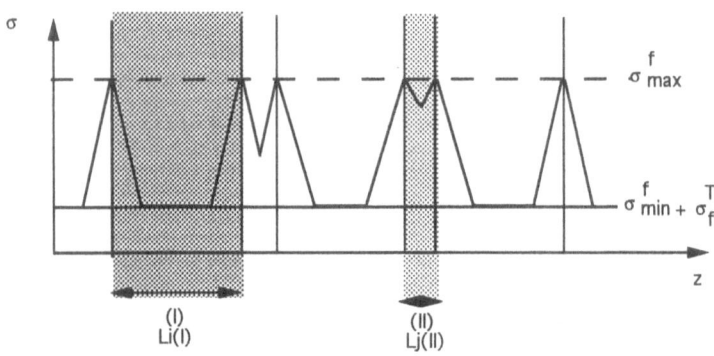

Fig.2 : Schematic diagram showing the stress fields induced by the matrix cracks in a minicomposite under tension and the associated type I and II fiber elements.

Incorporating the above equations of σ_M, and performing integration of equation (1) give the probability of failure of the ith type I matrix fragment:

$$P_{Mi} = 1 - \exp\left[-2S_M \cdot \left(\frac{\sigma}{\sigma_{0M}} \cdot \frac{a}{V_M \cdot (1+a)}\right)^{m_M} \cdot \left[\ell_i - \frac{\ell_{di}}{(m_M + 1)} \cdot (m_M + \beta)\right]\right] (5)$$

where l_{di} is the debond length in this i^{th} matrix fragment, and $\beta = \frac{l_o}{l_{di}}$ (figure 1).

This approach implicitly assumes that a single population of flaws is responsible for matrix cracking, and that the Weibull equation still describes scale effects as fragments become shorter and shorter. Other failure probability equations, such as the Weibull one as modified by Phani [14] could be used.

The strength σ_M^i of the ith matrix fragment of volume V_i is derived from equation (5).

It may be expressed as a function of a reference matrix strength σ_M^R that measures the strength of the initial volume of matrix (V_{OM}) in the uncracked minicomposite :

$$P_{MO} = 1 - \exp\left[-V_{OM} \cdot \left(\frac{\sigma_M^R}{\sigma_{OM}}\right)^{m_M}\right] \qquad (6)$$

Equating P_{Mi} to P_{MO} gives σ_M^i :

$$\sigma_M^i = \sigma_M^R \left(\frac{V_{OM}}{V_i}\right)^{1/m_M} \left(1 - \frac{l_{di}}{l_i} \frac{m_M + \beta}{m_M + 1}\right)^{-1/m_M} \qquad (7)$$

It is considered that the ith fragment fails if the stress acting on the matrix (σ_M) exceeds the fragment strength. A criterion for formation of a crack is thus:

$$\sigma_M^i \leq \sigma_M \qquad (8)$$

2.2 Fracture of fibers

The fracture of fibers involves the following features [11] :
(i) the individual fiber breaks occur at high loads near ultimate failure, and
(ii) the fibers fail only once. Once a fiber has broken anywhere in the gauge length of the specimen it is no longer capable of carrying the load [10, 11].
Probability of failure of the fibers is determined by the stress state induced by matrix cracks and also by the law of load sharing among the surviving fibers as fibers fail individually. A global load sharing, as observed in fiber bundles [15], is a realistic assumption for the examined minicomposites. The probability of failure of fibers within the minicomposites was determined as a function of the numbers of matrix cracks and individual fiber breaks. The failure probability was expressed in terms of an equivalent fiber length (denoted L_{equi}) reflecting the stress state induced by the matrix cracks. L_{equi} is defined as the length of a fiber subjected uniformly to the peak stress σ_{max}^f for the same failure probability. Failure probability for a fiber within the minicomposites is given by the following equation :

$$P = 1 - \exp\left[-A_f L_{equi} \left(\frac{\sigma_{max}^f}{\sigma_{of}}\right)^{m_f}\right] \qquad (9)$$

where

$$\sigma_{max}^f = \frac{F}{A_f^t (1 - \alpha)} \qquad (10)$$

F is the applied force, A_f is the cross sectional area of a single fiber, A_f^t is the total cross sectional area of all the fibers present in the minicomposites, m_f and σ_{of} are the statistical parameters pertinent to the fibers, and α is the fraction of individual fiber breaks.

148

α represents the failure probability of the Nth fiber. Therefore the number of fibers N that are broken under a given force F is related to the initial number of fibers N_0 by the following equation :

$$N = N_0 P \qquad (11)$$

Inserting the expression of P into equation (11) the fraction of surviving fibers is given by :

$$q = 1 - \alpha = \exp\left[-A_f L_{equi}(\alpha)\left(\frac{F}{A_f^t(1-\alpha)\sigma_{of}}\right)^{m_f}\right] \qquad (12)$$

The ultimate failure which may be described as an instability in the evolution of individual fiber failures is characterized by the following criterion :

$$\left.\frac{\partial F}{\partial \alpha}\right|_{\alpha=\alpha^*} = 0 \qquad (13)$$

α^* represents the critical value of α that causes ultimate failure.

2.3 Equivalent fiber length L_{equi}

The failure probability of a fiber within a minicomposite may be written as :

$$P = 1 - (1 - P_i)^k (1 - P_j)^t \qquad (14)$$

where $(1-P_i)^k$ and $(1-P_j)^t$ represent the survival probabilities respectively for the k type I and the t type II fiber elements defined in figure 2.

The stressfield along the fiber axis in a **type I fiber element** (distinct from minicomposite end) is described by the following equations :

$$\sigma_f(z) = \sigma_{max}^f\left[\frac{z}{l}B + 1\right] \qquad 0 < z < 1 \quad (15)$$

$$\sigma_f(z) = \frac{F}{A_f^t(1-\alpha)(1+a)} + \sigma_f^T \qquad z > 1 \quad (16)$$

now l denotes the debond length, that was previously referred to as l_{di}.

B is a constant : $B = \dfrac{-a}{(1+a)} + \dfrac{\sigma_f^T}{F}A_f^t(1-\alpha)$

Incorporating the stress state into the failure probability equation gives the following equation for the survival probability of a type I fiber element i of length $L_i(I)$:

$$1 - P_i = \exp\left[-2A_f \int_0^1 \left(\frac{\sigma_f(z)}{\sigma_{of}}\right)^{m_f} dz\right] \times \exp\left[-A_f\left(L_i(I) - 2l\right)\left(\frac{\sigma_{min}^f + \sigma_f^T}{\sigma_{of}}\right)^{m_f}\right] \quad (17)$$

After performing the integration, equation (17) becomes :

$$1 - P_i = \exp\left[-\left(\frac{\sigma_{max}^f}{\sigma_{of}}\right)^{m_f} A_f \left[\begin{array}{l} \dfrac{2l}{B(m_f + 1)}\left(-1 + (1 + B)^{m_f + 1}\right) + \\[2ex] \left(L_i(I) - 2l\right)\left(\dfrac{1}{1 + a} + \dfrac{\sigma_f^T}{F} A_f^t(1 - \alpha)\right)^{m_f} \end{array}\right]\right] \quad (18)$$

The equivalent fiber length in this type I element i is derived from equation (18) :

$$L_{equi}^{I, i}(\alpha) = L_i(I)\left[\begin{array}{l} \dfrac{2l}{L_i(I)}\dfrac{\left[(1 + B)^{m_f + 1} - 1\right]}{B(m_f + 1)} + \\[2ex] \left(1 - \dfrac{2l}{L_i(I)}\right)\left(\dfrac{1}{1 + a} + \dfrac{\sigma_f^T}{F} A_f^t(1 - \alpha)\right)^{m_f} \end{array}\right] \quad (19)$$

The total equivalent length of the k type I fiber elements is :

$$L_{equi}^k(\alpha) = \sum_{i=0}^{k} L_{equi}^{I, i}(\alpha) \quad (20)$$

The stress state in a **type II fiber element** (figure 2) is described by equation (15). The survival probability for the type II element j of length $L_j(II)$ is obtained by substituting the stress state into the following equation and integrating :

$$1 - P_j = \exp\left[-2A_f \int_0^{\frac{L_j}{2}} \left(\frac{\sigma_f(z)}{\sigma_{of}}\right)^{m_f} dz\right] \quad (21)$$

$$1 - P_j = \exp\left[-A_f\left(\frac{F}{A_f^t(1 - \alpha)\,\sigma_{of}}\right)^{m_f}\left[\frac{2l}{B(m_f + 1)}\left(-1 + \left(\frac{L_j(II)}{2l}B + 1\right)^{m_f + 1}\right)\right]\right] \quad (22)$$

The equivalent length for this type II element j is derived from equation (22) :

$$L_{equi}^{II, j}(\alpha) = L_j(II)\left[\frac{2l}{B(m_f + 1)}\left(-\frac{1}{L_j(II)}\right) + L_j^m(II)\left(\frac{1}{2l}B + \frac{1}{L_j(II)}\right)^{m_f + 1}\right] \quad (23)$$

The total equivalent length of the t type II fiber elements is :

$$L^t_{equi}(\alpha) = \sum_{j=0}^{t} L^{II,j}_{equi}(\alpha) \qquad (24)$$

The equivalent length of the fiber is :

$$L_{equi}(\alpha) = L^t_{equi}(\alpha) + L^k_{equi}(\alpha) \qquad (25)$$

3.1 Prediction of the mechanical behavior of SiC/SiC minicomposites

1- Simulation of matrix cracking and ultimate failure

The simulation of the matrix cracking phenomenon is based on the strengths of the matrix fragments generated by the transverse cracks. Due to size effects, matrix strength increases as the fragments become smaller and smaller. The successive fragment strengths were determined using equation (7). The simulation was conducted by iteratively incrementing the applied force F. At each step, the critical fragments were identified from the comparison of the applied stress to the fragment strengths. The critical fragments exhibit strengths smaller than the applied stress. Crack location in the effective volume of critical fragments was determined using the probability (λ_i) of location of the crack-inducing defect in the fragment: $\lambda_i = x_i/2(l_i-l_{di})$, where x_i indicates the location of the critical flaw (i.e. the distance from the debond tip), $2l_i$ is the fragment length, l_{di} is the debond length. Therefore, the lengths of both matrix fragments created at step i by failure of a fragment of length $2l_i$ are respectively (x_i+l_{di}) and ($2l_i- x_i-l_{di}$). λ_i data were given by computer generated random numbers ($0 < \lambda i < 1$).

The debond length associated to each matrix crack was calculated using equation (2). The element type was identified by comparing the debond length l_{di} to the element size $2l_i$: $2l_{di} < 2l_i$ for the type I elements, $2l_{di} > 2l_i$ for the type II elements. The equivalent length L_{equi} was computed using equation (25). The fraction of broken fibers α was determined by solving the equation (12) using an iterative method. The ultimate failure is indicated by the criterion given by equation (13).

Determination of residual stresses in minicomposites is not straightforward. Instead it requires computations based on the finite element method. Therefore, residual stresses were neglected in this paper.

The properties required for the computations are given in table 1.

Table 1: Mean characteristics of the SiC/SiC minicomposites [11].

V_f (%)	31.5
V_m (%)	61
Fiber Young's Modulus (GPa)	200
Matrix Young's Modulus (GPa)	400
Statistical parameters	
matrix : m_m	6.08
matrix : σ_{om} (Vo=1m3) (MPa)	10.51
fiber : m_f	5.45
fiber : σ_{of} (Vo=1m3) (MPa)	19.5
τ (MPa)	115

3.2 Force - deformation curves

The mean values of properties were used for the computations leading to the average force-deformation curves. Microcomposite deformations were derived from the stress field operating on the fibers (equations (15) and (16)) as a function of the number of matrix cracks and the fraction of broken fibers (α) (Appendix).

The computed curve coincides with the one obtained experimentally, for the particular value of τ = 80 MPa (Figure 3). Table 2 shows that the features of the minicomposites mechanical behavior that have been predicted for this particular τ, are in excellent agreement with the experimental data. This τ value is different from that one extracted from the experimental force-deformation curves [11, 15] (table 1), but the discrepancy falls within the scatter usually observed for this kind of data (around 50%).

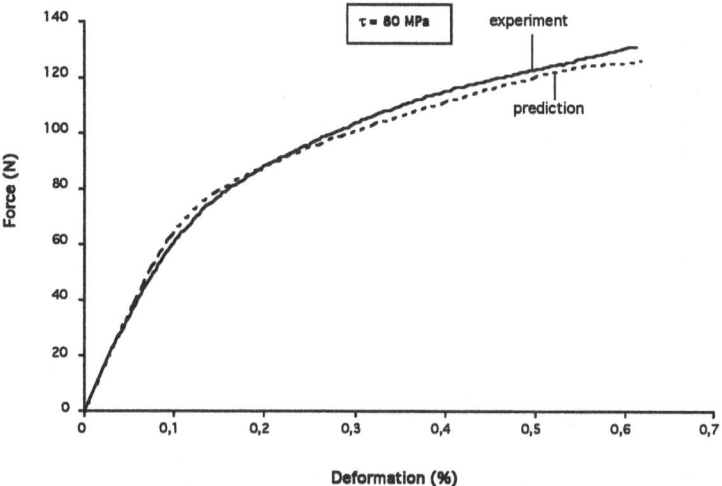

Fig.3 : Comparison of the force-strain behavior predicted from τ = 80 MPa with the experimental one for a SiC/SiC minicomposite.

The predicted critical fraction of individual fiber breaks was 16,76 % (table 2). As logically expected this value is similar to the one determined for bundles of fibers [15]. Computations confirmed that fiber breaks occur near ultimate failure and when the density of matrix cracks is significant (figure 4).

152

Table 2 - Features of the stress-strain behavior of the minicomposites :
predictions and experimental data.

	Forces at onset of matrix cracking (N)	Deformations at onset of matrix cracking (%)	Forces at ultimate failure (N)	Deformations at ultimate failure (%)	Matrix crack spacing distances (μm)	Critical fractions of individual fiber breaks (%)
Experiments	36,8	0,051	130	0,65	71	
Predictions (τ = 80 MPa)	35,78	0,05	125	0,61	79	16.76

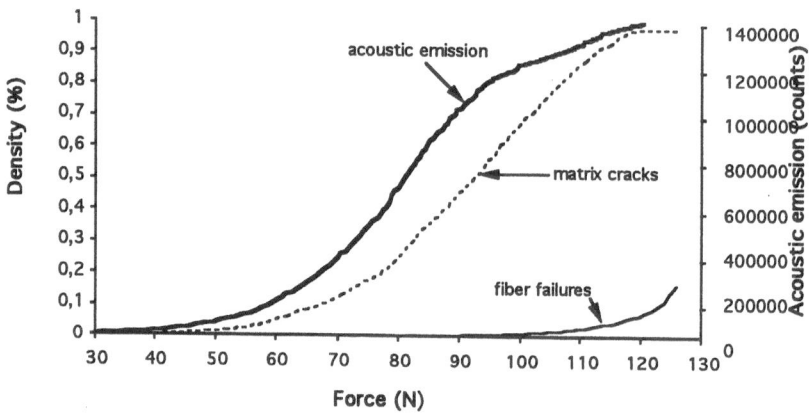

Fig.4 : Predictions of the densities of matrix cracks and fiber failures as a function of the applied forces, for a SiC/SiC minicomposite.

4- Prediction of microstructure - mechanical behavior relationships

4.1 Influence of the interfacial shear stress on the stress-strain behavior

The computations for various interfacial shear stresses τ ranging between 5 and 200 MPa, reproduced the trends which have been observed experimentally on practical CMC [2, 3].Thus it can be noticed from figure 5 that, when τ increases, the strain-to-failure tends to decrease whereas the corresponding applied force increases. On the other hand a plateau-like non linear domain appears for the low τ values.

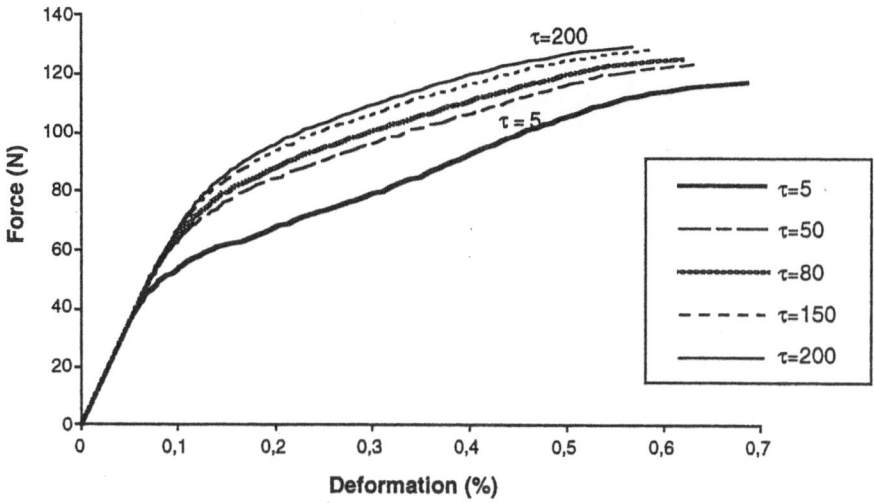

Fig.5 : Force-strain curves predicted for SiC/SiC minicomposites for various values of the international shear stresses τ.

4.2 Influence of fiber properties

The force-strain behavior was predicted for various values of fiber properties: i.e. (i) for various values of the average fiber strength with the Weibull modulus m_f remaining constant and (ii) then for various values of m_f at constant average fiber strength. The associated scale factors were derived from equation (26) :

$$\sigma_{of} = \frac{\overline{\sigma}_{fibre} V_f^{1/m_f}}{\Gamma\left(1 + \dfrac{1}{m_f}\right)} \qquad (26)$$

In figure 6 the average force-strain behavior of minicomposites was predicted for fiber strengths varying around the mean reference strength. A 10% variation around the reference leads to a significant scatter in the strain-to-failure (0.5-0.7%). Figure 6 also shows that fiber strength degradations strongly affect the ultimate failure of minicomposites: for a 30% degradation, the strain-to-failure diminishes from 0.6% to less than 0.4%. In practical composites, fiber strength degradations may be caused by processing or environment. On the contrary, stronger fibers improve the strain-to-failure. This may explain the scatter in the force-strain behavior that is observed with practical minicomposites [11, 15].

Figure 6 shows that for low Weibull's moduli the non linear domain of deformations becomes more pronounced and, as a logical consequence, the applied forces are lowered. This effect is caused by an increase in the fraction of individual fiber breaks. Thus it was calculated that when m_f decreases from 50 to 3, the critical fraction of individual fiber breaks increases from 2 to 29 %.

154

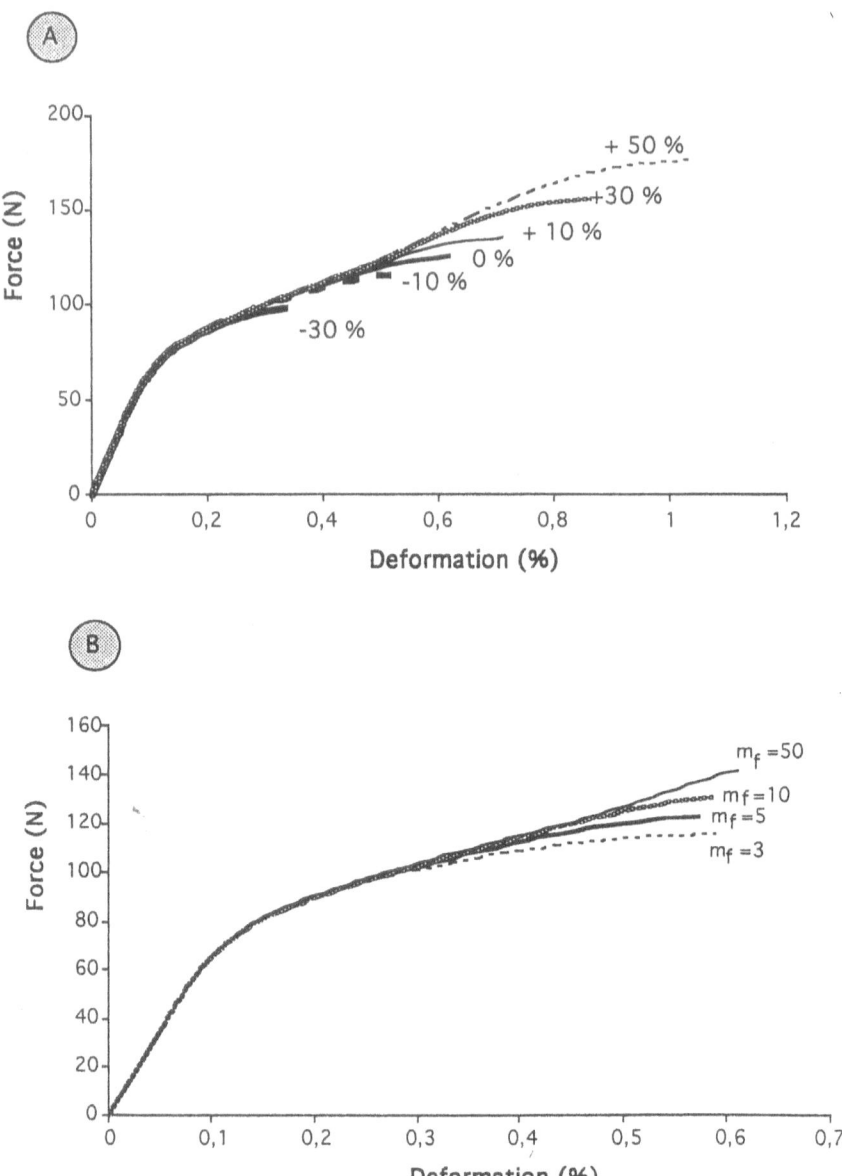

Fig. 6 : Prediction of the influence of fiber strength characteristics on the force-strain behvior of minicomposites from batch B (τ = 80 MPa) (A) for various ratios of fiber strength increase or decrease to the reference value ; (B) for various values of the Weibull modulus

Figure 7 shows the effect of fiber Young's modulus on the force-strain behavior. As logically expected, the fiber Young's modulus affects tremendously the strain-to-failure. Low moduli favor the ability to accomodate deformations whereas high moduli stiffen the material.

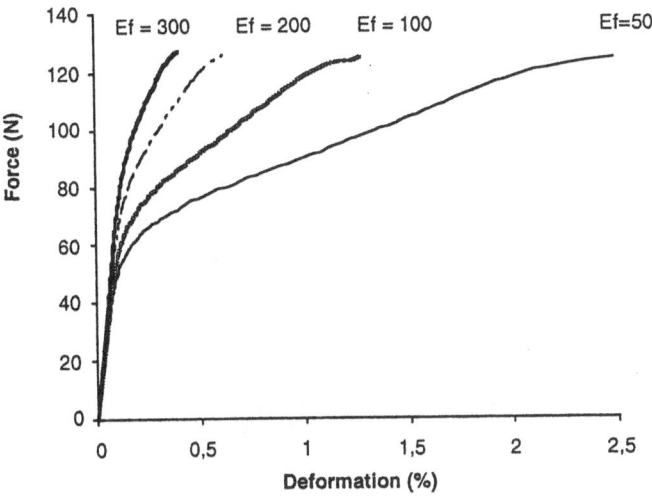

Fig.7 : Prediction of the influence of fiber Young's modulus (GPa) on the force-strain behavior of minicomposites (τ = 80 MPa).

4.3 Influence of matrix properties

Variations in the matrix properties were also described using the statistical parameters. The matrix strength was varied first around a reference value characterizing the actual SiC matrix, at constant m_m. Then, in a second step, it was set constant as computations were performed for various m_m values. The associated σ_{om} values were derived using equation (26). Figure 8 shows that the matrix strength exerts a fondamental influence on the force-strain behavior. Increasing the matrix strength elevates the stresses and more particularly the stress at onset of matrix cracking. By contrast, decreases in the matrix strength favor the onset of matrix cracking and a plateau-like domain of deformation. Computations indicated that the matrix crack spacing distance at ultimate failure is larger as the matrix is stronger : for a 50 % strength increase the spacing distance increased from 79 to 195 microns.

Similar effects are induced by the Weibull's modulus. Thus a pronounced non-linear domain of deformation is observed for low m_m values whereas a plateau-like domain was predicted for large m_m values. These trends are in agreement with logical expectation. Large m_m values characterize a nil scatter in strength data. As a consequence matrix cracking tends to be a deterministic phenomenon and the cracks form at a quite constant load. The associated mechanical behavior is the classical one that was modelled by Aveston, Cooper and Kelly (ACK) [16]. By contrast, low m_m values characterize a significant scatter in strength data and an associated sensitivity to size effects. The matrix cracks form at increasing stress levels as reflected by the non linear domain of deformation. The ACK model is deterministic in essence. It does not describe matrix multiple cracking in those CMC for which m_m < 10.

156

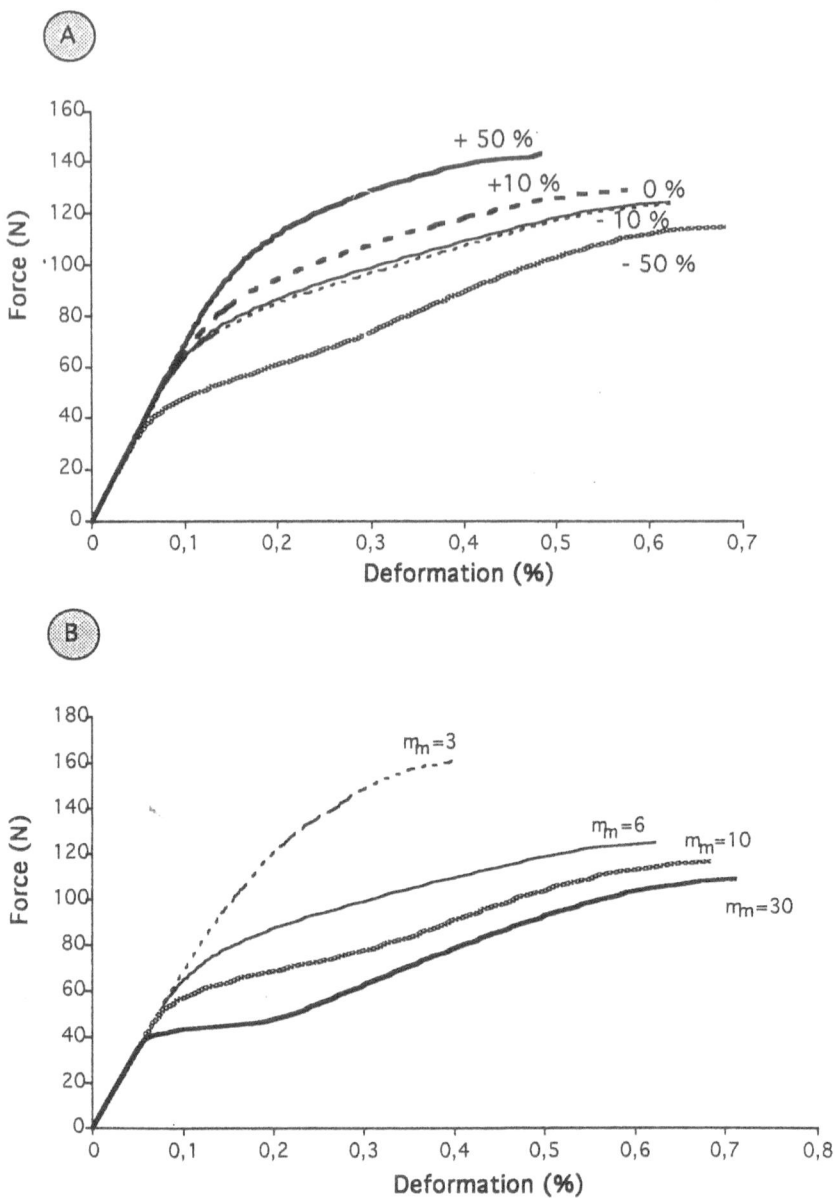

Fig. 8 : Prediction of the influence of matrix strength characteristics on the force-strain behavior of minicomposites from batch B (τ = 80 MPa) : (A) various ratios of strength increases or decreases to the reference values (B) for various values of Weibull's modulus

Computations indicated that the crack spacing distance at ultimate failure decreases from 1017 to 22 microns as m_m increases from 3 to 30.

The effect of the matrix Young's modulus (E_m) is described by figure 9 which shows that the force-strain behavior is dominated by the matrix at large E_m whereas the fiber becomes dominant at low Em. A pronounced non-linear domain of deformation was predicted for large E_m whereas the behavior tended to be linear for the low E_m values. Decreases in the Young's modulus led to enhanced decreases in the crack spacing distance at ultimate failure. These results are in agreement with the trends observed experimentally on practical CMC involving more or less stiff matrices (C, glass, glass-ceramics and SiC).

4.4 Size effects

Figure 10 represents the force-strain behaviors that were predicted for various minicomposite gauge lengths. It can be noticed that ultimate failure displays a significant sensivity to size effects as indicated by the strain-to-failure increase as gauge lengths decreases. This prediction is consistent with the damage mechanisms involved in minicomposite failure. It remains valid provided additional mechanisms (such as fiber multiple failure) do not operate. Experimental results on scale effects are not available yet.

5- Conclusions

The force-strain behavior of minicomposites was modeled on the basis of a fracture-statistics based description of matrix multiple cracking and fiber bundle failure. The minicomposites were assimilated to fiber bundles subjected to a typical stress-state induced by the presence of matrix cracks and associated debonding. A Weibull type statistical-probabilistic model was developed for the description of the matrix cracking process and fiber failure. This model provided stress-probability equations as a function of the number of matrix cracks and various constituent properties.
The force-strain behaviors that were predicted for SiC/SiC minicomposites were in excellent agreement with the experimental ones, thus validating the model. Estimates of the interfacial shear stresses were derived by fitting the predicted force-strain curves to the experimental ones.
Important trends in the influence of constituents properties on the mechanical behavior of minicomposites were anticipated. It was shown that matrix properties are preponderant factors to obtain high stresses and a pronounced non-linear domain of deformations. The matrix influence is similar to the interfacial shear stress one. Low stresses and a plateau-like domain of deformations are enhanced by a matrix displaying a low strength, a high Weibull's modulus, and a high Young's modulus. The fiber strength strongly affects the strain-to-failure. Thus degradation of fiber strength causes significant decreases in the minicomposite strain-to-failure whereas stronger fibers would lead to minicomposites able to accomodate higher deformations. Variability in fiber strength causes a variability in minicomposite strain-to-failure which may explain the scatter in force-strain curves observed with practical composites. Computations also showed that fibers with a low Young's modulus would increase the composite strain-to-failure whereas stiff fibers strongly decrease the capability to accomodate deformations. Finally the model showed that the minicomposites are significantly sensitive to size effects. The strength at ultimate failure was found in inverse proportion to the gauge length.

The proposed model thus appears to be an interesting tool for designing composite materials as a function of the desired properties. It provides quantitative guidelines for

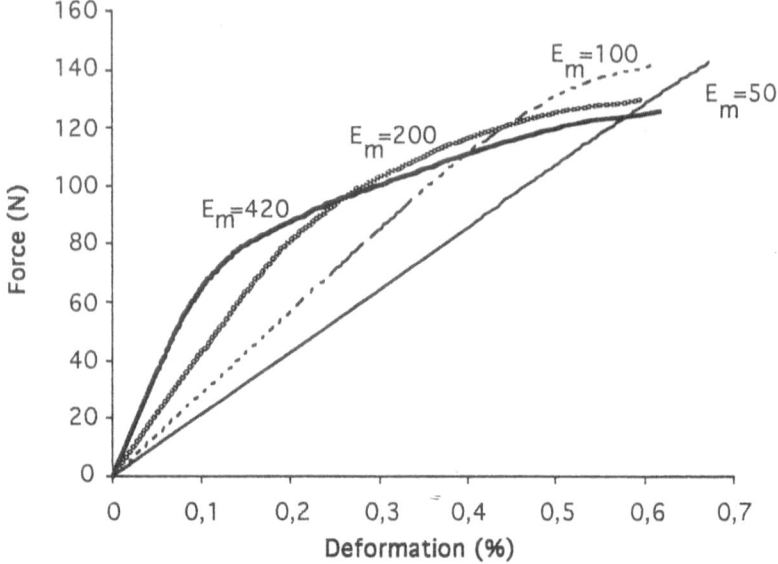

Fig.9 : Prediction of the influence of matrix Young's modulus (GPa) on the force-strain behavior of minicomposites from batch B (τ = 80MPa)

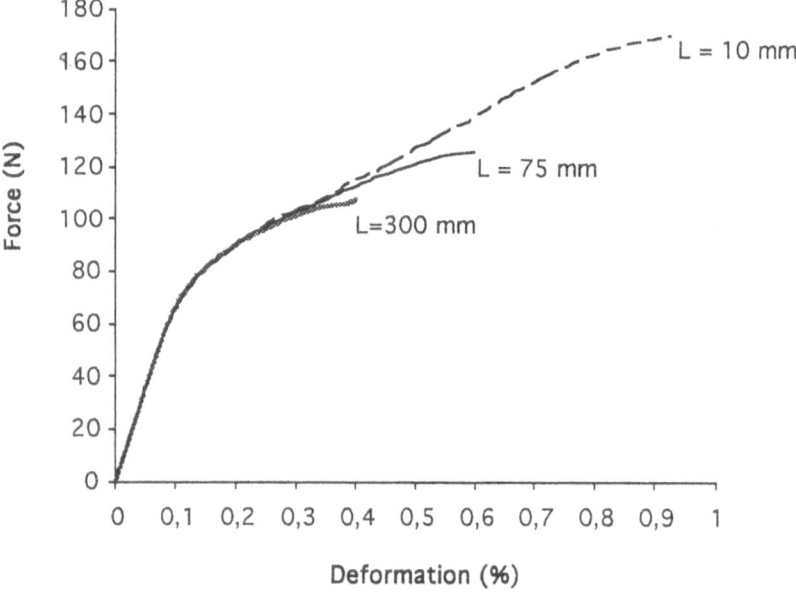

Fig. 10 : Prediction of the influence of gauge length on the force-strain behavior of minicomposites from batch B (τ = 80 MPa)

the selection of appropriate constituents. The effect of changes in constituent properties due to environmental conditions may also be predicted.

6- Appendix : Deformations of minicomposites

Minicomposite deformations are dictated by the fibers. Fiber elongation in the debonded portion of a **type I fiber element** is given by :

$$\Delta L_1 = \frac{2}{E_f} \int_0^1 \sigma_f(z) \, dz \qquad (A1)$$

$$\Delta L_1 = \frac{2l}{2E_f} \left(\frac{F}{A_f^t(1-\alpha)} \frac{2+a}{1+a} + \sigma_f^T \right) \qquad (A2)$$

Fiber elongation in the bonded portion of a type I fiber element is derived from the equation (16) of the stress field :

$$\Delta L_2 = \frac{1}{E_f} \int_0^{(L_i(I)-2l)} \sigma_f(z) \, dz = \frac{1}{E_m} \int_0^{(L_i(I)-2l)} \sigma_m(z) \, dz \qquad (A3)$$

$$\Delta L_2 = \frac{L_i(I) - 2l}{E_f} \left(\frac{F}{A_f^t(1-\alpha)(1+a)} + \sigma_f^T \right) \qquad (A4)$$

Fiber elongation for a type I element is :

$$\Delta L_i = \Delta L_1 + \Delta L_2 = \frac{L_i(I)}{E_f} \left(\frac{F\left(1+\dfrac{al}{L_i(I)}\right)}{A_f^t(1-\alpha)(1+a)} + \sigma_f^T\left(1-\dfrac{l}{L_i(I)}\right) \right) \qquad (A5)$$

The total elongation of the k type I fiber elements is :

$$\Delta L^I = \sum_{i=0}^{k} \Delta L_i \qquad (A6)$$

Similarly, deformations in the **type II fiber elements** are obtained by integrating equation (15) of the stress field for z between 0 and $L_j(II)/2$.

$$\Delta L_j = \frac{L_j(II)}{E_f} \left(\frac{F}{A_f^t(1-\alpha)} \left(1 - \frac{a}{1+a} \frac{L_j(II)/2}{2l} \right) + \sigma_f^T \frac{L_j(II)/2}{2l} \right) \qquad (A7)$$

160

The total elongation of the t type II elements is:

$$\Delta L^{II} = \sum_{j=0}^{t} \Delta L_j \qquad (A8)$$

The total deformation of the minicomposites is thus :

$$\varepsilon = \frac{\Delta L}{L} = \frac{\Delta L^I + \Delta L^{II}}{L} \qquad (A9)$$

7- Acknowledgements

The authors acknowledge support of CNRS and SEP one of them (NL) obtained a grant.

8- References

1. H.C. Cao, E. Bischoff, O. Sbaizero, M. Rühle, A.G. Evans, D.B. Marshall and J.J. Brennan, Effect of Interfaces on the Properties of Fiber-Reinforced Ceramics. J. Amer. Ceram. Soc. 73 [6] 1691-99 (1990).
2. R. Naslain, Fiber-matrix interphases and interfaces in ceramic matrix composites processed by CVI. Composites Interfaces, Vol 1, n°3, 253-286 (1993).
3. J. Lamon, "Interfaces and Interfacial Mechanics : Influence on the Mechanical Behavior of Ceramic Matrix Composites (CMC)". Journal de Physique IV, colloque C7, supplement au Journal de Physique III, volume 3, (novembre 1993) 1607-1616.
4. J. Lamon, N. Lissart, C. Rechigniac, D.M. Roach, J.M. Jouin, "Micromechanical and Statistical Approach to the Behavior of CMCs" pp 1115-1124 in Composites and Advanced Ceramics, Proceedings of the 17th Annual Conference and Exposition, (10-15 jan. 1993 - Cocoa Beach, Florida (USA). Ceramic Engineering and Science Proceedings, September-October 1993, The American Ceramic Society.
5. C. Droillard, J. Lamon, "Fracture Toughness of 2D Woven SiC/SiC CVI-Composties with Multilayered Interphases", J. Am. Ceram. Soc., 79 [4] 849-58 (1996).
6. L. Guillaumat, J. Lamon : "Multifissuration de composites SiC/SiC" (Multiple cracking in SiC/SiC composites), Revue des Composites et des Matériaux Avancés, Volume 3, numéro hors série, (1993) 159-171.
7. S. B. Batdorf, "Failure statistics of unidirectional long-fiber composites", Probabilistic Methods in the Mechanics of Solids and Structures, Proceedings of the IUTAM Symposium, Edited by S. Eggwertz and N.C. Lind, Springer-Verlag (1985)299-305.
8. M. G. Bader, "Tensile strength of uniaxial composites", Sciences and Engineering of Composite Materials, 1 [1] (1998).
9. R. E. Pitt, S. L. Phoenix, " Probability distributions for the strength of composite materials IV : localized load - sharing with tapering", Int. Journ. of Fracture, 22 (1983) 243 - 276.
10. H. Cao and M. D Thouless, "Tensile tests of ceramic-matrix composites : theory and experiment", J. Am. Ceram. Soc., 73 [7] (1990) 2091-94.
11. N. Lissart, J. Lamon, "Damage and Failure in Ceramic Matrix Minicomposites : Experimental Study and Model", Acta Mater. Vol. 45, N°3, pp. 1025-1044 (1997).
12. V. Arnault, "Relations entre microstructure et comportement mécanique des composites SiC/SiC : Analyse du rôle de l'interface dans le processus de fissuration matricielle dans des matériaux multifilamentaires", (Relationships between the microstructure and the mechanical behavior of SiC/SiC composites: influence of interfaces on matrix cracking) Ph.D.Thesis n°89 ISAL 0098, University of Lyon (1989).
13. L. Guillaumat and J. Lamon, "Fracture Statistics applied to modelling the non-linear stress-strain behavior in microcomposites : influence of interfacial parameters". Int. Journal of Fracture, (1997) in press.
14. K. K. Phani,"The Strength-Length Relationship for Carbon Fibers "Composites Science and Technology , 30 , 59-80 (1987).
15. N. Lissart, "Probabilité de rupture et fiabilité des composites à matrice céramique" (Failure probability and reliability of ceramic matrix composites) Ph. D. Thesis N° 1207, University of Bordeaux (1994).

KINETIC MODELS OF BRITTLE CRACK GROWTH: CRACK PATTERN STATISTICS AND LONGEVITY OF SOLIDS

A.I. MALKIN
Institute of Applied Mechanics Russian Academy of Sciences
117334, 32A Leninsky prosp., Moscow, Russia

1. Introduction

Brittle fracture as a kinetic process has been the subject of investigation for many years. However, the present notions on regularities and mechanisms of brittle crack growth is far from being complete. The main difficulties encountered in the description of crack growth are caused by an extremely complicated character of cooperative processes in a precracked zone of a solid. The state of the art of the kinetic theory is that physically meaningful, even oversimplified, models are of interest.

Early models of that type, both continual and lattice ones, deal with a one-dimensional rectilinear crack and treat the growth as a sequential dissociation of interatomic bonds [1-9]. These models are based on rough assumptions on the structure of the precracked zone as well as on the experimentally found regularities such as Jhurkov longevity equation [1, 5, 8]

$$t_s = t_0 \exp\left(\frac{E_0 - v\sigma_\infty}{kT}\right) \tag{1.1}$$

or its direct analog for crack growth velocity [5, 8, 10]

$$v_c = v_0 \exp\left(-\frac{E_0 - bK_I}{kT}\right) \tag{1.2}$$

(we use convential notations). Relations (1.1), (1.2) are commonly taken as the evidence of the thermofluctuation nature of crack growth at least at the initial stage of fracture.

Within the cited models crack growth is considered a deterministic process. However, it was found that crack growth generally shows the spasmodic stochactic process characteristics [1, 8, 10-14]. Prolonged intervals of static behaviour of a crack are alternated with the fast moving of the crack tip into its new position. The distance between the former and the latter positions of the crack tip is therewith much more than interatomic

161

G. N. Frantziskonis (ed.), PROBAMAT – 21st Century: Probabilities and Materials, 161–196.
© 1998 *Kluwer Academic Publishers.*

distance in a solid. It follows that the concept of sequential non-correlated rupture of interatomic bonds does not hold.

In line with commonly accepted notions [1, 5, 10], two causes should be recognized for the accidental nature of crack growth. These are the random structure of a solid (the random arrangement of structure defects) and thermal fluctuations (the accidental occupancy of the vibration modes localized in the precracked zone). It is reasonably to say that the role of these factors in fracture processes has attracted recently the most notice. A review of the experimental and theoretical results in this field is not presented herein. However, mention should be made of the important results, relevant to the effect of random structure of a solid on the crack patterns and propagation [10, 11, 15-20] and to the elucidation of crack growth regularities stemming from thermal fluctuations [21-25].

Clearly, to describe adequately growth processes the statistical approach have to be used. One possible way to the construction of semiempirical models of brittle crack growth was suggested in previous works [26, 27]. Here an attempt is made to promote the statistical approach [27].

We restrict the consideration below to the case of an individual two-dimensional crack in a widely used model geometry of "elastic plane". First we discuss a general scheme for the construction of statistical two-dimensional theory of brittle crack growth and suggest some specific models. We obtain the master equation in time-dependent distribution density of crack configuration and, then, the useful equations in marginal distributions. Finally we consider some results of the models with regard to experimentally observed regularities of brittle crack growth.

Before proceeding further some necessary remarks need to be made.

It should be remembered that empirical relations (1.1) and (1.2) are moderately accurate. In particular, preexponential factors are defined with an accuracy of the decimal order [5], so that their conceivable (comparatively weak) dependence on stress and temperature escapes detection. This must be taken into account in the comparison of theoretical results and empirical relations (1.1), (1.2).

One further remark concerns the validity of the above relations for the case of "microscopically" brittle fracture. Although physical intuition suggests that relation (1.1), (1.2) have to be valid with an ideal brittle fracture, only few experimental verifications of this statement are available. The reason is that an extremely unstable behaviour of "microscopically" brittle cracks leads to well-understood experimental problems [5]. Below we assume both relations are valid in this case.

A "macroscopically" brittle (quasi-brittle) cleavage is significantly more frequent type of fracture than "microscopically" brittle one and differs from it in that the structure of a precracked zone is produced by plastic deformations. When plastic deformations are small, we can assume, as with "microscopically" brittle fracture, that the macrostress tensor is presented by well known local solution of linear elasticity equations. Needless to say that such an assumption is not well substantiated and should be treated as a rough approximation. Nevertheless, it is hoped that an exessive stress in the vicinity of "dangerous" defects can be taken into account by the use of empirical coefficients. As for the time-dependence of the structure of a precracked zone, we assume the relaxation time is much less than the time scale of elementary growth event. If this is the case, the state of the precracked zone can be considered as stationary one.

2. Kinetic aproach to the description of brittle crack growth.

Let us assume that a crack can be represented by a chain of vectors (slots) $\vec{l}_i$ and hence is specified by the ordered set $\mathcal{L}_n = \left\{\vec{l}_i, \ i = 0,1,\dots n\right\}$. An elementary growth event involves the addition of random vector $\vec{l}_{n+1}$ to a crack tip at an arbitrary random instant of time (Fig.1).

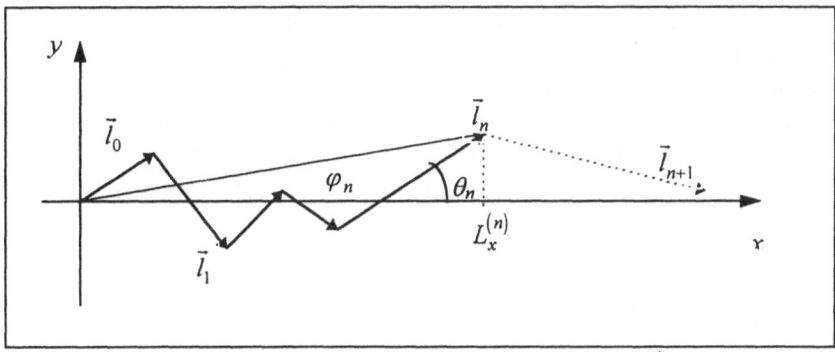

Fig.1. Geometry of the model

To formulate the kinetic model one should construct the probability density of $\mathcal{L}_n \rightarrow \mathcal{L}_{n+1}$ transition as a function of growth process prehistory. With this purpose let us consider an ideal experiment involving the observaton of the statistical ensemble of macroscopically identical strained samples with a given crack configuration. We assume the samples are thermostatically controlled. During the experiment time instants of the addition of the next microcrack and corresponding vector $\vec{l}_{n+1}$ are measured. It is apparent that the accumulation of the data from this experiment enables us to construct the probability density of the addition of microcrack $\vec{l}_{n+1}$ in a time t.

It seems reasonable to recognize two causes for the accidental nature of brittle crack growth [1]. These are the random arrangement of structure defects and the accidental occupancy of localized vibration modes in the vicinity of a crack tip.

Let us select from the ensemble the subset of elements with close atomic configurations of the initial and final (after the elementary growth event) states. Within this subset one can introduce, in principle, the reaction coordinate which is passed through the fixed saddle point on the energy surface in the configuration space of the system. In so doing, an elementary growth event can be related to overcoming of a definite energy barrier.

It should be recorded that there exists a direct analogy between brittle cleavage and unimolecular reactions of polyatomic molecules in gases. It seems to be evident if we assume that the occupancy of localized vibration modes is the controlling factor in the elementary cleavage event. If this is the case, the set of localized vibration modes corresponds to a polyatomic molecule, the energy exchange of localized modes with phonons corresponds to collisional activation in gases and the elementary cleavage event corresponds to the definite channel of unimolecular decomposition. From this standpoint the growth event is caused by the energy accumulation at one of the normalized vibration modes.

The considerations presented are moderately useful for the construction of the consistent theory because "ab initio" calculations for such complicated systems are practically impossible. Nevertheless, with the analogy outlined, the physically meaningful semiempirical models of crack growth can be constructed.

Within selected subset of microscopically identical elements, the distribution of random expectation time of a growth event may be assumed to have an exponential form. Such an assumption is verified by the well-known asymptotic theorems concerning outbursts of random processes [28, 29]. To specify the mean value of the growth event expectation time it is reasonable to adopt a so-called model of a rigid complex [30]. This model is appropriate in the case of small variations in angles and lengthes of interatomic bonds during the transition to the activated state. To put it in another way, the total number and the fundamental frequencies of vibration modes do not change appreciably during the moving along the coordinate of reaction. This assumption seems to be natural for the description of a brittle fracture which is characterized by small strains.

Furthermore, let the interaction between localized modes and phonons be intensive sufficiently to use Boltzmann statistics for the occupancy of vibrational states. This corresponds to high gas pressure limit of the theory of unimolecular decomposition [30]. Hence, the average expectation time of the elementary growth event becomes the form

$$\tau_d = \frac{2\pi\hbar}{L^+ kT} \cdot \frac{\exp(E / kT)}{1 - \exp(-\hbar\omega_e / kT)} \tag{2.1}$$

where E is activation energy, L^+ is the number of equivalent activated complexes, ω_e is the effective frequency depending on the ratio of the vibration statistical sums of the initial state and the activated complex [30]. With $\hbar\omega_e \ll kT$ pre-exponential factor does not depend on temperature; at normal temperature this is true when $\omega_e \leq 10^{13} s$.

We accept the details of the precracked zone microstructure to be not very significant, so that the average expectation time (2.1) depends on a limited number of macroparameters. It is reasonable to restrict our consideration to a minimal set of variables and parameters. Then, the average expectation time corresponding to the defined growth act $\bar{l}$ depends only on $\bar{l}$, temperature and local stresses. We suppose it to be Arrhenius function (2.1) with the activation energy linear-dependent on stresses.

The required invariance of the average expectation time in respect to coordinate system choice imposes the essential restrictions on the form of this function. For isotropic solids it can be written as

$$\tau_d(\bar{l}\,|\,\ell_n) = \tau_0 \exp\left\{\frac{E_0 - v_1(l)\sigma_{ii} - v_2(l)n_i n_k \sigma_{ik}}{kT}\right\} \tag{2.2}$$

where σ_{ik} are the components of the (macro)stress tensor, $l = |\bar{l}|$, $\bar{n} = \bar{l}/l$, $v_1(l)$ and $v_2(l)$ are certain functions of the added microcrack length, τ_0 and E_0 are assumed to be constant. Expression (2.2) is a general form of a scalar function fulfilling the requirments imposed.

The stress tensor entering expression (2.2) depends on the coordinates of the microcrack nucleation point. In this connection two generalized scenarios of crack growth should be recognized. In line with the first (microcrack addition) scenario the microcrack arises from a defect at the point $\bar{l}$ and then is added to the original crack tip. By the second (direct decohesion) scenario microcrack arises from the original crack tip and then grows until stopping at the point $\bar{l}$. In the latter case we assume the microcrack to originate at the point $\bar{\kappa} = \kappa\bar{n}$ with the effective curvature radius at the crack tip $\kappa << <l>$.

Stress distribution near the crack tip can be evaluated under the assumption that linear elasticity equations are applicable to the precracked zone description. For the case of isotropic solids it may be written in the well-known form [8]

$$\sigma_{ik} = \frac{K_I(\ell_n)}{\sqrt{2\pi\kappa}} f_{ik}^I(\bar{n}) + \frac{K_{II}(\ell_n)}{\sqrt{2\pi\kappa}} f_{ik}^{II}(\bar{n}) \tag{2.3}$$

where the stress intensity factors (SIF) K_I and K_{II} depend on the original crack configuration. The relationship (2.3) corresponds to the scenario of the direct decohesion at the crack tip. However, it is easy to check that, with minor reservations, the general form of $\tau_d(\bar{l}\,|\,\ell_n)$ does not depend on the scenario of crack growth. Notice that the insertion of parameter κ does not increase the number of empirical parameters as κ and $v_j(l)$ enter into relation (2.2) in the form $v_j(l)/\sqrt{\kappa}$.

In consistent (nonphenomenological) theory postulated relation (2.2) might correspond to a certain approximation of the mean field type. In this case the reasonable interpretation of expression (2.2) may be provided. This interpretation is based on the supposition that local stresses at the point of microcrack nucleation have a special form

$$s_{ik} = g_1(l)\sigma_{ik} + g_2(l)n_i n_k \sigma_{ik} \qquad (2.4)$$

where macrostress tensor σ_{ik} is assumed to result from the averaging over the statistical ensemble of macroscopically identical samples, functions $g_1(l)$ and $g_2(l)$ account for the excessive stresses in the vicinity of a "dangerous" defect. According to the commonly accepted viewpoint [1, 5, 20], the stress-induced reduction of the activation barrier is determined by the atomic arrangement in the neighborhood of this defect. Therefore, the coefficients in the linear dependence of activation energy on the components of tensor (2.4), i.e. activation volumes $v_1^{(0)}$ and $v_2^{(0)}$, should be independent on l. Clearly, each function $v_j(l)$ in (2.2) is a bilinear form of $v_j^{(0)}$ and $g_k(l)$.

Needless to say, the above interpretation is not the only one, and cannot be considered as a verification of relation (2.2) in terms of (2.3). These relations should be treated as a hypothesis, which is justified to a certain extent by the experimental data revealing the exponential dependence of crack growth velocity on SIF K_I [1, 8, 10].

As the next step, we introduce the state density function $\Omega(\bar{l})$, that is a number of microscopically distinguishable states, giving rise to the elementary growth event by vector $\bar{l}$. Let us discuss the virtual set of growth events in an arbitrary sample. Obviously, the event realized from this set is such that its random expectation time $t(\bar{l})$ has the minimum value. Thus, in order to construct the unconditional probability density of the expectation time it is sufficient to find the distribution of random variable

$$t_m = \min_{\{\bar{l}\}} t(\bar{l}) \qquad (2.5)$$

Therewith, the contribution of the growth event per vector $\bar{l}$ should be taken into account with weight $\Omega(\bar{l})$.

It is easy to demonstrate that the distribution of random value (2.5) also has an exponential form. The corresponding mean value can be determined from the expression

$$< t_n >^{-1} \equiv Z(\mathcal{L}_n) = \int\limits_{\{\bar{l}\}} \frac{\Omega(\bar{l})d\bar{l}}{\tau_d(\bar{l}|\mathcal{L}_n)} \qquad (2.6)$$

As a result, the distribution density of random vector $\bar{l}$ takes the form

$$w(\bar{l}|\mathcal{L}_n) = \frac{\Omega(\bar{l})}{Z(\mathcal{L}_n)\tau_d(\bar{l}|\mathcal{L}_n)} \qquad (2.7)$$

The above-written relations define completely the class of crack growth models, if functions $v_j(l)$ and state density function $\Omega(\bar{l})$ are prescribed.

3. Semiempirical models of crack growth

A practical implementation of the general kinetic scheme requires the concrete definition of functions $v_j(l)$ and $\Omega(\bar{l})$ introduced above. The calculations based on the more or less detailed description of the processes in a precracked zone seem to be too complicated to be done in the forseeable future. The more constructible approach to the statistical description of crack growth is the elaboration of semiempirical models derived from qualitative considerations. Here we discuss the features of the above functions and formulate some specific models.

3.1. SOME FEATURES OF FUNCTIONS $v_j(l)$ AND $\Omega(\bar{l})$

Assuming state density functions $\Omega(\bar{l})$ does not depend on stress, $\Omega(\bar{l})$ should be the function of the microcrack length but not its direction, i.e. $\Omega(\bar{l}) \equiv \Omega(l)$. This, as well as the form of expression (2.2), is dictated by the requirment of covariance.

The lack of the stress-dependence of the state density function means that the number of microscopically distinguishable states of the growth step is controlled by the original defect structure of a solid and does not vary with loading. By this means, the mechanical stress exclusevely affects the magnitude of the activation barrier which must be overcome when cleavage event is realized; new ways of reaction do not appear with the increase of stress.

Clearly, the assumption that the state density function is independent of stress can be justified in an ideal case of microscopically brittle fracture. The supposition that $\Omega(\bar{l})$ varies slightly with the direction of growth step seems to be less limiting. Note, that $\Omega(\bar{l})$ enters the relations (2.6), (2.7) coupled with the factor $\tau_d^{-1}(\bar{l}|\mathcal{L}_n)$. Thus, the presence of rapidly varying Arrhenius factor in $\Omega(\bar{l})$ leads to an additive renormalization of functions $v_j(l)$ only. This supposition provides the reason to adopt probability density (2.7) as the product of two factors: a rapidly varying exponential function of the growth step direction and a slightly varying factor which can be identified with a state density function without the loss of formal generality.

To agree with the experimentally observed regularities of crack growth it is necessary to impose specific restrictions on functions $v_j(l)$. In order to reveal these restrictions we rewrite expression (2.7) as

$$w(\bar{l} \mid \mathcal{L}_n) = \frac{\Omega(l)}{Z(\mathcal{L}_n)\tau_0} \exp\left(-\frac{E_0 - \Delta E}{kT}\right)$$

$$\Delta E = \frac{K_I}{\sqrt{2\pi\kappa}}\left[2(v_1 + v_2) - v_2 \cos^2\frac{\theta}{2}\right]\cos\frac{\theta}{2} + \qquad (3.1)$$

$$+ \frac{K_{II}}{\sqrt{2\pi\kappa}}\left[v_2 - 2v_1 - 3v_2 \sin^2\frac{\theta}{2}\right]\sin\frac{\theta}{2}$$

where angle θ is counted off the direction of the last unit of a crack. Function (3.1) peaks sharply in the most probable direction of vector $\bar{l}$. The most probable direction therewith must not coincide with the boundaries of the interval $(-\pi, \pi)$ at any load with nonnegative SIF K_I. Considering separately mode I and mode II crack, it is easy to obtain the necessary conditions

$$v_1(l) > 0$$

$$(3.2)$$

$$-2v_1(l) < v_2(l) < -v_1(l) / 4$$

Actually, the inequalities in $v_j(l)$ should be more violent than (3.2). As an example, let us suppose the average expectation time of growth step $\bar{l}$ to be nonincreasing function of hoop stress $\sigma_{\theta\theta}$ and strain $u_{\theta\theta}$. At given stress tensor component σ_{zz}, the stress-induced reduction of the activation barrier as a function of these arguments takes the form

$$\Delta E = \frac{v_2 + (1+v)v_1}{v}\sigma_{\theta\theta} - \frac{v_1 + v_2}{v}E_Y u_{\theta\theta} - (v_1 + v_2)\sigma_{zz} \qquad (3.3)$$

where v is Poisson's ratio, E_Y is Young modulus. It immediately follows that functions $v_j(l)$ should meet the inequalities

$$-(1+v)v_1(l) \leq v_2(l) \leq -v_1(l) \qquad (3.4)$$

For the solids with nonnegative Poisson's ratio inequalities (3.4) do not contradict each other and define considerably a narrower interval for $v_2(l)$ than (3.2).

Furthermore, let us consider the functions v_j dependence on the growth step length. If the original crack length is of the order of the growth step length, then the longevity of a sample will be defined by the expectation time of the first growth step. The reason is that the SIFs in this case rise sharply with a single growth step. So, the longevity is defined by the expression (2.6) wherein the integral over θ may be calculated, in view of the above suppositions, by the saddle point method. Basing on the need to obtain Jhurkov longevity equation (1.1), we can conclude that integration over l in (2.6) must lead to the exponential form of the stress dependence of function $Z(\mathcal{E}_n)$. In this case, it is logical to suppose that each v_j is a slightly varying function of growth step length. The alternative possibilities seem to be of little significance. By this means, v_j dependence on l can be neglected at least at the initial stage of growth, and v_j should be treated as parameters.

If the state density function does not depend on the growth step direction, then, since $v_j =$const, the length and the angle of a growth step appear to be independent random values. This is true with the same accuracy as relation (2.6) results in Jhurkov longevity equation.

In the general case of angle-dependent function $\Omega(\bar{l})$, the angle and the length distributions are interdependent. However, if the crack is close to rectilinear one, meaning that each unit direction is close to the direction of x axis, one can substitute $\theta = 0$ into $\Omega(\bar{l})$ and, in so doing, neglect the variations of this function with small angle variations. This corresponds to small-angle approximation discussed below.

3.2. SMALL-ANGLE APPROXIMATION

Let us assume that the original crack is close to rectilinear mode I one, so that $|K_{II}|/K_I \sim \varepsilon \ll 1$. While the parameters v_j satisfy inequalities (3.2), the most probable direction of the growth step will be close to the direction of the original crack. With an accuracy of the second order in ε, the stress-induced reduction of activation energy from (3.1) takes the form

$$\frac{\sqrt{2\pi\kappa}}{K_I}\Delta E = v_+ + \frac{1}{2}v_-\left(\frac{K_{II}}{K_I}\right)^2 - \frac{1}{8}v_-\left(\theta + 2\frac{K_{II}}{K_I}\right)^2 + O(\varepsilon^4) \qquad (3.5)$$

where $v_+ = 2v_- + v_2$, $v_- = 2v_1 - v_2$. Simultaneously, the state density function should be taken at $\theta = 0$. As a result, the distribution of the growth step direction takes the simple form

$$w_\theta(\theta|\pounds_n) = \frac{1}{\sqrt{2\pi <\delta\theta^2>}} \; \exp\left[-(\theta-\theta_m)^2/2<\delta\theta^2>\right]$$

$$\theta_m = -2\frac{K_{II}}{K_I}, \qquad <\delta\theta^2> = \frac{4kT\sqrt{2\pi\kappa}}{v_- \cdot K_I(\pounds_n)}$$

(3.6)

The validity conditions for distribution density (3.6) are $<\delta\theta^2> \; <<1$ and, at the same time, the smallness of the ratio of the forth-order terms omitted in (3.5) to $kT\sqrt{2\pi\kappa}/K_I$.

As is obvious from (3.6), the most probable direction of the growth step does not depend on parameters v_j associated with the defect structure of a solid. In contrast, dispersion is inversely proportional to the linear combination of these parameters. It increases linearly with temperature and decreases with SIF K_I. This fact seems to be in qualitative agreement with experimental observations [5].

Within adopted approximation, the distribution density of the growth step length can be given in terms of the state density function by

$$w_l(l|\pounds_n) = 2\pi l\Omega(l)/N$$

$$N = 2\pi\int_0^\infty \Omega(l)ldl$$

(3.7)

where N is the total number of states realizing the growth event. With the same accuracy, expression (2.6) becomes

$$Z = \frac{N}{\tau_0}\sqrt{\frac{<\delta\theta^2>}{2\pi}} \; \exp\left(-\frac{E_0-\Delta E_z}{kT}\right)$$

(3.8)

$$\Delta E_z = \frac{2v_+K_I + v_-K_{II}^2/K_I}{2\sqrt{2\pi\kappa}}$$

The models under discussion are distinguished, exclusive of numerical values of parameters, only by the form of function (3.7). To establish it in physically disparate situations, one should invoke some additional assumptions on the mechanisms of an elementary growth event.

3.3. DISTRIBUTION DENSITIES OF GROWTH STEP LENGTH

Let us consider the direct decohesion scenario of a crack growth. In line with it, a microcrack originates immediately at the crack tip and grows by the sequential disruption of strained interatomic bonds. We assume that the preceding desruption of each bond promotes the disruption of the next one, so that the sequence of bond disruptions consists of a chain reaction localized along the microcrack direction. Assuming the break of a disruption chain to be a rare event, we conclude that the typical growth step length is much longer than the interatomic distance in a solid. In this case it is reasonable to adopt the exponential distribution of the growth step length

$$w_l(l|\mathcal{L}_n) = \frac{\exp(-l/<l>)}{<l>}\eta(l) \qquad (3.9)$$

where $<l>$ is proportional to the product of the mean number of disruptions (until the disruption chain breaks) and interatomic distance in a solid. In general, the average step length dependence on the original crack contour is dictated by the SIF K_l dependence on $\mathcal{L}_n$.

The model (3.9) seems to be useful in the qualitative description of the low-temperature damage of "perfect" high-strength silicate, polymer and metallic glasses. It is widely believed that, for such solids, strong correlation of decomposition of neighbouring interatomic bonds is due to the chemical interaction between free valences and adjacent undecayed bonds. Consequently, the break of the disruptions chain is caused by the relatively infrequent events of the free valences blocking.

In line with microcavities addition scenario, a microcrack originates at some defect at a distance away from the crack tip and grows until adding to it. If the field of the "dangerous" defects may be approximated by the uniform Poisson's one, then the step length statistics will be defined by Rayleigh's distribution [29]

$$w_l(l|\mathcal{L}_n) = 2\pi\lambda l \cdot \exp(-\pi\lambda l^2)\eta(l)$$

$$<l>= 1/2\sqrt{\lambda} \qquad (3.10)$$

where λ is the mean density of defects.

Physically, the distribution (3.10) corresponds to an extremely rarefied field of pre-existing identical defects. Therewith, the mean density of defects must be taken as constant, so that distribution density (3.10) does not depend on the configuration of the original crack.

The distributions presented do not exhaust the possibilities and should be treated as extremely simplified examples. An important point is that the construction of physically meaningful model distributions of a step length is more convenient than that for the state density functions.

3.4. STRESS INTENSITY FACTORS WITHIN SMALL-ANGLE APPROXIMATION

Now, to formulate completely the models under discussion, it remains to determine SIFs as the functions of the crack configuration.

The two-dimensional problem on SIFs determination for an arbitrary, close to rectilinear mode I crack has been considered in [31-37]. The convenient perturbation procedure, suitable for the small deviations of the crack contour from a straight line, was suggested by Goldstein and Salganik [32]. The first order solution was obtained in [33] and, in an alternative form, in [34].

However, such an accuracy is insufficient for present purposes. While K_{II} should be known with an accuracy of ε, the sufficient accuracy for K_I, as evidenced by the expression (3.8), is ε^2. On the other hand, within small-angle approximation, not only deviations but also their derivatives (the left and the right ones) are small. This facilitates the problem to some extent.

Let us assume that a crack can grow only in the direction of increasing x. Then, SIFs must be calculated for the right endpoint of a crack only. Using the results of Goldstein and Salganik [33], we obtain

$$\frac{K_{II}}{K_I^{(0)}} = \frac{\theta_n - \varphi_n}{2} + \frac{2(1-\sigma)}{\pi L_x} \int_0^{L_x} dx \sqrt{\frac{x}{L_x - x}} \theta(x) + O(\varepsilon^3) \tag{3.11}$$

where $K_I^{(0)} = \sqrt{\pi L_x / 2} \cdot \sigma_{yy}$, $\sigma = \sigma_{xx}/\sigma_{yy}$, $\theta(x)$ is the step function defining the crack units direction, $\theta_n \equiv \theta(L_x)$ and φ_n are depicted in Fig.1. As the function of the elements of the set $\mathcal{L}_n$, K_{II} is given by

$$\frac{K_{II}}{K_I^{(0)}} \cong \frac{\theta_n - \varphi_n}{2} + \frac{2(1-\sigma)}{\pi} \sum_{i=0}^n R_i \theta_i$$

$$R_i = \arcsin\sqrt{S(0,i)/S(0,n)} \Big/ - \arcsin\sqrt{S(0,i-1)/S(0,n)} -$$

$$-\sqrt{S(0,i)S(i+1,n)/S^2(0,n)} + \sqrt{S(0,i-1)S(i,n)/S^2(0,n)} \tag{3.12}$$

$$S(i,j) = \begin{cases} \displaystyle\sum_{k=i}^j l_k, & i \le j \\ 0, & i > j \end{cases}$$

With the required accuracy of ε^2, function K_I is given by a further combersome expression.

Considerable simplification is possible for a lengthy crack with $<l>/L_x \ll 1$. Clearly, at the scale L_x the random function $\theta(x) - \varphi_n$ may be thought of as a rapidly oscillating one, since, in the first order in ε,

$$\int_0^{L_x} dx \left[\theta(x) - \varphi_n\right] \equiv 0$$

As $<l>/L_x \rightarrow 0$, this results, with minor reservations, in

$$\int_0^{L_x} dx \sqrt{\frac{x}{L_x - x}} \left[\theta(x) - \varphi_n\right] = 0\left(\varepsilon\sqrt{<l>/L_x}\right) \tag{3.13}$$

If the term (3.13) is neglected, the expression (3.11) becomes

$$\frac{K_{II}}{K_I^{(0)}} \cong \frac{\theta_n + (1 - 2\sigma)\varphi_n}{2} \tag{3.14}$$

It should be noted that, in general, the integral (3.13) and φ_n are of the same order of value in the sense that their dispersions are comparable. In this respect, formula (3.14) is not well founded. Nevertheless, we retain its form for the reason discussed below.

The approximation (3.14) coincides with the well-known solution for the straight crack with an infinitesimally small kink [34, 35]. It seems to be reasonable, since we assume the crack contour to be rapidly oscillating in the neighbourhood of the straight line. However, the question arises of whether the actual crack contour behaviour is compatible with this assumption.

To answer this question let us return to distribution density (3.6). The direction of the next growth step is defined by the angle $\theta_{n+1} = \theta + \theta_n$ counted off the x axis. At given φ_n, the mean (and the most probable) value of random θ_{n+1} satisfies the condition

$$< \theta_{n+1} > -\varphi_n = -2(1 - \sigma)\varphi_n$$

Hence, while $\sigma < 1$, the preferred growth step direction will bring the crack direction closer to x axis, and the above assumption does not contradict the actual behaviour of the crack contours.

The crack path stability condition $\sigma < 1$ coincides with that of Cotterell and Rice [34]. It seems to be natural as the most probable direction of growth step depends on SIFs ratio only. Note, that the condition presented is consistent, as stated in [34], with experimental results of Radon et al [38].

The above situation seems to be a good reason to adopt the simplified expression (3.14) at least for analytical studies. Evidently, this expression adequately represents K_{II} for big deviation angles of the crack tip. For the small deviation angles both φ_n and integral (3.13) do not affect significantly the growth direction distribution.

Now, to calculate function K_I in the limit $<l>/L_x \to 0$, one can immediatly use the results for an infinitesimally kinked crack [34, 35]. With an accuracy of ε^2 we have

$$
\frac{K_I}{K_I^{(0)}} = 1 - \frac{3}{8}(\theta_n - \varphi_n)^2 - \frac{3(1-\sigma)}{2}\varphi_n(\theta_n - \varphi_n) - (1-\sigma)\varphi_n^2 +
$$

$$
+ O\left(\varepsilon^2 \sqrt{<l>/L_x}, \; \varepsilon^4\right)
$$

(3.15)

The approximate expressions (3.14), (3.15) enable us to present the stress-dependent function ΔE_z from (3.8) as that of the crack tip coordinates and the direction of the last unit of the crack. In an explicit form

$$
\frac{\Delta E_z \cdot \sqrt{2\pi\kappa}}{K_I^{(0)}} \cong v_+ - \frac{v_1 + v_2}{2}\left[\theta_n + (1-2\sigma)\varphi_n\right]^2 + \frac{v_+(1-\sigma)(1-3\sigma)}{2}\varphi_n^2
$$

(3.16)

Function ΔE_z defines the reduction of the average life time of the crack in a given configuration. So, as it follows from (3.16), since the right inequality (3.4) is strictly satisfied, the greater is the direction deviation of the last crack unit from the most probable direction of the next step, the smaller is the time of static existence of the crack.

When the sum $v_1 + v_2$ is sufficiently small $\left(|v_1 + v_2| << v_-/2\right)$, the θ_n - dependence of (3.16) vanishes. Note, that if parameters v_j meet the inequalities (3.4), then, for solids with small Poisson's ratio (like some silicate glasses), this sum will necessarily be small. In this situation the life time of a given configuration of a crack decreases with φ_n while $\sigma < 1/3$ and increases while $1/3 < \sigma < 1$.

4. Kinetic equations

In the framework of the adopted approach, the crack growth is desribed, as noted above, by the probability density $V(\ell_n, t)$ of the transition from the initial state ℓ_0 to some arbitrary state ℓ_n in a time t. Here we obtain the equation for this function (master equation) and, thereupon, derive some useful equations in marginal distributions less informative than $V(\mathcal{L}_n, t)$.

4.1. MASTER EQUATION

The kinetic equation in function $V(\mathcal{L}_n, t)$ may be derived by a convential technique [29].

Let us consider the feasible changes of crack configuration in a time interval $(t, t + \Delta t)$ with $\Delta t \to 0$. If the crack configuration at the time instant t was $\mathcal{L}_n$, then the conditional probability to find the crack in the same configuration at the time instant $t + \Delta t$ is, clearly, the probability of the lack of any growth event in a time Δt. On the other hand, provided the configuration at time instant t was $\mathcal{L}_{n-1}$, the conditional probabililty to find the crack in $\mathcal{L}_n$ - configuration at $t + \Delta t$ is equal to the product $Z(\mathcal{L}_{n-1})w(\vec{l}_n | \mathcal{L}_{n-1})\Delta t$, where factors $Z(\mathcal{L}_{n-1})$ and $w(\vec{l}_n | \mathcal{L}_{n-1})$ are given by the general relationship (2.6), (2.7) with $\mathcal{L}_{n-1}$ in place of $\mathcal{L}_n$. The contribution from configurations with the number of units less than $n - 1$ is of the next order in Δt and should be neglected. So, with an accuracy to the first-order terms in Δt, we have

$$V(\mathcal{L}_n, t + \Delta t) = [1 - Z(\mathcal{L}_n)\Delta t]V(\mathcal{L}_n, t) +$$

$$+ Z(\mathcal{L}_{n-1})w(\vec{l}_n | \mathcal{L}_{n-1})\Delta t \cdot (1 - \delta_{no})V(\mathcal{L}_{n-1}, t), \quad n = 0, 1, 2 \ldots$$

where δ_{no} is Kronecker delta. In the limit $\Delta t \to 0$, it immediately follows

$$\frac{\partial V(\mathcal{L}_n, t)}{\partial t} = -Z(\mathcal{L}_n)V(\mathcal{L}_n, t) +$$

$$+ Z(\mathcal{L}_{n-1})w(\vec{l}_n | \mathcal{L}_{n-1})(1 - \delta_{no})V(\mathcal{L}_{n-1}, t) \tag{4.1}$$

This equation with the initial condition

$$V(\mathcal{L}_n, 0) = \delta\left(\mathcal{L}_n, \mathcal{L}_o^{(o)}\right)\delta_{no} \tag{4.2}$$

governs the evolution of the crack of the prescribed initial configuration $\mathcal{L}_o^{(o)}$.

Equation (4.1) presents an infinite system of ordinary differential equations depending parametrically on a crack configuration. This is caused by Markov property of the suggested scheme stemming from the ignoring of the kinetic prehistory impact upon the statistics of the growth elementary event. Configuration prehistory, in contrast, is built completely into the actual state and, therefore, does not result in statistical persistence.

The exact solution of initial value problem (4.1), (4.2) does not present and essential difficulties and may be written in the form

$$V(\mathcal{L}_n, t) = P(t| \mathcal{L}_n) V_n(\mathcal{L}_n)$$

$$V_n(\mathcal{L}_n) = \delta\left(\mathcal{L}_0, \mathcal{L}_o^{(o)}\right) \cdot \prod_{i=1}^{n} w\left(\bar{l}_i | \mathcal{L}_{i-1}\right) \tag{4.3}$$

$$P(t| \mathcal{L}_n) = \sum_{k=0}^{n} Z(\mathcal{L}_k) \exp\left[-Z(\mathcal{L}_k)t\right]/Z(\mathcal{L}_n) \cdot$$

$$\cdot \left\{ \delta_{no} + \prod_{\substack{j=0 \\ j \neq k}}^{n} Z(\mathcal{L}_j)/\left[Z(\mathcal{L}_j) - Z(\mathcal{L}_k)\right] \right\}$$

Here, function $V_n(\mathcal{L}_n)$ represents time-independent probability density of the configuration revealed after n growth step; $P(t| \mathcal{L}_n)$ is conditional probability of the fulfilment of inequalities

$$\sum_{j=0}^{n-1} t_j < t < \sum_{j=0}^{n} t_j$$

where t_j is random expectation time of the growth event from the state $\mathcal{L}_j$ to some next one. Function $P(t| \mathcal{L}_n)$ may be treated as the probability of n steps realization along the prescribed crack contour $\mathcal{L}_n$ in a time t .

It should be emphasized that the general form of the exact solution (4.3) is not connected with the specific form of functions $Z(\mathcal{L}_n)$ and $w\left(\bar{l} | \mathcal{L}_n\right)$. Nevertheless, to obtain the ultimate results without the above simplifications seems to be almost impossible.

To understand the statistical features of a crack growth, many-dimensional distribution (4.3) is not as needed as marginal distributions of the crack tip location, the total length of the crack contour, etc. From this standpoint, the exact solution is found to be moderately useful, since the analytical calculation of marginal distributions from (4.3) runs into invincible problems. An alternative way consists in the use of equation (4.1) to obtain the equations in required marginal distributions.

4.2. THE EQUATIONS IN MARGINAL DISTRIBUTIONS

As a preliminary, we obtain the kinetic equation in the joint distribution of the crack tip location and the direction of the last unit of a crack.

The crack tip x-coordinate and the total crack length coincide with an accuracy of $O(\varepsilon)$. With that accuracy, the current (random) values of the crack tip coordinates are introduced as

$$L_x^{(n)} = \sum_{i=0}^{n} l_i + O(\varepsilon^2)$$

$$\varphi_n = \frac{1}{L_x^{(n)}} \cdot \sum_{i=0}^{n} l_i \theta_i + O(\varepsilon^3)$$

(4.4)

where the initial crack is assumed to be a straightline segment of length l_0 and direction $\varphi_0 \equiv \theta_0$.

Then, distribution density $V_{l\theta}(L_x, \varphi, \theta, t)$ should be introduced in the following way:

$$V_{l\theta} = \sum_{n=0}^{\infty} \int d\mathcal{L}_n \delta\left(L_x - L_x^{(n)}\right)\delta(\varphi - \varphi_n)\delta(\theta - \theta_n)V(\varphi_n, t)$$

(4.5)

Hereinafter, θ denotes the angle, counted from the direction of x axis.

To obtain the equation in function $V_{l\theta}$, one should apply the projection operator (4.5) to the equation (4.1). As a result, we have

$$\frac{\partial V_{l\theta}}{\partial t} = -Z(L_x, \varphi, \theta)V_{l\theta} + \int_{0}^{L_x - L_x^{(o)}} dl \int_{-\infty}^{\infty} d\theta' \frac{w_l(l) \exp\left[-\left(\theta + (1 - 2\sigma)\varphi'\right)^2 / 2 < \delta\theta^2 >\right]}{\sqrt{2\pi < \delta\theta^2 >}} \cdot$$

$$\cdot \frac{L_x}{L_x - l} Z(L_x - l, \varphi', \theta')V_{l\theta}(L_x - l, \varphi', \theta', t)$$

(4.6)

$$\varphi' = \varphi - \frac{l(\theta - \varphi)}{L_x - l}$$

$$V_{l\theta}(L_x, \varphi, \theta, 0) = \delta(\varphi - \varphi_0)\delta(\theta - \theta_0)$$

where, in the integrand, function $< \delta\theta^2 >$ depends on $L_x - l$ in line with relationship (3.6) upon the substitution of $K_I^{(0)}$ for K_I; function $Z(L_x, \varphi, \theta)$ is given by (3.8) in terms of (3.16).

In the above mentioned special case that $|v_1 + v_2|$ is small, the θ-dependence of function Z can be neglected. If so, the equation in marginal distribution of the crack tip location $V_l(L_x, \varphi)$ is obtainable from (4.6). Upon integrating (4.6) over θ, we have the initial value problem

$$\frac{\partial V_l}{\partial t} = -Z_2(L_x, \varphi)V_l + \int\limits_0^{L_x - L_x^{(0)}} dl \int\limits_{-\infty}^{\infty} d\theta \frac{w_l(l) \exp\left[-(\theta + (1 - 2\sigma)\varphi')^2 / 2 < \delta\theta^2 >\right]}{\sqrt{2\pi < \delta\theta^2 >}} \cdot$$
$$\cdot \frac{L_x}{L_x - l} Z_2(L_x - l, \varphi')V_l(L_x - l, \varphi', t)$$

(4.7)

$$\varphi' = \varphi - \frac{l(\theta - \varphi)}{L_x - l}$$

$$V_l(L_x, \varphi, 0) = \delta\left(L_x - L_x^{(0)}\right)\delta(\varphi - \varphi_0)$$

where $Z_2(L_x, \varphi) \equiv Z(L_x, \varphi, 0)$.

Further reduction leads to rectilinear growth approximation, wherein the crack contour is represented by an intercept of x axis. For lengthy cracks, since $< l > / L_x << 1$, one might expect the characteristic magnitudes of the angle φ is much less than $\sqrt{< \delta\theta^2 >}$, at least when the values of parameter σ is not close to unity (recall that $\sigma < 1$ is the condition of macroscopic stability of the growth direction). If this is the case, one can neglect the angle-dependence both of function Z_2 and exponential term in the integrand of equation (4.7). Upon integrating equation (4.7) over φ, we obtain the following initial value problem for the distribution density of the x-coordinate of crack tip $V_1(L_x, t)$:

$$\frac{\partial V_1}{\partial t} = -Z_1(L_x)V_1 + \int\limits_0^{L_x - L_x^{(0)}} dl w_l(l)Z_1(L_x - l)V_1(L_x - l, t)$$

(4.8)

$$V_1(L_x, 0) = \delta\left(L_x - L_x^{(0)}\right)$$

where $Z_1(L_x) \equiv Z(L_x, 0, 0)$.

The equations presented describe the crack in terms of time-dependent statistical distributions. Another way of a crack examining is to describe the statistical features of a crack contour disregarding the growth kinetics. The latter approach is less informative but in some respects more readily available.

Two-dimensional crack patterns statistics is defined by function $V_n(\mathcal{L}_n)$ from (4.3). This function presents the distribution density of a crack configuration at a given number of growth steps. Since the future growth events do not affect the crack sections which have originated earlier, it may be treated also as the distribution density of the crack patterns in the section with n kinks.

The "structural" description, as well as the kinetic one, face the problem of the extraction of useful information from many-dimensional distribution density $V_n(\mathcal{L}_n)$. The more readily available approach is based, as before, on the equations in marginal distributions.

Let us consider the equation for n-dependent joint distribution of crack tip location coordinates and the total crack length. The function sought is imposed as

$$V_s(L_x, \varphi, L|n) = \int d\mathcal{L}_n \delta\left(L_x - L_x^{(n)}\right)\delta(\varphi - \varphi_n)\delta(L - L_n)V(\mathcal{L}_n) \qquad (4.9)$$

where the second-order terms should be retained. In contrast to (4.4), we must use now the relationships

$$L_x^{(n)} = \sum_{i=0}^{n} l_i\left(1 - \theta_i^2/2\right) + O\left(\varepsilon^4\right)$$

$$\qquad (4.10)$$

$$L_n = \sum_{i=0}^{n} l_i$$

whereas the expression for φ_n is still valid.

As is clear from (4.3), the original function $V_n(\mathcal{L}_n)$ satisfies the recursion relation

$$V_n(\mathcal{L}_n) = w\left(\vec{l} \mid \mathcal{L}_{n-1}\right)V_{n-1}(\mathcal{L}_{n-1}) \qquad (4.11)$$

So, by applying the operator (4.9) to the relation (4.11), we obtain the sought-for integro-difference equation

$$V_s(L_x, \varphi, L|n) = \int_{\gamma_s} dl d\theta \, \frac{w_l(l)\, \exp\left[-\left(\theta + (1 - 2\sigma)\varphi'\right)^2/2 < \delta\theta^2 >\right]}{\sqrt{2\pi < \delta\theta^2 >}} \cdot$$

$$\cdot \frac{L_x}{L_x - l\left(1 - \theta^2/2\right)} V_s\left(L_x - l\left(1 - \theta^2/2\right), \varphi', L - l \mid n - 1\right) \qquad (4.12)$$

$$\varphi' = \frac{L_x \varphi - l\theta}{L_x - l\left(1 - \theta^2/2\right)}$$

To specify the integration domain γ_s, an additional remark is to be made. The obvious pitfall of equation (4.12) is the formal possibility of turnings and self-intersections of a crack contour. The probability of such configurations is exponentially small $\left(\sim e^{-const/\varepsilon^2}\right)$ and, with adopted accuracy, of no significance. The probability of configurations with negative L_x is therewith still less, since the consideration here has been provided for the lengthy cracks. However, the function $< \delta\theta^2 >$ in the integrand of (4.12) is defined for the positive value of its argument $L_x - l\left(1 - \theta^2/2\right)$ exclusevely. Thus, formally, we need to restrict the integration domain to exclude this indeterminacy or, else, to complete appropriately the definition of $< \delta\theta^2 >$. In particular, it may be assumed that $< \delta\theta^2 > \rightarrow\infty$ at negative arguments and integration domain presents the strip $l \in \left(0, L - L_0\right), \theta \in \left(-\infty, \infty\right)$.

Note, that within "structural" description any suppositions on function Z is unnecessary, since it does not appear in the above consideration.

In the subsequent treatment of the problem we shall not solve the equation (4.12) as it stands. Instead of this, we shall consider the simpler partial differential equations corresponding to so-called diffusion models.

4.3. DIFFUSION MODELS

With the additional suppositions, the above nonlocal equations may be substituted by the more readily solved diffusion ones. The conditions providing diffusion approximation validity are well understood [29]. Primarily, the variation of the mean expectation time of growth event with a single growth step has to be small. In the case under consideration this condition takes the form

$$\frac{v_+ K_l^{(0)}}{\sqrt{2\pi\kappa kT}} \cdot \frac{< l >}{L_x} \ll 1 \tag{4.13}$$

Note, that inequality (4.13) is much stronger than the validity condition of lengthy crack relationship (3.14), (3.16).

Another necessary condition (the rapid decreasing of function $w_l(l)$ with l tending to infinity) is met automatically, while we adopt this function in the form (3.9) or (3.10).

For simplicity, we restrict the consideration to the special case of parent equation (4.7) with $K_l^{(0)}$ - independent function $w_l(l)$. Expanding the integral operator (4.7) in terms of growth step statistical moments, we have

$$\frac{\partial V_l}{\partial t} = - <l> \frac{\partial Z_2 V_l}{\partial L_x} + \frac{2(1-\sigma) <l>}{L_x} \cdot \frac{\partial \varphi Z_2 V_l}{\partial \varphi} + \frac{<l^2>}{2} \frac{\partial Z_2 V_l}{\partial L_x^{\,2}} -$$

$$- \frac{2(1-\sigma) <l^2>}{L_x} \cdot \frac{\partial \varphi Z_2 V_l}{\partial \varphi \partial L_x} + \frac{<l^2>}{2L_x^2} \cdot \frac{\partial^2}{\partial \varphi^2} \Big[<\delta\theta^2> +4(1-\sigma)^2 \varphi^2 \Big] Z_2 V_l \quad (4.14)$$

$$V_l(L_x, \varphi, 0) = \delta\big(L_x - L_x^{(0)}\big)\delta(\varphi - \varphi_0)$$

When the angle-dependence of function $Z_2(L_x, \varphi)$ may be neglected, the one-dimensional diffusion model can be immediately obtained by integration (4.14) over φ.

It should be remembered that diffusion models describe the crack growth at gross scales of time and length, when the observation time far exceeds the characteristic expectation time of the elementary growth event.

Within "structural" approach, as with kinetic one, the more readily solved diffusion equations are of practical interest. These equations are derivable from (4.12) on the assumption that function (4.9) depends on n smoothly, so that the difference $V_s(L_x, \varphi, L|n) - V_s(L_x, \varphi, L|n-1)$ may be approximated by the first derivative with respect to n. Therewith, the dependence of function $<\delta\theta^2>$ on L_x has to be smooth too. This latter condition

$$<l> \frac{\partial <\delta\theta^2>}{\partial L_x} << 1$$

coincides with the above lengthy crack one and hence does not lead to any new restriction.

The diffusion equation resulting from (4.12) is too unwieldy. So, we shall restrict ourselves to the simpler case of the equation in joint distribution of the crack tip x-coordinate and the total length of the crack contour $V_l(L_x, L|n)$. Leaving out the details of straightforward derivation, we present this equation in the ultimate form

$$\frac{\partial V_l}{\partial n} = - <l> \frac{\partial}{\partial L_x}\big(1 -<\delta\theta^2>/2\big)V_l - <l> \frac{\partial V_l}{\partial L} + \frac{<l^2>}{2} \frac{\partial^2 V_l}{\partial L^2} +$$

$$+ <l^2> \frac{\partial}{\partial L_x}\big(1 -<\delta\theta^2>/2\big)\frac{\partial V_l}{\partial L} + \qquad\qquad (4.15)$$

$$+ \frac{<l^2>}{2} \frac{\partial^2}{\partial L_x^{\,2}}\big(1 - <\delta\theta^2> +3<\delta\theta^2>^2/4\big)V_l$$

Equation (4.15) is used below to obtain the conditional distribution of the total length of crack contour at a given crack tip coordinate.

It is easy to check that, within diffusion approximation, there is a one-to-one correspondence between kinetic equation and "structural" ones. The "structural" equations can be obtained from the corresponding kinetic ones by the substitutions of unity for the function Z and n for t. In particular, this enables to obtain immediately the "structural" analog of equation (4.14).

5. Crack patterns statistics

To discuss the statistical properties of crack patterns we now turn to "structural" equations. Let us suppose that in the above-mentioned ideal experiment the geometric parameters of the crack contour can be measured only after the complete decomposition of a sample. In practice this corresponds to a quantitative fractographic analysis of crack patterns. In this situation the local properties of crack patterns, i.e. the distributions of a single growth step at the given configuration prehistory, are determined completely by the relationship (2.7) in terms of (3.6), (3.14) and (3.9) or (3.10). These properties are obvious from the previous consideration. It is of more interest to discuss the statistical features of crack patterns at a gross scale of length.

Macroscopically, the quantitutive fractography involves the detection of empirical distributions of crack contour parameters at a given sufficiently stretched section of a crack. Here, we obtain the distributions of a crack contour deviation from a straight line and the total length of a crack contour. The latter distribution allows to discuss the influence of stress and temperature on the actual work of brittle fracture.

5.1. STATISTICS OF CRACK CONTOUR DEVIATIONS FROM A STRAIGHT LINE

To obtain the probability distribution of crack contour deviations we shall use the "structural" analog of the initial value problem (4.14). To formulate this new problem one should substitute $Z_2 \equiv 1$ and $t = n$ into equation (4.14). Since both equations are adequate only for lengthy cracks within small-angle approximation, it is evident that they possess two small parameters. These are the ratio of the average growth step length and the x-coordinate of the initial crack tip location $\nu_0 = <l>/L_x^{(o)}$, and the angle dispersion of the single growth step at $L_x = L_x^{(o)}$ denoted by ε^2. Using the natural scales for the variables, the "structural" problem may be written in the form

$$\frac{\partial V_d}{\partial n'} = -\frac{\partial V_d}{\partial x} + \frac{2(1-\sigma)}{x}\frac{\partial \varphi' V_d}{\partial \varphi'} + \frac{1}{2x^2}\frac{\partial^2}{\partial \varphi'^2}\Big[h(x) + 4(1-\sigma)^2 \mu^2 \varphi'^2\Big]V_d -$$

$$-\frac{2(1-\sigma)\mu^2}{x}\frac{\partial^2 \varphi' V_d}{\partial \varphi' \partial x} + \frac{\mu^2}{2}\frac{\partial^2 V_d}{\partial x^2} \tag{5.1}$$

$$V_d(x,\varphi',0) = \delta(x-1)\delta(\varphi' - \varphi'_0)$$

where the new variables are designed by $n' = <l> n\big/L_x^{(0)}$, $x = L_x\big/L_x^{(0)}$, $\varphi' = \varphi <l>\big/\varepsilon\sqrt{v_0 <l^2>}$; $\mu^2 = <l^2>\big/<l> L_x^{(0)} \ll 1$. As for the function $h(x) = <\delta\theta^2>\big/\varepsilon^2$, the relation (3.6) gives $h(x) = x^{-1/2}$. Below we omit primes in designations, since this cannot lead to a misunderstanding.

Equation (5.1) is derived for the joint distribution of crack tip coordinates at a given number of growth steps $V_d(x,\varphi|n)$, whereas we would like to have the distribution of angle φ (or, else, of deviation $y = x\varphi$) at given x. However, if the solution of equation (5.1) is available, the required distribution can be obtained by the following straightforward procedure. At first, upon integrating the solution over φ we obtain the probability density of x -coordinate at a given number of growth steps. The division of the original distribution by this one gives the conditional distribution of angle φ at given x and n. In what follows we obtain the distribution of a random number of growth steps until the intersection of given boundary x. In the case under consideration the probability of the repeated intersections is exponentially small, therefore we can impose this distribution by simple relation

$$V_n(n|x) = \frac{\partial}{\partial n}\int_x^\infty dx' \int_{-\infty}^\infty d\varphi V_d(x',\varphi|n) \tag{5.2}$$

Finally, forming the convolution of (5.2) and conditional distribution of φ at given x and n, we have the desired distribution of angle φ at given x.

Although $\mu^2 \ll 1$, one cannot eliminate the terms of the order of μ^2 from (5.1). The reason is that the solution of the resulting equation is not adequate at wide changes of n. However, using the multiple-scales method [39] one can easily obtain the zero-order solution, which is uniformly valid at least within the domain $o < n \le \mu^{-2}$. Since this solution is represented by the product of desired distribution $V_d(\varphi|x)$ and distribution of x at given n, the above procedure appears to be trivial. The distribution sought is determined ultimately by

$$V_d\left(\varphi|x\right) = \frac{\exp\left[-\left(\varphi - <\varphi>\right)^2 / 2 < \delta\varphi^2 >\right]}{\sqrt{2\pi < \delta\varphi^2 >}}$$

$$<\varphi> = \varphi_o x^{-2(1-\sigma)} \tag{5.3}$$

$$<\delta\varphi^2> = \frac{1 - x^{-(5-8\sigma)/2}}{2(5-8\sigma)x^{3/2}}$$

Probability distribution (5.3) reveals the stability of a crack growth direction with $\sigma < 1$ in terms of the mean value of angle φ. To this must be added that angle dispersion decreases along the direction of growth when $x > \left(8(1-\sigma)/3\right)^{2/(5-8\sigma)}$. In contrast, the dispersion of transverse deviation $y = x\varphi$ increases with x in any case, so that the amplitude of random oscillations about an average crack line increases along the growth direction. Note, that the stability condition $\sigma < 1$ coincides with the above one obtained from the rough qualitative considerations.

Let us discuss the statistical properties of the transverse deviation of a crack contour in relation to stress ratio σ.

When $\sigma < 1/2$ the average deviation $< y > = x < \varphi >$ tends to zero with $x \to \infty$. In other words, the statistical persistence is damped out with a distance from the initial point in a crack contour. With $\sigma > 1/2$, on the contrary, the average deviation increases along the growth direction, even though the direction stability condition is satisfied. So, we can conclude that, when the stress ratio goes through the critical value $\sigma_* = 1/2$, the transverse deviation gains the property of statistical persistence.

The next critical value of the stress ratio is associated with the asymptotic behaviour of the dispersion for $x \to \infty$. As is obvious from (5.3), when $\sigma < 5/8$ the asymptotics takes the form $< \delta y^2 > \sim x^{1/2}$ and does not depend on the stress ratio, whereas with $\sigma > 5/8$ we have $< \delta y^2 > \sim x^{-2+4\sigma}$. Thus, the asymptotics of the transverse deviation is changed qualitatively when the stress ratio goes through the critical value $\sigma_{**} = 5/8$.

It seems to be important that, while $1/2 < \sigma < 5/8$, the increasing of $< y >$ is accompanied by the decreasing of the ratio $< y > / < \delta y^2 >^{1/2} \sim x^{-(5-8\sigma)/4}$. The fluctuations should thus be expected to damp the trend in deviation of the crack contour. While $5/8 < \sigma < 1$ that ratio tends to a constant, and one might expect the unstable behaviour of the deviation. Of course, the direction of a crack growth therewith remains stable.

As we have restricted the consideration to the one-point distribution (5.3), the detailed analysis of these situations is beyond the scope of the present work.

It remains to point out that deviation dispersion is directly proportional to temperature and inversely proportional to the principal stress. As for the dependence on the initial crack length, the dispersion appears to be asymptotically independent on it with $\sigma < 5/8$; otherwise, it is proportional to this parameter to the power $(5 - 8\sigma)/2$.

5.2. STATISTICS OF CRACK CONTOUR LENGTH

To derive the probability distribution of the total crack contour length in the intercept $\left(L_x^{(0)}, L_x^{(0)} + \Delta L_x\right)$ we shall use equation (4.15). Preparatory to this, we will first discuss the accuracy of that equation.

Recall that we retain purposely the term of the order of ε^4 on the right hand side of (4.15). It is easy to check that this term results in variation from unity of the coefficient of correlation between the length and the endpoint location of a crack. So, the nonsingular conditional distribution of the crack length may be exclusively obtained with the terms of the order of ε^4. At first glance it seems to be necessary in such situation to retain the next term in the expansion (4.10). However, the straightforward calculation reveals that the refinement of this sort leaves the principal term of the coefficient of correlation invariant, and results only in small corrections to the parameters of the ultimate distribution. Note, that independence of the principal term of correlation coefficient expansion over ε from the accuracy of expansion (4.10) is governed by the normal form of growth step direction distribution (3.6).

Let us restrict the consideration to the case of the small length of given intercept $\Delta L_x \ll L_x^{(0)}$. In this case we can assume L_x-dependent function $< \delta\theta^2 >$ to be constant and denote it, as previously, by ε^2. Then, the initial value problem for equation (4.15) can be written in the dimensionless form as

$$\frac{\partial V_l}{\partial n} = -\frac{\partial V_l}{\partial x_\delta} - \frac{\partial V_l}{\partial z} + \frac{D}{2}\frac{\partial^2 V_l}{\partial z^2} + D\frac{\partial^2 V_l}{\partial z \partial x_\delta} + \frac{D\left(1 + \varepsilon^4/2\right)}{2}\frac{\partial^2 V_l}{\partial x_\delta^2}$$

$$(5.4)$$

$$V_l(x, z, 0) = \delta(x)\delta(z)$$

where $x_\delta = \Delta L_x / < l > \left(1 - \varepsilon^2/2\right)$, $z = \Delta L / < l >$, $D = < l^2 > / < l >^2$. The solution of problem (5.4) presents a two-dimensional normal distribution with the coefficient of correlation $\rho \cong 1 - \varepsilon^4/4$.

According to the above-mentioned procedure, to derive the sought - for conditional distribution we should first obtain the distribution of the number of crack contour kinks in the intercept ΔL_x. While the solution of (5.4) is available, one can immediately use the relation of type (5.2). As a result we have

$$V_n(n|x_\delta) = \frac{n + x_\delta}{2\sqrt{2\pi D}\, n^{3/2}} \; \exp\left[-(n - x_\delta)^2/2Dn\right] \tag{5.5}$$

where the correction of the order of ε^4 to the parameter D is neglected.

The expressions for the average value and the dispersion of the random number of kinks in dimension variables are presented by

$$<n> \; \cong \; \frac{\Delta L_x}{<l>}\left[1 + \frac{2kT\sqrt{2\pi\kappa}}{v_- K_I^{(0)}\left(L_x^{(0)}\right)}\right]$$

$$\tag{5.6}$$

$$<\delta n^2> \; \cong \; \frac{<l^2>}{<l>^2}<n>$$

It is clear from (5.6) that the first-order correction both to the average value and to the dispersion of number of kinks is decreased with local stress and increased with temperature.

Note, that since each growth step is accompanied by the irradiation of acoustic pulse, relationships (5.5), (5.6) may be also applied to relate the acoustic emission characteristics to the elongation of the crack.

Furthermore, forming the convolution of (5.6) and conditional distribution of random z at given x_δ and n, we obtain

$$V_Z(Z|x_\delta) = \frac{e^{x_\delta/D}}{\pi\sqrt{2}D\varepsilon^2} \; K_0\left(\sqrt{\varepsilon^4 x_\delta^2 + 2(z - x_\delta)^2}\Big/D\varepsilon^2\right) +$$

$$+ \frac{\varepsilon^2 x_\delta \; e^{x_\delta/D}}{\pi\sqrt{2}D\varepsilon^2 \sqrt{\varepsilon^4 x_\delta^2 + 2(z - x_\delta)^2}} \; K_1\left(\sqrt{\varepsilon^4 x_\delta^2 + 2(z - x_\delta)^2}\Big/D\varepsilon^2\right) \tag{5.7}$$

where $K_0(...)$ and $K_1(...)$ are modified Hankel's functions. Function (5.7) presents the dimensionless sought - for distribution of the total length of a crack contour between the boundaries $L_x^{(0)}$ and $L_x^{(0)} + \Delta L_x$. The exact expressions for the first statistical moments of (5.7) have the form

$$< z > \ = x_\delta$$

(5.8)

$$< \delta z^2 > \ = \frac{1}{2} D \varepsilon^4 x_\delta + \frac{1}{2} \sqrt{\frac{\pi}{2}} D^2 \varepsilon^4$$

Since the dispersion is of the order of ε^4, distribution (5.7) peaks sharply at $z = x_\delta$. Clearly, the fact that the root-mean square deviation of the crack contour length is much less than the average length at given ΔL_x is a direct consequence of the small deviations of growth step direction from a straight line.

Note, that the second term in the dispersion in relations (5.8) is relatively small with $x_\delta \gg 1$ and should be neglected. The origin of this term is determined by the inconsistency of the diffusion model with $x_\delta \leq 1$. This remark relates as well to expressions (5.6) in which such terms have been omitted.

With $x_\delta \gg D$ and $(z - x)/x_\delta \varepsilon^2 \ll 1$ distribution (5.7) can be substituted by the normal one. Needless to say, that relations (5.8) therewith are still valid.

Macroscopically, the relations presented refer to the local properties of lengthy cracks at given SIF. So, it makes sense to turn to the specific value, i.e. to the crack contour length in the interval ΔL_x divided by ΔL_x. We denote this value by A.

The average value $< A >$, as is clear from (5.8), does not depend on the length of the interval along the macroscopic growth direction and takes the form

$$< A > = 1 + \frac{2kT\sqrt{2\pi\kappa}}{v_- K_I^{(0)}}$$

(5.9)

Value (5.9) is evidently proportional to the specific work of (brittle) fracture. Thus, we can conclude that the first-order geometric correction to the actual specific work of fracture is proportional directly to the temperature and inversely to SIF.

The fluctuation of the specific work of fracture in the interval ΔL_x is therewith defined by relation

$$\sigma_A \equiv \sqrt{< \delta A^2 >} = \frac{2kT\sqrt{2\pi\kappa}}{v_- K_I^{(0)}} \cdot \sqrt{\frac{2 < l^2 >}{< l > \Delta L_x}}$$

(5.10)

When (5.10) is compared with (5.9), it is apparent that the ratio of geometric correction $< A > -1$ and root-mean square deviation σ_A depends neither on temperature nor on SIF. Within the context of the model presented, this ratio, besides the dependence on ΔL_x, is defined only by the structure parameter of a solid $< l^2 >/< l >$. Note, that with $\Delta L_x >> < l >$, where only the diffusion model is valid, the magnitude of σ_A is much less than geometric correction. So, within the above-mentioned ideal experiment, the geometric correction is to be an easy-to-measure characteristic of a crack.

6. Crack growth kinetics and longevity of solids

Most of the above-written equations in time-dependent distribution densities of geometric parameters of a crack are too complicated to be solved analytically . Because of this, to discuss the qualitative features of growth process we would do well to turn to the more readily solved equations. First we consider a quasisteady growth, when the mean velocity of a crack is approximately constant in time. Thereupon we discuss the problem on accelerating growth of rectilinear crack and, within rectilinear growth approximation, set up the longevity distribution of a solid in respect to an individual crack.

6.1. QUASISTEADY GROWTH OF A CRACK

The variations of the mean expectation time of growth event can be considered as small so long as the increment of crack tip coordinate ΔL_x meets inequality

$$\frac{v_+ K_I^{(0)}\left(L_x^{(0)}\right)\Delta L_x}{2\sqrt{2\pi\kappa}\,kTL_x^{(0)}} << 1 \qquad (6.1)$$

In that case, the mean growth velocity is approximately constant. Clearly, within statistical description inequality (6.1) is violated necessarily in the tail of distribution. So, the statistical meaning of inequality (6.1) implies that the probability of a crack tip to be find exterior to the interval $\left(L_x^{(0)}, L_x^{(0)} + \Delta L_x\right)$ must be small.

Let us consider the crack growth within one-dimensional approximation. Since in this approximation $Z_1(L_x) \cong Z_1\left(L_x^{(0)}\right)$, initial value problem (4.8) can be rewritten as

$$\frac{\partial V_1}{\partial \tau} = -V_1 + \int_0^{x_\delta} dx' w_1(x') V_1(x_\delta - x', \tau)$$

$$V_1(x_\delta, 0) = \delta_+(x_\delta)$$

(6.2)

where $\tau = Z_1\left(L_x^{(0)}\right) t$ and $x_\delta = {}_\Delta L_x / <l>$. With growth step length distribution (3.9), the exact solution of problem (6.2) takes the form

$$V_1(x_\delta, \tau) = e^{-\tau} \delta_+(x_\delta) + \sqrt{\tau/x_\delta} I_1\left(2\sqrt{x_\delta \tau}\right)$$

(6.3)

where $I_1(\ldots)$ is modified Bessel function. The first term in the right hand side of equation (6.3) corresponds to the initial state of a crack, whereas the second one describes the growth with the constant mean velocity. The dispersion of distribution (6.3) is linear in time.

With step length distribution (3.10), the solution can be represented by the convolution integral or by the series. In that case the cumulants of the crack length distribution are presented by

$$\gamma_{2k+1} = \frac{\tau}{\sqrt{\pi} \pi^k k!}, \quad \gamma_{2k+2} = \frac{2^{k+1} \tau}{\pi^{k+1}(2k+1)!!}$$

(6.4)

In both cases the crack length distribution tends to the normal one as would be expected in terms of central limit theorem [28, 29]. This raises the question of when the exact solution of problem (6.2) can be replaced by the normal distribution. The simple evaluation gives that the latter is acceptable with $Z_1\left(L_x^{(0)}\right) t \geq 10^2$. Since the normal distribution is the exact solution of the corresponding diffusion equation, we can conclude this condition to be the validity one of the diffusion approximation.

Using expressions (6.3), (6.4) to obtain the mean velocity of a crack as a function of SIF, we have

$$< \dot{L}_x > = v_0 \exp\left[-\frac{E_0 - bK_I^{(0)}\left(L_x^{(0)}\right)}{kT} \right]$$

$$v_0 = \sqrt{\frac{2}{\pi}} \frac{N <l>}{\tau_0} \left[\sqrt{2\pi\kappa} kT / v_- K_I^0\left(L_x^0\right) \right]^{1/2}$$

(6.5)

$$b = v_+ / \sqrt{2\pi\kappa}$$

The dispersion $< \delta L_x^2 >$ is linear with time, so that the ratio of the root-mean square deviation $\sqrt{< \delta L_x^2 >}$ and the average crack length is small when $Z_1 t \gg 1$. As soon as the inequality of type (6.1) in $\sqrt{< \delta L_x^2 >}$ (in place of ΔL_x) is met, one can substitute $< L_x >$ for $L_x^{(0)}$ into relation (6.5). The relation resulting from this substitution should be treated, with minor reservations, as the equation in average crack length.

Note, that relation (6.5), in view of the empirical one (1.2), allows to evaluate roughly the parameters of the model. Therewith, since pre-exponential factor in (1.2) is defined with an accuracy of decimal order, the variation of v_0 with $L_x^{(0)}$ in (6.5) seems to be moderately important.

Let us discuss the interdependence of crack growth velocity and actual work of brittle fracture. With this purpose we should obtain the conditional distribution of a crack contour length at a given time interval of crack growth between the boundaries $L_x^{(0)}$ and $L_x^{(0)} + \Delta L_x$.

As noted above, for a lengthy crack the typical magnitudes of the angle φ have to be much less than ε. While $|\varphi|/\varepsilon \ll \varepsilon$, the angle-dependence of function Z_2 from equation (4.7) can be ignored. So, to obtain the conditional distribution of the crack contour length we can use, as before, equation (4.15) with the substitution $Z_1 t$ for n. The crack tip velocity is therewith defined as $v_c = \Delta L_x / t$.

Clearly, the sought - for distribution is the normal one. The minor deviation from the previous consideration is in the need for inclusion of forth-order terms in the expansions over ε. The reason is that the velocity-dependence of the average crack contour length appears first in the terms of this order.

Denoting the specific length of a crack contour by A_v, we have

$$< A_v > = 1 + \frac{\varepsilon^2}{2} - \frac{3\varepsilon^4}{8}\left(1 - \frac{4Z_1 < l >}{3v_c}\right) \tag{6.6}$$

As one might expect, the average length of a crack contour is the greater, the smaller is crack tip velocity. As quantity (6.6) is proportional to the actual work of brittle fracture, the same is true for the latter one. However, from relation (6.6) it follows that variations of the average work of brittle fracture stemming from fluctuations of the crack growth velocity is small when compared to the root-mean square fluctuations of the work of a fracture. So, in the case considered the interdependence of growth velocity and work of fracture is weak. It must be emphasized that this conclusion is true exclusevely for the cracks with $|\varphi| \ll \varepsilon^2$, i.e. in a limiting case of a very lengthy crack.

The dispersion of the specific length of a crack contour at given velocity is presented by

$$< \delta A_v^2 >= \frac{< l^2 > \varepsilon^4}{2 < l > \Delta L_x} \cdot \frac{Z_1 < l >}{v_c} \tag{6.7}$$

It is obvious from (6.7) that, at given growth velocity, the fluctuations of a specific length of a crack contour is the greater, the smaller the growth velocity is. It seems to be natural since the velocity decreasing is caused by the deviations of growth step directions from the macroscopic direction of growth and results in the increasing number of steps in the intercept ΔL_x.

The growth velocity and the specific length of a crack contour were considered here as local characteristics of macroscopic (rectilinear) crack. Such an approach is meaningful only when fluctuations are small in comparison with corresponding mean values. This places one necessary constraint on the minimal crack length which follows from (6.1) in view of the above remark on the adequacy domain of the diffusion description, namely

$$\frac{L_x^{(0)}}{< l >} \geq 10^3 \frac{< l >}{\kappa} \left(\frac{v_+ \sigma_{yy}}{kT} \right)^2 \tag{6.8}$$

Inequality (6.8) should be treated as the necessary condition of turning from the statistical description of crack growth to thermodynamic one.

6.2. NONSTATIONARY CRACK GROWTH AND LONGEVITY OF SOLIDS

To discuss characteristic features of nonstationary growth of brittle crack and longevity of solids we shall restrict our consideration to the case of rectilinear crack. Preparatory to this, two remarks need to be made. First, to consider the longevity we pass on to the edge crack which enables us to eliminate the formal necessity of the examination of crack growth in two opposite directions. Within rectilinear growth approximation this makes no significant modifications, exclusive of new expression for SIF $K_I^{(0)} \cong 1{,}12\sqrt{\pi L_x} \sigma_{yy}$ [8]. Second, it should be emphasized that we consider here the conditional longevity, i.e. the sample life time relative to the growth of the prescribed crack. Generally, this quantity is not the direct analog of the experimentally defined one. To obtain the actual longevity, even though the interactions in the ensemble of growing cracks can be neglected, one should construct a distribution of minimum from a set of conditional longevities. This needs to invoke the additional considerations and goes beyond the scope of the present work.

Instead of initial value problem (6.2) we should use now (4.8) in the form

$$\frac{\partial V_1}{\partial \tau} = -u^{-1}(x_\delta)V_1 + \int_0^{x_\delta} dx\omega_1(x_\delta - x)u^{-1}(x)V_1(x,\tau)$$

$$u(x_\delta) = (1 + x_\delta/x_0)^{1/4} \exp\left[-\beta\left(\sqrt{x_0 + x_\delta} - \sqrt{x_0}\right)\right] \qquad (6.9)$$

$$V_1(x_\delta,0) = \delta_+(x_\delta)$$

where the variables are the same as before, $x_0 = L_x^{(0)}/< l >$ and

$$\beta = \frac{1.12v_+\sigma_{yy}}{kT}\sqrt{\frac{< l >}{2\kappa}}$$

Since the crack healing in the considered models is neglected, the crack tip can intersect an arbitrary boundary $x_\delta = x_*$ only once. Thus, the integral distribution of the intersection time instant is defined by the simple relation

$$P(\tau|x_*) = 1 - \int_0^{x_*} dx'V(x',\tau) \qquad (6.10)$$

It is important that $P(\tau|x_*)$ do not necessarily equal to the integral of $V_1(x_\delta,\tau)$ between the limits x_* and ∞. The equality takes place exclusively when normalizing integral of problem (6.9) is conserved. Integrating equation (6.9) between the limits 0 and x_*, in view of (6.10) we have

$$\frac{\partial P(\tau|x_*)}{\partial \tau} = \int_0^{x_*} dxV(x,\tau)u^{-1}(x)R(x_* - x) \geq 0 \qquad (6.11)$$

where $R(y) = \exp(-y)$ or $R(y) = \exp\left(-\pi y^2/4\right)$ for growth step distributions (3.9) and (3.10) respectively. If the limit of the right hand side of relation (6.11) with $x_* \to \infty$ exists and distincts from zero, normalizing integral decreases monotonically with τ. Clearly, the right-hand side of (6.11) presents, with $x_* \to \infty$, the probability flux to infinity, whereas the left-hand side is the distribution density of the life time of semi-infinite sample.

Nonconservation of normalizing integral means that, at any (nonzero) instant of time, the probability of complete decomposition of a sample is nonzero. It is precisely the situation that is the case in problem (6.9). To make this statement apparent, let us consider the case of the exponential distribution of growth step length (3.9). Then, the Laplace transform of the solution of equation (6.9) is given by

$$\tilde{V}_1(x_\delta, p) = \frac{\delta_+(x_\delta)}{1+p} + \frac{u(x_\delta)}{(1+p)(1+pu(x_\delta))} \exp\left[-p\int_0^{x_\delta} dx \frac{u(x)}{1+pu(x)}\right] \quad (6.12)$$

Using relation (6.12), it is easy to check that the limit of (6.11) is bound to be nonzero. Evidently, this situation is governed by the rapid decreasing of function $u(x_\delta)$ with $x_\delta \to \infty$. It should be recalled also that the time interval of the crack tip moving between consequent states has been neglected in the assumption scheme.

In so doing, the Laplace transform of the longevity distribution takes the form

$$\tilde{\psi}(p|x_0) = \frac{1}{1+p} \exp\left[-p\int_0^\infty dx \frac{u(x)}{1+pu(x)}\right] \quad (6.13)$$

where $u(x) \equiv u(x, x_0)$ is presented by (6.9). It immediately follows that the average life time and its dispersion can be written as

$$<\tau> = 1 + \frac{3\sqrt{\pi}e^{\beta/\sqrt{x_0}}}{2\beta^{5/2}x_0^{1/4}}\left[1 - \frac{4}{3\sqrt{\pi}}\Gamma_{\beta\sqrt{x_0}}(5/2)\right] \quad (6.14)$$

$$<\delta\tau^2> = 1 + \frac{2\sqrt{x_0}}{\beta}\left[1 - \frac{1}{2\beta\sqrt{x_0}} + \frac{1}{2\beta^2 x_0}\right] \quad (6.15)$$

where $\Gamma_{\beta\sqrt{x_0}}(5/2)$ is partial gamma-function. In the limit $\beta\sqrt{x_0} \gg 1$, $x_0 = const$ the right-hand sides of relations (6.14) and (6.15) coincide with an accuracy of $O(1/\beta^2 x_0)$.

The longevity distribution cannot be expressed in an explicit form. However, the asymptotic relation in the above limit is given by Γ-distribution

$$\psi(\tau|x_0) = \frac{\tau^{2\sqrt{x_0}/\beta} \exp(-\tau)}{\Gamma(1+2\sqrt{x_0}/\beta)} \quad (6.16)$$

When $\beta/\sqrt{x_0} \gg 1$, longevity distribution (6.16) is close to the exponential one of expectation time of the first growth event.

The longevity statistics in the case of step length distribution (3.10) was examined numerically in the previous work [40]. It was revealed that distribution densities are qualitatevely the same, although quantitative characteristics significantly disagree.

Note, that Γ-distribution of the longevity of silicate glass filaments was obtained experimentally with a high degree of assurance [24, 25].

To discuss the features of time-dependent crack length statistics we should use, as it follows from the preceeding observation, the conditional distribution of the form

$$V_c(x_\delta, \tau) = \frac{V_1(x_\delta, \tau)}{1 - P(\tau|\infty)} \qquad (6.17)$$

Relation (6.17) implies that the crack length statistics should be defined on the subset of unruptured elements of the above statistical ensemble. The characteristic property of the model under consideration is that the evolution of statistical ensemble with $\tau \to \infty$ consists only in the increasing of number of ruptured samples. It means that, with $\tau \to \infty$, distribution (6.17) tends to some time-independent function of x_δ. In the case of step length distribution (3.9) and $\beta\sqrt{x_0} \gg 1$ it can be easily shown analytically by using relations (6.12) and (6.16). In general case the numerical verification of this property was obtained in [40]. It was also shown that, in the most interesting case of $\beta\sqrt{x_0} \gg 1$, both the average value and the root-mean square deviation of the limit distribution are close to the initial crack length.

So, while we keep track of the individual sample for a long time (within the above ideal experiment), the result will be following: the sample will be ruptured completely or, alternatively, it will have the static crack with a length close to the initial one. Clearly, this is the result of sharp decreasing of the activation barrier of elementary growth event with the increasing crack length.

7. Conclusion

In the above discussion we formulated the semiempirical approach to the statistical description of jump-like growth of a two-dimensional crack. For the sake of simplicity we limited ourselves to the minimum number of statistical degrees of freedom and ignored the stress relaxation processes. Apparently, the models considered are oversimplified to describe adequately the growth of real crack. However, certain of the above results were found to be in a qualitative agreement with the experimentally observed regularities.

It is important that certain generalizations which do not change radically the suggested procedure are possible. In particular, we can introduce additional statistical variables characterising the state and even the evolution of a precracked zone.

References

1. Yokobori, T. (1966) *An Interdisciplinary Approach to Fracture and Strength of Solids*, Wolters-Noordhot Scientific Publications Ltd, Groningen
2. Barenblatt, G., Entov, V., and Salganik, R (1966) On the Kinetics of Crack Propagation, *Izv. AN SSSR, MTT*, 5, 82-92; 6, 76-80 (in Russian).
3. Barenblatt, G., Entov, V., and Salganik, R (1967) On the Kinetics of Crack Propagation, *Izv. AN SSSR, MTT*, 1 122-128; 2, 148-150 (in Russian).
4. Bartenev, G., and Tulinov, B. (1977) Kinetic Theory of Brittle Fracture of Polymer Glasses, *Mechanika Polymerov*, I, 3-11 (in Russian).
5. Regel', V., Slutzker, A., and Tomashevskii (1974) *Kinetic Nature of Strength of Solids*, Nauka, Moscow (in Russian).
6. Irwin, G., and Paris, P. (1971) Fundamental Aspects of Crack Growth and Fracture, in H. Leibovitz (ed.) Fracture, v. III, *Engineering Fundamentals and Environmental Effects*, Academic Press, New-York, pp. 1-46.
7. Cherepanov, G. (1966) On Crack Propagation in Solids, *Int. J. Solids & Structures*, 5, 863-871.
8. Cherepanov, G. (1974) *Mechanics of Brittle Fracture*, Nauka, Moscow (in Russian).
9. Thomson, R., Hsieh, C. and Rana, V. (1971) Lattice Trapping of Fracture Cracks. *J. Appl. Phys.*, 42, N 8, 3154-3160.
10. Thomson, R. (1983) Physics of Fracture, in R. Latanision and J. Pickens (eds) *Atomistic of Fracture*, pp. 167-204.
11. Vladimirov, V. (1984) *Physics of Metal Fracture*, Metallurgia, Moscow (in Russian).
12. Albrecht, R., Shmidt, V., and Betechtin, V. (1977) The Damage Processes Preceding Semi-Brittle Fracture in *Dependence on Deformation*. *Phys. Stat. Sol. (a)*, 39, 621-630; 40, 147-152.
13. Ravi-Chandar, K., and Knauss, W. (1986) An Experimental Investigation into Dynamical Fracture: I. Crack Initiation and Arrests, *Inv. J. of Fract*, 25, 247-262.
14. Shokey, D., Seamon, L., and Curran, D. (1985) The Microstatistical Fracture Mechanics Approach to Dynamic Fracture Problem, *Int. J. of Fract*, 27, 145-158.
15. Mandelbrot, B., Passoja, D., and Paullay, A. (1984) Fractal Character of Fracture Surfaces of Metals, *Nature*, 308, 721-722.
16. Herrman, H. (1989) Fractal Deterministic Cracks, *Physica D*, 38, 192-197.
17. Louis, E., and Guinea, F. (1989) Fracture as a Growth Process, *ibid*, 235-241.
18. Herrmann, H., and Roux, S. (eds) (1990) *Statistical Models for Fracture of Disordered Materials*, North-Holland, Amsterdam.
19. Frantziskonis, G. (1994) Crack Patterns Related Universal Constants, in D. Breysse (ed) *Probability and Materials*, Kluwer Academic Publishers, Netherlands.
20. Frantziskonis, G. (1994) On the Possibly Multifractal Properties of Dissipated Energy in Brittle Materials, *Appl. Mech. Rev.*, 47, N 1, part 2, 5132-5140.
21. Vladimirov, V. (1975) A Kinetic Approach to Theory of Fracture of Crystalline Solids, *Int. J. of Fract.*, 11, 869-874.
22. Orlov, A., Petrov, V., and Vladimirov, V. (1970) A Kinetic Approach to Fracture of Solids. I. General Theory, *Phys. Stat. Sol.*, 42, 197-205.
23. Orlov, A., Petrov, V., and Vladimirov, V. (1972) A Kinetic Approach to Fracture of Solids II. The Time to Fracture, *ibid*, 47, 293-299.
24. Petrov, V., Bashkarev, A., and Vetlegren', V. (1993) *Physical Foundation of Longevity Prediction for Structural Materials*, Polytechnika, Sankt-Peterburg (in Russian).
25. Tsoi, B., Kartashov, E., Shevelev, V., and Valishin, A. (1997) *Damage of Thin Polymer Films and Fibers*, Chimia, Moscow (in Russian).

26. Malkin, A. (1995) On Statistical Theory of Brittle Crack Growth, in A. Bakker (ed) *VII-th Int. Cont. on Mechanical Behavior of Materials. Book of Abstr.*, Int. Congress Centre, Hague, pp. 857-858.

27. Malkin, A. (1995) Statistical Approach to the Theory of Brittle Crack Growth, *Physics - Doklady, 40,* N 7, 319-322.

28. Tikhonov, V. (1970) *Outbursts of Random Processes*, Nauka, Moscow (in Russian).

29. Rytov, S. (1976) *Introduction to Statistical Radiophysics: Part I. Random Processes*, Nauka, Moscow, (in Russian).

30. Kuznetsov, N. (1982) *Kinetics of Unimolecular Reactions*, Nauka, Moscow (in Russian).

31. Banichuk, N. (1970) Determination of the Form of a Curvilinear Crack by Small Parameter Technique, *Izv. AN SSSR, MTT, 7,* N 2, 130-137 (in Russian).

32. Goldstein, R., and Salganik, R. (1970) Plane Problem of Curvilinear Cracks in an Elastic Solid, *Izv. AN SSSR, MTT, 7,* N 3, 69-82 (in Russian).

33. Goldstein, R., and Salganik, R. (1975) Brittle Fracture of Solids with an Arbitrary Crack, in I. Vorovich et al (eds) *Advances in Mechanics of Deformable Media*, Nauka, Moscow, pp. 156-170 (in Russian).

34. Cotterell, B., and Rice, J. (1980) Slightly Curved or Kinked Cracks, *Int. J. of Fract.,* 16, 155-169.

35. Savruk, M. (1980) Stress Intensity Factors for a Curvilinear Crack Little Differing from a Bow-Shaped or Rectilinear one, *Fisiko-Khimicheskaya Mekhanika Materialov*, 2, 57-62 (in Russian).

36. Savruk, M. (1988) *Stress Intensity Factors in Bodies with Cracks*, Naukova Dumka, Kiev.

37. Romalis, N. (1982) On the Equivalence of Energetic and Dynamic Approaches to the Crack Theory under Biaxial Loading Conditions, *Fiziko-Khimicheskaya Mekhanika Materialov*, 3, 12-15.

38. Radon, J., Lever, P., and Culver, L. (1977) Fracture Toughness of PMMA Under Biaxial Stress, *Fracture, v. 3,* Univ. of Vaterloo, 1113-1118.

39. Nayfeh, AA. (1976) *Perturbation Methods*, Wiley-Interscience Publications, New-York.

40. Malkin, A., Dyachkin, A., and Nikitin, N. (1997). Kinetic Models of Crack Growth and Longevity Statistics of Solids, Physics - Doklady, 42, N 5 (in press).

FROM CLASSIC TO FRACTAL MECHANICS OF DISORDERED MEDIA: SELF-CONSISTENCY VERSUS SELF-SIMILARITY

T. CHELIDZE*, Y. GUEGUEN**, LE RAVALEC, M***

* Institute of Geophysics, 1. Alexidze str, 380093, Tbilisi,
Georgia
** Ecole Normale Superieure, 24, rue Lhomond, 75231, Paris,
France
***Geosciences Rennes, University Rennes1, Campus Beaulieu, Bat.15,
35042, Rennes, France

1. Introduction

A classical approach to the mechanics of heterogeneous (porous) media, such as rocks, ceramics and other composites is based on the concept of the representative elementary volume (REV). It is believed that as far as the REV is larger than the size of the elementary heterogeneity (pore), the object can be considered as a quasihomogeneous one. In other words, if the size of the sample L is much larger than the characteristic size of an elementary heterogeneity (grain of rock, pore) the condition of quasihomogeneity (or statistical homogeneity) is fulfilled. It follows that quasihomogeneity can be defined as a translational invariance (TI) of a physical property of material, if the sample is equal or larger than REV. This approach can be applied to any physical property, which belongs to the generalized conductivity (GC) class. In following we will use the well known principle of physical analogies, which postulates that as far as different transport

G. N. Frantziskonis (ed.), PROBAMAT – 21st Century: Probabilities and Materials, 197–231.
© 1998 *Kluwer Academic Publishers.*

process obey the same mathematical formalism, formulas for such generalized conductivities as dielectric constant, conductivity, scalar elasticity, permeability are essentially identical. These expressions, which often have been obtained independently for various properties [28, 7, 20, 22, 35, 25], are very similar mathematically.

The theories, founded on the principle of self-consistent field, can be considered as a basic tools for a quantitative analysis of heterogeneous media. A lot of experiments prove that these theories work quite well in the case of weak heterogeneity (small contrast in properties of components, small concentration of heterogeneities).

In the last years, howerver, both experimental and theoretical studies confirme the reality of a functional dependence of transport properties of heterogeneous media on the size of the system. This means in turn that in some particular conditions the principle of translational invariance breaks down together with the effective medium approach (EMA) which is founded on this principle. These and some other data gave birth to a new approach to the mechanics of heterogeneous materials which can be defined as a fractal mechanics [5, 8, 9, 16, 21, 32, 41]

The goal of this paper is to study the constraints for application of EMA, to clear up, what is a substitute of it and to show, how the nature dictates one to change ones approach depending on the type of symmetry of system. We will try to show that the fundamental transition in the physics of disordered media is a transition from self-consistency to self-similarity (SS) , or from the translational invariance (TI) to the scaling, or, lastly, from a simmetry in relation to translation of REV to a simmetry in relation to scaling procedure.

2. Examples

There are many examples of a good fit of experimental data on physical properties of composite media with effective medium theory. Fig. 1, for example, shows that theoretical and measured bulk moduli of model porous systems may be quite close to each other in a wide enough range of porosities.

At the same time there are numerous examples of unusual behaviour of porous and granular media, which cannot find explanation in the framework of classic EMA.

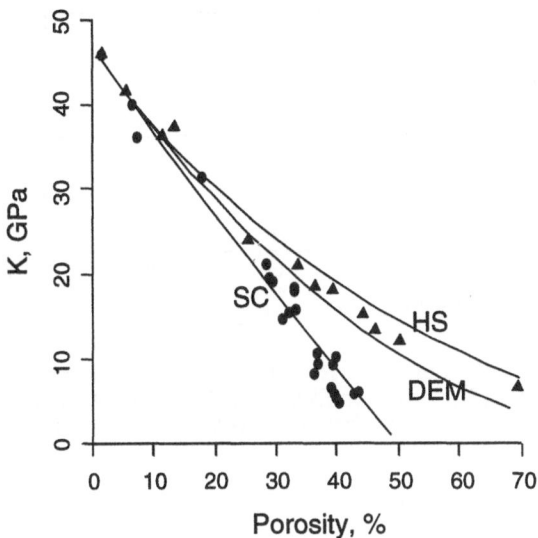

Fig. 1. Estimated and observed bulk moduli for dry synthetic sandstone   and glass foam samples. Dots indicate dynamic moduli of sintered glass beads, triangles indicate bulk moduli of glass foam samples Walsh et al.[38]. HS line shows Hashin-Shtrikman upper bound on bulk moduli for various porosities; DEM curve shows differential self-consistent estimates of bulk moduli; SC curve shows self-consistent estimates [6]

Fig. 2 shows the acceleration time history from June 1, 1964 Niigate (Japan) earthquake [2]. It is evident from the record that soil changes abruptly at 8 to 10 seconds from elastic medium, responding to a high frequency perturbation, to an unconsolidated (liquified) one, responding to low frequency waves. This is an example of a mechanical transition from consolidated to unconsolidated state under relatively weak perturbation, which do not affect at all elastic properties of massive rocks.

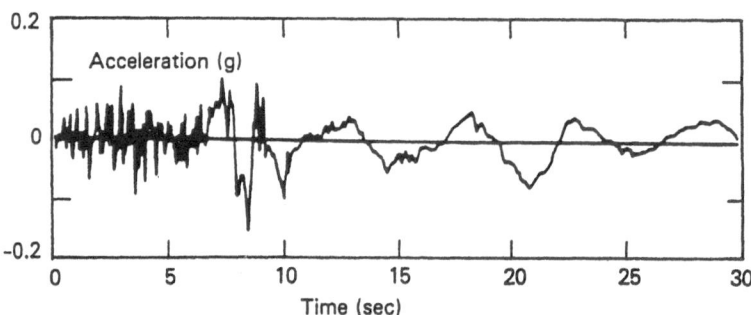

Fig. 2. Acceleration time hystory from the June 16, 1964 Niigata, Japan earthquake, recorded in an appartament house [2]. Transition from consolidated to an unconsolidated state (mechanical percolation), known as liquefaction is evident after 8-10 sec of oscillations.

Fig 3 a, b, c illustrate the dependence of elastic properties of a square lattice of plastic on the damage level (a), of velocity of acoustic waves propagating trough a pack of acrilic plastic plates on the number of plates (b) and velocity of elastic waves in natural rocks, both for laboratory and field scale (c), on the size of system [10]. All these data confirm scale-dependence of elastic properties of heterogeneous system. Evidently, this contradicts to one of the basic concepts of EMA - independence of physical parameters of the size of sample, as far as the latter is much larger than REV, namely, the size of grain of rock or the size of elementary crack (bond of network).

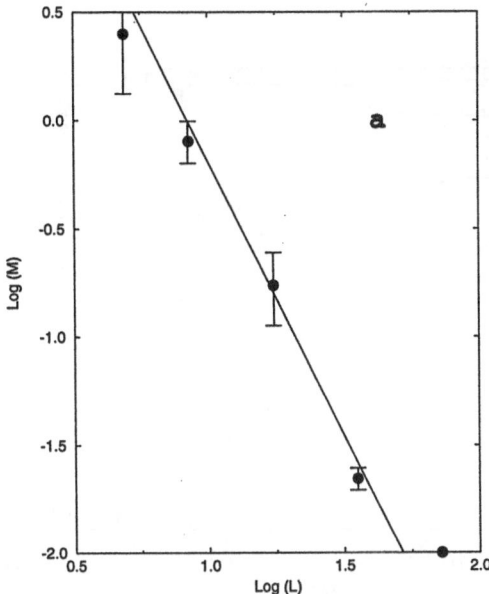

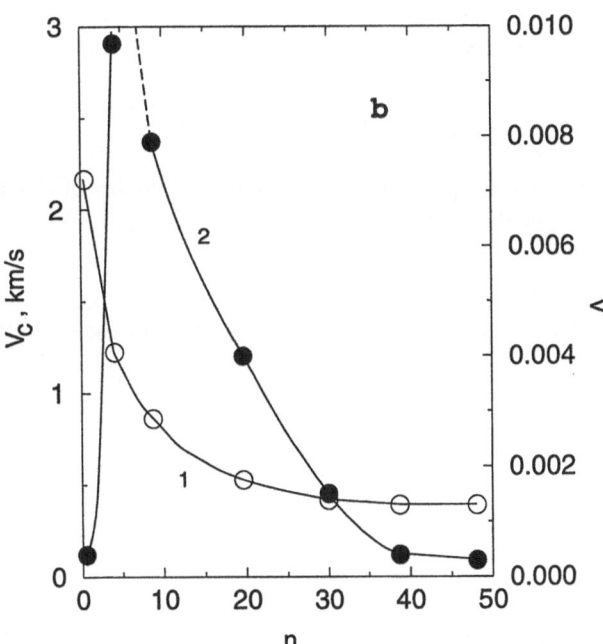

202

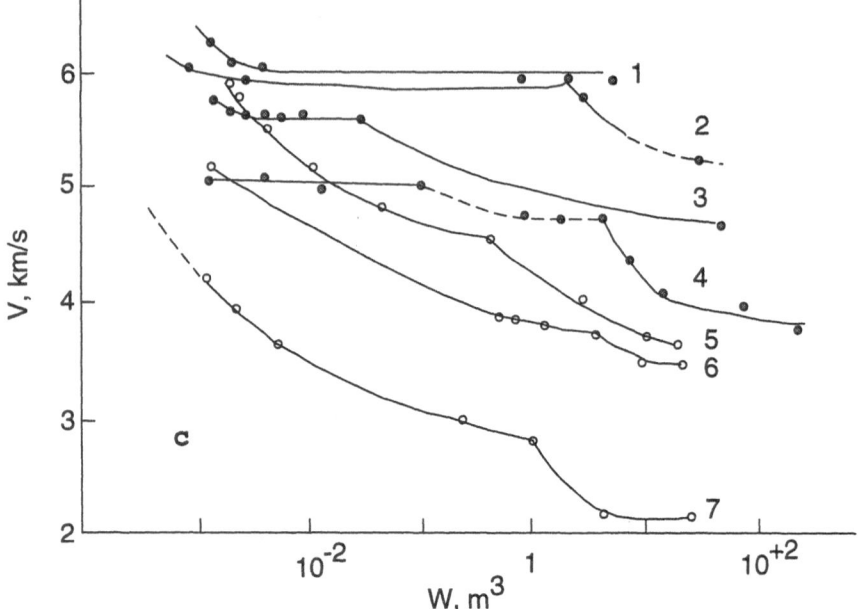

Fig. 3.a. Elastic modulus M of damaged (diluted) plastic lattice versus lattice size L (in units of elementary bond's length l) for samples, cut from the large lattice, damaged up to the mechanical percolation threshold ($p = p_c$), according to Chelidze [10]. The slope of the straight line lgM(lgL) is approximately 3, which is close enough to the theoretical values of mechanical percolation critical exponent $e_m = 3,6$.

   b. Elastic wave velocity v_p and its variance V versus number of plates in the layered model (relative pile thickness) n at constant load. 1 - v_p; 2 - V. Note maximal scatter of experimental data at n = 8 [11]

   c. Elastic wave velocity v_p versus the volume of sampled material [33]. 1 - diabase, 2 - granites, 3 - basalts, 5 - gneisses, 6 - limestones Toctogul, 7 - limestones Inguri.

Fig. 4 a from Shankland and Waff [34] presents the conductance G of the network of resistors versus fraction of the breaked bonds p (breaking makes the resistance of a bond infinite) for one experimental run. The more or less smooth dependence of G on p (for one run) is observed only for small values of damage p . At high damage level the $G(p)$ curve became step-wise which implies that the same elementary act , namely, destruction of one bond of lattice, may result quite different response in effective property. Similar results are obtained by Thompson et al [37] for the resistance of a sandstone as a function of injection pressure of mercury (Fig. 4 b). In mechanics the phenomenon of step-wise increment of deformation at monotonous loading after some critical value of the load is known as the Savar-Masson effect [4]. The typical example is shown on Figure 4c. Such nonregular dependence of generalized conductivity on the concentration p of inclusions (defects)   contradicts to another basic assumption of EMT, namely that GC is a smooth differentiable function of p . Thus EMA is valid only at small damage level, where this function is monotonous (Fig. 4 a, b).

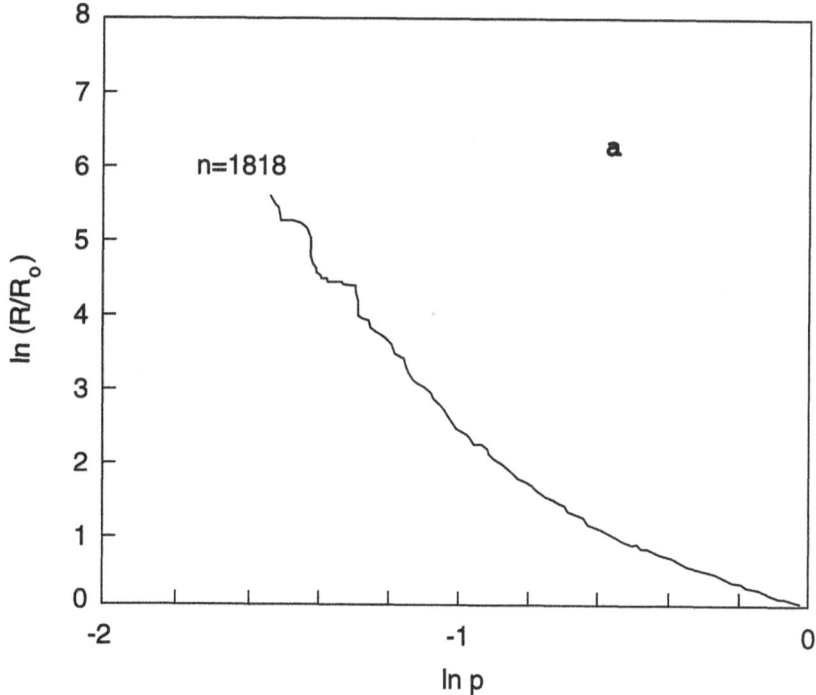

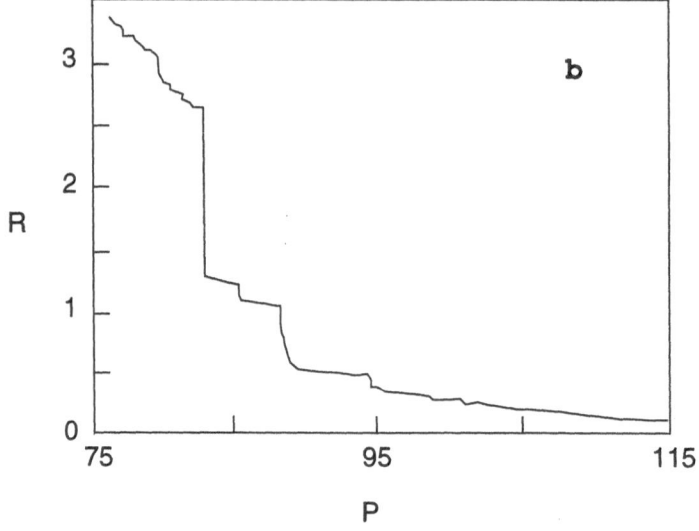

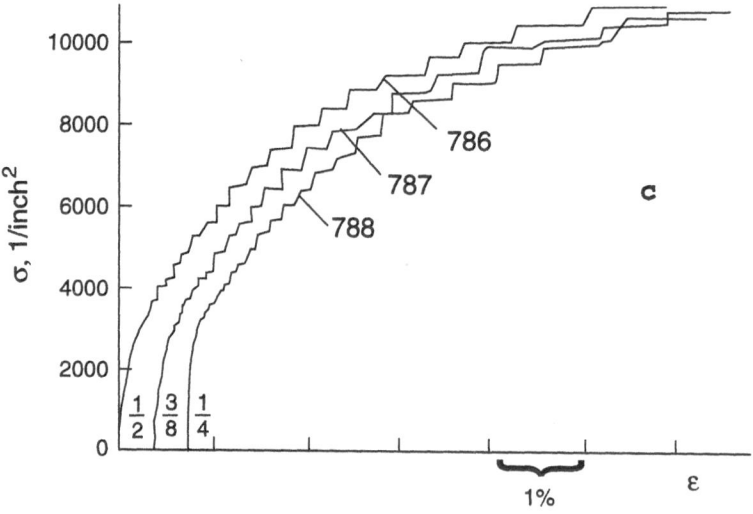

Fig. 4.a. Natural logarithm of normalized resistivity of the 3D (physical) network of resistors as a function of natural logarithm of p_i , number of intact (conducting) bonds.

Note, that as ln p_i decreases below -1 , both the number and magnitude of resistivity jumps increases drastically [34].

b. Resistance R of a sample sandstone as a function of the injection pressure P at mercury porosimetry [37]. At pressures close to percolation threshold of mercury (P from 75 to 95) the R(P) dependence experiences strong jumps, and only at P > 95 the curve became smooth

c. Stress σ (l/inch^2) versus strain ε at extension for 3 cylindrical samples of aluminium with various cross-sections (1/2 ; 3/8 and 1/4 "); results obtained by Sharp in 1966, after Bell [4]; at large enough strain the $\varepsilon(\sigma)$ became stair-case like.

Resuming, one should admit, that experimental evidence confirms the effectiveness of EMA in many cases, and, at the same time, shows its limitations when the system is close to critical state. It seems instructive to analyse the theoretical and experimental constraints on application of EMT, the reasons of its breakdown in certain situations and to find out, what in this case should be used instead of EMT. We try to show in the following that all above mentioned anomalies find their natural explanation in theories of percolation and fractals.

3. When the effective medium theory became uneffective

Effective medium theory (EMT) has been derived for very small perturbations of homogeneity of the embedding medium. It is believed that EMT itself contains its own constraint, in other words there is some critical (threshold) concentration of component, and extension of the theory above the threshold results in an (to us imaginary) transition of conductor-insulator type, whoch should indicate that the EMT is not more valid. In order to analyse this conjecture we will relate to EMT of electrical conductivity of composite, keeeping in mind that the behaviour of any other generalized conductivity, including elastic moduli, will be the similar.

Effective conductivity g of bicomponent mixture, consisting of embedding medium with conductivity g_1 and spherical inclusions with conductivity g_2 , according to Reinolds and Hough [31] is:

$$\frac{g - g_1}{3g} = \frac{g_2 - g_1}{g_2 + g_1} p_2 \qquad (1)$$

where p_2 is a partial volume of inclusions.

In the limiting case, when $g_2 \gg g_1$ from (1) it follows that

$$g = \frac{g_1}{(1 - 3p_2)} \qquad (2)$$

It can be imajined that equation (2) contains some threshold, namely that at $p_2 = 1/3$ the conductivity g jumps from some finite value to $g = \infty$, that is transition metal-insulator occurs.

It seems nevertheless that this threshold cannot be considered as a physical one. The matter is that in principle it is possible to fill the insulating matrix with conducting polydisperse spheres (recall that according to the expression (1) conductivity g does not depend on the size of particles) up to $p_2 \approx 1$, that is to much higher values of p_2, than 1/3, without any contact between them which means that metal-insulator transition at $p_2 = 1/3$ can be avoided. On the other hand classic EMT ignore the possibilty of clustering of inclusions; they are tacitly assumed to be distributed equidistantly. Meanwhile, physical and numerical experiments show that owing to clustering of conductive inclusions transition metal-insulator or percolation transition for spherical particles is achieved at much lower values of p_2, than is predicted by EMT namely at p_2 of order of 0.17 ; it can be still smaller for elongated particles or in case of three-dimensional continuum percolation [30,9].

Thus, it seems that threshold in EMT is fictituos and results from an attempt to apply it beyond its limiting conditions, which demand small concentation of inclusions and small contrast in physical properties of components. Let us consider, for example, the expression for electrical field of spherical inclusion E_i of dielectric constant ε_2 and radius a on the distance r from its center in the embedding media of dielectric constant ε_1 under applied external electric field E_0:

$$E_i = E_0 \frac{a^3}{r^3} \frac{\varepsilon_2 - \varepsilon_1}{\varepsilon_2 + 2\varepsilon_1} \qquad (3)$$

In EMT it is assumed that electrical fields of particles do not interfere significantly; only in this case dielectric constant of embedding medium does not deviate much from the value ε_1. From (3) it follows that for an interparticle distance r the condition $E_i \gg E_0$ is fulfilled if

$$\frac{a^3}{r^3} \frac{\varepsilon_2 - \varepsilon_1}{\varepsilon_2 - 2\varepsilon_1} \ll 1 \qquad (4)$$

From (4) it follows that to have the ratio E_i / E_0 less than 0.01 (week constraint!) at $\varepsilon_2 - \varepsilon_1 > 1$ the partial volume of inclusions should not exceed 0.25. Arbitrarily, to be true for any significant concentration of inclusions the allowed contrast in generalized conductivitties should be very small, less than 0.025.

This simple example shows that expression (3) for systems with strong contrast in properties of components became incorrect much earlier than a critical value of $p_2 = 1/3$ is attained.

Thus it seems that thresholds in EMT are rather mathematical than physical ones [25].

4.Differential self-consistent approach (DSCA)

Differential self-consistent approach , or DSCA [7, 9, 18, 20] allows to calculate effective properties of composite at large concentration of inclusions using procedure of integration of effects of infinitely small additions of inclusions, for which the expression (1) is still valid. Experimental data on Fig.1 show that in many cases this approach gives quite satisfactory results up to the considerable values of p_2. At the same time it is essential to note, that in DSCA additional inclusions are supposed to be embedded in an effectively homogeneous medium. This implies that some rearranging (annealing, stirring) procedure is needed after each act of addition of inclusions in order to avoid clusering and to homogenize mixture. This can be a case in such disordered systems as emulsions, which are stabilized by some repulsive interactions, or in foamy structures, where clustering is excluded. In such systems DSCA is valid even at high concentration of inclusions (Fig 1).

But quite often nature can refuse to rearrange heterogeneities. The matter is that some processes have memory, so that inclusions are actually frozen in their spatial positions. In the latter case addition of inclusions inevitably leads to their clustering (the inclusions belong to a cluster, if an interparticle distance r is less than some concrete value r_0; in "diluted" lattices r_0 is the lattice constant or the interaction range).

We can consider cluster (aggregate, pocket) of inclusions as one large inclusion with properties, defined by EMT for concentrated mixture and in this manner extend its applicability range to the aggregated mixtures [25]. In this case the REV is of order of the aggregate size L_a and EMA can be applied for $L >> L_a$, as the medium on this scale is again a quasihomogeneous one.

5. Effective network approximation (ENA)

In 1973 S. Kirkpatrick [23]advanced a new approach to a modelling of physical properties of composite by considering it as a randomly damaged network of conductances, which we refer to as an effective network approximation (ENA) in orded to stress its difference from EMA. Kirkpatrick's model also implies some threshold value of damage of network,at which the effective conductance of the network g_n falls to zero:

$$g_n = g_0 \frac{p - 2/z}{1 - 2/z} \tag{5}$$

where g_o is the conductance of intact network, p is the probability of a bond to be broken and z is the coordination number for a given network. According to (5) at $p = 2/z$ the conductivity became of network zero. Again, as in the case of EMA , the ENA predictions both for the threshold value of p and behaviour of g near actual threshold p_c are wrong: numerical experiments show that percolation transition happens at smaller values of p, than predicted by ENA , i.e. $p = 2/z$ and values of g_n from (5) deviate from experimental ones more and more as the system approaches p_c.

Again these deviations are explained by the extension of theory beyond its validity limit. ENA as well as EMA, is founded on the assumption of homogeneity of embedding media (network) which fails close to p_c .

Near p_c breaking of a single red bond [12] can lead to isolation of a big block of intact bonds and to strong step-like change of g_n in contrast with the homogeneous domain far from p_c , where each elementary act produces infinitely small variation of g_n (Fig. 4 a,b).

Thus ENA as well as EMA is only valid far enough from p_c , that is for very small percent of damage (inclusions).

6. Mechanical percolation threshold

If the thresholds, appearing in EMA, seem to be ficticious, there are real, experimentally documented critical transitions for a mechanical percolation process. Experimental data on typical (sol-gel) transition are presented in , Figure 5 from [1]

These results are useful for understanding of other kinds of mechanical percolation, such as liquefaction (Figure 2), damage fracture of solids , cementation of sediments or sintering of grains at high temperatures. Parameter p of Figure 5 can be interpreted as a partial volume of elastic elements in viscoelastic mixture: monomers in sols, number of cemented contacts , or just number of intact elements of a damaged solid. Below p_c the system is viscous (or liquified) and its viscosity scales as $(p - p_c)^{-0.8}$; thus at p_c viscosity

210

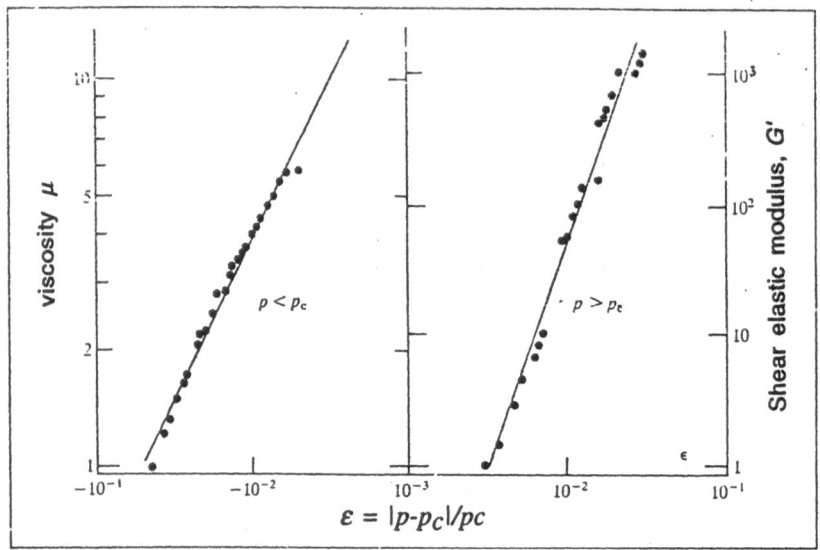

Fig. 5. Mechanical properties of a colloid at zero frequency below ($p<p_c$) and above ($p<p_c$) the gelation (percolation) threshold. Below the gelation threshold the system behaves like a liquid; its viscosity μ diverges at the threshold as $(p-p_c)^{-0.8}$ (left straight line). Above the gelation threshold the system manifests elastic behaviour ; the shear elastic modulus G' increases like $(p-p_c)^{3.2}$ (right straight line)

diverges. Above p_c the system is an elastic body with an elastic modulus M, which scales as $(p -p_c)^{3.2}$. Dependence of M on $(p -p_c)$ for gel is the same as for a diluted elastic lattice (Figure 4 b.). This proves the universal nature of mechanical percolation in various heterogeneous materials.

7. Transition from translational to scale invariance (homogeneous and fractal regimes)

Above we have shown, that refined EMT or DSCA works quite well in many cases. This means that hypothesis about self-consistency of the field is valid and the heterogeneous system which is larger than its REV may by safely replaced by a homogeneous medium without any significant change in a local field around the given REV.

But any refinement of EMT is useless, if the heterogeneous system loses its property of translational invariance (TI) , that is when there is not any relevant REV, for which the system can be considered as a homogeneous one. This happens when, due to clustering of elementary heterogeneities of size l, the structure of system became self-similar (fractal) and thus obeys the principle of scaling; see Mandelbrot, [19, 27]. It is evident that in a fractal regime any homogenization procedure is impossible.

The transition from translational invariance to self-similarity (TI-SS transition) in disordered systems is the most important type of transitions caused by proliferation and clustering of inclusions and it can be considered as a hallmark, limiting application of EMA: indeed, the system cannot be both quasihomogeneous and fractal.

In infinite self-similar systems there are only two characteristic length scales - the size of elementary object l (that of an inclusion, a bond, a defect, etc) and the correlation lenghth L_C , which loosely can be defined as a size of the largest cluster . Correlation length depends on the concentration of defects; it diverges at a critical concentration p_c when the infinite cluster (IC) appears in the system; thus at $p = p_c$ there is not any characteristic length in it (except l).

Any transport property g of heterogeneous system in a fractal regime obeys percolation theory [36] and near to p_c follows a power law

$$g \propto g_0 (p - p_c)^d \qquad (6)$$

where g_0 is some amplitude constant, and d is a fractal dimension for a given property. The power law dependence on the difference $(p-p_c)$ is a signature of self-similarity.

Of course some refinements are needed to take into account pecularities of electrical [23], mechanical [16, 21,32] and other percolation processes, which lead to the different values of fractal dimension for various physical parameters.

Far enough from critical concentration p_c $(p << p_c)$ the correlation length and, hence, cluster sizes are not infinite. That means, that for scale $L >> L_c$ the system is homogeneous and EMA approach holds again, but in this case the REV is that of the largest cluster L_c and the properties of inclusions are thouse of cluster. In contrast with above EMA-model of homogeneous aggregates the clusters on the length scales $L < L_c$ have fractal structure and naturally, properties of clusters depend on their fractal dimension; thus, generalized conductivity of system though is independent of the scale (sample size) for $L >> L_c$, may depend on the fractal dimension of clusters.

Fig. 6a illustrates the behaviour of a generalized conductivity g when concentration of inclusions (damage of lattice) changes from 0 to 1: there is a homogeneous regime domain for $L >> l$ and $l << L >> L_c$ and a fractal regime domain when L is of order of L_c. Some generalized conductivities , for example, viscosity and dielectric constant, have negative exponent in (6); they diverge at $p \approx p_c$.

8. Finite size transitions

In finite systems a specific kind of transition can be observed. Let us consider finite pieces of size L of an infinite system in which $p \geq p_c$. For small enough L , namely when L is comparable with a correlation length L_c the "finite size"transition may occur. Let us consider, for example, conductor-insulator mixture in a conducting state at $p > p_c$, which is beeng divided into smaller parts. When the scale of partition became less than the correlation length, it is highly probable that in the separated pieces of a system the spanning conductivity path of a large sample will be destroyed; decimation of a conducting system causes its transition to an insulating state.Similarly, decimation of mechanically disjointed large system leads to appearance at L $< L_c$ of many intact blocks; we have finite size transition from uncosolidated to consolidated state. This is an exellent example of breakdown of the ergodic approach - small parts of the infinite system behave themselves in a quite different way, than the system as a whole.

The schematic behaviour of generalized conductivity of a finite system versus sample size is presented on the Fig. 6b. Here also, like on Fig. 6a, there are domains of homogeneous regime at $L >> l$ and
$l << L >> L_c$ and the fractal regime domain at $L \approx L_c$. Fig. 7 shows the (experimental) dependence of the resistivity of a two-dimensional system on its size, where domains of fractal and homogeneous regimes are clearly seen. Example of size-dependent mechanical properties are presented on fig. 3b, c.

Finite size scaling is a powerful method of percolation theory as it allows to obtain fractal dimension of process from the scale dependence of generalized conductivity [32].

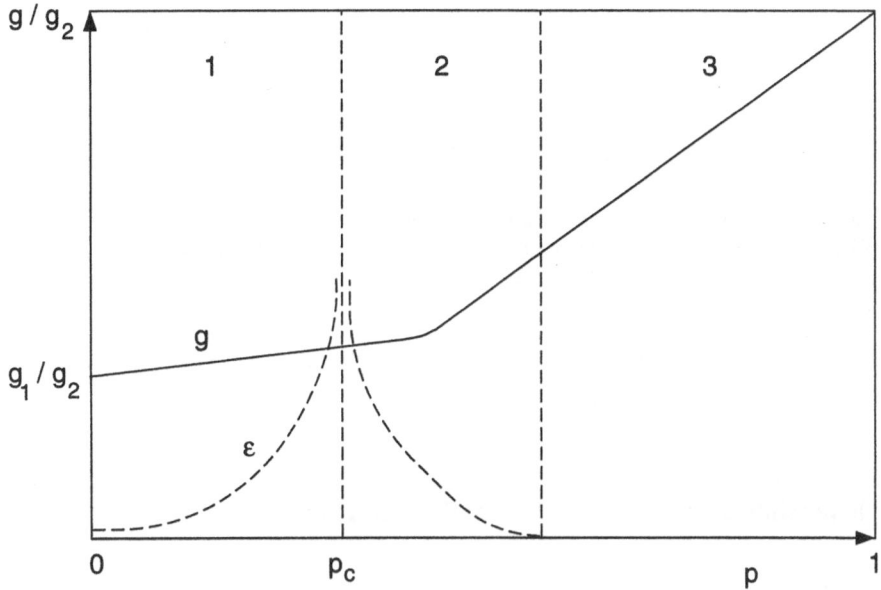

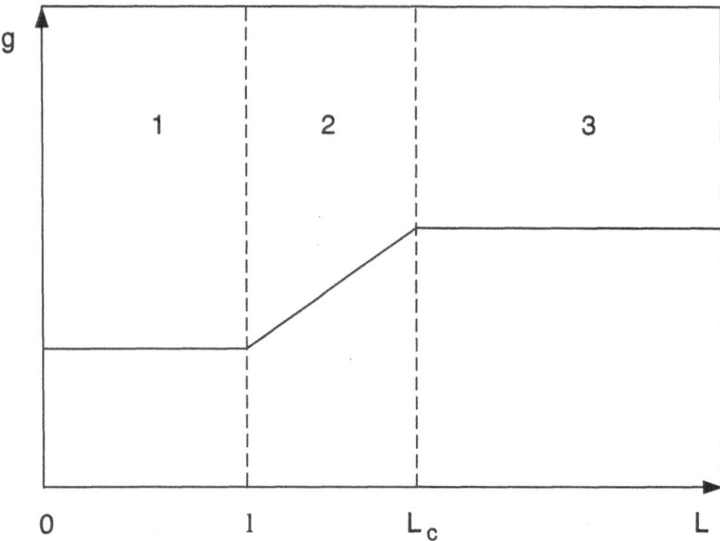

Fig.6 a Schematic behaviour of normalized nondivergent (g/g2) and divergent ($\varepsilon/\varepsilon_1$) generalized conductivity of heterogeneous material versus partial volume of inclusions p_2; p_c corresponds to the percolation threshold.

   b. Schematic behaviour of generalized conductivity g of a finite heterogeneous system system versus sample size L. 1 and 3 correspond to the domain of homogeneous regime; 2 corresponds to a fractal one.

9. Is it possible to converge EMA and percolation?

It may be thought that incorporation of the actual percolation threshold values as a limiting concrntration into EMT may inprove the performance of the latter. This idea seems be not very promisive, because, as it has been shown above, EMA and percolation theory are incompatible, as they are founded on mutually excluding basic assumptions.

   Of course, taking into account the excluded insulating volume, that is volume, which cannot be filled in by the solid grains, may improve the fit of experimental data with EMT and that is actually done Mukerji et al for the conductivity of insulator-conductor mixture [39]. Nevertheless, this procedure cannot be considered as a convergense of EMA and percolation, as it is claimed in [39]; it is just unification of sand

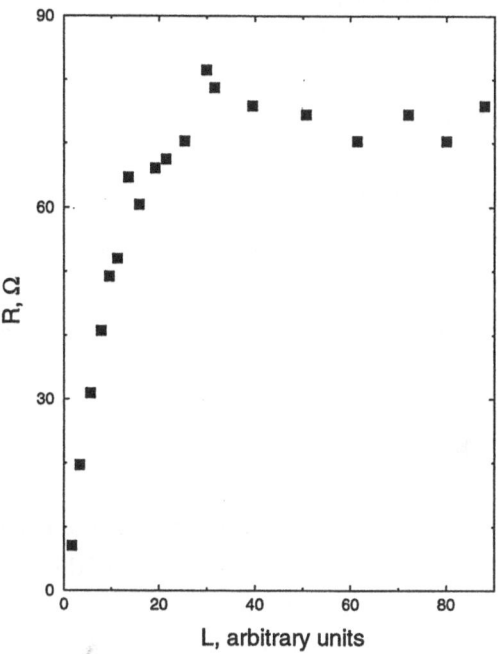

Fig. 7. Resistivity of a 2D conductor-insulator system versus sample size (experimental data of Palevski and Deutsher, [29]. For L less than L_c,where L_c is the correlation lenght R depends on the size of sample (fractal domain); for $L>L_c$ resistivity is not scale-dependent (homogeneous domain). Thus, system is simultaneously quasihomogeneous for the large scale measurments and fractal for small-scale tests.

grains and air phase into one nonconducting phase (see also [40]. Moreover, as will be shown later on, percolation for conductivity or permeability in porous rocks and mechanical transition from consolidated to unconsolidated state may happen at quite different porosities, contrary to conjecture of Mukerji et al [39].

It seems also that lawfullness of application of such methods, as renormalization group (RG) for calculation of physical properties by EMA [41] near percolation threshold, also is doubtful because RG is a method for analysis of self-similar systems and EMA is valid for quasihomogeneous media.

10. Experimental criteria of TI-SS transitions.

It is useful to formulate expetimental criteria of diagnostics of TI-SS transition, which would allow one to know which regime the system is in and conseqeantly which approach should be used in calculations.

    i. If a physical property g reveals dependence on the sample size L and obeys power law

$$g \sim L^x ,$$

it is very likely reflecting fractal process with a fractal dimension x. In homogeneous regime g does not depend on L and it may be calculated by EMT (Fig. 3a, b, c).

    ii. The main postulate of the best EMA technique, namely the differential self-consistent method impose the condition, that function $g(p)$ is differentiable, hence, the infinitesmal increase of p should produce the infinitesmal change of g. Hence, $g(p)$-curves should be smooth for each (separate) experimental run. At p , close enough to p_c , even smallest change of p may induce step-like variation in non-averaged generalized conductivity g which means that EMA is not any more valid. The matter is that the additional defect , generated in a blob (dense cluster of defects) causes much less effect than is predicted by EMT, whereas breaking of a solitary (so called red) bond between blobs causes much greater response (step-like increase) in g than given by EMT. (Fig. 4a)

    iii. Both physical and numerical experiments show that near percolation threshold ($p = p_c$) as well as at finite size transition ($L = L_c$) the scatter in measured (calculated) values of generalized conductivities increased dramatically for seemingly identical series of experiments (Fig; 3 b). The maximum in the scatter is caused by very high sensitivity of physical property of system, which is near to a critical point, to the smallest changes in p or L ; these smallest changes in p may be caused by infinitesmal variations in experimental conditions at various runs. For finite size transitions the scatter is minimal in homogeneous domains where $L\gg l$ and $l\ll L\gg L_c$ and maximal for $L\approx L_c$. This effect can be used for experimental pinning of L_c.

iv. Experiments also show high sensitivity of generalized conductivity g to the smallest external perturbation ΔP (stress, temperature) when the system is close to the percolation threshold.

Thus, plotting, for example, the strain-sensitivity of g or $\Delta g/\Delta P$ as a function of concentration of component p makes it possible to define p_c (Fig. 8).

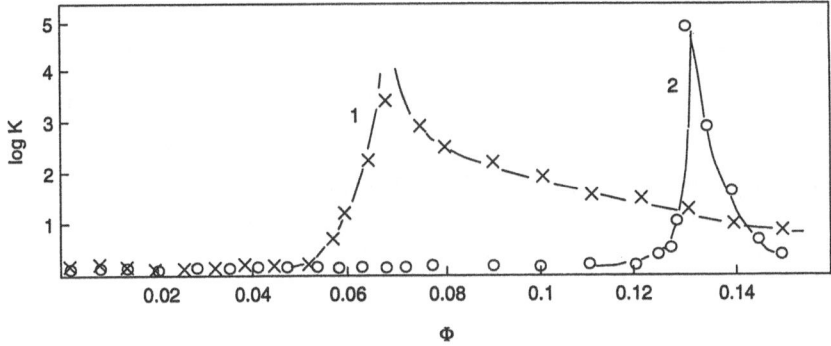

Fig. 8.Strain-sensitivity K of a conductor-insulator mixture for different
concentrations Φ of the conducting phase; conductor in mixture: 1-
graphite, 2 - thin copper rods.

v. There is a striking detail in frequency dependence of complex elastic properties of heterogeneous system, which can be considered as a signature of fractal behaviour: namely, near the percolation threshold real M' and imaginary M" parts of elastic moduli reveal quite identical power law dependence on frequency : $M'\approx M''\approx f^s$ (Figure 9). That, in turn, means that their ratio, or dissipation factor $Q^{-1} = M''/M'$ should be independent of frequency (constant Q model). Experiments confirm such type of behaviour of electrical and mecanical loss factors in the fractal regime [41].

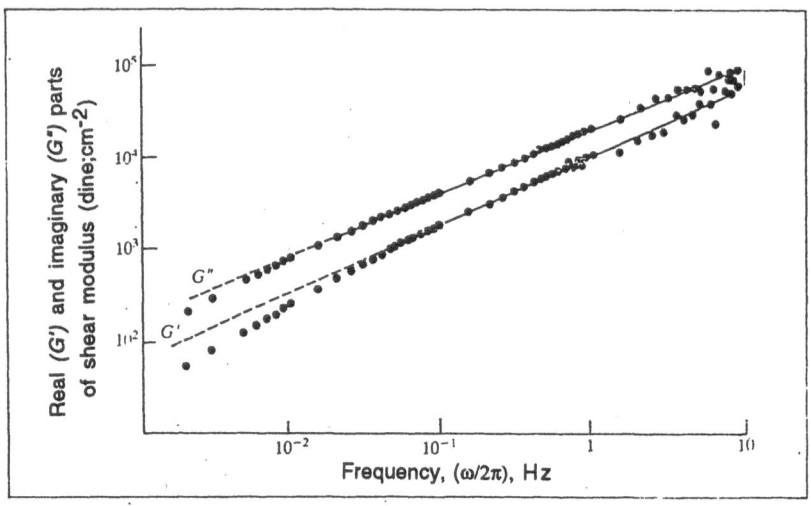

Fig. 9. Frequency dependence of the real G' and imaginary G'' parts of the complex shear modulus for a colloid sample at concentration $(p - p_C) = 5.36 \ 10^{-4}$. The straight lines correspond to a power law dependence $G' \approx G'' \approx f^{0.72}$,which means that shear dissipation factor $Q = G''/G'$ is constant [1]

11. Percolation thresholds for various processes and structures.

We showed above that physically well grounded thresholds for such transitions as conductor-insulator, consolidated-unconsolidated solid, permeable-unpermeable medium are given by percolation theory. Naturally, a question arises: are these thresholds invariant or they differ for various processes and various structures?

Both exact calculations and numerical experiments prove that percolation threshold is different for various structures (lattices); the larger is the coordination number of bonds z in the lattice the less is p_c.

The question of identity of p_c for various processes on the same structure is a delicate one. It seems that due to a purely geometrical definition of percolation problem p_c should be the same for any transport process, if the structure of system is identical (say, triangular lattice). For such processes as conductivity and permeability of a porous dielectric this may indeed be a case (except systems with very fine pores), but for the mechanical percolation threshold of such system p_c is not the same as for electrical percolation. Feng and Sen [16] show that both the percolation threshold and the fractal dimension for elasticity of randomly damaged (diluted) solid with central (interparticle) forces are quite different from these of electrical percolation, in contrary to de Gennes conjecture [19]. On the other hand Sahimi and Arbabi [32] suggested that for elastic percolation with bond-bending interparticle forces the threshold for elastic moduli is the same as in ordinary percolation both in 2D and 3D systems, if only each site of network has at least $D:(D-1)/2$ bonds with the nearest site.

Now let us consider the results of physical experiments, which are devoted to a study of the critical porosity of solid p_p (that is such porosity when the solid cannot support any finite stress and actually a transition consolidated-unconsolidated solid occures) and critical porosity for permeability (conductivity) of porous materials.

Experimental data, show that typically at 20-25% porosity most of porous solids both with closed and open porosity retain approximately half of their (intact) strength . On the other hand, permeability of porous rocks (sandstones) may be of order of hundred millidarsy, that is very high, for samples with connected porosity 10-20% .

We can conclude from analysis of these data that a porous rock with connected porosity of order of 20 % is quite permeable for brine and has high conductivity if saturated; hence this porosity is high enough to ensure percolation transition for these transport properties. At the same time it is also quite evident that materials with such porosity may have large enough elastic modulus M and high mechanical strength as the solid phase is still consolidated.

Thus connectivity for the transport does not mean at all that the porous material has zero elastic moduli and can be disjointed mechanically. In other words, the critical porosities for transport and elasticity are not the same.

These experimental data seem to contradict conclusions of Sahimi et al [32] about identity of mechanical and ordinary (conductivity) percolation, but the more detailed analysis prove that the contradiciton is actually fictitious.

The matter is that Sahimi and Arbabi [32] consider desintegtration of *conducting* solid (network).In this case the critical concentration $p_c = 0.25$ for mechanical percolation is that of *intact conducting* bonds; that means that in order to have zero elastic moduli you need to destroy 75 % of bonds in the intact cubic *conducting* lattice. Thus in such network presence of 25% of *intact* bonds guaranties appearance of global conductivity as well as finite elastic modulus of system.

The situation is quite different in porous *dielectrics and rocks:* the value of critical porosity for mechanical percolation remains the same as in porous metals, namely around 80 % [43], but the critical (brine-saturated) porosity for conductivity and permeability is much less: to make the insulating cubic lattice conductive only 0.25 of all bonds should be conductive. According to Pike and Sieger [30], in insulator-conductor powder mixtures approximately 15 % of the whole volume should be conductive to reach global conductivity.

Thus, it seems that the critical porosity values for 3D elastic (damage) percolation p_{cep} and for transport p_{ctp} in a porous dielectric are complementary, that is [9]

$$p_{cep} = 1 - p_{ctp} \qquad (7)$$

The mechanical connectivity threshold p_{cep} should be considered as a lower bound for decoupling of solid; the matter is that solid (intact) phase may be disconnected, but existence of anchor-like configurations on the interface between blocks can prevent mechanical decoupling [9].

12. Dynamical percolation transitions

Some of percolation transitions are strongly affected by dynamical processes. Let us consider, for example, the plain elastic wave of a wavelenght λ, propagating in a fractal heterogeneous system with the elementary heterogeneity size l and the correlation lenght L_C. We can imagine fractured media, containing fractal network of elementary cracks of

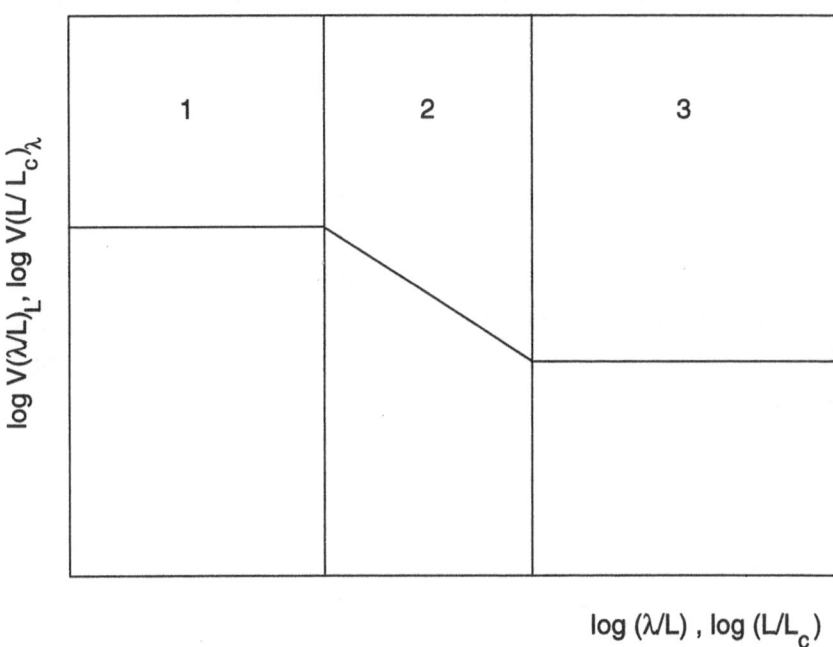

Fig.10. Schematic behaviour of the velocity of elastic waves versus relative size of system in units of correlation length L_C) at fixed λ or versus relative wavelength λ/L (in units of system size L) at fixed L.

size l. For $\lambda/l \gg 1$ the medium is effectively homogeneous and the velocity is constant. The same is true for $\lambda/L_C \gg 1$, when the wavelength is much larger than correlation length of crack clusters. Between these two homogeneous domains the velocity is not stable and depends on the wavelength λ (Fig. 10). It is evident that as the parameter of transition is ratio λ / L_c, the same kind of transition can be realized at constant λ by variation of the size of the sample L (Fig. 3 b).

In classic percolation approach the occupied (damaged) site or bond cannot be healed. Introduction of some rules for local healing of damage (restoring initial state of element) leads to a new domain of dynamics of disordered media, including cellular automata models and the theory of self-criticality [3]. Dynamic models much better correspond to a complicated, dynamical character of seismic processes in the Earth's interior than classic percolation, which , nevertheless, is a quite satisfactory tool for interpretation of laboratory data [9, 10, 21] and is incorporated into dynamical models as a special case.

13. Variability of thresholds for TI-SS transition.

We have seen in previous sections some examples of the variability of thresholds depending on the structure of system and the character of the process. Here we would like to discuss some other factors, which can significantly shift the position of TI-SS transition threshold.

i. Microstrucuture of components. The percolation thresholds as a rule are calculated for mixtures with homogeneous components. In practice, however, the component (inclusion) may be heterogeneous itself: examples are (secondary) porosity of inclusions and existence of a layer (shell) with different properties on the surface of inclusion.

Experiments on various sintered powders [14, 26] show that the mechanical percolation threshold for such systems is extremely low : the system became elastic and may sustain some finite stress at a critical volume fraction of solid p_S such as $p_S = 0.04$ (metal powder sinter) , $p_S = 0.07$ (sintered perlite), $p_S = 0.066$ (sintered submicron silwer beam). We presume that these anomalously low thresholds are due to porosity of grains in the sinter.The critical volume fraction (CVF) of mutually inpenetrable grains for appearence of infinite elastic cluster p_{cvf} in sinters is at least that of ordinary 3D percolation $p_{cvf} = 0.16$ (see [30]), then in order to have above experimental values of critical partial volume of sinter, namely, $p_S = 0.06$ we should assume grain porosity $p_g = p_S / p_{cvf} \approx 0.06 / 0.16 = 0.38$.

Generally, for systems with porous component (grain) of porosity p_g the critical volume fraction of solid p_S is

$$p_S = p_{cvf} p_g \tag{8}$$

The same arguments hold for grains with shells, for example, insulator grains with a conducting shell, where the conductor-insulator transition may happen at very small CVF of conductor p_{csh}

$$p_{csh} = p_{cvf} p_{sh} \tag{9}$$

where p_{sh} is a volume fraction of conducting envelops of a grain.

ii. Hard core (interaction range). Regular lattice models of disordered media imply some minimal inpenetrable distance (hard core) between sites of order of a lattice constant. If this restriction is released and the sites are allowed to be overlap each other (zero hard core) then CVF increases significantly, from ~ 0.15 to ~ 0.3. Concept of hard core of various size can help to understand differences in threshold values for seemingly similar processes, like conductivity and permeability: hard core or immobilized layer of liquid on the surface of pores can prevent flow in narrow gaps, but allow transport of charges.

Percolation with a varying interaction range can be considered as a special case of hard core models. Indeed, percolation may be a result of increased interaction range R of sites even at relatively small concentration of occupied sites (R-percolation). R-percolation reflects the variation of a local field, associated with a given site under action of some extenal field [30].For example, according to "forced" percolation model of fracture the solid may fail at much lower damage level , then it is needed for critical porosity model, due to the expansion of local stress field range under action of external stress [9]. In this model the critical stress σ_c for a given damage level p is a relevant percolation parameter and not the critical porosity, because forced model takes into account long-range stress-dependent interactions unlike pure damage model, where only nearest neighbours can interact. Other fracture criteria, including "damage"* analogs of Griffith's critical crack length , may be envisioned for accumulated (fractal)damage mechanisms [9, 21]

iii. Anisometric particles. In mixtures with randomly arranged elongated particles (sticks) the percolation threshold may be very low: sticks percolate, when the mean distance between their centers is approximately half of their length [30]. Thin copper wire pieces in an insulating powder percolate at $p_c = 0.07$ [9].

iv. Finite size effects. (see section 7).

v. Correlation. Many natural phenomena are correlated or anisotropically correlated. Correlation causes at first decrease of p_c for small values of correlation factor and then significant increment of p_c for strong correlation [15]. In anisotropically-correlated models p_c is a function both of direction and strength of correlation, as shows Fig. 11 from Chelidze [8, 9].

vi. Frequency shift. Geometric or ordinary percolation thresholds correspond to the static or stationary problems and do not depend on the frequency of excitation. On the other hand, any transport property has its own characteristic relaxation time. Thus if the alternating or transient field is applied to a fractal system, the dynamical aspect of percolation may reveal itself in the effect of frequency shift of the threshold. Fig. 12 a illustrates this effect for a.c. conductivity of a conductor-insulator mixture; the p_c shifted from ~ 0.8 at high frequency to ~ 0.14 at low frequency. The underlying physics can be easily understood: at high enough frequencies the thin insulating gaps between conducting inclusion are shunted by reactive currents and became effectively

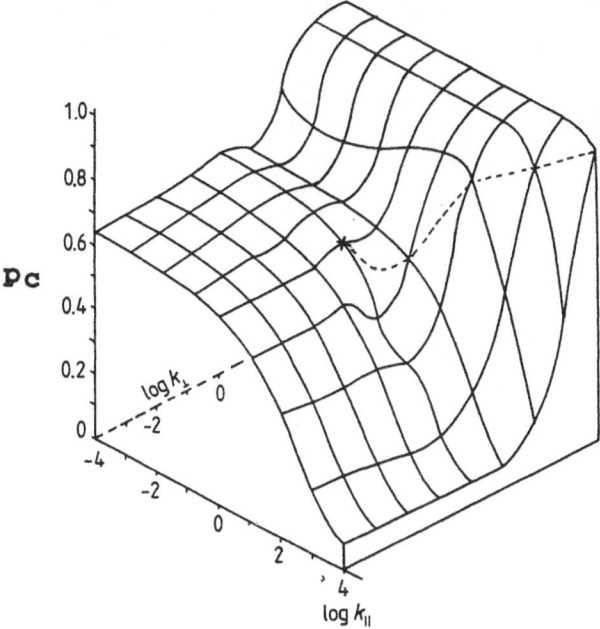

Fig. 11. Variation of the percolation threshold p_c of a square lattice as a function of anisotropic correlation factor (K_p/K_n) or ratio of correlation strengths in directions, parallel (K_p) and normal (K_n) to the external (correlating) field [9]. For very large K_p/K_n the network of cracks losses its ramified structure; in the limiting case there is left only one straightly propagating linear crack.

conductive. As a result, effective volume fraction of conductive phase increases with frequency even at fixed concentration of ohmic conductor p_0 and at high enough frequency the global conductor-insulator transition may occur at the concentration of conductor p_0 which is less than p_c, according to the expression [9]:

$$p_0 = p_c(1 - k_f \Delta f_c) \qquad (10)$$

where k_f is some factor and Δf_c is the frequency increment which compensates exactly the deficit of conductor $(p_c - p_0)$ at "dynamical" transition as compared to a static one.

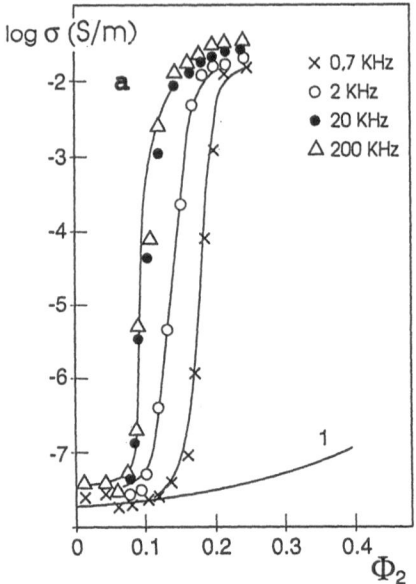

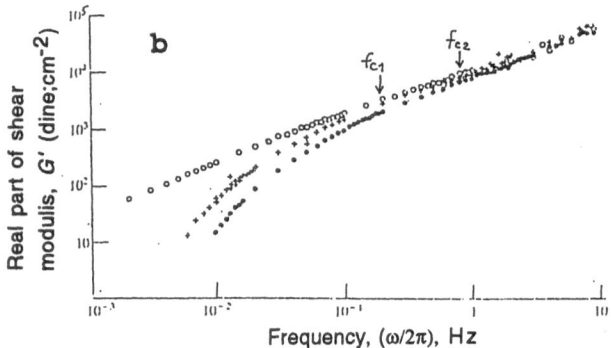

Fig. 12. a. Dependence of conductivity σ of powder mixture of quartz and graphite on partial volume of conductive component Φ_2 at various frequencies: 1- calculated values

of σ according to MBH-theory, 2 - 5 -measured values (kHz): 2 - 0.7, 3 - 2.0, 4 - 20, 5 - 200. Note, that percolation threshold shifts from ~ 0.12 for LF to ~ 0.08 for HF. [9].

b. Frequency dependence of the real part of shear modulus of colloid system for various concentations of elastic phase below gelation threshold p_c.
o, +, • correspond to samples with $(p - p_c)$ equal to 5.36 10^{-4}, 3.75 10^{-3} and 7.33 10^{-3} respectively. Note that we need less and less concentration of elastic elements at higher frequencies to obtain the same shear modulus [1].

The similar effect can be predicted for the so called superelastic percolation (Fig. 12b), that is transition from unconsolidated (viscous) to consolidated state in a system, consisting of viscous and rigid elements [9, 41]. It is evident that at high enough (angular) frequencies, larger than relaxation time of viscous element, the superelastic system became elasic at smaller concentration of rigid elements, than it is needed for purely geometrical (static) transition. The effect should be described by the same equation (10) where p_c is now the critical concentration for static superelastic transition and k_f is the frequency stiffening factor.

Experiments on mechanical properties of colloids near gelation point confirm these conclusions: data, presented on the Figure 12 b show that in order to have the same shear modulus G' of system, say 10^2 dyne.cm^{-2} you need much less concentration of elastic elements for higher frequencies than for lower ones.

It seems interesting to calculate k_f from experimental data (Figure 12 b), assuming that the upper curve corresponds to the shear modulus of gelated system, that is to $(p - p_c) \approx 0$. Then, inserting into (10) the frequency increments $\Delta f_{c1} = 0.11$ Hz for $(p - p_c) = 3.75 \ 10^{-3}$ and $\Delta f_{c2} = 0.8$ Hz for $(p - p_c) = 7.33 \ 10^{-3}$ (at these frequency increments shear moduli of system, which is still below static percolation (gelation) threshold , became equal to shear modulus of gelated system) we obtain for the frequency-stiffening factor values correspondingly $k_1 = 3.3 \ 10^{-2}$ and $k_2 = 0.9 \ 10^{-2}$. Thus it seems that for a given system k is of order of 10^{-2}. From $(p - p_c)$ values it is possible in principle to assess the volume of the stiffened liquid near the colloidal particle and its dependence on the frequency.

As far as the shunting and stiffening happen preferably in the narrowest gaps between clusters the fractal dimension of the dynamical infinite cluster may be less than that of infinite cluster for the random percolation; we guess that it may be close to the fractal dimension of the backbone.

Frequency stiffening is predicted also by poroelasicity theory for solid with fluid-bearing pores [42].

14. Implications of TI-SS transitions to some practical problems

TI-SS transitions may be important for practical applications. For example, any theory of geophysical prospecting uses the concept that a given geological formation (layer, block, etc) is a quasihomogeneous one, that is, the physical properties of formation does not depend on scale : say, on electrode spacing or explosion-registration point separation, if only the geophysical field does not exceed the borders of (homogeneous) formation.

Nevertheless, experiments show, that physical properties of apparently homogeneous rocks or rock massives may change in regular manner with variation of the scale of sampling (Fig. 3 a,b,c).There is not anything striking in these results from the point of view of TI-SI transitions. For the data, presented on Fig. 3c, the sample size L is always much larger, than grain size of rock; thus the most probable type of heterogeneity in these apparently homogeneous geological formations is the correlation length of crack network. As far as crack network may be ierarchical, several TI-SS transitions may occur in the same geological formation, when the sample size increases (Fig. 3c).

It should be stressed, that the phenomenon of scaling of physical properies in fractal geological medium is practically ignored in geophysical interpretation, though possibility of scale-dependence cannot be excluded .

For example, electrical vertical sounding curve, similar in form to that in Fig.7 will doubtlessly be interpreted as a sign of two-layered structure with different resistivities of layers. This will be correct in the great majority of cases, but it cannot be excluded that the curve of this form is a result of TI-SS transition within the same geological formation.

The separate problem is how to handle the phenomenon of scaling of physical properties in the classic mathematical formalism of geophysics: the traditional material relations, used as an complementary addition to the differential equations of mathematical physics, do not provide for scaling.

Thus, scaling and fractality do not promise easy life to mechanicians, geophysicicts and generally to all, who are involved in the physics of disordered media, but, on the other hand, they offer a new look of traditional mechanical and geophysical problems and new tools to solve them.

Acknowledgements

This research has been supported by the Ministry of Higher Education and Research of France (program PAST). T. Chelidze (Georgia) is extremely grateful for this support.

REFERENCES

1. Adam, M., Delsanti, M. (1989). Percolation-type growth of branched polymers near gelation threshold. *Contemporary Physics*, **30**, 203-218.
2. Aki, K. (1988) in T. Reiner , *Seismic Hazard Assessment,* 1990
3. Bak, P., Tang, C., Wiesenfeld,K. (1988). Self-organized criticality. *Phys. Rev.A.*, **38**, 364-374.
4. Bell, J. F. (1973). *Mechanics of Solids* M.1., In: Encyclopedia of Physics. S. Flugge (Ed), v. V/1,Springer-Verlag.
5. Benguigi, L. (1984).Experimental study of the elastic properties of percolating system. *Phys.Rev.Lett.* **53**, 2028-2032.
6. Berge, P. A., Bonner, B. P. Berryman, J. G. (1995). Ultrasonic velocity-porosity relationships for sandstone analogs made from fused glass beads. *Geophysics*, **60,** 108-119.
7. Bruggeman D.Berehnung (1935)Verschiedeneren Physicalisher Konstanten von Heterogenen Substanzen. *Ann.Phys.*, **24**, 636-679.
8. Chelidze, T. L. (1982). Percolation and Fracture. *Phys. Earth Plan. Int.*, **28**, 93-101
9. Chelidze, T. L. (1987). *Percolation Theory in the Mechanics of Geomaterials*.Nauka, Moscow, 136 p (in Russian).
10. Chelidze, T. L. (1993). Fractal damage mechanics of geomaterials. *Terra Nova*, **5**, 421-437.
11. Chelidze, T. L., Chergoleishvili T. T., Manjgaladze P. V. (1987).Elastic properties of loosely-jointed layered media and fractal mechanics. *Geophysical Journal (Kiev)*, **9**, 25-30 (in   Russian).
12. Coniglio, A. (1982). Cluster structure near the percolation threshold. *J. Physics A*, **15**, 3929-3844.
13. De Gennes P. G. (1976). On a relation between percolation theory and elasticity of gels.*J. Physique Lett.* **37**. 213-233.
14. Deptuck, D., Harrison, J. P., Zavadski, P. (1985). Measurment of elasticity and conductivity of a three-dimensional percolation system. *Phys. Rev. Lett.*, **54**, 913-916.
15. Duckers, L. J. (1978). Percolation with nearest neigbour interaction. *Phys. Lett. A.*, **67**,93-94.
16. Feng, S., Sen, P. (1984). Percolation on elastic network: new exponent and threshold. *Phys. Rev. Lett.*, **52** , 216-219.
17. Gueguen Y., Chelidze, T., and Le Ravalec, M. Microstructure, Percolation Thresholds and Rock mechanical Properties (*Tectonophysics*, in press)
18. Gueguen Y. and V. Palciauskas. (1994) *Introduction to the Physics of Rocks*. Princeton, 294 p.
19. Gouet J.-F. (1992). *Physique et structures fractales*. Masson.
20. Hanai T. (1968)Electrical Properties of Emulsions; in : *Emulsion Science*. (Ed. P. Sherman) p.p. 354-478. London, N.Y. Academic   Press,
21. Herrmann, H. J.., Roux, S. (Eds). 1990. *Statistical models for the fracture of disordered media*. North-Holland.
22. Hill, R. A. (1965). A self-consistent mechanics of composite materials. *J. Mech. Phys. Solids.*, **13**, 213-223.
23. Kirkpatrick, S. (1973). Percolation and conduction. *Rev. Mod. Phys.*, **45**, 574-605.

24. Kolesnikov, Yu. M., Chelidze, T. L. (1985). The anisotropic correlation in percolation theory. *J. Phys. A.*,**18**, L273- L275.

25. le Ravalec, M., Gueguen, Y. and Chelidze, T. Elastic wave velocities in partially saturated rocks: saturation hysteresis. *J.Geophys.Res.* **101**, 837-844.

26. Maliepaard, M. C., Page, J. R., Harrison, J.P.,Stubbs, R.J. (1985). Ultrasonic study of the vibrational modes of sintered metal powders. *Phys. Rev.B.*, **32**, 6261-6271.

27. Mandelbrot, B. (1982). *The Fractal Geometry of Nature*. Freeman.

28. Maxwell J.C. 1891) *A Treatise on Electricity and Magnetism*. Dover Publ.

29. Palevski, A., Deutsher, G. (1984). Conductivity measurments of percolation fractal. *J.Physics A.*, **17**, L895-L898.

30. Pike, G. E., Seager, C. H. (1974). Percolation and conductivity. *Phys. Rev.*, **10**, 1421-1435.

31. Reynolds, I.A. and Hough,I.M. (1957) Formulae for dielectric constant of mixtures.*Proc.Phys.Soc.***70**, 769-775

32. Sahimi, M., Arbabi, S. (1993). Mechanics of disordered solids. II. *Phys. Rev. B.*,**47**, 703-712.

33. Savich, A. (1983). Geophysical monitoring of basements of hydroelectric stations. In: *Earthquake Prognosis* (Moscow-Dushanbe), No. 4, pp.273-288.(in Russian).

34. Shankland, T.G., Waff, H. S.(1974). Conductivity in fluid-bearing Rocks. *J. Geoph. Res.*, **79**, 4863-4868.

35. Shermergor, T. D. (1977)*Theory of elasticity of microheterogeneous media* . Moscow. Nauka Publ. House.(in Russian).

36. Stauffer, D. and A.Aharony.(1994).*Percolation Theory*, Taylor&Francis

37. Thompson, A. H., Katz, A. J., Rashke, R. A.(1987). *Phys. Rev. Lett.*,**58**, 29.

38. Walsh, J. B., Brace, W.F., England, A. W. (1965). Effect of porosity on compressibility of glass. *J.Am.Ceramic.Soc.* **48**, 605-608.

39. Mukerji,T., Berryman,J., Mavko,G.,Berge,P.(1995) Differential effective medium modelling of rock elastic moduli with critical porosity constraints. *Geoph.Res.Lett.* **22**, 555-558.

40. Chelidze T, Derevijanko,A., Kurilenko,O.(1977).*Electrical spectroscopy of heterogeneous systems.* Naukova Dumka, Kiev (in Russian)

41. Sahimi, M. 1994) *Applications of Percolation Theory*. Taylor&Francis.

42. Le Ravalec, M., Gueguen, Y. High and low frequency elastic moduli for a saturated porous cracked rock. *Geophysics*, in press

43. Kendall, K.(1984) Connection between structure and strength of porous solids , in D.Johnson and I.Sen (eds), AIP Conf. Proc. N107, *Physics and Chemistry of porous media*, AIP, N.Y. pp 79-87.

PROBABILISTIC MODELS OF STRUCTURES

DOMINIQUE JEULIN

Centre de Morphologie Mathématique
Ecole des Mines de Paris
35 rue Saint Honoré, F77300 Fontainebleau, FRANCE

Abstract. We consider the construction and properties of some basic random structure models (point processes, random sets and random function models) for the description and for the simulation of heterogeneous materials. The use of random structure models in the physics of the change of scale is introduced for the estimation of effective properties and for fracture statistics modeling.

1. Introduction

Materials usually present heterogeneous properties at various scales. This is certainly even more common in the case of natural materials like rocks, soils. The heterogeneity is a difficulty to design parts with given properties, and with a controlled dispersion. In order to manage it, it is necessary to develop a probabilistic approach.

In this paper, we review some morphological models of random media, that may be useful on two different levels: they provide a description of the heterogeneous structure, and sometimes enable us to predict some macroscopic properties of materials.

In what follows, the first level is illustrated by different classes of models of random media, and by their morphological characterization. The second level is illustrated by the introduction of models in the prediction of the macroscopic properties of random media from their microstructure, and by fracture statistics models.

G. N. Frantziskonis (ed.), PROBAMAT – 21st Century: Probabilities and Materials, 233–257.

2. Types of random structure models

The principal types of regionalized data that are used in materials, and the corresponding types of models can be commonly classified as follows:

- i) Dispersions of small particles in a matrix (defects like non metallic inclusions in steel,...), modelled by realizations of stochastic point processes.
- ii) Granular structures (polycrystals), assimilated to random tessellations of space.
- iii) Two phase (porous media) or multiphase structures (composite materials with several components) may be simulated by random sets (binary), or multicomponent random sets.
- iv) Rough surfaces (steel plate, fracture surface,...), chemical concentration mappings, and more generally grey level images can be represented by random functions.
- v) Multivariate data (multi species chemical mappings, components of a vector or of a tensor in every point x of space) are modelled by multivariate random function models.
- vii) Data on a network connecting vertices and their properties depending on connectivity, can be modelled by random graphs.

3. Basic properties of random media

When deciding to use a probabilistic approach, it is important to be able to characterize a random medium by appropriate tools. A generalization of the notion of random variable can be proposed for random sets, through the Choquet capacity [31], which can be physically interpreted according to its morphological meaning as illustrated below. We will also recall some morphological criteria for the quantitative analysis of microstructures.

3.1. THE CHOQUET CAPACITY

To characterize a set A in a space E (for instance the grains of a porous medium seen as a subset of the euclidean space R^3), we can select a set $K \subset E$, and examine the mutual locations of K and of A, by means of the answer to the two following exclusive questions:

$$\text{i) is } K \text{ disjoint from } A? \ (K \cap A = \emptyset)?$$
$$\text{ii) } K \text{ hits } A \ (K \cap A \neq \emptyset)? \tag{1}$$

For instance if $E = R^3$ and if K is the point x, the test 1 enables us to test whether $x \notin A$ (i) or $x \in A$ (ii), i.e. to which component of the two-phase medium (A, A^c) the point x belongs. The same process can be repeated for K made of the pair of points $\{x, x+h\}$, or more generally of a denumerable set of points $\{x_1, x_2, ..., x_n, ...\}$. By increasing the number of points, the number of positive answers to questions i) and ii) decreases, and a richer information is obtained about the structure of A. Let us assume now that $x \notin A$ and that we want to estimate the distance separating the point x from the set A, noting $\vee$ by and $\wedge$ the supremum and the infimum:

$$R(x, A) = \wedge_{y \in A}\{d(x, y)\} \tag{2}$$

Let $B_x(r)$ the closed ball with radius r and with center x: $y \in B_x(r) \Leftrightarrow d(x, y) \leq r$.

$$R(x, A) = \vee\{r; B_x(r) \subset A^c\} \tag{3}$$

Equation (3) involves sets $B_x(r)$ containing a non denumerable set of points.

Models derived from the theory of Random Closed Sets (RACS) by G. Matheron [29, 31] are characterized by their Choquet capacity $T(K)$ defined on the compact sets K. For a RACS noted A,

$$T(K) = P\{K \cap A \neq \emptyset\} = 1 - P\{K \subset A^c\} = 1 - Q(K) \tag{4}$$

In practice, various geometrical figures are used for the compact set K, in order to test various morphological properties of a heterogeneous structure.

For upper semi continuous random functions (usc RF) $Z(x)$, a generalization of the Choquet capacity, $T(g)$ can be defined [16, 24] on the lower semi continuous functions (lsc) g with a compact support K

$$T(g) = P\{x \in D_Z(g)\}; D_Z(g)^c = \{x, Z(y) < g(y - x), \forall y \in K\} \tag{5}$$

A particular and usual case is the spatial law defined on a finite number of points $x_1, x_2, ..., x_n$,

$$F(x_1, x_2, ..., x_n, z_1, z_2, ..., z_n) = P\{Z(x_1) < z_1, Z(x_2) < z_2, ..., Z(x_n) < z_n\} \tag{6}$$

Other particular cases give the changes of supports by $\vee$ (supremum) or by $\wedge$ (infimum), which generate new RF:

$$Z_\vee(K) = \vee_{x \in K}\{Z(x)\}$$

$$Z_\wedge(K) = \wedge_{x \in K}\{Z(x)\}$$

We must stress the following points:

- The Choquet capacity of a RACS is equivalent to the distribution function of a random variable.

- Two models (RACS, usc RF) with the same functional $T(K)$, $T(g)$, cannot be distinguished (theoretically as well as experimentally).
- The functional $T(K)$, (or $T(g)$), connects theory and experiments; it is used to estimate the parameters of a model and to test its validity.

3.2. CALCULATION OF THE FUNCTIONAL $T(K)$

The functional $T(K)$ (or $T(g)$) has to be determined as a function of the assumptions of the model, of its parameters, and of the compact set K or the function g. For a given model, the functional T is obtained:

- by theoretical calculation
- by estimation on simulations of the random structure, or on samples of the real structures. In the last case, it is possible to estimate the parameters from the "experimental" T , and to test the validity of assumptions. The estimation of the functional $T(K)$ is easily performed from the implementation of the basic operations of mathematical morphology, namely erosions or dilations.

The functions $T(K)$ (K being variable) obtained for a given model are coherent (which is not the case of any prior analytical model, as often used in the literature).

After specification and validation of the model from available data, it is possible to use a predictive implementation of its properties (such as $T(K)$ for compacts K not used during the identification step). Practical examples are the following: 3D properties deduced from 2D observations (stereology); change of support by $\vee$ or $\wedge$ in the case of a change of scale in fracture statistics (section 6.2).

3.3. MORPHOLOGICAL CONTENT OF THE CHOQUET CAPACITY

In the euclidean space R^n, with translations by vector x, the Choquet capacity of A for K at the origin of coordinates O can be expressed by:

$$T(K) = P\{K \cap A \neq \emptyset\} = P\{O \in A \oplus \check{K}\} \qquad (7)$$

where $A \oplus \check{K}$ is the result of the dilation of A by K:

$$A \oplus \check{K} = \{x, K_x \cap A \neq \emptyset\} = \cup_{-y \in K} A'_y = \cup_{x \in A, y \in K} \{x - y\} \qquad (8)$$

For K translated at x :

$$T(K_x) = P\{K_x \cap A \neq \emptyset\} = P\{x \in A \oplus \check{K}\} \qquad (9)$$

The functional $Q(K)$ is connected to the erosion by:

$$Q(K_x) = P\{K_x \subset A^c\} = P\{x \in A^c \ominus \check{K}\} \qquad (10)$$

with

$$A \ominus \check{K} = \{x, K_x \subset A\} = \cap_{y \in K} A_{-y} \qquad (11)$$

The dilation and erosion are the basic operations of Mathematical Morphology, introduced by G. Matheron and J. Serra. An experimental estimation of $T(K)$ can be obtained by an image analysis performed on realizations of A, involving the dilation operation. In the general case, several realizations of A are required, and $T(K_x)$ is estimated from a frequency for every point x. For a **stationary** random set, $T(K_x) = T(K)$ ($T(K_x)$ is invariant by translations of K). If in addition the RACS A is **ergodic**, $T(K)$ can be estimated from a single realization of A with:

$$T(K)^* = P\{x \in A \oplus \check{K}\}^* = V_V(A \oplus \check{K})^* \qquad (12)$$

where V_V is the volume fraction (for a RACS in R^3). Eq. (12) provides a practical process to estimate $T(K)$ from image analysis on samples of a microstructure.

Every compact set K brings its own information on the set A, considered as a realization of a random set [29, 31, 38, 15]. For instance, we consider the following types of compact sets:

- $K = \{x\}$

$$T(x) = P\{x \in A\} \qquad (13)$$

For a stationary RACS in R^3, we get the volume fraction V_v from $T(x) = p = V_V(A)$
- $K = \{x, x + h\}$

$$T(x, x + h) = P\{x \in A \cup A_{-h}\}$$

$$Q(x, x + h) = P\{x \in A^c \cap A^c_{-h}\} \qquad (14)$$

$Q(x, x+h)$ is the covariance of A^c. It depends only on h for a stationary random set.

- $K = B(r)$ (closed ball with radius r)

$$T(B_x(r)) = P\{x \in A \oplus B(r)\} \qquad (15)$$

$T(B_x(r))$ enables us to estimate the distribution of the random variable $R(x, A)$ defined in Eqs (2, 3):

$$F_x(r) = P\{R(x, A) < r \mid x \in A^c\} = \frac{T(B_x(r)) - T(x)}{1 - T(x)} \qquad (16)$$

The distribution function $F_x(r)$ does not depend on x for a stationary RACS.

The covariance and the distribution $F(r)$ are useful for studying the spatial distribution of A. The covariance enables us to measure the notion of scale by its range, defined as the length a for which the correlation between $k(x)$ and $k(x + a)$ vanishes, $k(x)$ being the indicator function of the set A at point x ($k(x) = 1$ when $x \in A$ and $k(x) = 0$ when $x \in A^c$). The covariance is a convenient tool to study the same materials at different scales [38]. In addition, as it depends on the orientation in space of the vector with origin x and extremity $x + h$, it is sensitive to anisotropy. This is the most convenient way to characterize anisotropic textures.

3.4. MORPHOLOGICAL CRITERIA

The main morphological types of data to quantitatively describe random structures can be classified according to the main following morphological criteria:

- **Basic measures** (volume fraction, surface area, integral of mean curvature,...) are global quantitative measurements whith suitable stereological properties (they can be estimated for tridimensional objects from lower dimensional sections (1D or 2D such as images).
- **Sizing** of objects in a microstructure can be obtained as a size distribution (with some restrictions on the shape, such as a limitation to spheres, concerning estimation from 1D or 2D to 3D); sizing of media of any shape (including connected networks) can be performed by the opening (erosion followed by a dilation) or closing (dilation followed by an erosion) morphological operations by convex structuring elements [31, 38].
- **Distribution in space** is very important to account for the presence of **scales**:

  - Clustering of objects may be studied by the probability distribution of distances $F_x(r)$ (estimated from volume fraction measurements after dilations with increasing sizes, as in Eq. (16))
  - Scales and their superimposition can be quantified from second order statistics, based on the covariance, from which nested structures such as clusters, clusters of clusters, repulsion effects, as well as periodicity in images are easily detected; the integral range is a standard measure of the size of a representative volume element of the microstructure described by a stationary and ergodic random structure: in the space, a volume V is made of $n = \dfrac{A_3}{V}$

volume elements inside which the average values of a RF $Z(x)$ are uncorrelated random variables, and therefore the integral range A_3 is a good measure of the notion of scale.

- **Anisotropy** is accessed from directional measurements, such as the variation of the covariance or of the chord size distribution with their orientation.

— **Connectivity** has a major incidence on the physical properties of composites made of phases with a high contrast. The **connectivity numbers** N in R^2 and in R^3 are topological properties describing the overall connectivity of an object. To estimate N in R^3, serial sections in 3D, or a random set model must be implemented. **Propagation phenomena** (light in optics, sound in acoustics, fluid in a porous medium,...) with different propagation velocities in heterogeneous media, involve the existence of paths (and of percolation) across a specimen; for a valued graph, one can estimate (in 2D or in 3D) the distance to a source, usually called the **geodesic distance** (namely the length of shortest paths), and its probability distribution function, characterizing the tortuosity of a network. This was applied to fracture of polycrystalline graphite, diffusion in polymers, fracture of simulated random media at different scales (porous media, polycrystals) [17], fluid flow between rough surfaces [10].

4. Point processes

Point processes are probably the most simple kind of random structure that we could imagine [8]. They have their own interest, since they can reproduce the occurrence of very small defects isolated in a matrix. In addition, they are often the first step in the construction of a more complex random medium.

Point processes can be studied as particular RACS with the Choquet capacity $T(K)$. They can also be characterized by means of the probability generating function $G_K(s)$ of the random variable $N(K)$ representing the number of points of the process contained in K.

4.1. THE POISSON POINT PROCESS

The non homogeneous Poisson point process in R^n with a regionalized intensity $\theta(x)$ $(x \in R^n; \theta \geq 0)$ is a point process with the following properties: the numbers $N(K_i)$ are independent random variables for any family of disjoint compact sets K_i. Moreover, the number of points of the process contained in a compact set K is a Poisson random variable with parameter

$\theta(K)$:

$$\theta(K) = \int_K \theta(dx) \tag{17}$$

$$P_n(K) = P\{N(K) = n\} = \frac{\theta(K)^n}{n!} \exp(-\theta(K)) \tag{18}$$

The probability generating function $G_K(s)$ of the random variable $N(K)$ is

$$G_K(s) = \sum_{n=0}^{n=\infty} P_n(K)s^n = \exp(\theta(K)(s-1)) \tag{19}$$

For the volume element dx centered in x, we have:

$$P_1(dx) = \theta(dx) \text{ and } P_0(dx) = 1 - \theta(dx) \tag{20}$$

When the intensity θ remains constant in space, $\theta(K) = \theta\mu_n(K)$, where μ_n is the Lebesgue measure in R^n (volume in R^3), the process is stationary, and $P_n(K) = P_n(K_x)$ for any translation of K by a vector x. From a physical point of view, this kind of random process is the prototype model of structure without any order.

As a random set, the Choquet capacity of the Poisson point process is obtained from

$$T(K) = 1 - Q(K) = 1 - P_0(K) = 1 - \exp(-\theta(K)) \tag{21}$$

In the stationary case

$$T(K) = 1 - \exp(-\theta\mu_n(K)) \tag{22}$$

5. Random sets and Random functions models

Starting from a point process, it is possible to build more general models, that can be called grain models. A part of such models are presented in [22].

5.1. THE BOOLEAN MODEL

The Boolean model [29, 31, 38] is obtained by implantation of random primary grains A' (with possible overlaps) on Poisson points x_k with the intensity θ: $A = \cup_{x_k} A'_{x_k}$. For this model, we have in the stationary case, with $q = P\{x \in A^c\}$:

$$T(K) = 1 - Q(K) = 1 - \exp\left(-\theta\overline{\mu}_n(A' \oplus \check{K})\right) = 1 - q^{\frac{\overline{\mu}_n(A' \oplus \check{K})}{\overline{\mu}_n(A')}} \tag{23}$$

This expression results from the fact that the number of primary grains hit by K follows a Poisson distribution with average $\theta\overline{\mu}_n(A' \oplus \check{K})$. Particular cases of Eq. (23) give the covariance $Q(h) = P\{x \in A^c, x+h \in A^c\}$ and the three point probability $Q(h_1, h_2) = P\{x \in A^c, x + h_1 \in A^c, x + h_2 \in A^c\}$ entering into the calculation of third order bounds (subsection 6.1). We have

$$Q(h) = P\{x \in A^c, x + h \in A^c\} = q^2 \exp(\theta K(h)) \tag{24}$$

with the geometrical covariogram

$$K(h) = \overline{\mu}_n(A' \cap A'_{-h}) \tag{25}$$

$$\begin{aligned} Q(h_1, h_2) &= P\{x \in A^c, x + h_1 \in A^c, x + h_2 \in A^c\} \\ &= \exp\left(-\theta\overline{\mu}_n(A' \cup A'_{-h_1} \cup A'_{-h_2})\right) \end{aligned} \tag{26}$$

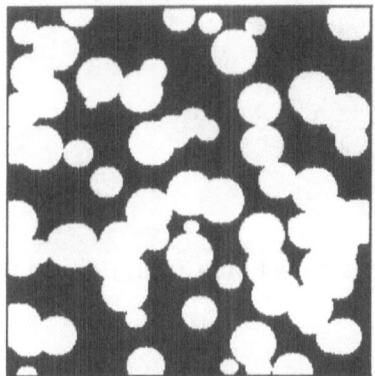

Figure 1. Boolean model of spheres; $V_v = 0.5$.

Figure 2. Boolean model with Poisson polyhedra.

Any shape (convex or non convex, and even non connected) can be used for the grain A'. Most often in the literature Boolean models of spheres are considered (figure 1); in [37], a Boolean model with Poisson polyhedra was found appropriate for WC-Co composites, as illustrated by a simulation in figure 2. Other examples are of interest:

- Using as a primary grain A' a point process with a limited extent, the resulting Boolean model is the Neyman-Scott point process, which reproduces clusters.
- Replacing the Poisson points by Poisson varieties [31], enables us to generate random sets models with fiber or strata textures [16, 24] (figure 3).
- Anisotropic versions of this model can be easily proposed, from anisotropic primary grains (for instance ellipsoids in R^3).

Figure 3. Boolean model built from Poisson planes (2D section).

5.2. THE COLOR DEAD LEAVES MODEL

The dead leaves model [15, 16, 24, 22] is obtained sequentially by implantation of random primary grains $A'(t)$ on a Poisson point process: in every point x is kept the last occurring color during the sequence. In this way, non symmetric random sets are obtained if two different families of primary grains are used for A and for A^c. When using the same family of primary grains for the two sets, a mosaic model built on a random tessellation is obtained (figure 4). The shape of the resulting cell is non convex (and even non connected!), due to the overlaps occurring during the construction of the model. Using random primary grains of various colors $A_i'(t)$, results into a multi component random sets, useful to simulate random media with crystals of different phases. For these models, and for their generalization to the RF case, up to five points statistics were derived in some particular cases [24]. The covariances $C_{ii}(h)$ $(i = 1, 2)$ of the two phase model, obtained in the case of a primary grain A_i' non depending on the time t, and implanted with a constant intensity θ_i, are given by

$$C_{ii}(h) = \frac{P_i r_i(h) + 2P_i^2(1 - r_i(h))}{2 - r(h)} \tag{27}$$

with $r_i(h) = K_i(h)/K_i(0)$, $\theta K(h) = \sum_{i=1}^{i=2} \theta_i K_i(h)$, and the volume fraction P_i given by

$$P_i = \frac{\theta_i K_i(0)}{\theta K(0)} \tag{28}$$

From the color dead leaves, it is possible to derive a model of non overlaping particles (for instance spheres), by considering the 'intact' grains (which were not covered by other grains during the sequence). The size distribution of the non overlaping particles, as well as their volume fraction can be estimated for this model. In the case of a single type of grain the resulting area fraction in R^2 and volume fraction in R^3 are given for the homogeneous model by 0.25 and 0.125 respectively.

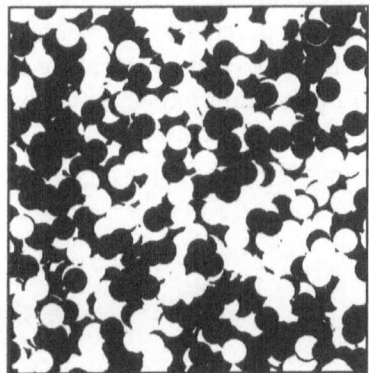 

Figure 4. Dead leaves models of discs ($P_i = 0.5$) and of Poisson polyhedra ($P_i = 0.5$).

5.3. THE MOSAIC MODEL

The mosaic model [30, 15, 16, 24] or "cell" model [30, 32] is built in two steps: starting with a random tessellation of space into cells, every cell A' is affected independently to the random set A with the probability p and to A^c with the probability q. The medium is symmetric in A and A^c, which may be exchanged (changing p into q).

A particular random mosaic can be obtained from a Poisson tessellation of space by Poisson random planes in R^3 (with the intensity λ, which is a scale parameter) [31, 38]. The cells of these tessellations are Poisson

244

polyhedra and Poisson polygons. These can be used as a primary grain in the Boolean model [37].

5.4. COMBINATION OF BASIC RANDOM SETS

Starting from the basic models, it is possible to generate more complex structures, presenting for instance a superposition of various scales. For instance, fluctuations of morphological properties (such as the local volume fraction of one phase, p) may exist in real materials.

5.4.1. *Union and intersection of random sets*
A first way to combine random sets is to consider the union or the intersection of two independent random sets A_1 and A_2. Since we have $(A_1 \cup A_2)^c = A_1^c \cap A_2^c$, we can limit our purpose to the intersection. For this model, $p = p_1 p_2$ and $P(K) = P\{K \subset (A_1 \cap A_2)\} = P\{K \subset A_1\}P\{K \subset A_2\}$.

In figure 5a from [23] is given a simulation of a composite material with a particle volume fraction equal to 0.49. The particle phase percolates in this case, while it does not percolate at the same volume fraction for a standard Boolean model (as in figure 5b), for which the percolation threshold is close to 0.6. It is therefore easy to generate in this way (and by iterating the same process) interesting random media with performant expected properties at low volume fractions, due to a low percolation threshold.

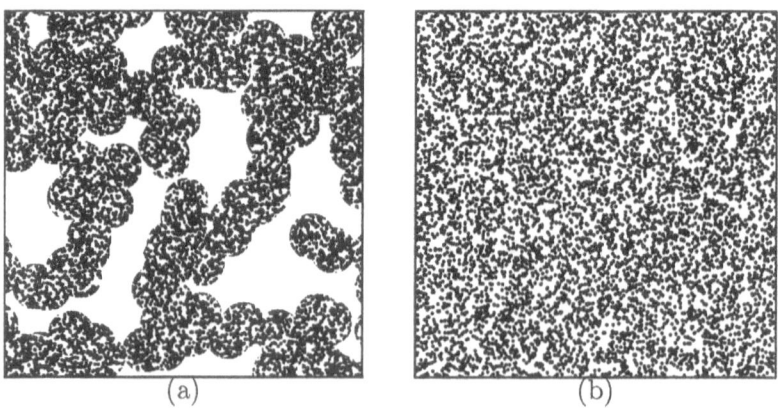

(a)   (b)

Figure 5. Simulation of a two scale Boolean model (a) and of a sphere Boolean model (b) with the same volume fraction (0.49).

5.4.2. *A hierarchical model*
A simple hierarchical model with two separate scales is built in two steps [24]:

— we start with a primary random tessellation of space into cells,

— every cell is intersected by a realization of a secondary random set (with random parameters). Realizations in separate cells are independent.

Any random tessellation can be used in this construction. Any type of random set (mosaic model, Boolean model, dead leaves model, etc...) can be used in the second step.

An example of simulation of a hierarchical model is presented in figure 6. This model starts from a Poisson tessellation of the space into cells. Each cell is intersected by a Boolean model of discs, with the density θ (with the probability 0.5) or 5θ (with the probability 0.5). The lines of the tessellation are absent from the final structure.This simulation of a two scale random set can represent a two phase material with an overall particle volume fraction equal to 0.3 presenting local fluctuations of the volume fraction. These fluctuations are generated by random changes in density (average number of Poisson points) between cells of the tesselation.

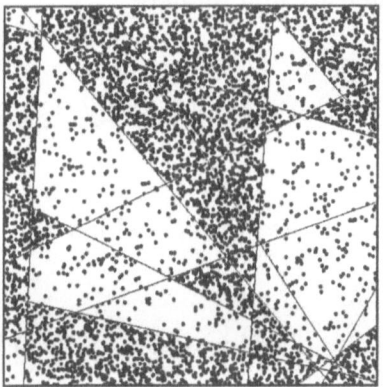

Figure 6. Simulation of a hierarchical model on a Poisson tesselation.

5.5. RANDOM FUNCTION MODELS

The previous random sets models have their continuous version, as random functions: the Boolean RF considers the random implantation of primary random functions on points of a Poisson point process. In that case the $\cup$ operation for overlapping grains is replaced by the supremum ($\vee$) or by the infimum ($\wedge$). This type of model is very useful for applications to fracture statistics (section 6.2, and [16, 24, 19]), and to simulate rough surfaces. Examples of simulations are given in figure 7.

246

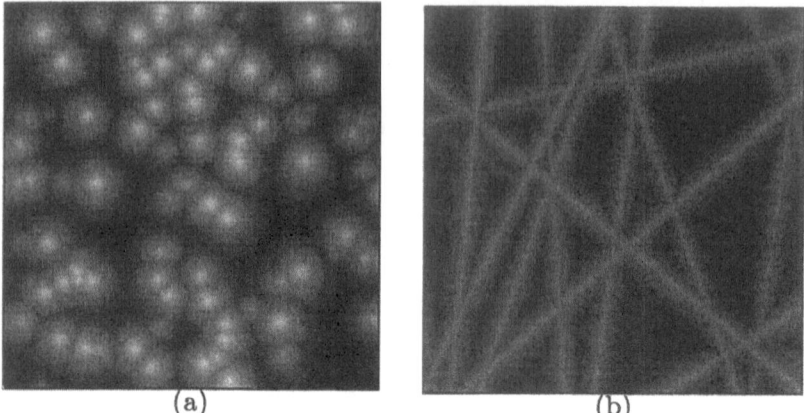

(a)   (b)

Figure 7. Boolean RF with cone primary grains (a) and Boolean Variety RF (b).

Starting from a sequence of primary RF Z'_t, implanted according to the supremum with the intensity $\theta(t)$ we have

$$1 - T(g) = Q(g) = \exp\left(-\int_R \overline{\mu}_n(D_{Z'_t}(g))\,\theta(dt)\right) \qquad (29)$$

As particular cases are obtained the spatial law and the change of support by the supremum:

$$
\begin{aligned}
1 - T(g) &= P\{Z(x_1) < z_1, ..., Z(x_n) < z_n\} \\
&= \exp\left(-\int_R \overline{\mu}_n(A_{Z'_t}(z_1)_{x_1} \cup ... \cup A_{Z'_t}(z_n)_{x_n})\,\theta(dt)\right)
\end{aligned} \qquad (30)
$$

If $Z_\vee(x) = \vee_{x \in K}\{Z(x)\}$ and if $A_{Z'_t}(z) = \{x,\ Z'_t(x) \geq z\}$ we have

$$P\{Z_\vee(K) < z\} = \exp\left(-\int_R \overline{\mu}_n(A_{Z'_t}(z) \oplus \check{K})\,\theta(dt)\right) \qquad (31)$$

Similarly, the dead leaves and the mosaic random sets models have their counterpart as random functions (figures 8, 9). The dead leaves RF is used to reproduce scanning electron micrographs, and provides algorithms to estimate the morphological properties of powders [18].

Figure 8. Dead Leaves RF with cones.   *Figure 9.* Poisson Mosaic

6. Change of scale in random media

Models of random media can be efficiently applied to the prediction of the macroscopic behavior of a physical system from its microscopic behavior. Firstly they are used to **estimate the effective properties** (namely the overall properties of an equivalent homogeneous medium) of random heterogeneous media from their microstructure. The approach, using variational principles, provides bounds of the effective properties for linear constitutive equations. It is illustrated by third order bounds (from third order central correlation functions) derived for some of the random sets models. Secondly they provide **fracture statistics models.**

6.1. PREDICTION OF THE EFFECTIVE PROPERTIES

The use of materials as components in manufactured parts is based on their physical properties, resulting from their microstructure. An usual approach to connect the properties to the microstructure is to look for correlations. A more efficient approach is to use predictive tools based on some morphological information and on random texture models.

An estimation of the overall properties of random composites can be obtained from perturbation expansions [29, 30], or from bounds based on a limited amount of statistical information [4, 26]. The commonly used bounds for isotropic media are Hashin and Shtrikman (H-S) bounds based on the volume fractions [13]. However many different morphologies can be encountered for a given volume fraction. Tighter bounds can be derived from additional statistical information, such as the infinite set of their correlation functions [4, 26, 14]. In practice it is difficult to obtain results beyond the third order correlation functions, which already provide interesting bounds that are useful for the comparison of microstructures according to their incidence on the behaviour of composite media.

For illustration, we consider random composites made of two phases A_1 (with fraction p) and A_2 (with fraction $q = 1 - p$) having a scalar dielectric permittivity ϵ_1 and ϵ_2 (with $\epsilon_1 > \epsilon_2$) (the same approach could be followed for other physical properties such as the thermal or electrical conductivity, the permeability of porous media,...). The composite is modelled by a stationary and isotropic random set A (with $A = A_2$ and $A^c = A_1$). The third order bounds (upper bound ϵ_+ and lower bound ϵ_-) are derived from a perturbation expansion of the electric field and a variational principle on the stored electrostatic energy [4]. They are expressed in R^d as a function of ϵ_1, ϵ_2 and of the three-point probability $Q(h_1, h_2)$. In R^d ($d = 2, 3$), using G. Milton [33, 34] and S. Torquato [40] notations,

$$\epsilon_-/\epsilon_1 =$$
$$\frac{1 + ((d-1)(1+q) + \zeta_1 - 1)\beta_{21} + (d-1)(((d-1)q + \zeta_1 - 1))\beta_{21}^2}{1 - (q + 1 - \zeta_1 - (d-1))\beta_{21} + ((q - (d-1)p)(1 - \zeta_1) - (d-1)q)\beta_{21}^2} \tag{32}$$

$$\epsilon_+/\epsilon_2 =$$
$$\frac{1 + ((d-1)(p + \zeta_1) - 1)\beta_{12} + (d-1)(((d-1)p - q)\zeta_1 - p)\beta_{12}^2}{1 - (1 + p - (d-1)\zeta_1)\beta_{12} + (p - (d-1)\zeta_1)\beta_{12}^2} \tag{33}$$

$$\beta_{ij} = \frac{\epsilon_i - \epsilon_j}{\epsilon_i + (d-1)\epsilon_j} \tag{34}$$

where the function $\zeta_1(p)$ introduced by G. Milton is obtained from the probability $P(h_1, h_2)$ that the three points $\{x\}$, $\{x + h_1\}$, $\{x + h_2\}$ belong to A_1.

$P(h_1, h_2) = P(|h_1|, |h_2|, \alpha)$, with $u = \cos\alpha$ (α: angle between the vectors h_1 and h_2)

$$\zeta_1(p) = \frac{9}{4pq} \int\limits_0^{+\infty} \frac{dx}{x} \int\limits_0^{+\infty} \frac{dy}{y} \int\limits_{-1}^{+1} (3u^2 - 1)P(x, y, \alpha)du \quad \text{in } R^3 \tag{35}$$

$$\zeta_1(p) = \frac{4}{\pi pq} \int\limits_0^{+\infty} \frac{dx}{x} \int\limits_0^{+\infty} \frac{dy}{y} \int\limits_0^{\pi} P(x, y, \alpha) \cos(2\alpha)\, d\alpha \quad \text{in } R^2 \tag{36}$$

This function was calculated for the different types of random sets introduced in section 5 [32, 39, 40, 27, 21, 28].

If the two phases A_1 and A_2 are symmetric, the case of $p = 0.5$ produces an autodual random set (the two phases having the same probabilistic properties, as for the mosaic model), for which the third order central correlation function equal to zero. Therefore in two and three dimensions, the third order bounds of a symmetric medium present a fixed point at $p = 0.5$. In addition, it is known [29, 30], that for an autodual random set in two dimensions the effective permittivity is equal to the geometrical average of the two permittivities.

Contrary to the mosaic model, the Boolean model is not symmetric (and not autodual for $p = 1/2$). Therefore, different sets of bounds are obtained when exchanging the properties of A and A^c (which is not the case for the second order H-S bounds). This is illustrated for the case of the thermal conductivity λ of ceramic materials modelled by two Boolean models of spheres with a constant radius (in figures 10a and 10b, the morphology of the low conducting bright phase is exchanged): here, the third order bounds increase when $\lambda(x) = \lambda_1 > \lambda_2$ for $x \in A^c$(figure 11). This is due to the fact that it is easier for the "matrix" phase A^c to percolate than for the overlapping inclusions building A. We see here that third order bounds are sensitive to the morphology in a more accurate way than the H-S bounds. In [21, 27, 28], various morphologies are studied in two and three dimensions. Two-scale hierarchical models accounting for local fluctuations of the volume fraction of a Boolean model can be optimized with respect to the third order bounds.

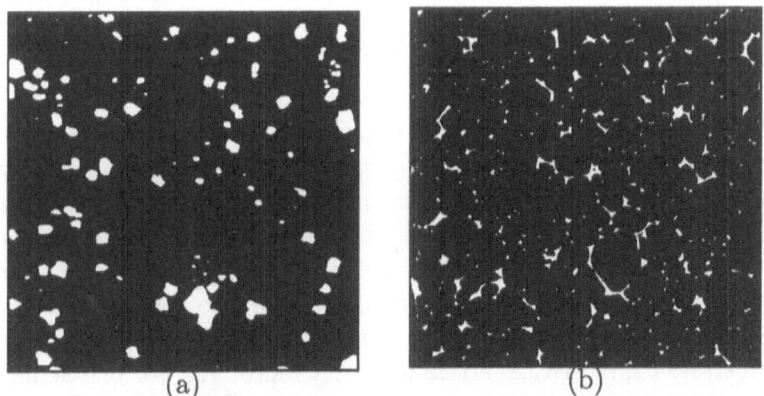

(a)   (b)

Figure 10. AlN textures (AlN in black; Y rich phase in white).

Similar derivations can be given for the elasticity case, following the developments given by Milton [34], as was made for some basic models [39, 40], and even for nonlinear physical properties [1, 36]. The more general approach considers multiphase and even continuous models (scalar or multivariate) [16, 24], for which the calculation of third order bounds can be made using the general derivation based on the third order correlation function [4, 32, 26, 14].

250

An extension of bounds to the complex dielectric permittivity was developped by D. Bergman [7] and by G. Milton [33]. It was applied to various types of random textures [23], showing that third order bounds could generate separate domains in the complex plane.

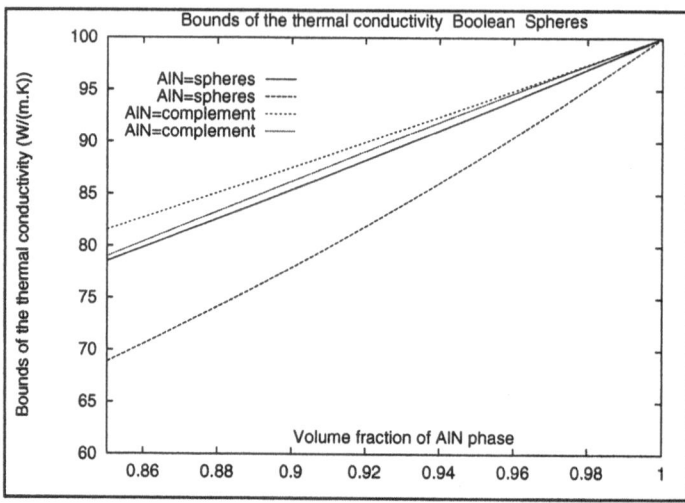

Figure 11. Bounds of the thermal conductivity of AlN materials ($\lambda_1 = 100$ W/(m.K) for AlN $\lambda_2 = 10$ W/(m.K) for the second phase)

6.2. FRACTURE STATISTICS MODELS

Fracture statistics models based on random structures [16, 24, 19] can be derived. The main purpose of this approach is to estimate the probability of fracture of a specimen under a given solicitation. Differently from the case of effective properties, there is in fracture a great sensitivity to local defects (there is a large effect of tails of the distributions on the macroscopic fracture behavior). One of the main points of fracture statistics approaches deals with size effects. For this, it is necessary to use models based on the reproduction, at a point scale, of the variations of a fracture criterion (critical stress σ_c, or critical stress intensity factor K_{Ic} for brittle materials; local fracture energy γ). These models are based on different macroscopic fracture criteria: the weakest link assumption is suitable for the sudden fracture of brittle materials; the models with a damage threshold allow several sites with a crack initiation before fracture; models based on the fracture energy (Griffith's criterion) account for the propagation and arrest of cracks in a random medium. For every family of models the probability of fracture is worked out as a function of the loading conditions and of the parameters of the selected random structure models. Some interesting aspects for applications, such as the prediction of expected scale effects, are derived. The proposed models can be tested at different scales (including the microscopic scale, by use of image analysis). The diversity of the obtained theoretical distributions for fracture statistics offers new possibilities for the microstructure based interpretation and modelling of mechanical data obtained on materials.

6.2.1. *Weakest link model*
Consider first the case of the weakest link model, where there is a sudden propagation of a crack after its initiation. In this case, the fracture statistics is governed by the most critical defects. The fracture probability is expressed by

$$P\{\text{non fracture}\} = P\{\inf(\sigma_c(x) - \sigma(x)) \geq 0\} \qquad (37)$$

where $\sigma_c(x)$ is the random critical stress, $\sigma(x)$ is the applied stress field, and inf is the minimal value over the sample.

Explicit expressions of the fracture probability are obtained for uncoupled critical and applied stress fields and for specific models (like Boolean random functions). They are more general than the standard Weibull model. The expected scale effect is a decrease of strength (to a constant for large samples) with the size of specimens.

For illustration, consider a medium where defects are introduced in a matrix with an infinite strength. For instance defects with a critical stress σ_c being a random variable Z' inside a closed random set A_0', implanted in space on Poisson points with the intensity $\theta(\sigma)$. We take for σ_c the minimum of the values of Z' on overlapping grains, building so a Boolean RF. The resulting field $\sigma_c(x)$ is a mosaic with domains where σ_c is constant. In brittle tensile fracture, we can restrict the purpose to the scalar case with $\sigma > 0$ (σ being the maximal main stress). For a uniform stress applied field, the probability distribution of the overall fracture stress σ_R of a specimen B is obtained from Eq. (38), generalizing Eq. (23) obtained for the Boolean random set:

$$P\{\sigma_R \geq \sigma\} = P\{\text{non fracture of } B\} = \exp(-\bar{\mu}(A_0' \oplus \check{B})\Phi(\sigma)) \qquad (38)$$

where $\bar{\mu}$ is given by averaging over the realizations of the random set A_0', and

$$\Phi(\sigma) = \int_0^\sigma \theta(t)dt \qquad (39)$$

A model of this kind was used for fracture of ceramics, the function $\Phi(\sigma)$ being estimated by image analysis from the distribution function of defects (inclusions and porosities ranging from $20\mu m$ to $70\mu m$) [5].

When applying a non uniform stress field $\sigma(x)$ and for point defects,

$$P\{\text{non fracture of } B\} = \exp(-\int_B \Phi(\sigma(x))dx) = \exp(-\Phi(\sigma_{eq})) \qquad (40)$$

with $\Phi(\sigma_{eq}) = \int_B \Phi(\sigma(x))dx$. This formulation with the equivalent stress σ_{eq} makes possible to put together results of different types of fracture tests, for the identification of a model, as made in [6] for the Weibull statistics, σ_{eq} being the so-called Weibull stress.

For defects with $\theta(\sigma) = m\theta(\sigma - \sigma_0)^{m-1}$ and $m > 1$ and $\sigma > \sigma_0$, we have $\Phi(\sigma) = \theta(\sigma - \sigma_0)^m$, and in the case of a homogeneous stress field, σ_R follows a **Weibull distribution** with $\sigma_u^m = 1/\theta$:

$$P\{\sigma_R \geq \sigma\} = \exp(-\bar\mu(A_0' \oplus \check B)\left(\frac{\sigma - \sigma_0}{\sigma_u}\right)^m) \qquad (41)$$

This distribution is well known in the practice of fracture statistics, in the case of point defects [12, 25]. When $\sigma_0 = 0$, the coefficient of variation of σ_R, $D[\sigma_R]/E[\sigma_R]$ does not depend on V, which can be easily checked from data. Scale effects can be seen from the variation of the median strength σ_M, as a function of the specimen volume V:

$$\sigma_M = \sigma_0 + KV^{-1/m} \qquad (42)$$

For a mixture of two populations of defects following Weibull distributions with parameters $(\theta_1, \sigma_{01}, m_1)$ et $(\theta_2, \sigma_{02}, m_2)$, resulting into a bimodal Weibull distribution when a uniform stress field is applied.:

$$\Phi(\sigma) = \theta_1(\sigma - \sigma_{01})^{m_1} + \theta_2(\sigma - \sigma_{02})^{m_2} \qquad (43)$$

Other distributions functions can be derived from other functions $\theta(\sigma)$, like the Pareto distribution, the sigmoidal distribution, and so on [16, 24, 19].

6.2.2. Critical defect fraction

We can relax the weakest link assumption in two different ways, letting the defects (where a potential crack can initiate) reach a critical volume fraction of the domain where $\sigma_c(x) < \sigma(x)$ or a critical density (number of defects per unit volume) [16, 19]. In both cases, no size effect is expected for the median strength, while the dispersion of the observed strength decreases with the size of the specimen.

6.2.3. *Griffith crack arrest criterion*

Fracture statistics models were derived for two dimensional media with a random fracture energy $\Gamma(x)$ [19]. The following situations were considered: straight crack propagation according to the mode I opening under various loading conditions, resulting in stable or unstable crack propagation; the case of a microcrack initiation on defects, followed by an advance or arrest of the crack, was also examined. For a given model of RF concerning the fracture energy $\Gamma(x)$, the following types of probability distributions can be expressed as a function of the model and of the loading conditions: fracture stress σ_R, toughness (standard G_c, and at crack arrest G_a), critical length of defects (unstable microcrack a_c or crack after arrest l_a). Since these probability distributions are related in a coherent way, it is possible to use them in practice to test models from data at different scales:

- mechanical data $(\sigma_R, G_c, \sigma_a, G_a)$,
- size distribution of cracks observed for given loading conditions (a_c, l_a).

Calculations made on a Poisson mosaic, and on a Boolean random mosaic, with the distribution function $F(\gamma) = P\{\Gamma < \gamma\}$ provide various scale effects: even an increase of the strength with the size of specimens can be expected for very slowly decreasing tails of the distribution $F(\gamma)$; this situation corresponds to rare reinforcements with a high fracture energy appearing along the crack path.

6.2.4. *Fracture statistics models and simulations*

When considering damaging materials at different scales, like fiber composites or metals under a ductile fracture, the fracture process must be studied by means of numerical simulations [2, 20, 3, 9, 11], since it is difficult to account for the complex stress field resulting from the interaction between various damage sites. In the case of composite materials, we used Monte Carlo simulations for the fracture criterion, combined with finite elements (FE) calculations. The following methodology was developed to study the fracture statistics unidirectional composites for fiber fracture [2, 3], for transverse fracture, and finally for the fracture of laminate composites. The first step is based on the experimental identification of the population of defects (point defects are considered here), by appropriate mechanical tests and by calculation of the local stress field seen by the microstructure; in this experimental part, a statistical volume element (SVE) is determined, as the elementary volume element broken during the progression of damage. Then, the statistical information and the fracture criterion are introduced in a FE calculation: to every SVE is attributed a random strength σ_R corresponding to the population of defects; during the calculation, it is broken if the average stress in the SVE, or more rigorously the equivalent stress

σ_{eq} defined by Eq. 40 is larger than σ_R. Monte Carlo simulations enable us to study the fracture behaviour on a first scale (RVE). If necessary, the material can be considered as a set of RVE to study its behaviour on the next scale. The main difficulty in running simulations is to determine the appropriate microstructural element converted into a SVE, and to know its fracture statistics. When operating on scales with increasing lengths, the statistical models proposed in the previous sections can be used to generate the necessary random variables. However, the correct corresponding type of assumption (weakest link, critical density,...) has to be introduced on a physical basis corresponding to the behavior of the material, as known from experiments.

Similar approaches are developped for ductile fracture statistics in steels:

- in [9], the fracture of a C-Mn steel containing clusters of MnS inclusions, which are germs for cavities during the fracture process, was simulated by axisymmetric FE calculations, using a continuum damage mechanics approach based on the Gurson potential controlling the growth of cavities. At the begining of the simulation, an initial random distribution of cavities on the SVE mesh (with size $250\mu m$, defined by the average distance between clusters) is generated by means of its local volume fraction; a lognormal distribution of volume fractions is used after fitting the data on MnS volume fraction obtained by quantitative image analysis on fields with size $250\mu m$.

- In [11], the statistics of the fracture toughness of duplex stainless steels (austeno-ferritic) is simulated in two dimensions to study the effect of the embrittlement of the ferrite by aging. In this case again, the Gurson plastic criterion is used to predict fracture. However in these simulations the random variable of interest is the local damage rate for the growth of cavities. Its probability distribution was measured by quantitative metallography on specimens, and then introduced in the generation of random damage rates on the SVE corresponding to the mean size of austenitic grains which generate clusters of microcracks ($1mm^2$).

In these two last types of simulations, the decrease of ductility and of its scatter (size effect) is well predicted by the model.

From these examples of simulations, general guidelines can be derived. Firstly, a damage parameter (density of point defects, microcrack network parameter, cavity volume fraction, cavity growth rate,...), connected to the microstructure, should be selected at a given scale, the SVE. Secondly, statistical properties of this parameter should be obtained by image analysis (or by micromechanical tests); until now this information was limited to the probability distribution function over domains with a given size, but higher order information can be estimated to recover probabilistic information on the damage parameter, considered as the realization of a random function (section 3.1). Finally, simulations of the damage parameter as initial conditions for the prediction of its evolution by means of finite elements have to be performed; for this step, morphological models of random media, as well as change of scale models to generate correct simulations on different scales, can be useful. These last points need more efforts in the future.

7. Conclusion

Random models of structures may be of interest to simulate the complex morphology of materials at a microscopic level. Our approach, based on measurements obtained by image analysis, makes possible to test and select appropriate models, and to estimate their parameters. Moreover, beyond this descriptive power, they enable us to predict overall properties of the material, to model their fracture statistics behaviour, and to provide ways of simulation. There are still connexions to make for a better integration of such models in reliability calculations at a much larger scale, as required in the mechanics of structures. These last points will help to make the random structure models a useful tool for the design of microstructures by the materials scientists.

References

1. K.E. Barett, D.R.S. Talbot (1996) Bounds for the effective properties of a nonlinear two-phase composite dielectric, *Proceedings of the CMDS8 Conference (Varna, 11-16 June 1995)*, ed K.Z. Markov, World Scientific Company, 92-99.
2. Baxevanakis C., Jeulin D., Renard J. (1995) Fracture Statistics of a Unidirectional Composite, *International Journal of Fracture*, **73**, 149-181.
3. Baxevanakis C., Jeulin D., Lebon B., Renard J. (1997) Fracture statistics modeling of laminate composites, accepted for publication in *International Journal of Solids and Structures*, N-02/97/MM, Paris School of Mines Publication.
4. M. J. Beran (1968) *Statistical Continuum Theories*, J. Wiley, New York.
5. C. Berdin, G. Cailletaud, D. Jeulin (1993) Micro-Macro Identification of Fracture Probabilistic Models, *Proc. of the International Seminar on Micromechanics of Materials, MECAMAT'93*, Fontainebleau, 6-8 July 1993, 499-510, Eyrolles, Paris.
6. F.M. Beremin (1983) A local criterion for cleavage fracture of a nuclear pressure vessel steel, *Metall. Trans. A.* 14A, 2277-2287.
7. D. Bergman (1978) The dielectric constant of a composite material: a problem in classical physics, *Phys. Rep. C*, **43**, 377-407.
8. D. R. Cox, V. Isham (1980) *Point Processes*, Chapman and Hall, New York.
9. K. Decamp, L. Bauvineau, J. Besson, A. Pineau (1997) Specimen size and geometry effects on ductile fracture of a C-Mn steel, submitted to *Int. J. of Fracture*.
10. Demarty C.H., Grillon F., Jeulin D. (1996) Study of the contact permeability between rough surfaces from confocal microscopy, *Microscopy, Microanalysis, Microstructure*, **7**, 505-511.
11. L. Devillers-Guerville, J. Besson, A. Pineau (1997) Notch fracture toughness of

a cast duplex stainless steel: modeling of experimental scatter and size effect, to appear in *Nuclear Engineering & Design.*

12. H. L. Harter (1978) A bibliography of extreme-value theory, *International Statistical Review*, **46**, 279-306.

13. Z. Hashin and S. Shtrikman (1962) A variational approach to the theory of the effective magnetic permeability of multiphase materials, *J. Appl. Phys.*, **33**, 3125-3131.

14. M. Hori (1973) Statistical theory of the effective electrical, thermal, and magnetic properties of random heterogeneous materials. II. Bounds for the effective permittivity of statistically anisotropic materials, *J. Math. Phys.* **14** , 1942-1948.

15. D. Jeulin (1987) Random structure analysis and modelling by Mathematical Morphology, *Proc. CMDS5*, ed. A. J. M. Spencer, Balkema, Rotterdam , 217-226.

16. D. Jeulin (1991) *Modèles morphologiques de structures aléatoires et de changement d'échelle*, Thèse de Doctorat d'Etat, University of Caen.

17. Jeulin D. (1993) Damage simulation in heterogeneous materials from geodesic propagations, *Engineering computations*, **10**, 81-91.

18. Jeulin D. (1993) Random models for the morphological analysis of powders", *Journal of Microscopy*, **172**, Part 1, 13-21.

19. D. Jeulin (1994) Random structure models for composite media and fracture statistics. In: *Advances in Mathematical Modelling of Composite Materials*, K.Z. Markov (ed), 239-289, World Scientific Company (Advances in Mathematics for Applied Sciences, vol 15).

20. D. Jeulin, C. Baxevanakis, J. Renard (1995) Statistical modelling of the fracture of laminate composites. In: *Applications of Statistics and Probability*, Lemaire M., Favre J.L., Mébarki A. (eds), 203-208, Balkema, Rotterdam.

21. D. Jeulin, A. Le Coënt (1996) Morphological modeling of random composites, *Proceedings of the CMDS8 Conference (Varna, 11-16 June 1995)*, ed K.Z. Markov, World Scientific Company, 199-206.

22. D. Jeulin (ed) (1997) Proceedings of the Symposium on the *Advances in the Theory and Applications of Random Sets* (Fontainebleau, 9-11 October 1996), World Scientific Publishing Company.

23. Jeulin D., Savary L. (1997) Effective Complex Permittivity of Random Composites, accepted for publication in *Journal de Physique I*, section Condensed Matter, September 1997.

24. D. Jeulin, *Morphological models of random structures*, CRC press, in preparation.

25. P. Kittl, G. Diaz (1986) Weibull's fracture statistics, or probabilistic strength of materials: state of the art, *Res. Mechanica*, **25**, 99-207.

26. E. Kröner (1971) *Statistical Continuum Mechanics*, Springer Verlag, Berlin.

27. A. Le Coënt (1995) *Observations et modélisation statistique de matériaux composites à agrégats*, PhD thesis, Paris School of Mines.

28. A. Le Coënt and D. Jeulin (1996) Bounds of effective physical properties for random polygons composites, *C.R. Acad. Sci. Paris*, **t. 323**, Série II b, 299-306.

29. G. Matheron (1967) *Eléments pour une théorie des milieux poreux*, Paris.

30. G. Matheron (1968) Composition des perméabilités en milieu poreux hétérogène: critique de la règle de pondération géométrique, *Rev. IFP*, **23**, 201-218.

31. G. Matheron (1975) *Random sets and integral geometry*, J. Wiley, New York.

32. M. Miller (1969) Bounds for the effective electrical, thermal and magnetic properties of heterogeneous materials, *J. Math. Phys.* **10**, 1988-2004.

33. G. Milton (1980) Bounds on the complex dielectric constant of a composite material, *Appl. Phys. Lett.*, **37**, 300-302.

34. G. Milton (1982) Bounds on the elastic and transport properties of two component composites, *J. Mech. Phys. Solids* **30**, 177-191.

35. Pélissonnier-Grosjean C., Jeulin D., Pottier L., Fournier D., Thorel A. (1997) Mesoscopic modeling of the intergranular structure of Y_2O_3 doped aluminium nitride and

application to the prediction of the effective thermal conductivity, to appear in *Proc. Conference of the European Ceramic Society, Versailles, June 22-26, 1997.*

36. P. Ponte Castaneda (1996) Variational methods for estimating the effective behavior of nonlinear composite materials, *Proceedings of the CMDS8 Conference (Varna, 11-16 June 1995)*, ed K.Z. Markov, World Scientific Company, 268-279.

37. J. L. Quenec'h, J.L. Chermant, M. Coster and D. Jeulin (1994) Liquid phase sintered materials modelling by random closed sets, *Mathematical morphology and its applications to image processing*, eds J. Serra and P. Soille, Kluwer Academic Pub., Dordrecht, 225-232.

38. J. Serra (1982) *Image analysis and mathematical morphology*, Academic Press, London.

39. S. Torquato and F. Lado (1986) Effective properties of two phase disordered composite media: II Evaluation of bounds on the conductivity and bulk modulus of dispersions of impenetrable spheres, *Phys. Rev. B* **33**, 6428-6434.

40. S. Torquato (1991) Random heterogeneous media: microstructure and improved bounds on effective properties, *Appl. Mech. Rev.* **44**, 37-76.

A NEW MODEL FOR INTER-FIBRE-FAILURE OF HIGH STRENGTH UNI-DIRECTIONALLY REINFORCED PLASTICS AND ITS RELIABILITY IMPLICATIONS

R. RACKWITZ
Technical University Munich
Arcisstr.21, D-80290 Munich, Germany

S. GOLLWITZER
RCP-GmbH, Barerstr. 48, D-80799 Munich, Germany

1. Introduction

Fiber-reinforced plastic laminates with unidirectionally reinforced layers can be used efficiently in many fields of application because of their extremely favorable strength properties combined with little weight. Yet, a number of failures have been observed when the laminate is also subject to inter-fiber failure in the various layers, a failure mode in which such materials are much weaker than if the stresses are primarily in the direction of the fibers. Stress states in which interfiber failures may occur originate in combined loading and when non-zero multidimensional stresses are likely to occur. There is a long history of formulation and improvement of both fiber and interfibre failure criteria. At present, the so-called Tsai-Wu-criterion (Tsai/Wu, 1971) for static, short time loading favors most attention among practitioners although its theoretical basis is somewhat inadequate for the material under question. There is also clear experimental evidence that the failure mechanism is different from the one assumed in Tsai-Wu's criterion. Accordingly, many attempts have been made to improve the failure criterion for interfibre failure. Hashin (1980) first put forward the idea that failure occurs in a plane where its strength is weakest as compared to the combined stresses acting in that plane. Later, Puck (1992) took up this idea and developed the concept of maximum in-plane stresses based on Mohr's ideas for the failure of brittle materials. Several forms of failure criteria have been proposed based on those ideas but experimental verification was missing. In a major joint research project an attempt has been made to verify experimentally Mohr's hypothesis and to fit several special alternative forms of the strength criterion (see Puck, 1996, and Cuntze et al., 1997). It was demonstrated that the new failure criterion in fact is able to represent the data quite well.

259

G. N. Frantziskonis (ed.), PROBAMAT – 21st Century: Probabilities and Materials, 259–274.
© 1998 *Kluwer Academic Publishers.*

In the following only one of those strength criteria, the simplest, is briefly reviewed and their support by experimental data is discussed. Then, detailed numerical studies are reported if the primary strength parameters in those criteria are uncertain as well as the stresses and some implication for practical design rules are discussed. The discussions are limited to stresses in the (σ_{22}, τ_{21})- plane which is the most important in practical applications.

2. New Strength Criterion due to Jeltsch-Fricker and Experimental Verification

It is first assumed that there is no interaction between fiber failure and inter-fiber failure. For interfibre failure it is assumed that

    i. a interfibre failure occurs always in a failure plane parallel to the fibers

    ii. for interfibre failure only the stresses in the failure plane are relevant

Let $\sigma = (\sigma_{11}, \sigma_{22}, \sigma_{33}, \tau_{23}, \tau_{31}, \tau_{21})^T$ be the stress state in a Cartesian coordinate system (x_1, x_2, x_3) where the x_1-axis is oriented parallel to the fibers (see figure 1). Then, the three stresses in a plane parallel to the fibers but rotated by the angle θ around the x_1-axis are:

$$\sigma_n(\theta) = \sigma_{22}\cos^2(\theta) + \sigma_{33}\sin^2(\theta) + 2\tau_{23}\sin(\theta)\cos(\theta)$$
$$\tau_{nt}(\theta) = -\sigma_{22}\sin(\theta)\cos(\theta) + \sigma_{33}\sin(\theta)\cos(\theta) + \tau_{23}(\cos^2(\theta) - \sin^2(\theta)) \qquad (1)$$
$$\tau_{nl}(\theta) = \tau_{31}\sin(\theta) + \tau_{21}\cos(\theta)$$

as can be verified by simple geometrical considerations.

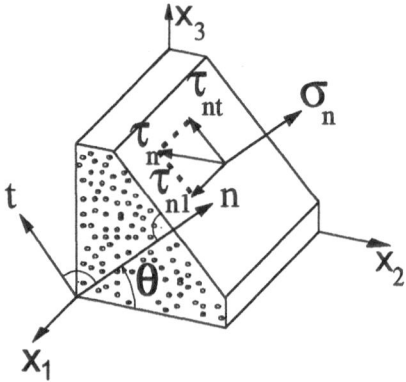

Figure 1: Stresses in failure plane

According to Hashin/Mohr the failure criterion then has the form

$$1 - F(\sigma_n(\theta_U), \tau_{nt}(\theta_U), \tau_{nl}(\theta_U)) \le 0 \qquad (2)$$

where θ_U is the angle which maximizes the term $F(\sigma_n(\theta), \tau_{nt}(\theta), \tau_{nl}(\theta)) \leq 1$. With $p := p^t$, $p^c = p R_\perp{}^c/R_\perp{}^t$ a simple model is (Jeltsch-Fricker, 1996; Puck, 1996)

$$1 - \sqrt{(1-p)^2 \left(\frac{\sigma_n}{R_\perp^t}\right)^2 + \left(\frac{\tau_{nt}}{R_{\perp\perp}^M}\right)^2 + \left(\frac{\tau_{nl}}{R_{\perp\parallel}}\right)^2} - p\frac{\sigma_n}{R_\perp^t} \leq 0 \quad \text{for } \sigma_n \geq 0 \quad (3a)$$

$$1 - \sqrt{p^2 \left(\frac{\sigma_n}{R_\perp^t}\right)^2 + \left(\frac{\tau_{nt}}{R_{\perp\perp}^M}\right)^2 + \left(\frac{\tau_{nl}}{R_{\perp\parallel}}\right)^2} - p\frac{\sigma_n}{R_\perp^t} \leq 0 \quad \text{for } \sigma_n < 0, \quad (3b)$$

where

$$0 < p < \ 1 - \left(\frac{R_\perp^t}{R_{\perp\perp}^M}\right)^2 < \ 1 \quad (3c)$$

This implies $R_\perp{}^t < R_{\perp\perp}{}^M$ and failure occurs for pure shear τ_{23} at an angle of $\theta_U = 45°$. Further there is

$$R_{\perp\perp}^M = \frac{R_\perp^c}{1 + \sqrt{1 + 2p\, R_\perp^c / R_\perp^t}} = R_\perp^c \cos^2\left(\theta_U^c\right) \quad (3d)$$

$$p = \frac{R_\perp^t}{2R_\perp^c}[\tan^4\left(\theta_U^c\right) - 1] \quad (3e)$$

The three basic strengths, the transverse tensile strength $R_\perp{}^t$, the transverse compression strength $R_\perp{}^c$ and the shear strength $R_{\perp\parallel}$, respectively, and p are free parameters; $R_{\perp\perp}{}^M$ is determined from eq. (3d). $R_\perp{}^t$, $R_\perp{}^c$, $R_{\perp\parallel}$ and p need to be determined experimentally.

Another suitable criterion derived on the basis of the same ideas has been investigated by Plica (1995)

$$1 - \sqrt{\left(\frac{\max\{\sigma_n, 0\}}{R_\perp^t}\right)^2 + \left(\frac{\tau_{nt}}{R_{\perp\perp}^M - \mu_{\perp\perp}\sigma_n}\right)^2 + \left(\frac{\tau_{nl}}{R_{\perp\parallel} - \mu_{\perp\parallel}\sigma_n}\right)^2} \leq 0$$

where $R_\perp{}^t < R_{\perp\perp}{}^M$, $0 \leq \mu_{\perp\parallel} \leq \dfrac{R_{\perp\parallel}}{R_\perp^t}$ and $0 \leq \mu_{\perp\perp} \leq \dfrac{R_{\perp\perp}^M}{R_\perp^t} - 1$ must be fulfilled.

Further there is

$$R_{\perp\perp}^{M} = \frac{R_{\perp}^{c}}{2}\left(\sqrt{1+(\mu_{\perp\perp})^{2}} - \mu_{\perp\perp}\right) = \frac{R_{\perp}^{c}}{2}\cot\left(\left|\theta_{U}^{\circ}\right|\right) \quad \text{and} \quad \mu_{\perp\perp} = -\cot\left(2\left|\theta_{U}^{\circ}\right|\right)$$

Here, the quantities in the denominators of the second and third term under the square root allow the interpretation of corresponding to Mohr-Coulomb's strength hypothesis. In this case the most dangerous failure angle must be determined numerically and the estimation of the parameters is somewhat more involved. Therefore, this model is no more considered in the following although it might describe the physical facts a little better having in mind that the "friction parameters" $\mu_{\perp\parallel}$ and $\mu_{\perp\perp}$ should be different for stresses in nt- and the nl-direction.

For the simple model eq. (3) and plane stresses (σ_{22}, τ_{21}), the failure angle θ_U and the failure curve can be determined analytically in terms of the acting stresses (σ_{22}, τ_{21})

$$1 - \sqrt{(1-p)^{2}\left(\frac{\sigma_{22}}{R_{\perp}^{t}}\right)^{2} + \left(\frac{\tau_{21}}{R_{\perp\parallel}}\right)^{2}} - p\frac{\sigma_{22}}{R_{\perp}^{t}} \leq 0 \quad \text{for } \sigma_{22} \geq 0 \quad (\theta_U = 0) \quad (4a)$$

$$1 - \sqrt{p^{2}\left(\frac{\sigma_{22}}{R_{\perp}^{t}}\right)^{2} + \left(\frac{\tau_{21}}{R_{\perp\parallel}}\right)^{2}} - p\frac{\sigma_{22}}{R_{\perp}^{t}} \leq 0 \quad \text{for } -R_{\perp\perp}^{M} \leq \sigma_{22} < 0, \quad (\theta_U = 0) \quad (4b)$$

$$1 + \frac{\sigma_{22}}{R_{\perp}^{c}} + \left(\frac{\tau_{21}}{R_{\perp\parallel}(1+\sqrt{1+2pR_{\perp}^{c}/R_{\perp}^{t}})}\right)^{2} \cdot \frac{R_{\perp}^{c}}{\sigma_{22}} \leq 0$$

$$\text{for } -R_{\perp}^{c} \leq \sigma_{22} < -R_{\perp\perp}^{M} \quad \text{and} \quad \cos^{2}(\theta_{U}) = \frac{R_{\perp\perp}^{M}}{|\sigma_{22}|} \quad (4c)$$

Other similar criteria can be set up (see Cuntze et al., 1997) but it was found that they differ only slightly from the simple model just described. Moreover, almost all of the alternatives require comparatively heavy numerics already in the two-dimensional (σ_{22}, τ_{21})- case because the maximization with respect to the fracture angle must be carried out numerically. Figure 2 shows the failure curves for different two-dimensional stress states. It is noted that the failure surface in any stress state can be inferred from the failure criteria in the failure plane. This criteria (3) and (4) have at most 4 parameters which all can be determined from simple tests in the (σ_{22}, τ_{21})-plane. The criterion is no more valid and not meant for large bi-axial compression stresses $(\sigma_{22}, \sigma_{33})$. For very large bi-axial compression stresses one has to expect that the transverse deformation of matrix is constrained by the fibers. The applied compression stresses will stretch them and ultimately there will be tensile failure of the fibers if matrix collapse is not occurring before. There are experimental indications that for up to $\sigma_{22} = \sigma_{33} = 600$ [MPa] no failure can be observed. For fiber reinforced plastics this stress state is of minor interest, however.

The parameters $R_\perp^t$, $R_\perp^c$ and $R_{\perp\|}$ can be determined directly from tensile, compression and shear (torsion) tests, respectively, at standard tubular test specimens in the (σ_{22}, τ_{21}) -plane. p can only be assessed indirectly. Usually it is more satisfying to use all tests in the method of least squares for

$$g(s, R) = 1 - F(s, \theta_U, R) \qquad (5)$$

where R is the unknown parameter vector with dimension q so that an estimate $\hat{R}$ can be found from

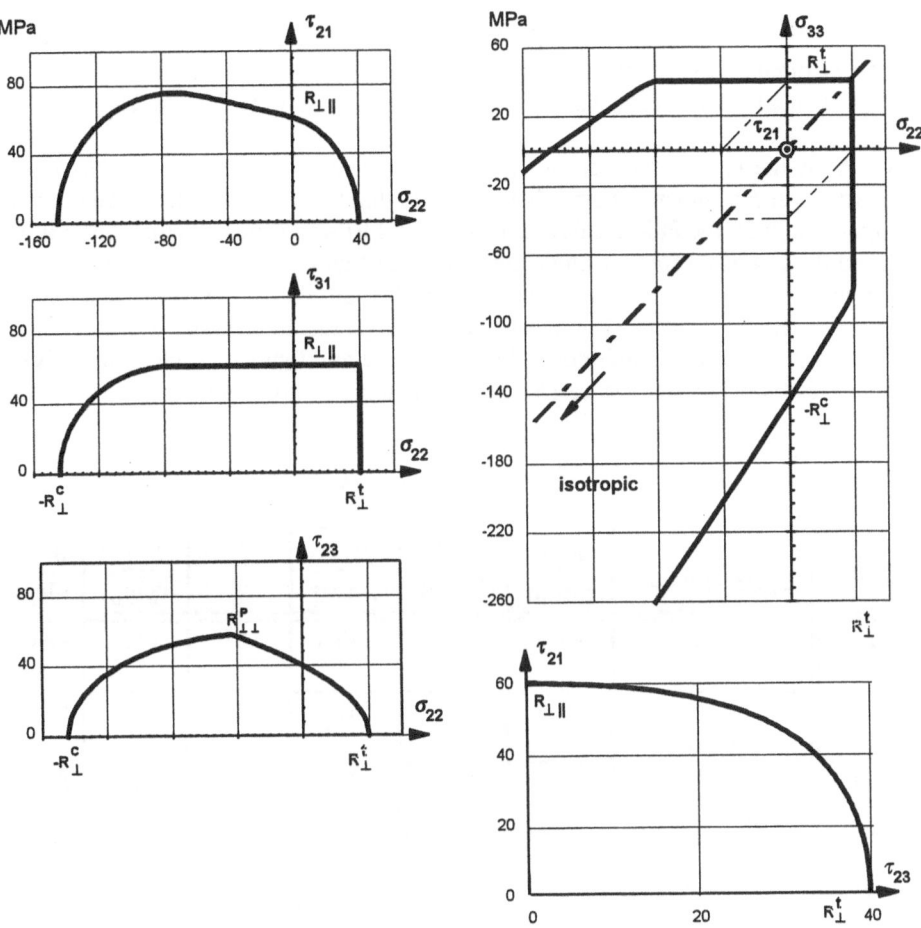

Figure 2: Cuts of failure curves through the $(\sigma_{22}, \sigma_{33}, \tau_{23}, \tau_{31}, \tau_{21})$-failure surface

$(R_\perp^t = 40$ MPa, $R_{\perp\|} = 60$ MPa, $R_\perp^c = 145$ MPa, $p = 0.152; \Rightarrow R_{\perp\|}^M = 59$ MPa$)$

$$\hat{R} = min\{\sum_{i,j} g(\sigma^{i,j}, R)^2\} \qquad (6)$$

Here i runs over the "rays" along which the tests have been performed and j over the tests along a specific ray. The quadratic residuum then is:

$$s^2_R = \frac{1}{N}\sum_{i,j} g(\sigma^{i,j}, R)^2 \tag{7}$$

where N may be replaced by N - q in order to account for the loss of "degrees of freedom" when q parameters are to be determined. The use of function (5) is appropriate as it takes account of slope of the failure function.

The following tests have been performed at three different but typical materials:

A1: T/C/T tubular test specimens, GFRP: EP-resin, LY556/HY917/DY070 F, N=140
A2: T/C/T tubular test specimens, GFRP: EP-resin, LY556/HT976, N=61
A3: T/C/T tubular test specimens, CFRP: T300: EP-resin, LY556/HT976, N=72

Series A1 is the largest data set and especially interesting as it has many data in the compression-shear domain where it is most likely that the above model can be verified. In figures 3a to 3b the failure curves for the two GFRP-test series together with all test data and the computed fracture angle are plotted. In figure 4 the CFRP-tests are shown. The fracture angle could be measured only rather unreliably and only at a limited number of test specimens. The measured values agree by far and large with the predicted values. Note the different scales in figures 3 and figure 4. Table 1 presents the summary statistics of the test series. The residuum in parenthesis is the root mean square of eq. (7) multiplied by N/(N-4).

Table 1: Summary statistics of tests

Test Series	$N s_R^2$ $s_R \sqrt{N/(N-4)}$	$R_\perp^t$ [MPa]	$R_{\perp\parallel}$ [MPa]	$R_\perp^c$ [MPa]	p	$R_{\perp\perp}^M$ [MPa]	$\theta^c U$ [deg]
A1	0.73 (0.073)	44.0	68.9	150.4	0.096	65.7	48.6 °
A2	0.38 (0.082)	53.9	69.5	162.7	0.248	63.0	51.5 °
A3	0.67 (0.099)	59.1	98.4	231.2	0.169	91.5	51.0 °

For all test series the root mean square residuum is of the same order of magnitude. It is seen that the proposed failure criterion can be fitted excellently to the data of all three materials. The same is true when the few other data from the literature are used. The other similar models described in Cuntze et al. (1997) lead to approximately the same degree of goodness-of-fit and the same root mean square residuum. Finally, strength parameters and fracture angle have been verified by tests for other stress combinations (see Cuntze et al. 1997).

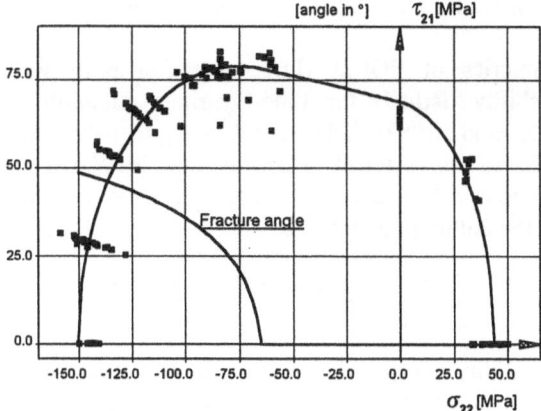

Figure 3a: GFRP Standard Tests A1 (N=140)

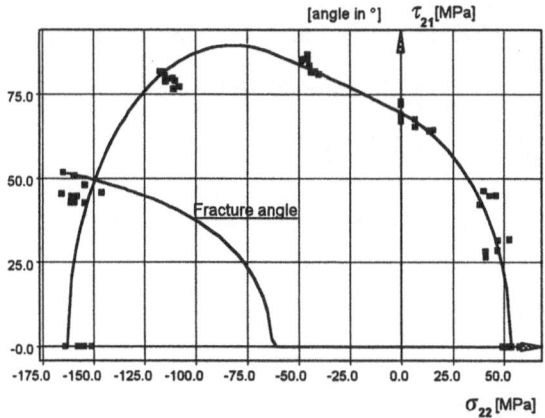

Figure 3b: GFRP- Standard Tests A2 (N=61)

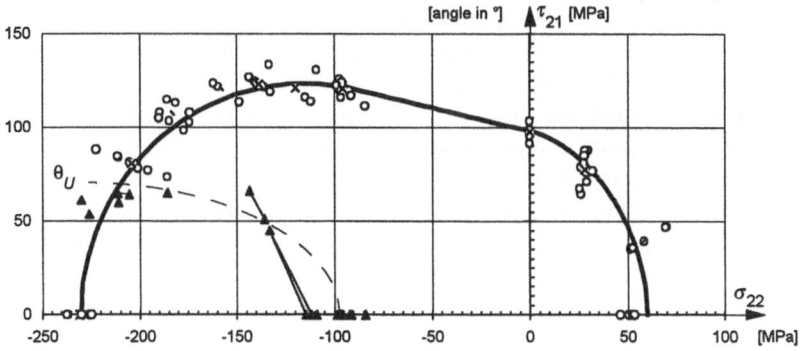

Figure 4: CFRP- Standard-Tests A3 ▲——▲ measured failure angles (N=72)

3. First- and Second Order Reliability Methods - Definitions

Theory and numerics of FORM (First-Order-Reliability-Method) and SORM (Second-Order-Reliability-Method) for time-invariant reliability analysis are well-known (Hohenbichler et al., 1987). Subsequently we give only some definitions which will be used later. Given the vector of uncertain variables X and a state function $g(x)$ such that $g(x) > 0$ denotes safe states, $g(x) = 0$ the limit state of failure surface and $g(x) < 0$ the failure state the failure probability is

$$P_f = \int_{g(x)\leq 0} f_X(x)dx = \int_{g(T(u))\leq 0} \varphi_U(u)du \tag{8}$$

where $f_X(x)$ the probability density of X and $\varphi_u(u)$ the standard normal density. For convenience, it is assumed that $f_X(x)$ is continuous and $g(x)$ is at least twice differentiable. If a probability distribution transformation $U = T^{-1}(X)$ exists (see, Rackwitz/Fiessler, 1978; Hohenbichler/Rackwitz, 1981; Der Kiureghian/Liu, 1986; Winterstein/Bjerager, 1987) the second integral can be approximated to first order by

$$P_f \approx \Phi(-\beta) \tag{9}$$

with $\beta = ||u^*||$ and u^* as the solution of

$$u^* = min \; ||u^*|| \quad for \; \{u: g(u) \leq 0\} \tag{10}$$

$\Phi(.)$ is the standard normal integral. β is the so-called geometrical reliability index. A second order approximation is (Breitung, 1984)

$$P_f \sim \Phi(-\beta) \prod_{i=1}^{n-1}(1-\beta\kappa_i)^{1/2} \tag{11}$$

which can be shown to be asymptotically exact ($\beta \to \infty$ or $P_f \to 0$). Here, the κ_i are the main curvatures of the failure surface in u^*. This theory together with a large number of distribution models and the corresponding probability distribution transformations has been implemented in several computer codes, for example in COMREL (COMREL, 1996) which will be used for the reliability analyses to follow. Alternatively or in addition, failure probability computations can be performed by Monte Carlo methods preferably with importance sampling.

Important information is also contained in several importance and sensitivity measures. For example, with $\beta_E = -\Phi^{-1}(P_f)$ the generalized reliability index

$$\alpha_{E,i} = \frac{\partial \beta_E}{\partial u_i} \frac{1}{\|\alpha_E\|} \approx \alpha_i \tag{12}$$

it can be shown that asymptotically there is $\beta_E \approx \beta$. Those α-values ($-1 \leq \alpha \leq 1$) are measures for the importance of a variable with respect to the reliability index. More precisely, they measure the sensitivity of the reliability index with respect to small changes of the mean of a variable in the independent standard space. Similarly, one can define dimensionless elasticities for deterministic parameters p_j

$$e_{p,j} = \frac{\partial \beta_E}{\partial p_j} \frac{p_j}{|\beta_E|} \tag{13}$$

Of great importance is the fact that the β-point (design point, most likely failure point) can be related to classical partial safety factors γ_i

$$\gamma_i = \frac{x_i^*}{x_{c,i}} \tag{14a}$$

where $x_i^* = F_i^{-1}[\Phi(u_i^*)] = F_i^{-1}[\Phi(-\alpha_i \beta)]$ is the design point and $x_{c,i}$ some characteristic value (usually a quantile) of the quantity X_i . $F_i^{-1}(.)$ is the inverse distribution function of the i-th variable. Definition (14a) is valid for "loading" variables, i.e. for variables for which the derivative of the limit state function is negative and the design value usually is larger than the characteristic value. In the contrary case we have

$$\gamma_i = \frac{x_{c,i}}{x_i^*} \tag{14b}$$

The relation $x_i^* = F_i^{-1}[\Phi(u_i^*)] = F_i^{-1}[\Phi(-\alpha_i \beta)]$ also allows to define a representative $a_{r,i}$ which is valid for general dependent variables and reduces to the sensitivity given in eq. (12) for independent variables.

$$\alpha_{r,i} = \frac{-\Phi^{-1}[F_i(x_i^*)]}{\beta} \tag{15}$$

If necessary, β can be replaced by β_E. to account for larger curvatures of the failure surface.

4. Reliability Analyses

In practice, the strength parameters as well as the loads must be considered as random variables. The following marginal distributions are proposed. The Weibull distribution for all three strength parameters is chosen because failure is brittle in all three regions of eq. (4) and, although not uniquely, a size effect can be observed (Puck/Schürmann, 1982). No statistical verification tests were possible due to limited data. Those data which were suitable for such analysis at least showed negative skewness typical for the Weibull distribution in contrast to the log-normal distribution which is frequently used for strength quantities. In fact, whereas one should expect the Weibull distribution for tensile strength from weakest link and fracture mechanics considerations and, to a certain extent, also for shear strength, a tendency to the normal distribution should be observable in the range where the failure plane rotates around the x_1-axis. This is because the Daniels model as a system of elements in parallel appears more appropriate in this range. As can be seen in table 2 the means are close to the values observed for the test series A1. The variability has been chosen larger than observed in test series A1 because larger variability must be expected in practical production than in highly controlled laboratory experiments. The parameter p varies very little. Therefore, a uniform distribution will be chosen. No residuum is taken into account because it is assumed that the observed residuum of the failure surface originated from measurement errors (see table 1) which may not entirely be true.

For most of the analyses to be discussed below the means of the stress components (σ_{22}, τ_{21}) are determined by an affine shrinking of the failure curve, calculated for the means of the strength parameters, by a factor of 1.75. Then, the mean of σ_{22} can be varied between $(-144/1,75) = -82$ and $(40/1,75) = 22$. The corresponding mean of τ_{21} has to be determined numerically.

Table 2: Stochastic model for standard study case

Variable	$R_\perp^t$	$R_{\perp||}$	$R_\perp^c$	p	σ_{22}	τ_{21}
Distribut.	Weibull	Weibull	Weibull	Uniform	Normal	Normal
Mean	40	61	144	0.150	see text	see text
St. Dev.	5,85	5,85	10,1	0.006	-	-
C.o.V.	14,6%	9,6%	7,0%	4,0%	15%	15%

The standard deviations $S[X_i] = V[X_i] \cdot E[X_i]$ for the stress components are determined to correspond to a constant coefficient of variation of 15%. Both stress components (σ_{22}, τ_{21}) are likely to originate from the same load or loads in most cases. Therefore, a not too small correlation should be assumed, for example $\rho[\sigma_{22}, \tau_{21}] = \pm 0.80$. The correlation coefficient is taken positive for positive stresses σ_{22} and negative for negative stresses σ_{22}. The actual correlation between stresses depends very much on the type of use and can easily be determined for each specific application. Also, the strength parameters must be stochastically dependent on each other because all three strength parameters are affected by the same material characteristics, but to a different degree. Unfortunately, such correlations are extremely difficult to verify experimentally. It would be necessary to determine all three strength parameters at the same test specimen. Correlations between resistance parameters simply express the fact that those strength parameters are not the real basic uncertainty variables. Only micro-mechanical considerations could help to assess these correlations on a sound basis which, however, are not yet available. For subsequent numerical analysis we have chosen the following standard values:

$$\rho[R_\perp^t, R_\perp^c] = +0,30; \quad \rho[R_\perp^c, R_{\perp||}] = +0,50; \quad \rho[R_\perp^t, R_{\perp||}] = +0,80$$

taking into account that all R-values depend on matrix strength and fiber contents. The dependency between $R_\perp^t$ and $R_\perp^c$ should be relatively small due to the significantly different fracture modes. The very similar fracture modes for $R_\perp^t$ and $R_{\perp\|}$ dictate a relatively large coefficient of correlation. The stochastic model in terms of marginal distributions and correlations as defined before requires a special transformation technique, the so-called Nataf-transformation (Der Kiureghian/Liu, 1986) which is only approximate but can be shown to be sufficient for the case under consideration.

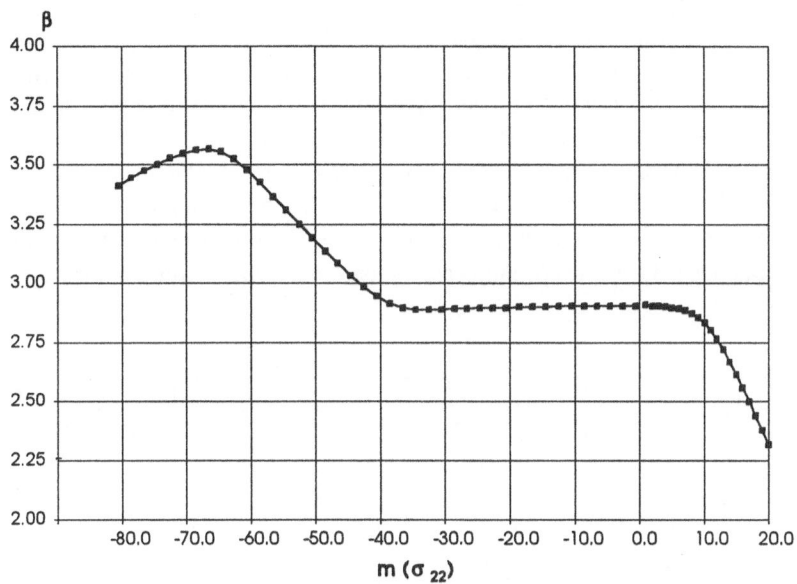

Figure 5: Reliability index β versus mean of load σ_{22}

In figure 5 the reliability index is shown for variation of the mean of σ_{22}. The "bulge" at $m(\sigma_{22}) \approx -65$ indicates the change from eq. (4b) to eq. (4c). For larger, negative $m(\sigma_{22})$ β rapidly approaches zero. The transition at $\tau_{21} = 0$ from the shear mode to the tensile mode is smooth. For larger positive σ_{22} the reliability index decreases rapidly to zero. The variations of the reliability index with running $m(\sigma_{22})$ and corresponding $m(\tau_{21})$ can be explained by the fact that the various strength parameters have different importance in the three domains defined in eq. (4). The reliability index curve is also affected by the different variabilities of the strength parameters $R_\perp^t$, $R_{\perp\|}$ and $R_\perp^c$ given in table 2.

Figure 6 shows the representative α-values according to eq. (15). Here again, one can identify the three domains of eq. (4). Note also, that for larger positive $m(\sigma_{22})$ the influence of $R_\perp^c$ does not vanish due to the correlations. Similar observation can be made for other variables. As expected the influence of the parameter p is very small.

In Table 3 the reliability indices, representative α- values, partial safety factors and elasticities are listed for some $m(\sigma_{22})$-$m(\tau_{21})$-combinations and the standard correlations given above. It is first noted that FORM and SORM results are very close to each other. The same also holds for other reliability levels, variabilities, correlations and distribution assumptions as can be demonstrated easily. The SORM result in turn is very accurate as can be verified by importance sampling according to Hohenbichler/Rackwitz (1988). For the stochastic model described before there is only one unique β-point for each $m(\sigma_{22})$-$m(\tau_{21})$-combination. This is different when no correlations are assumed and/or the stresses have much larger spread as will demonstrated below. Then, due to the convexity of the failure surface (in the original space, see figure 2) it may happen that there are two β-points which need to be located by special methods. These two β-points correspond to different failure modes.

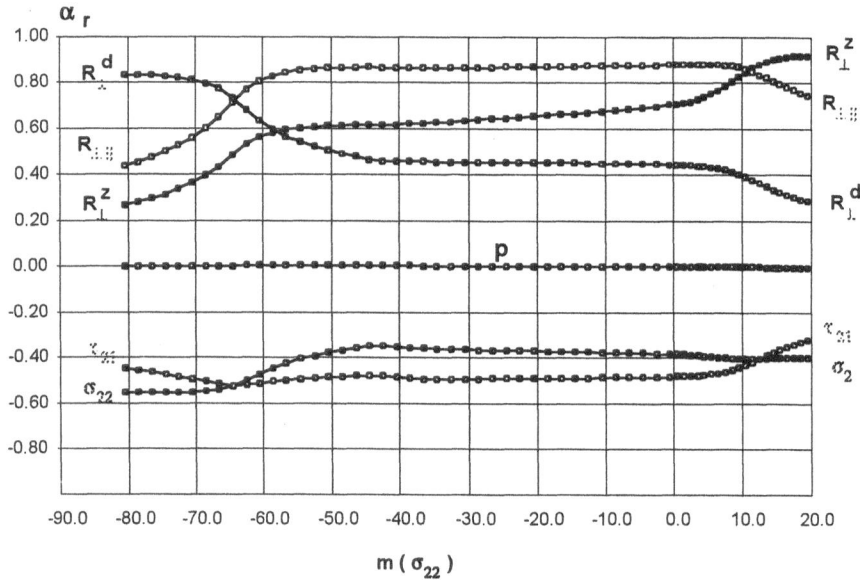

Figure 6: Representative sensitivity a_i versus $m(\sigma_{22})$

For reliability indices around 3 the specific distribution assumptions for the resistance parameters are already quite important. If a joint lognormal distribution would be assumed reliability indices of around 4.5 are obtained for the same sets of statistical moments. The influence of the particular stochastic model for strength parameters and stress components becomes even more significant for parameter sets resulting in larger reliability indices. This large sensitivity against distribution assumptions needs further study. For the present reliability level of about $\beta \approx 3$ (corresponds to $P_f \approx 10^{-3}$) the partial safety factors related to the mean are also

Table 3: Representative α-values, partial safety factors γ and elasticities e for selected m(σ₂₂)-m(τ₂₁)-combinations

$\sigma_{22}; \tau_{21}$ (1)	-130 ; 45	-70 ; 75	-20 ; 65	+5 ; 60	+20 ; 45	+35 ; 2
$\sigma_{22}/1.75$	-75	-40.0	-11	+3	+11	+20
β_{FORM}	3.49	2.90	2.93	2.86	2.88	2.29
β_{SORM}	3.44	2.91	2.93	2.84	2.82	2.28
$\alpha_r(R_\perp{}^t)$	+0.32	+0.62	+0.68	+0.72	+0.88	+0.92
$\alpha_r(R_\perp\|\|)$	+0.51	+0.86	+0.87	+0.88	+0.85	+0.74
$\alpha_r(R_\perp{}^c)$	+0.82	+0.46	+0.45	+0.44	+0.37	+0.28
$\alpha_r(\sigma_{22})$	-0.55	-0.36	-0.37	-0.39	-0.39	-0.40
$\alpha_r(\tau_{21})$	-0.48	-0.49	-0.48	-0.48	-0.40	-0.32
$\gamma(R_\perp{}^t)$ (2)	1.20	1.41	1.49	1.52	1.78	1.55
$\gamma(R_\perp\|\|)$	1.24	1.43	1.45	1.43	1.41	1.22
$\gamma(R_\perp{}^c)$	1.38	1.11	1.11	1.10	1.08	1.04
$\gamma(\sigma_{22})$	1.29	1.16	1.16	1.17	1.17	1.14
$\gamma(\tau_{21})$	1.25	1.21	1.21	1.21	1.17	1.11
$e(R_\perp{}^t)$	-0.04	-0.36	-0.13	+0.07	+0.90	+1.94
$e(R_\perp\|\|)$	+0.28	+2.20	+1.97	+1.82	+0.86	+0.01
$e(R_\perp{}^c)$	+1.62	+0	+0	+0	+0	+0
$e(p)$	+0.03	+0.26	+0.09	-0.03	-0.10	-0.15
$e(\sigma_{22})$	-1.18	+0.26	+0.09	-0.05	-0.55	-1.31
$e(\tau_{21})$	-0.23	-1.59	-1.41	-1.32	-0.63	-0.01

(1) Failure stress combinations on mean value failure curve

(2) Partial safety factors related to the mean

272

collected in table 3. For stresses with coefficient of variation $V = 15\%$ are between 1,1 and 1,3 - on the average 1,2. The partial safety factors for the resistance parameters vary quite considerable, that is from 1,1 to 1,5 and in extreme cases even up to 1,7. They are largest when they are the corresponding resistance quantity dominates the failure curve, i.e. for predominant compression there is $\gamma(R_\perp^C) \approx 1,4$ and else $\gamma(R_\perp^C) \approx 1,1$. For predominant shear it is $\gamma(R_\perp ||) \approx 1,45$ and else $\gamma(R_\perp ||) \approx 1,25$

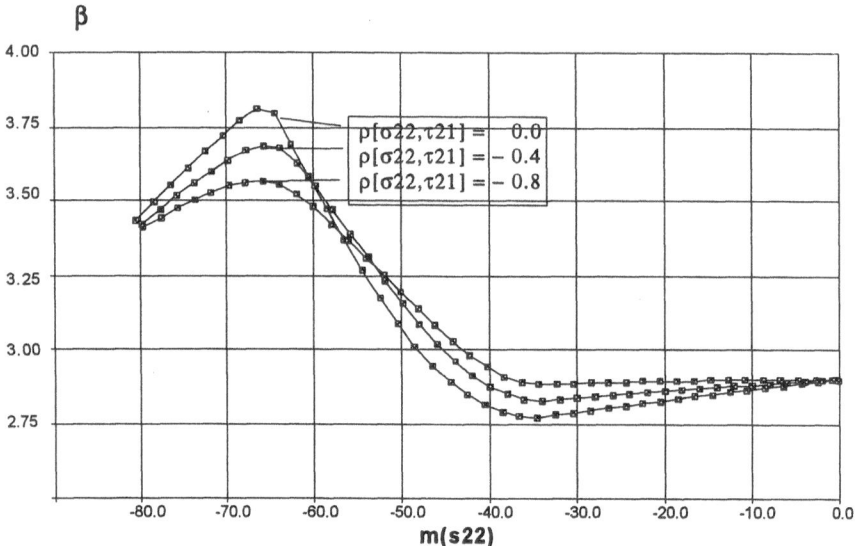

Figure 7. Reliability index versus $m(\sigma_{22})$ for different correlations of stress components

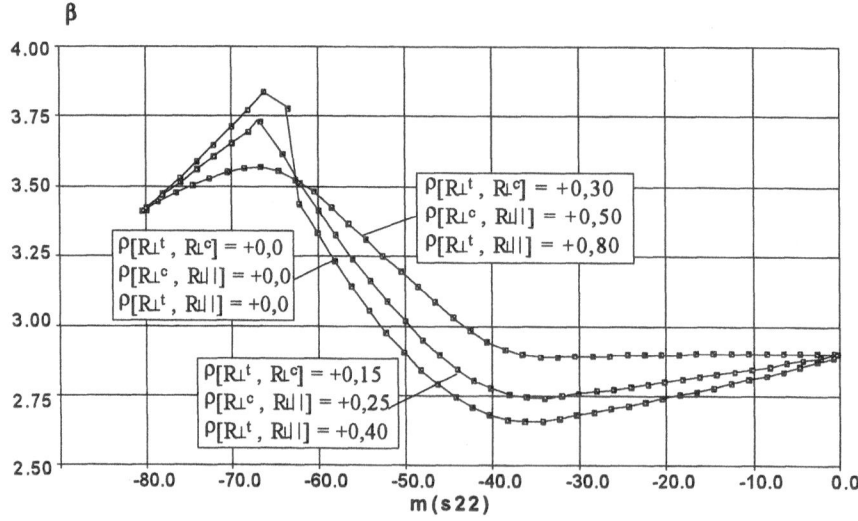

Figure 8. Reliability index versus $m(\sigma_{22})$ for different correlations of strength components

whereas for predominant tensile stress it is $\gamma(R_\perp^t) \approx 1,7$ and else $\gamma(R_\perp^t) \approx 1,4$. These numbers, of course, depend on the chosen reliability level. Partial safety factors related to characteristic values (so-called R-values) are somewhat smaller. Table 3 clearly shows that constant reliability cannot be achieved by partial safety factors related to the corresponding means. If they are related to suitable characteristic values some improvement is possible. While the existence of more or less large correlations between the stress components must be expected in many practical applications, the above assumptions on correlations between the resistance parameters may appear arbitrary. Therefore, it is interesting to study their influence on the reliability. This is shown in figure 7 where the correlation coefficient between applied stresses is varied, and in figure 8 where the correlations between the strength parameters are varied. For stresses $m(\sigma_{22}) \approx -65$ there is, in fact, a switch from one β-point to the other (figure 8). Both figures indicate that the effect of correlations is present but not excessively large.

5. Discussion, Summary and Conclusions

A new macro-mechanical, multi-axial criterion for interfibre failure of fiber reinforced plastics by Jeltsch-Fricker based on ideas of Hashin and Puck is presented and experimentally verified. Reliability analysis for stresses in the (σ_{22}, τ_{21}) -plane are performed. It is pointed out that both stress components as well as the resistance parameters should be dependent variables. Dependencies of the resistance parameters originate from common factors not captured by the experiments. Depending on the stress combination the reliability index varies slightly. It is largest for stresses where the failure surface starts to rotate about the x_1-axis. Sensitivities and partial safety factors are calculated for typical parameter variations. The reliability calculations can also be performed for general stress states at the expense of some more numerical effort and for time-variant stresses leading to essentially the same general insights.

Further research is necessary with respect to two aspects. At the one hand distributional assumptions, especially for the resistance parameters, can affect reliabilities quite significantly. On the other hand, the true uncertainty variables remain unknown in a purely macro-mechanical consideration. Therefore, the macro-mechanical findings have to be supported by micro-mechanical considerations.

6. Acknowledgments

Financial support of the project (see Cuntze et al., 1997) by the German Ministry of Education, Science, Research and Technology is gratefully acknowledged.

274

7. References

Breitung, K. (1984) Asymptotic Approximations for Multinormal Integrals, *Journ.Eng. Mech.*, ASCE, **110, 3,** 357-366

COMREL Users Manual, (1996) RCP GmbH Munich,

Cuntze, R., Deska. R., Szelinski, B., Jeltsch-Fricker, R., Meckbach, S., Huybrechts, D., Kopp, J., Kroll, L., Gollwitzer,S., Rackwitz, R. (1997) Neue Bruchkriterien und Festigkeitsnachweise für unidirektionalen Faserkunststoffverbund unter mehrachsiger Beanspruchung - Modellbildung und Experimente, *VDI-Fortschrittsberichte*, Reihe 5, No. 506, VDI-Verlag GmbH, Düsseldorf, 1997

Der Kiureghian, A., Liu, P.-L. (1986) Structural Reliability under Incomplete Probability Information. *Journal of Engineering Mechanics*, ASCE, **112,** No. 1, 85-104

Hohenbichler, M., Rackwitz, R. (1981) Non-Normal Dependent Vectors in Structural Safety, *Journ. Eng. Mech.*, ASCE, **107,** 6, 1227-1249

Hohenbichler, M., Gollwitzer, S., Kruse, W., Rackwitz, R. (1987) New Light on First- and Second-Order Reliability Methods, *Struct. Safety*, **4,** 267-284

Hohenbichler, M.; Rackwitz, R. (1988) Improvement of Second-order Reliability Estimates by Importance Sampling. *Journal of Eng. Mech.*, ASCE, **114,** 12, , 2195-2199

Hashin, Z. (1980) Failure Criteria for Unidirectional Fiber Composites, *Journ. Appl. Mech.*, **47,** 329-334

Jeltsch-Fricker, R. (1996) Bruchbedingungen vom Mohrschen Typ für transversal-isotrope Werkstoffe am Beispiel der Faser-Kunststoff-Verbunde, *ZAMM*, **76,** 505-520

Plica, S. (1995) Zum Einfluß der Bauteilgröße auf die Zuverlässigkeit von Bauteilen aus Faser-Kunststoff-Verbund, *Berichte aus dem Konstruktiven Ingenieurbau*, Technische Universität München, **7,**

Puck,A., Schürmann, H. (1982) Die Zug/Druck-Torsionsprüfung an rohrförmigen Probekörpern. *Kunststoffe*, **71,** 554-561

Puck, A. (1992) Praxisgerechte Bruchkriterien für hochbeanspruchte Faser-Kunststoff-Verbunde, *Kunststoffe*, **82,** 149-155

Puck, A., Festigkeitsanlayse von Fase-Matrix-Laminaten Modelle für die Praxis), Hanser Verlag, München/Wien, 1996

Rackwitz, R., Fiessler, B. (1978) Structural Reliability under Combined Random Load Sequences. *Comp. & Struct.*, **9,** 484-494

Tsai, S.W., Wu, E.M. (1971) A General Theory of Strength for Anisotropic Materials, *Journ. Composite Materials*, **5,** 58-80

Winterstein, S., R., Bjerager, P. (1987) The Use of Higher Moments in Reliability Estimation. *Proc. ICASP 5, Int. Conf. on Appl. of Statistics and Prob. in Soil and Struct.*, **2,** Vancouver, 1027-1036

PROBABILISTIC APPROACH TO DIMENSIONING OF STRUCTURES EXPOSED TO STOCHASTIC OPERATING LOADS

M. BILY
Institute of Materials and Machine Mechanics of the Slovak Academy of Sciences
Racianska 75, P.O. Box 95, 830 08 BRATISLAVA 38, Slovakia

1. Introduction

There is no doubt that the nature is of a stochastic multifactorial character manifesting itself by various excursions, short and long term trends and generally speaking by a scatter of any observed item. It is therefore natural that all our activities should take this fundamental fact into account and should apply methods and models exploiting the probabilistic language. Dimensioning as the process of adjusting a structure to its expected use is not an exception and consequently one should also adopt procedures reflecting the stochastic character of this problem.

In order to create a structure satisfying requirements of its users and environment a few groups of knowledge are required concerning (Fig. 1):
- use conditions,
- operating loads,
- properties of materials used,
- design philosophy,
- computational models, and/or
- experimental models.

In what follows these groups of knowledge are characterized and compared with the present practice, and proposals for further research are formulated. This field of interest is of extreme importance if it is realized that by spending in the EU countries about 2×10^8 ECUs annually in research and development of problems preventing fractures the potential savings could reach $(8 \pm 2) \times 10^{10}$ ECUs per year [1].

Although the views presented concern dimensioning against all damage mechanisms, nevertheless, the cyclic character of use conditions and use loads

G. N. Frantziskonis (ed.), PROBAMAT – 21st Century: Probabilities and Materials, 275–298.

276

predermine that most (may be 70 - 90 percent) failures are due to fatigue. For this reason dimensioning against fatigue will be the main focus of this paper.

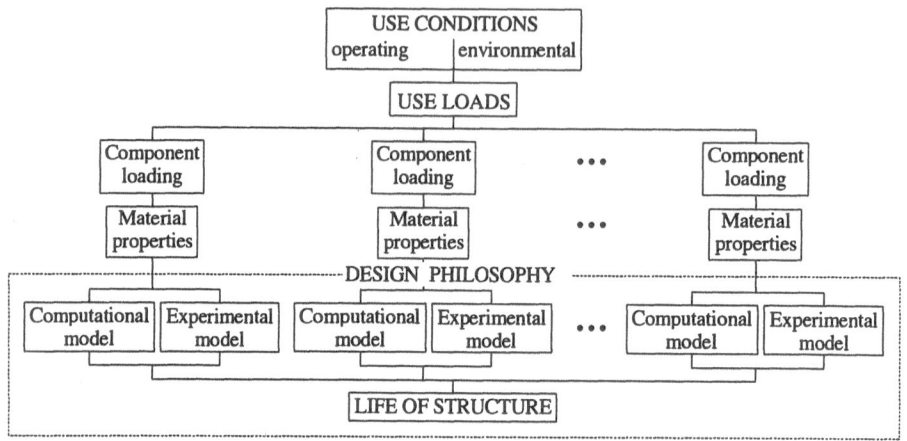

Figure 1. Chain of steps to be taken when dimensioning a structure

2. Use conditions

Knowledge of use conditions of a given structure is the first prerequisite for a successful design. They are the source of use loads and if they are not assessed with a reasonable confidence, all the subsequent computational/experimental steps may be subject to rather significant errors.

According to the IEC recommendations [2] use conditions consist of operating and environmental conditions, appearing simultaneously or sequentially, being mutually dependent, mutually exclusive, dependent or independent.

The *operating conditions* are related to functioning of the structure and comprise a combination of single *operating parameters* and their *severities* (e.g. functional modes, input signals, loads, manipulation by operator, various supplies).

The *environmental conditions* represent physical and chemical conditions external to the structure to which it is subjected at a certain time and also comprise a combination of single *environmental parameters* and their *severities* (e.g. climatic parameters such as outdoor and indoor conditons, earthquake shocks, free falls, radiation, biological agents, atmospheric corrosion).

From this brief description it should be obvious that the problem of use conditions and especially their quantification is far from simple. One should consider two successive steps:

(a) Ascertainment and analysis of true use conditions of an existing structure representing typical operating and/or environmental profiles. The purpose of this step is to break down the totality of use conditions into constituents which can be conveniently handled separately and to determine the duration of each activity by relation to the total time of the structure life.

(b) Formulation of an equivalent model of the use conditions of a structure considered as a basis for all subsequent experimental and theoretical works during development and testing of the structure or its parts. To make such a model useful it should naturally contain all important combinations of operating and environmental parameters as well as their durations and severities.

The use conditions may be described by a set of n generally stochastic characteristics $x_1, x_2, ..., x_n$ (parameters and their severities). Then the phenomenon that the structure is exposed to the jth use conditions is equivalent to the phenomenon that the characteristics $(x_1, x_2, ..., x_n)$ belong to the n-dimensional space D_j, or in the mathematical form

Pr{use conditions are in the state j} =

$$\text{Pr}\{ x_1, x_2, ..., x_n \in D_j \} = \iint_{(D_j)} ... \int f(x_1, x_2, ..., x_n) \, dx_1 dx_2 ... dx_n , \quad (1)$$

where Pr denotes the probability and $f(x_1, x_2, ..., x_n)$ is the multidimensional joint probability density function of the stochastic vector $(x_1, x_2, ..., x_n)$.

Nevertheless, this approach does not seem to be practically feasible for the time being because this multidimensional probability density function of n stochastic variables cannot be generally determined from the marginal probability density functions $f_1(x_1), f_2(x_2), ..., f_n(x_n)$, as the stochastic quantities $x_1, x_2, ..., x_n$ may be dependent; the alternative approach based on the conditional probability densities of each variable does not offer any simplification either. Thus one should adopt another way for determination of the true use conditions and can try to quantify the hypothetical life-time unit as the substitution for the true use conditions. This unit represented by the matrix

$$
k = \begin{vmatrix}
p_1 & p_2 & \cdots & p_j & \cdots & p_q \\
x_{11} & x_{12} & \cdots & x_{1j} & \cdots & x_{1q} \\
\cdot & \cdot & \cdots & \cdot & \cdots & \cdot \\
\cdot & \cdot & \cdots & \cdot & \cdots & \cdot \\
\cdot & \cdot & \cdots & \cdot & \cdots & \cdot \\
x_{n1} & x_{n2} & \cdots & x_{nj} & \cdots & x_{nq}
\end{vmatrix} \quad (2)
$$

is the choice of the use characteristics $x_{1j}, x_{2j}, ..., x_{nj}$ and their corresponding probabilities of occurrence p_j (j = 1,2,...,q) designed on the basis of a suitable description of the use conditions. The *j*th vector may represent, for example, the operation of a fully loaded lorry delivering goods, moving on the second gear on the country road with a very rough surface during a severe frost, etc.; the probability of occurrence of such a vector or the percentage of its appearance during the life-time unit (say one month) makes, for example, 3 percent. Other vectors designed in a similar way will form the full set of use conditions, i.e. 100 percent.

The criterion of the choice and so the formulation of matrix (2) follows from the assumption that the hypothetical life-time unit application yields either the theoretical life-time estimation coincident (in the probabilistic sense) with the life-time corresponding to the true use conditions or the degree of damage during the laboratory verification is equivalent with a sufficiently high probability to the degree of damage caused by the true use conditions after the same time of use (considering calender time compression, if applicable). The life-time (use time) of the structure is understood as

$$L = \lambda k = \lambda \sum_{j=1}^{q} k_j, \tag{3}$$

where k_j is a part of the life-time unit **k** when the structure operates in the use conditions described by the jth elementary combination of use characteristics x_i, and λ is a number of repetitions of the life-time unit during the total life-time L.

In order to make a practical use of the proposed approach it is indispensable to observe operation of a given structure, determine its activities and their duration, idetify operating and environmental parameters and their relationships, evaluate their severities and then assess the combinations of parameters and severities and estimate their simultaneous durations (probabilities of occurrence). Although it may seem to be a troublesome procedure it has been successfully applied, for example, to a skidder used for logging trees. The typical (hypothetical) life-time unit equal to one working day made 4.15 hours of the actual working day time. One must admit, however, that this kind of structure is exploited in a fairly monotonic way and the working day description of this skidder comprises a relatively few parameters and their severities (about 45).

3. Operating loads

The use conditions generate operating loads which can in fact represent forces, moments, pressures, deformations, velocities, accelerations and other quantities characterizing behaviour of structures (in practical cases of measuring, recording, analysis and simulation we usually deal with non-dimensional signals, anyway).

In order to assess the use loads (or better, the typical use loads characterizing the life-time unit) a few partial steps are to be undertaken.

3.1. ANALYSIS OF LOADS

Theoretically the monitored, measured and recorded loads may not need to be analysed at all and their realizations may be used directly in computational/ experimental procedures. For many reasons this is not reckoned, however, to be the optimal way and this is why some kind of analysis is required. Depending on the subsequent application of the results obtained, three different approaches seem to be feasible.

3.1.1. *Correlation theory statistical characteristics*

The extent of the correlation theory statistical analysis depends on its aim. In dimensioning tasks it is usual sufficient to compute a probability density function of process ordinates $p(x)$ and/or power spectral density $S(f)$, in dynamics of structures also the cross-correlation characteristics are of use. Suitable software makes this computation an easy job. Nevertheless, not always it is so straightforward way because most use loads (processes) possess evident non-stationary (time-dependent) properties. A strongly non-stationary process (e.g. with a stepwise varying mean value or variance or both of them as illustrated in Fig. 2) is usually detected during its visual inspection but if the statistical process characteristics are not so clearly time-dependent some tests for stationarity should be carried out [3].

If the process is proved to be non-stationary, the question arises how to carry its statistical analysis or what to do with the obvious non-stationarity in the frame of the correlation theory.

The answer is not satisfying, unfortunately, as the correlation theory of non-stationary processes can be hardly used in practical applications despite that some efforts have been published (see, e.g., the Priestly's approach based on the evolutionary spectrum [4]). Thus in the most correct mathematical and at the same time practical way the non-stationary process characteristics should be evaluated as ensemble averages in discrete time moments t_i from more process realizations (Fig. 3a). For $i = 1,2,...$ we can get a set of

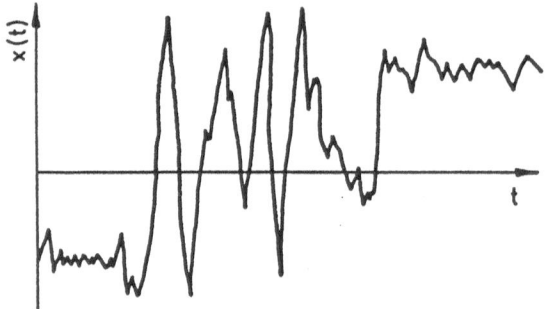

Figure 2. An example of a non-stationary process with time-varying mean value and variance, resp.

probability density functions f(x, t_i),
autocorrelation functions R(t_i , τ) and
power spectral densities S(f, t_i).

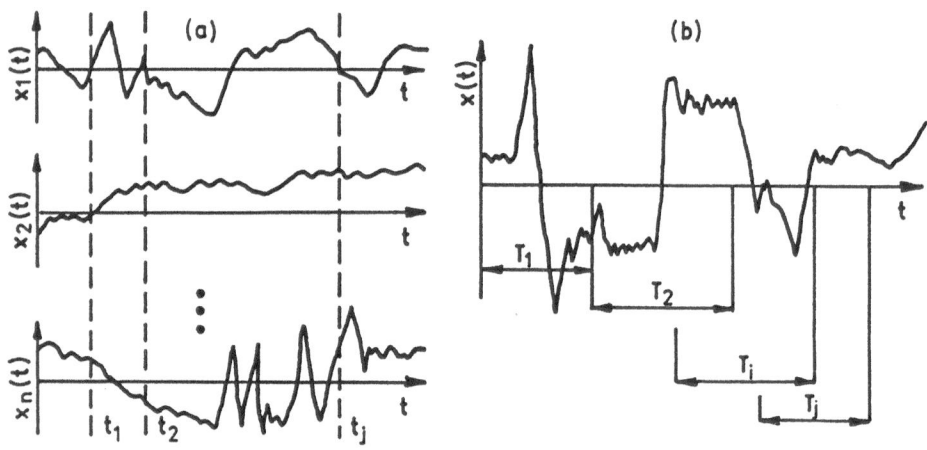

Figure 3. Scheme of calculation of ensemble averages (a) and parametric (b) statistical characteristics

Unfortunately, in practical applications a sufficient number of process realizations obtained in identical conditions is not usually available. Moreover, these statistical characteristics can hardly be directly used for dimensioning purposes as no adequate formulae which could exploit them are available as yet. This is why the so-called parametric evolutionary statistical characteristics are preferred defined for various segments of a long non-stationary process with the length T_i, $i = 1,2,...,j$ (Fig. 3b). Thus we get the

parametric evolutionary mean value

$$[m(t)]^T = \frac{1}{T} \int_{t-T}^{t} x(\xi)d\xi, \tag{4}$$

parametric evolutionary variance

$$[s^2(t)]^T = \frac{1}{T} \int_{t-T}^{t} \{x(\xi) - [m(t)]^T\}^2 d\xi, \tag{5}$$

parametric evolutionary probability density function

$$[f(x,t]^T, \tag{6}$$

parametric evolutionary autocorrelation function

$$[R(t,\tau)]^T = \frac{1}{T} \int_{t-T}^{t} x(\xi)x(\xi+\tau)d\xi \quad \text{and} \tag{7}$$

parametric evolutionary power spectral density

$$[S(t,f)]^T = \int_{-\infty}^{\infty} [R(t,\tau)]^T \exp(-i2\pi f\tau)\, d\tau \tag{8}$$

where f stands for frequency.

The idea of this step is analogous as for stationary ergodic processes except that for a non-stationary process more than one statistical characteristic are obtained.

3.1.2. *Markov chain analysis*
As known, the Markov chain analysis is based on counting transitions between the subsequent discrete process ordinates i and j and dividing them by the total number of process occurrences at the corresponding levels yielding the transition probability densities $p_{i,j}$ (as usually the process ordinates are classified into a finite number of K intervals). If only two adjacent process ordinates are taken into account then the Markov chain of the first order results characterized by the transition probability density matrix shown in Fig. 4.

Our experience has shown, however, that this kind of analysis is not decriptive enough as it cannot reflect a pronounced non-stationary behaviour (e.g., sudden changes of mean process level as shown in Fig. 2). This is why it is recommended to use the transition probability density matrices between three

$p_{i,j}$	1	2	. . .	K-1	K
1	$p_{1,1}$	$p_{1,2}$	. . .	$p_{1,K-1}$	$p_{1,K}$
2	$p_{2,1}$	$p_{2,2}$	. . .	$p_{2,K-1}$	$p_{2,K}$
. . .	. . .	. . .	. . .	. . .	. . .
K-1	$p_{K-1,1}$	$p_{K-1,2}$	. . .	$p_{K-1,K-1}$	$p_{K-1,K}$
K	$p_{K,1}$	$p_{K,2}$	. . .	$p_{K,K-1}$	$p_{K,K}$

Figure 4. The first order Markov probability density matrix of process ordinates classified into K intervals

adjacent ordinates (2nd order chain). This means that the probability density of the transition from the *i*th to the *j*th level is determined exactly in the same way as above, but instead of one matrix, K matrices of dimension K x K are formed, depending on the preceeding levels.

In an analogous way one could consider four or more ordinates should it be physically meaningful. In our verifications we analysed a number of non-stationary processes [13] and found that even in strong non-stationary cases the 2nd order chain provided very good results. A partial proof offers Fig. 5 comparing the macroblock shapes (obtained by the Rain Flow Method - see further) of the original stochastic strongly non-stationary process and of the same process analysed and simulated by means of the 1st order and 2nd order matrices, resp. It is obvious that the 2nd order chain yields results which are practically identical with those obtained from the original process; this cannot be said about the 1st order chain analysis.

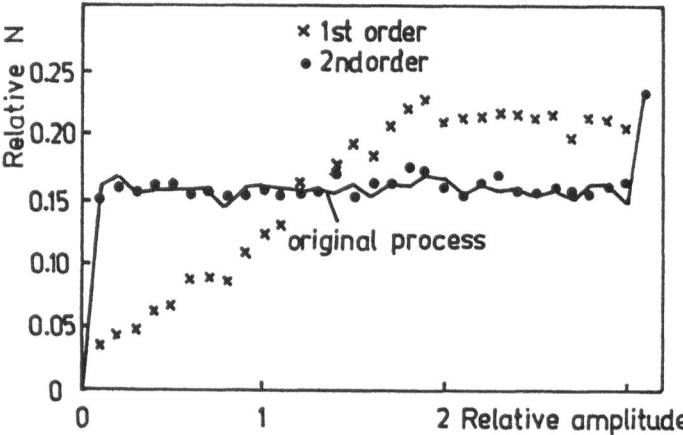

Figure 5. Comparison of macroblocks obtained by the RFM from the original non-stationary process and from the processes simulated by the 1st order and 2nd order Markov chains

3.1.3. *Rain Flow Method (RFM)*

This is a special single-purpose procedure used in dimensioning against fatigue when a stochastic process is analysed in such a way that its stochastic sequence of ordinates is according to specified rules transformed into groups of harmonic amplitudes (called a macroblock of amplitudes); the criterion for this transformation are the closed material hysteresis loops as illustrated in Fig. 6.

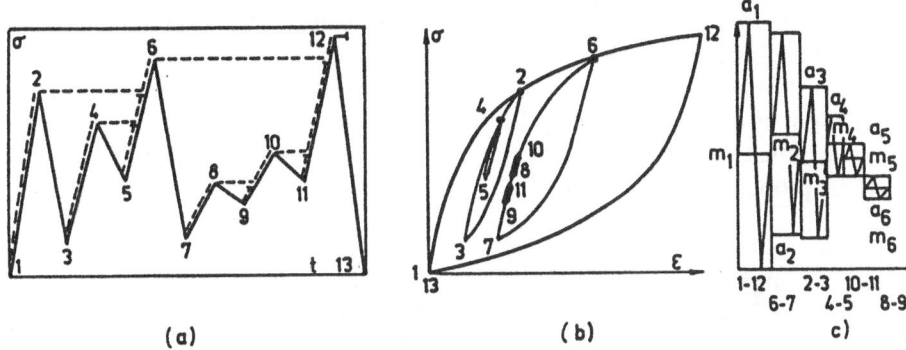

Figure 6. Illustration of the RFM yielding a macroblock of harmonic cycles

Although this kind of analysis is more deterministic than probabilistic, one should be aware that similarly as for the parametric characteristics obtained from segments of one long process (Fig. 3) here also more segments are to be analysed yielding a scatter of macroblocks.

3.2 SIMULATION OF LOADS

A situation may occur when the original recorded operating stochastic process is to be reproduced (not replayed) or, in other words, its certain characteristics are to be simulated.

Three possible ways of analysis provide the basis for three kinds of simulation algorithms.

3.2.1. *Simulation of correlation theory characteristics*

Basically one can simulate a probability density function of white noise, an arbitrary power spectral density of a Gaussian process, and an arbitrary probability density function together with an arbitrary power spectral density. The corresponding algorithms for the stationary processes are well elaborated (see, e.g., [6]) and so no principal problems are to be anticipated in practice.

Nevertheless, one word of caution is worth mentioning here.When simulating the pseudo-random nubers representing the process ordinates one should be aware that theoretically they range from minus to plus infinity or from zero to plus infinity. Even if in practice the simulation algorithms have certain limitations, the maximum generated number may exceed, say, the ultimate tensile strength of a loaded component (although with a very low probability). This naturally does not correspond to real operating conditions and so to avoid it, a certain maximum limit is to be prescribed. For the Gaussian process it usually makes three root mean square values s because the probability of its exceeding is negligible; but for other types of processes such a fixed limit is obviously incorrect as the probability of appearance of amplitudes above it is fairly high, as obvious from Fig. 7. This is why any practical choice of the cut-off level must take into account not only the load process itself but also the material properties.

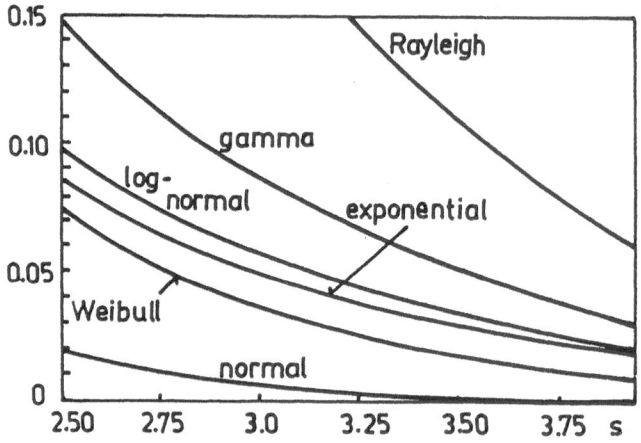

Figure 7. Tail parts of various probability density functions

The situation becomes more complicated when the simulated process is to possess non-stationary properties. The principal approaches to the simulation concide with those applied to stationary processes except that both the probability density function and power spectral density are time-dependent, i.e. $f(x,t)$ and $S(f,t)$.

As an example let us present the simulation algorithm for a non-stationary Gaussian stochastic process $x(t)$ (white noise) whose probability density function has the form

$$f(x,t) = \frac{1}{\sqrt{2\pi}\sigma(t)} \exp\left\{ -\frac{[x-\mu(t)]^2}{2\sigma^2(t)} \right\}, \tag{9}$$

where $\mu(t)$ and $\sigma(t)$ are deterministic functions representing the mean value and standard deviation, resp.

This non-stationary process can be simulated according to the formula [6]

$$x_i(t) = \sigma(t_i)\alpha_i + \mu(t_i) \tag{10}$$

where α_i are random numbers with a normalized Gaussian distribution and $\sigma(t_i)$, $\mu(t_i)$ are the discrete time interval values of $\sigma(t)$ and $\mu(t)$, resp.

In order to simulate the time-dependent power spectral density S(f,t) one can exploit a transformation through a linear dynamic system with the time-dependent trasfer function H(t,f) excited by a non-stationary stochastic process x(t) with a given time-dependent power spectral density $S_x(f,t)$. In the convolution form it can be expressed as

$$y(t) = \int_{-\infty}^{\infty} h(\xi) x(t-\xi) d\xi, \tag{11}$$

where $h(\xi)$ is the unit impulse response function. After rather tedious rearrangement [6] we finally get

$$S(f,t) = H(f,t)S_x(f,t). \tag{12}$$

Although these approaches are mathematically rather complicated they are, nevertheless, realizable. From the practical point of view there is, however, one principle obstacle here which for the time being can be hardly overcome. In theory we can assume and simulate practically any non-stationary process property (say, time-dependent mean value and/or variance and/or power spectral density). But there is no generally applicable tool how to obtain the corresponding time-dependent functions from realizations of non-stationary operating processes (except, perhaps, the subjective assessment of the mean value and variance trends in so called separable processes) and neither a tool for comparing the simulated and original characteristics. Should, however, such a tool be derived the correlation theory approach to the non-stationary process simulation becomes reasonable.

Nevertheless, if for some reasons we want to exploit in dimensioning tasks the present state of the correlation theory of non-stationary processes the only feasible way seems to use the parametric evolutionary statistical characteristics (4) - (8). This in fact leads to the philosophy that the long non-stationary process is understood to be part by part stationary, each stationary segment

286

having its own stationary statistical characteristics as schematically illustrated in Fig. 8.

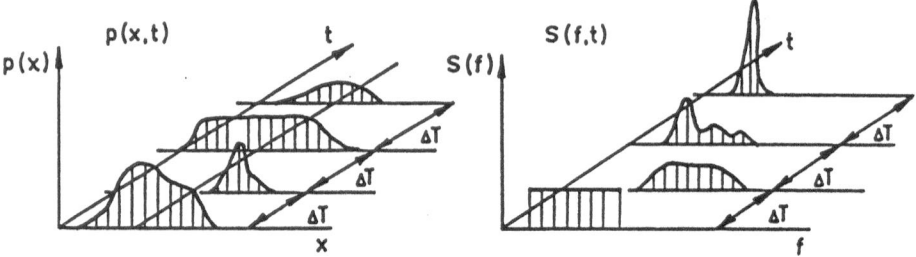

Figure 8. Part by part stationary probability density functions p(x) and power spectral densities S(f) of a non-stationary process

3.2.2. *Simulation of Markov chains*

As shown above the 2nd order Markov chain analysis applied to K discrete ordinate levels provides K x K matrices. These matrices created by the transition probability densities similar as shown in Fig. 4 then serve for the simulation purposes. According to our experience with various stationary and non-stationary processes this approach is descriptive enough to characterize the non-stationary process behaviour.

Nevertheless, one objection could be raised against this conclusion, viz. the comparison of the original and simulated process properties is based on the evaluated macroblock (the result of the RFM - Fig. 5), i.e. a single purpose characteristic used in designing against fatigue. But there is probably no other way how to prove this good engineering coincidence.

3.2.3. *Macroblock simulation*

Stochastic process simulation based on the evaluated macroblock is a method of simulation of stochastic processes used in fatigue applications for more than a half of century. The simulated process is composed of blocks of harmonic cycles or even of individually randomly distributed harmonic cycles with varying amplitudes and mean levels. Nevertheless, even this simplest way of simulation contains queries as, for example, the influence of amplitude sequence (load history) and small amplitudes on fatigue life, etc.

In conclusion to this part on simulation it is worth mentioning that some structural elements are affected by more than one damage mechanism (say, fatigue and creep, fatigue and corrosion) and because their action may be synergetic the corresponding operating loads are to be taken as one composed loading.

4. Material properties

The main material properties related to dimensioning of structures and components under stochastic loading are
- modulus of elasticity,
- yield and ultimate tensile strength,
- plastic properties,
- fatigue strength (including temperature, corrosion, irradiation),
- cyclic material properties,
- fatigue crack propagation characteristics,
- notch sensitivity, and
- some other special properties (e.g., balistic resistance, hardness, resistance to wear, etc.).

Which of these data are important depends on the way of dimensioning adopted, on the operating loading, on the environmental conditions, on the design concept, etc. But in all cases we have to deal with the selected material properties inherently exhibiting a certain scatter. Even if the manufacturers' aim is to reduce it as economically possible it is reckoned to be unavoidable and so its statistical nature should be included in the computational and/or experimental estimations.

Most good material manufacturers control quality of their products and are able to provide the minimum value of a certain characteristic or at least its median value (corresponding to 50 percent of appearance); rarely the whole field of values for various loads (as in the case of the fatigue curves described by the probability - stress - number of cycles to fracture lines). If one considers that good structures are related to the probability of failure of about 10^{-5} and less then the scatter of the material property selected may be more decisive than its mean value.

Because practically all dynamically loaded structures (especially machines) are prone to fatigue damage (it is estimated that 70 - 90 percent of all operating failures are due to this mechanism), designing against fatigue has very high priority and except some rather special applications it is the prevalent way of dimensioning. From this point of view all material characteristics related to cyclic loading are of importance and should be obtained with high accuracy despite the costs of their experimental determination. Specifically it concerns the S/N (Wöhler) curve and Manson-Coffin curve, resp. (Fig. 9) which should be always presented with their confidence limits (say, 2.5 and 97.5 percent) and/or with the confidence bands for the median (50 percent) curve. Because all these curves are quite well described mathematically in the form

$$\sigma_a = \sigma_f'(2N_f)^b \quad \text{and} \quad \varepsilon_a = \frac{\sigma_f'}{E}(2N_f)^b + \varepsilon_f'(2N_f)^c, \tag{13}$$

where σ_a and ε_a are the stress and strain harmonic amplitudes, resp., and N_f is the number of cycles to fracture, it is sufficient to present the corresponding coefficients only: σ'_f and ε'_f - fatigue strength and fatigue ductility coefficients, resp., b and c - fatigue strength and fatigue ductility exponent, resp., and E - modulus of elasticity.

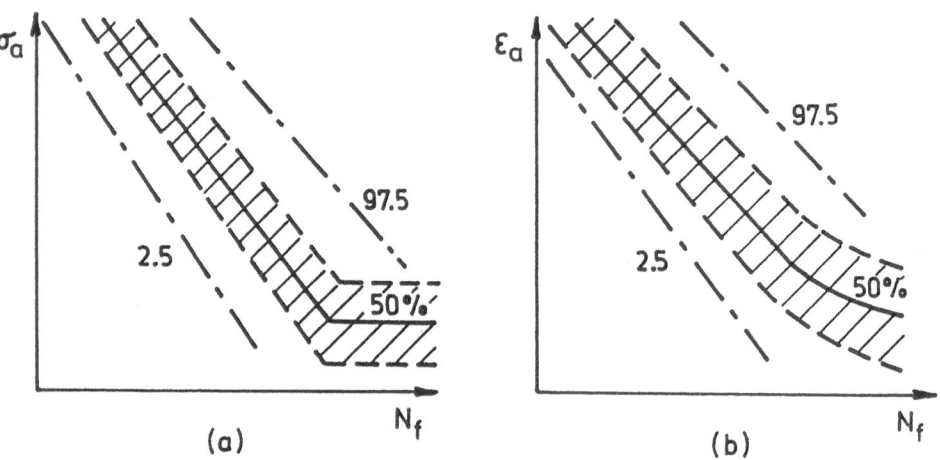

Figure 9. S/N (Wöhler) and Manson-Coffin curves with their scatter bands

Furthermore, it has been proved during the last decade that these characteristics may not be always sufficient for the fatigue life assessment and that also the cyclic stress-strain curve in the form

$$\varepsilon_a = \frac{\sigma_a}{E} + \left(\frac{\sigma_a}{k}\right)^{1/n} \tag{14}$$

is of interest (here k is the cyclic strength coefficient and n is the cyclic strain hardening exponent). Because the operating fatigue life estimation based on these characteristics is very sensitive to the slopes b, c and n of these curves, their scatter is also to be assessed with high accuracy, if possible with the corresponding scatter.

In connection with these statements one word of caution is appropriate here. Sometimes high operating stresses and requirements on low weights force designers to use steels with high yield and ultimate tensile strength. But as usually such materials rapidly decrease their properties under cyclic loading (so called cyclic softening materials) and the result may be astonishing: the resulting fatigue life is not much better than for ordinary low carbon steels (especially when welded).

5. Design philosophy

Structures are usually so complicated that the designer has to adopt a certain philosophy how to create a functioning reliable and safe item considering the specified time interval, limited costs, given staff and technical facilities. A good design in the early stage of the design process is mainly the result of the component design. Thus if the designed structure has no critical components which are potentional sources of premature damage limiting the operating life it will become reliable (as usually also safer) and will possess a higher quality at lower costs.

5.1. SYSTEM APPROACH

The designing tasks to avoid the critical locations is in the first step reduced to finding them and determining their importance for the successful operation of the structure. In order to succeed a few possible ways can be adopted. The simplest and most straightforward approach is to rely on the past experience with similar exploited structures. Considering that new innovated structures may contain even 70 - 90 percent of the previously used components, the importance of the carry over is obvious, as the slogan states: New designs, new problems with reliability!

However, a more objective way is to use tools offerred by the Failure Modes and Effect Analysis (FMEA), Failure Modes, Effects and Criticality Analysis (FMECA) and Failure Tree Analysis (FTA). All these methods perform a qualitative system reliability (or safety) analysis from a lower to a higher structural level in all five life-time stages but most importantly in the design and development stage.

Details of these procedures can be found in numerous standards and papers (see, e.g., [7,8]).

The basic questions which are to be answered by the FMECA are as follows:
- In what way can each item of the structure conceivable fail?
- How can these failures be detected?
- What mechanisms might produce these modes of failures?
- Are these failures on the safe or unsafe side?
- What are the consequences of such failures?
- What inherent provisions are made in the design to compensate for these failures?

If also the criticality analysis of failures is to be undertaken then each failure effect is classified according to its consequences for the structure and its criticality number FCR is computed [9]:

The (FTA) was originally used as a technique for the qualitative analysis only but later on its quantitative analysis has also been developed. Two different approaches are possible as schematically shown in Fig. 10:

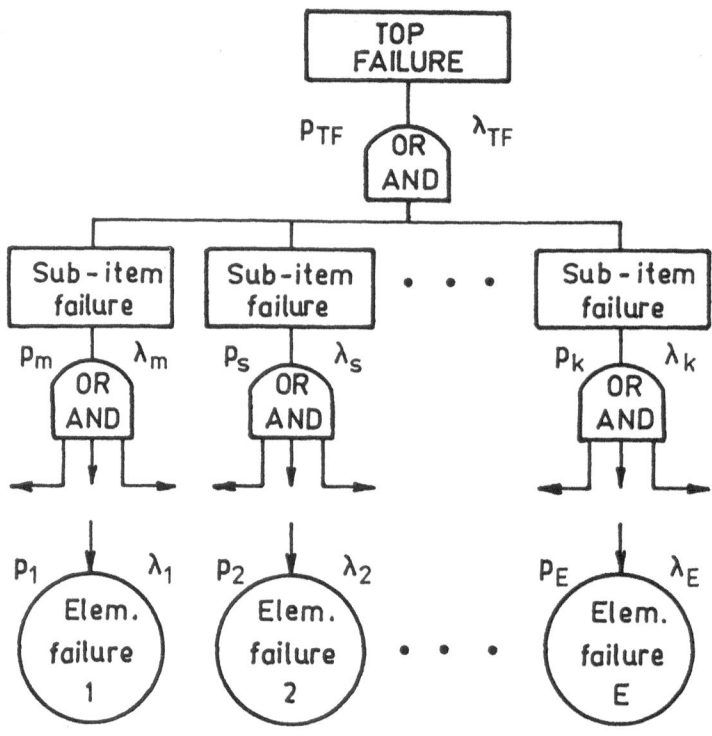

Figure 10. Qualitative and quantitative scheme of the FTA

(a) The calculation is based on the failure intensity data starting from the failure intensities of the elementary failures $\lambda_1, \lambda_2, ..., \lambda_E$. Going upwards the tree through the logic symbols (gates) OR or AND the next upper branch failure intensities are calculated according to the elementary theory of reliability rules, so we get λ_s. Thus moving higher and higher along the tree finally the top failure intensity λ_{TF} for the case analysed is determined.

(b) The calculation is based on the probability of failure data starting from the probability of failures of the elementary items $p_1, p_2, ..., p_E$. Analogously as in the previous case the upper sub-item probability of failure p_s is computed according to the same probability rules as above and so on up to the probability of the top failure p_{TF}.

It is now the designer's problem to assess whether this result is satisfactory in two ways: either knowing the elementary data (probabilities or intensities of failures), the top failure probability or intensity of the whole structure (usually its part only) should correspond to the probability or intensity of failure required for the structure (or its part). *Vice versa*, knowing the required overall probability or intensity of the top failure it is the designer's duty to split it into its sub-items and finally elementary items; this may lead even to changes of the structural set up and/or to new structural development.

In any case this kind of analysis provides the quantified information about the importance of every failure considered and so about the criticality of component failures. It is natural that the more critical components should then attract the subsequent attention of the working team and evoke more "accurate" computational and/or experimental methods. The whole quantitative analysis of Failure Trees can nowadays be performed on computers [7,8].

5.2 COMPONENT APPROACH

Successful component design as usually means keeping the stresses (strains) caused by external loads below a reasonably justified level in all component sections. In the simplest case it means trying to avoid stress/strain concentrators (notches) by using large radii, ground welds, etc. On the other hand, the stress/strain concentrators cannot be avoided. Because their roots are potential sources of microcracks it is important to determine their stresses/strains knowing stresses/strains of the surrounding (smooth) volumes (which can be measured).

The simplest approach utilizes the static stress concentration factors and strength reduction factors (given for elastic stresses/strains in handbooks), by which the "smooth stresses/strains" are multiplied. Nevertheless, often the notch roots undergo plastic deformation substantially changing the whole picture (due, e.g., to residual stresses) and so such a step should be understood as a coarse approximation.

Another approach is based on the stochastic process transformation into a macroblock (say, using the RFM as mentioned hitherto) and then determination of the "cyclic" stress/strain concentration factors - functions of the corresponding stress/strain levels, as derived by Polák [5].

The situation becomes, however, even more misty when taking stochastic loading. The possibility to monitor continuously the hysteresis loops under stochastic loading and so the actual smooth stresses/strains [11] has generated an idea not to seek for the separate notch stress σ_n and strain ε_n values but directly determine the notch hysteresis energy according to equation

$$\int_0^{\varepsilon_n} \sigma \, d\varepsilon = K_f^2 \int_0^{\varepsilon_s} \sigma \, d\varepsilon, \tag{15}$$

where ε_n and ε_s are the notch and smooth strain values, resp., and K_f is the strength reduction factor. This approach seems to be promising and may yield a more descriptive fatigue life estimation of notched components. The work is now in progress.

6. Computational model for components

A computational model for the operating life estimation of a stochastically loaded component is to be structured for it could exploit the input information on operating loads and material properties and yield the fatigue life (time, number of repetitions, number of take-offs, etc.). A long evolution of various views, ideas and opinions caused that a large variety of models had been developed, verified and "proved", providing a number of formulae. Two fundamentally different approaches are feasible.

6.1. INTERFERANCE THEORY OF RELIABILITY (ITR)

At first glance the ITR model seems to be the most suitable model for dimensioning as it covers both the classical design philosophies (including the old-style safety factor) as well as the probabilistic nature of operating loads and material properties, offerring some estimator of the risk of failure.

Let C stands for the load capacity (limit stress, limit strain, characteristic strain, fatigue limit, etc.) and L characterize the operating load; both quantities are stochastic and have their corresponding probability density functions f and distribution functions F. Then the ITR model can be formulated as follows:

$$P_f = \Pr[C \leq L] \tag{16}$$

yielding the probability of failure [12]

$$P_f = 1 - \int_0^\infty F_L(x) \, f_C(x) \, dx \text{ or reliability } \Pr = 1 - P_f, \tag{17}$$

where Pr stands for the probability.

Yet other models are also feasible, e.g. the ITR(C - L) model, ITR(lnC/L) model and especially the ITR[g(C,L,...)] model, where g(C,L,...) is a function of operating loads, material properties, load capacity, environment, dimensions, design philosophy, etc., defining the failure surface and separating all possible combinations of the stochastic variables which cause failure from all possible combinations which do not.

Thus the probability of failure may be expressed as

$$P_f = \iint \cdots \int f(x_1, x_2, \ldots, x_n) \, dx_1 dx_2 \ldots dx_n \,, \qquad (18)$$

where $f(x_1, x_2, \ldots, x_n)$ is the joint probability density function of the variables.

From this expression it is immediately aparent that in practice there are never enough data to define this joint probability density function, despite that the ITR[g(C,L,...)] models have been thoroughly examined and various solutions for numerous combinations and properties of stochastic variables C, L,... were derived.

As an example, using eq. (17) we computed reliability of a shaft loaded by the 8 level macroblock shown in Fig. 11 (obtained by the RFM) representing the probability density function of operating stresses $f_L(s)$. The fatigue limit CC of the material used has the probability density function $f_C(\sigma_C)$ with the mean value $\overline{\sigma}_C = 50 \, \text{MPa}$ and root mean square value $s = 10$ MPa. The resulting reliability Pr = 0.711 = 71.1 %.

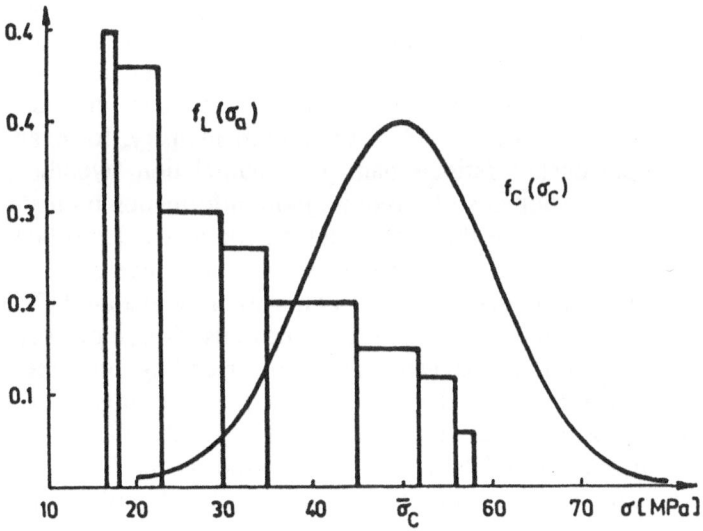

Figure 11. Practical example of the ITR model computation

The present knowledge of material fatigue does not seem, however, to favour this kind of calculation. First of all, the fatigue damage process is a multifactorial phenomenon and to attempt to create the joint probability density function $f(x_1, x_2, ..., x_n)$ for these factors (amplitudes, mean levels, fatigue limit, statistical characteristics of loads, stresses/strains, fatigue curve slope, strain hardening exponent, load history, etc.) is hardly resonable, especially when realizing their correlations. Further, the relation (17) does not seem to be descriptive enough as it is known that also other statistical characteristics may influence the fatigue damage accumulation (e.g. a power spectral density).

6.2 FATIGUE LIFE ASSESSMENT

Fatigue life assessment utilizes the fatigue damage hypotheses which have been under development for almost 80 years starting from the first linear damage rule proposed by Palmgren and in 1945 rediscovered by Miner. Since that tens of various ideas have appeared exploiting the S/N or Manson-Coffin curves and operating load charactersistics and yielding the operating life assessment. Because in those times computers were not available engineers tried to find an alternative way of the stochastic process analysis and developed so called counting methods, counting various characteristic process parameters (local amplitudes and mean levels, level transitions, etc.) and thus transforming the analysed stochastic process into a macroblock of harmonic amplitudes [12] (example see in Fig. 6 obtained by the RFM, which is also one of these methods). Many successful applications of the fatigue life calculation based on the RFM, Manson-Coffin curve and the Palmgren-Miner linear damage rule have proved that this approach is fairly accurate and this is why it is recommended by the American SAE and aircraft industry, for example. Most other more sophisticated fatigue damage accumulation hypotheses on the macroblock basis developed so far require more information on material - load properties and generally speaking have not proved to yield substantially more accurate results in standard cases. Nevertheless, large divergencies compared with experimental results may arrise for non-stationary loading [13] and this is why hypotheses based on other principles come to use (see, e.g., [14]).

It is natural that the stochastic nature of operating processes has kept attracting specialists trying to develop a method of the fatigue life estimation using certain stochastic process characteristics (say, probability density function or power spectral density). The result is then either in the form of the time to fracture with its probabilistic scatter band or the probability of fracture. One can mention the successful effort due to Rajkher [12] or Kliman [14], both exploiting the power spectral density of a stationary stochastic process, and some others. Nevertheless, for the time being there is not enough data and experimental evidence to assess the accuracy of these approaches and further investigations are required to include the most relevant material and load parameters and provide the probability of failure or time to failure in the probabilistic sense.

In conclusion to this part let us mention the Kliman's effort [15] to estimate the whole fatigue life distribution function for the median and extreme values (2.5 and 97.5 %) of the input S/N fatigue curves, resp., and a non-stationary process analysed by the RFM. The result shown in Fig. 12 illustrates the following: when the calculation is based on the median S/N curve, the shortest fatigue life makes 99 hours and the longest life 135 hours (curve M_m). When also its 95 percent probability interval is considered, the minimum life for the lower band is 60 hours (curve M_l) and the maximum life for the upper band is 166 hours (curve M_u). When also the 2.5 and 97.5 % reliability scatter bands for the S/N curves are considered then the minimum life may reach 20 hours (curve E_l) and the maximum life 544 hours (curve E_u).

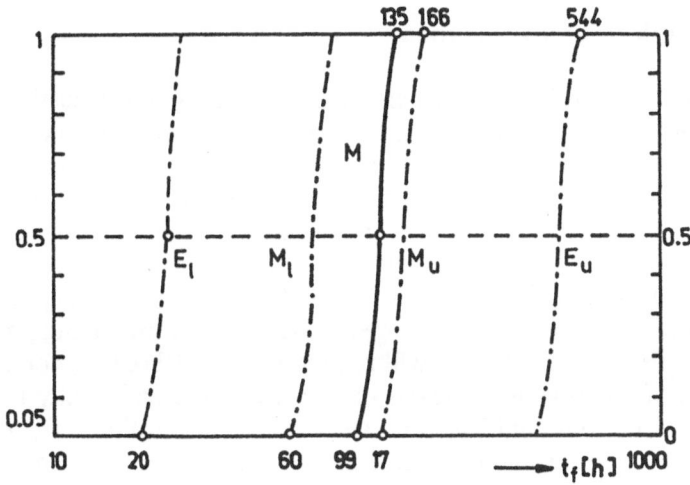

Figure 12. Fatigue life distribution functions

This result is obviously rather conflicting. One one hand we are able to take into account the probabilistic nature of the material and load properties but, on the other hand, this approach yields such a large scatter of lives that it is very problematic to make any conclusion on the suitability of the calculated component. It is natural that if the component is very critical then the E_l curve will cover a large safety and will be the most appropriate. If, however, the failed component is not supposed to cause serious consequences the median M_m curve (may be with its boundaries) will satisfy.

Yet another problem concerning the operating fatigue life estimation is worth mentioning, viz. the multiaxiality of operating loads (stresses/strains).

Considering that the uniaxial operating loads are rather rare (most relevant stresses/strains are biaxial) on should try to include this fact both into the computational and experimental procedures.

The situation is relatively clear for static loading thanks to various strength hypotheses and relatively simple experiments. A more complicated case occurs for harmonic biaxial loading where the life depends on whether the loading is in phase or out of phase. And practically nothing is known about multiaxial stochastic loading effect on operating life where the resulting stress/strain changes not only its amplitude but also its direction. Investigations of these problems obviously contain combined difficulties compared with the uniaxial loading.

7. Future subjects of investigations

The review of probabilistic approaches to dimensioning of stochastically loaded structures presented above can serve as a basis for the assessment of knowledge of the whole subject as well as of "white spots" asking for clarification.

(a) According to our experience the weakest link of all designing procedures is the knowledge of use conditions, corresponding operating loads and environmental effects. The question consists not only in defining the operating and environmental parameters but mainly in defining the column vectors of their mutual appearance together with the corresponding probabilities as shown in the matrix (4). This will require further elaboration for every kind of structures, collection and analysis of data on its use and finally formulation of the typical life-time unit. It is no sense to try to develop a more accurate computational/experimental method of the structural life assessment if the input information is insufficient.

(b) Various possibilities and problems of the operating load statistical description seem to lead to one practical and realizable conclusion: considering that more than fifty percent of all operating loads exhibit a non-stationary character the most appropriate method of their analysis is the 2nd order Markov chain and the K x K matrices for K discrete process levels. Nevertheless, these matrices can be used for the simulating purposes only but not as a direct tool for the operating life estimation. The future therefore lies either in developing a practical method for the life assessment based on the Markov matrices, or in a wider use and verification of the evolutionary statistical characteristics (4) - (8). Which of them or whether all of them is also questionable. It will hardly be the probability density function only as some fatigue experiments clearly show that this single characteristic does not sufficiently well characterize the fatigue damage accumulation process and that also the power spectral density has a remarkable influence. From this point of view the application of the multiparametric ITR models is problematic and certainly the dimensioning trend in mechanical engineering does not point to it (remarkable effort is obvious in civil engineering, however).

(c) More experimental data are required on the qualitative and quantitative influence of various non-stationarities on the operating life. On one hand, if the operating process is long enough then even long time trends are smoothed and the process looks to be stationary. But on the other hand, a limited experimental evidence proves that, e.g., sudden rapid changes and overloads do effect the fatigue damage accumulation and should be taken into account. This is especially true when a stochastic process of a limited length is applied and abrupt events occurring with very low probability (although important) are to be included.

(d) In order to promote a certain round robin effort in the field of non-stationary process application better descriptive tools for the assessment of their time-dependent statistical properties are required. The correlation theory can be quite successfully applied in simulation of time-dependent characteristics but their mathematical description derived from a measured process is practically impossible.

(e) The literature presents tens of fatigue damage accumulation hypotheses based on various assumptions and phenomena. But in order to approach the calculation to the probabilistic nature of damage accumulation under stochastic loading, more descriptive formulae are required including all relevant load characteristics (even time-dependent), material characteristics and physical phenomena (e.g. the hysteresis energy evolution).

(d) All material data should be provided with their probabilistic limits (scatter). This will help reducing safety factors and estimating the criticality of components. Because many structures are exposed to more than one damage mechanism acting synergetically, this kind of material data is also required. It would be a great help for designers if apart from this basic information also data for typical components are available (e.g., screwed connections, components with typical notches, reveted joints, etc.).

(e) Although the FMECA and FTA approaches are becoming widely used their further effective application depends to a large extent on the quality of input information, viz. on the elementary failure intensities or probabilities of failure. If these data for typical elementary components are available the FMECA and especially FTA would become more accurate and easier to be used.

(f) Because the decisive limitation of structural lives stem from notches (stress/strain concentrators) it is desitable to elaborate a method which could transfer stresses/strains from their vicinity (as usually measured) to their roots.

(g) Further investigations of the correlation of multiaxial (biaxial) stresses/strains and corresponding operating life will undoubtedly help decreasing the probability of operating failures.

As seen, the whole subject of dimensioning is really very wide and offers numerous opportunities for both research - development as well as for practical applications. Although some practicians claim that the up-to-date understanding of the designing procedures (mainly deterministic) and data available are sufficient to create reliable structures, two conclusions from the final report [1] are worth noting here:

(a) The improvement of codes, standards and design rules, the training of designers and skilled people, the transfer of existing knowledge developed by research and development, and the introduction of total quality concepts in companies could reduce the costs due to fracture by about 50 %.

(b) The investment in research and development, and in knowledge transfer, on structural integrity represents a very good strategy.

8. References

1. Faria, L. (1991) *The Economic Effects of Fracture in Europe*, Commission of the European Communities, Study Contract No. 320105, Brussels.
2. *Equipment Reliability Testing, Part 2: Guidance for Design of Test Cycles, Part 3: Preferred Test Conditions* (1982) IEC Technical Committee No. 56, Publication 605, Geneve.
3. Bendat, J. and Piersol, A.G. (1980) *Engineering Application of Correlation and Spectral Analysis*, J. Wiley, N. York.
4. Priestly, M.B. (1981) *Spectral Analysis and Time Series*, Academic Press, N. York.
5. Bílý, M. (ed.) (1993) *Cyclic Deformation and Fatigue of Metals*, Elsevier, Amsterdam.
6. Čačko, J., Bílý, M. and Bukoveczky, J. (1988) *Random Processes: Measurement, Analysis and Simulation*, Elsevier, Amsterdam.
7. *Analysis Techniques for System Reliability - Procedure for Failure Mode and Effect Analysis (FMEA)* (1985), IEC Standard, Publication 812, IEC, Geneve.
8. Limnios, N. (1987) Event Trees and their treatment on PC computers, *Reliability Engng* **18**,197-204.
9. Jordan, W.E. (1972) Failure Modes Effects and Criticality Analysis, in *Proceedings 1972 Annual Reliability and Maintainability Symposium*, San Francisko,30-41.
10. Mingxiang, J. (1985) Theory and algorithm of the quantitative analysis of Fault Trees, *Reliability Engng* **12**,241-257.
11. Petrovič, M., Horanský, P. and Bílý, M. (1996) Material fatigue via continuously monitored hysteresis energy, in *13th Danubia - Adria Symposium*, Rajecke Teplice,43-44.
12. Bílý, M. (1989) *Dependability of Mechanical Systems*, Elsevier, Amsterdam.
13. Bílý, M. and Prohácka, J. (1996) Fatigue life estimation under non-stationary random loading, in D.M.R. Taplin et al. (eds.) *Advances in Fracture Resistance in Materials,2*, Tata McGraw Hill Publ. Co.,New Delhi,143-151.
14. Kliman, V. (1984) Fatigue life estimation under the random behaviuour of the loading process, in C.J. Beevers (ed.) *Fatigue 84*, Engng Materials Advisory Services Ltd., Birmingham,903-913.
15. Kliman, V. (1993) Prediction of the random load fatigue life distribution, in J. Solin et al. (eds.) *Fatigue Design, ESIS*, Mech. Engng Publications, London,241-255.

MODELING OF MATERIAL UNCERTAINTIES USING THE NONLINEAR STOCHASTIC FINITE ELEMENT METHOD

A. HALDAR and J. HUH
University of Arizona
Department of Civil Engineering and Engineering Mechanics
University of Arizona, Tucson, AZ 87521, U.S.A.

1. Introduction

Safety evaluation of structural systems with realistic consideration of their behavior is becoming increasingly important to the engineering profession. Most of the planning and design of engineering systems is accomplished under conditions of uncertainty. Three major sources of uncertainty are (1) environmental uncertainty, (2) structural uncertainty and (3) modeling uncertainty. Structural uncertainty can be attributed to structure-related parameters such as the structural geometry, material properties, and boundary conditions. They are difficult to predict because of the influence of numerous factors during manufacturing, fabrication and construction. Structural uncertainty, specifically material uncertainty, is emphasized in this paper.

2. Modeling of Material Uncertainty

The modeling of material uncertainty consists of two items: the method used for the probabilistic analysis, and the stochastic description of the material parameters to be used in the numerical algorithm. Both items need comprehensive evaluation and are the subjects of the paper.

2.1 STOCHASTIC FINITE ELEMENT METHODS

Most of the commonly used structural reliability analysis procedures require that the response of the structure be known explicitly in terms of input variables. For most practical structures, the responses are not available in closed form, thus making them unsuitable for the reliability analysis. On the other hand, it is also known that if the basic variables are stochastic, every quantity computed during the deterministic analysis is also stochastic, being a function of the basic variables. The currently available reliability methods can still be used if the stochastic variation of the response can be

G. N. Frantziskonis (ed.), PROBAMAT – 21st Century: Probabilities and Materials, 299–310.

300

tracked in terms of the stochastic variation of the basic variables at every step of the deterministic analysis. The classical deterministic finite element method (FEM) provides such an opportunity, and this concept forms the basis of the stochastic finite element method (SFEM).

To evaluate the safety of complicated structural systems, a finite element-based representation is very desirable, since it is also the first step in a conventional deterministic analysis. Furthermore, failure limit states, i.e., the condition of the system just before failure, need to be investigated to estimate the reliability or probability of failure of a structural system. As the structural reliability area matures, a realistic representation of a structure is becoming essential, in terms of loading, geometric arrangement, material properties, behavior (linear or nonlinear), boundary conditions, etc. Since these features are routinely incorporated in a typical deterministic finite element method, their use in the context of SFEM will make the reliability method more desirable for practical implementation.

The basic concept can be described in the following way:

$$K \, U = F \qquad (1)$$

where K is the global stiffness matrix, U is the vector of nodal displacements and F is the global load vector. Due to the uncertainty in the basic variables, the precise values of the quantities in this equation are uncertain. Hence, an error term has to be added to the deterministic value used for each of these quantities, as:

$$K = K_0 + \Delta K, \quad U = U_0 + \Delta U, \quad F = F_0 + \Delta F \qquad (2)$$

Alternatively, one may think of the quantities K, U and F as having deterministic parts K_0, U_0, and F_0, respectively, and stochastic parts ΔK, ΔU, and ΔF, respectively.

Neglecting the product $[\Delta K] \, [\Delta U]$ and separating the deterministic and stochastic parts, it can be shown that:

$$U_0 = K_0^{-1} \, F_0 \qquad (3)$$

and:

$$\Delta U = K_0^{-1} \, [\Delta F - \Delta K \, U_0] \qquad (4)$$

Thus, the variation in the response U is obtained knowing the variation in K and F. The matrices can be obtained as:

$$\Delta K = \frac{\partial K}{\partial X} \Delta X \qquad\qquad (5)$$

$$\Delta F = \frac{\partial F}{\partial X} \Delta X \qquad\qquad (6)$$

This is simply the application of the chain rule of differentiation to finite element analysis to compute the sensitivity of the response with respect to that in the basic variables.

The error terms introduced above can be obtained alternatively using a Taylor series expansion, paving the way for a generalized formulation and higher-order approximations. The computation of response statistics using an expansion about a deterministic state is referred to as the perturbation approach [9]. The Neumann expansion approach has also been used instead of a Taylor series expansion to treat the random field of material properties. The other two commonly used solution stategies for SFEM are the approximate limit state function method and the reliability method [7].

The reliability approach is used in this study. It is based on the basic concept of reliability analysis methods, i.e., First Order Reliability Method (FORM) or Second Order Reliability Method (SORM) [8]. In the context of FORM/SORM, a performance function can be expressed by an n-dimensional standard Gaussian random variable space, by transforming the basic random variables into the uncorrelated standard normal space. An iteration algorithm is used to locate the design point (the most likely failure point) on the limit state function using the first or second order approximation. In each iteration, the structural response and the response gradient vectors are calculated using the deterministic finite element models. Once the design point is located, the distance from this point to the origin in the standard normal space is the reliability index β. This will be discussed further later.

2.2 DISTRIBUTED STATE OF MATERIAL

The general consensus in the profession is that the parameters related to the material properties need to be modeled as spatially varying random processes or random fields. Depending upon the type of structure, these parameters may have random fluctuations in a space of one, two, or three dimensions, in addition to variation accross different samples or realizations. In the context of SFEM, the random field can be represented in a continuous or discontinuous way. In the continuous approach, the random field is expanded in a sum of continuous deterministic functions with random coefficients [1, 4, 10, 12, 15, 16]. In the discontinuous approach, the distributed material properties are discretized into sets of spatially correlated random variables. In other words, the structure is divided into several elements, and a random variable is used to represent the random field over each element. The statistical correlation between the random variables for different elements are derived based upon the correlation characteristics of the corresponding random field. The theoretical basis for such a representation has been provided by Vanmarcke [13].

Several methods have been proposed for the discretization of random fields into random variables [11]. These determine the value of the random field over an element based on (i) the values at one or more points of the element, e.g., the midpoint method, the nodal point method, and the interpolation method, or (ii) the spatial average of the random field over the element. As discussed earlier, the random field may also be discretized by expansion into a series using basis random variables [11], Karhunen-Loeve expansion [12], or Neumann expansion [15]. The emphasis here is not on the relative merits and demerits of these methods, but on questions relating to practical implementation such as selective representation of a few of the distributed parameters as random fields and the number of elements for discretization.

To further elaborate the subject, the midpoint and the spatial everage methods can be considered. The midpoint method represents a random field $X(t)$ over an element i as $X_i = X(t_i)$, where t_i is the location of the centroid of the ith element. For such representation, the means and the covariance matrix of the resulting random variables can be readily estimated. For a one-dimensional random field, the spatial averaging method represents the random field $X(t)$ over an element i as:

$$X_i = \frac{1}{T} \int_{-\frac{T}{2}}^{\frac{T}{2}} X(t) \ dt \qquad (7)$$

where T is the averaging (temporal or spatial) interval. Equation 7 can be generalized to higher dimensions. The mean value of the random variable X_i is not affected by the averaging operation. For a wide-sense homogeneous random field $X(t)$ with mean μ and variance σ^2, the mean of X_i is simply:

$$E[X_i] = E[X(t)] = \mu \qquad (8)$$

However, the variance of X_i is not obtained as a simple average and has to be expressed as:

$$Var[X_i] = \gamma(T) \ \sigma^2 \qquad (9)$$

where $\gamma(T)$ is defined as the variance function of $X(t)$ which measures the reduction of the point variance σ^2 under local averaging. Vanmarcke and Grigoriu [14] defined the properties of the variance function and established a methodology to compute the covariance matrix of the resulting random variables. They also presented several models of the correlation functions for random fields (such as triangular, exponential and Gaussian) and the corresponding variance functions. They observed that the variance function becomes inversely proportional to T at large values of T for each model, and referred to the proportionality constant as the scale of fluctuation θ:

$$\gamma(T) = \frac{\theta}{T} \quad as \quad T \to \infty \qquad (10)$$

The scale of fluctuation can be viewed as the approximate length over which strong correlation persists in the random field. The information can be used advantageously in determining the size of the random field element mesh. Haldar [6] commented that when a single type of test was conducted at all locations, the sampling distances should be chosen to be larger than the scale of fluctuation in order to avoid redundancy in the gathering of information. However, if different tests were to be performed at two locations, the distance between the two locations should be chosen to be smaller than the scale of fluctuation for maximum effectiveness.

The discretization of a random field $X(t)$ into a set of random variables X_i also raises questions about the probability distribution functions for X_i. In the midpoint discretization method, the distribution of the random variables remains the same as that of the underlying random field. However, for the spatial averaging method, this is true only if the underlying random field is Gaussian, since the integration operation is linear. For non-Gaussian random fields, the distribution of X_i as defined by Eq. 7 is very difficult to obtain. Der Kiureghian [2] presented approximate descriptions of the distribution function for such cases.

3. Practical Random Field Discretization

The theoretical basis for considering distributed structural parameters as random fields and their use in stochastic finite element analysis are discussed in Section 2. However, to implement these concepts, two issues of practical importance need to be addressed, considering the efficiency and accuracy in the computations. They are: (i) should all the distributed parameters be considered to be random fields?, and (ii) if a parameter is considered to be a random field, what is the optimal random field mesh to be used?

The first issue can be addressed by using sensitivity indices. A unit sensitivity vector (γ) with respect to the stochastic variations in the basic random variable X can be defined as [3]:

$$\gamma = \frac{D \, B^T \, \alpha}{|D \, B^T \, \alpha|} \qquad (11)$$

where D is the diagonal matrix of standard deviations of the random variables X, B is a transformation matrix [11], and α is the unit vector normal to the limit state surface away from the origin. During SFEM-based reliability analysis, if certain components of γ remain consistently small through the first few iterations, then the randomness in the corresponding variables may be ignored in all subsequent iterations. These

304

variables may be approximated at the latest iteration point and considered to be deterministic for future computations. This significantly reduces the size of the problem.

Mahadevan and Haldar [11] addressed the issue of stochastic mesh refinement in great detail. They observed that the choice of the appropriate random field element mesh was strongly dependent on the correlation characteristics of the random field, such as the scale of fluctuation and correlation length. They suggested that an element size equal to the scale of fluctuation might in general be considered adequate.

4. A Unified Stochastic Finite Element Method for Nonlinear Problems

The discussion on SFEM so far has been mainly applicable to linear problems, where the uncertainty associated with the problem is relatively small, and is based on the displacement-based finite element method. However, as stated earlier, to estimate the reliability or risk of failure of a structure, the performance functions need to be formulated just before failure, when the structure is expected to behave nonlinearly. The nonlinear SFEM-based approach is expected to be computationally more demanding. The efficiency and robustness of the deterministic FEM is very important, particularly in capturing the nonlinear behavior of the structure. The displacement-based FEM may not be attractive in this capacity. The other alternative may be more desirable, i.e., the assumed stress-based FEM. In this approach, the tangent stiffness can be expressed in explicit form, the stresses of an elelment can be obtained directly, fewer elements are required in describing a large deformation configuration, and integration is not required to obtain the tangent stiffenss. It is very accurate and efficient in analyzing the nonlinear responses of frame structures. Thus, stress-based SFEM is developed further here.

The method is based on the first-order reliability method (FORM) and necessitates the definition of a limit state function $G(x,u,s)$, where vector x denotes the set of basic random variables pertaining to a structure (e.g., loads, material properties and structural geometry), vector u denotes the set of displacements involved in the limit state function, and vector s denotes the set of load effects (except the displacement) involved in the limit state function (e.g., stresses, internal forces). The displacement u can be expressed as $u = QD$, where D is the global displacement vector and Q is a transformation matrix. In general, x, u and s are related in an algorithmic sense, e.g., a finite element code. In the SFEM developed by the authors, nonlinearities due to geometric, material, and partially restrained (PR) connections can be incorporated. The algorithm evaluates the performance function deterministically, with the corresponding gradients at each iteration point. It converges to the most probable failure point or design point, and calculates the corresponding reliability index β. The following iteration scheme can be used for finding the checking point [5]:

$$y_{i+1} = \left(y_i^t \alpha_i + \frac{G(y_i)}{|\nabla G(y_i)|} \right) \alpha_i \qquad (12)$$

$$\nabla G(y) = \left[\frac{\partial G(y)}{\partial y_1}, \ldots, \frac{\partial G(y)}{\partial y_n} \right]^t \qquad (13)$$

$$\alpha_1 = - \frac{\nabla G(y_i)}{|\nabla G(y_i)|} \qquad (14)$$

$$\nabla G = \left[\frac{\partial G}{\partial s} J_{s,x} + \left(Q \frac{\partial G}{\partial u} + \frac{\partial G}{\partial s} J_{s,D} \right) J_{D,x} + \frac{\partial G}{\partial x} \right] J_{y,x}^{-1} \qquad (15)$$

where J_{ij} are the Jacobians of transformation and the y_i's are statistically independent random variables in the standard normal space. The evaluation of the quantities in Eq. 15 will depend on the problem under consideration (linear or nonlinear, 2-D or 3-D, etc.) and the performance functions used. These quantities need to be derived as described briefly below.

4.1. EVALUATION OF JACOBIANS AND ADJOINT VARIABLE METHOD

To evaluate the gradient ∇G, the four Jacobians in Eq. 15 need to be computed properly. Because of the triangular nature of the transformation, $J_{y,x}$ and its inverse are easy to compute. Since s is not an explicit function of the basic random variables x, $J_{s,x} = 0$. $J_{s,D}$ and $J_{D,x}$, however, are not easy to compute, since s, D and x are implicit functions of each other. The adjoint variable method [5,7] is used in this study to compute the product of the second term in Eq. 15 directly rather than evaluating its constituent parts. The method is very accurate and efficient. It can not be shown here due to lack of space, but will be elaborated during the presentation.

It is important to point out that all the quantities required for the computation of $\nabla G(y)$ in Eq. 15 can be obtained in a simple explicit form considering different sources of nonlinearities, including material nonlinearities. Material nonlinearity arises when yielding occurs or if the stress-strain behavior exhibits a nonlinear constitutive relationship. The key to solving material nonlinearity problems is to properly define a constitutive reltionship and the yield criterion in tracing a stress strain path.

Since the methodology is verified using steel frame structures, a brief discussion on the modeling of its nonlinear behavior is necessary at this stage. The three most common assumptions for steel material behavior are elastic-perfectly-plastic, isotropic strain hardening, and kinematic strain hardening models. Elastic-plastic behavior is characterized by an initial elastic material response on which a plastic deformation is superimposed after a certain level of stress has been reached. In this study, the discussion is restricted to elasto-plastic material nonlinearity.

For steel structures, material nonlinearity arises when yielding spreads through the cross-section (plastification) and along the member length (plastic zone) as the moment in the cross-section increases from the initial yield moment M_y to the full plastic moment M_p. Depending upon the degree of accuracy desired, the concentrated plasticity (plastic hinge) model or the distributed plasticity (plastic zone) model can be used in the algorithm.

Various yield criteria have been suggested for metals. The most simple and frequently used yield criteria are the Thresca criterion and Von Mises criterion. The Von Mises concept is used in this study in the form of an interaction equation. For the reliability analysis, the corresponding performance function can be expressed as [5]:

$$G(x, u, s) = 1.0 - \frac{P_u}{P_n} - \frac{8}{9}\left(\frac{M_{ux}}{M_{nx}} + \frac{M_{uy}}{M_{ny}}\right); \quad if \quad \frac{P_u}{\phi P_n} \geq 0.2$$

$$(16)$$

$$G(x, u, s) = 1.0 - \left(\frac{P_u}{2P_n} + \frac{M_{ux}}{M_{nx}} + \frac{M_{uy}}{M_{ny}}\right); \quad if \quad \frac{P_u}{\phi P_n} < 0.2$$

$$(17)$$

where ϕ is the resistance factor; P_u is the required tensile/compressive strength; P_n is the nominal tensile/compressive strength; M_{ux} and M_{uy} are the required flexural strength with respect to the x-axis and y-axis, respectively; and M_{nx} and M_{ny} are the nominal flexural strength with respect to the x-axis and y-axis, respectively. As stated in the Load and Resistance Factored Load manual published by the American Institute of Steel Construction (AISC, 1994), P_n can be evaluated as $P_n = AF_{cr}$ (compression) or $P_n = AF_y$ (tension), $M_{nx} = Z_x F_y$ and $M_{ny} = Z_y F_y$. They cannot be discussed further due to lack of space. It is important to point out that the interaction equation will contain the information on the material nonlinearity as well as the uncertainty in the parameters describing the material behavior. The basic random variables considered in the interaction equation are the Young's modulus, yield stress, area, plastic modulus, the moments of inertia, and all the applied loads.

5. Numerical Example

To elaborate the algorithm further, a single story frame shown in Fig. 1, is considered. The statistical properties of the basic random variables are given in Table 1. Using the nonlinear SFEM discussed above, the reliability of the frame is estimated for the strength limit state represented by either Eq. 16 or 17 as appropriate for the following

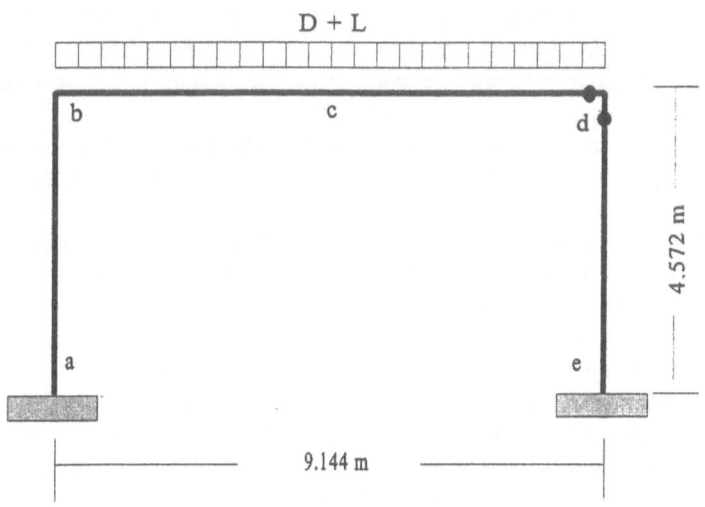

TABLE 1. Description of Basic Random Variables

Variables	Mean Values	COV	Distribution		
			Case 1	Case 2	Case 3
E (kN/m²)	2.000×10^8	0.06	Lognormal	Lognormal	Lognormal
F_y (kN/m²)	2.604×10^5	0.10	Lognormal	Lognormal	Lognormal
A (m²)	1.270×10^{-2}	0.05	Lognormal	Lognormal	Constant
I_x (m⁴)	3.970×10^{-4}	0.05	Lognormal	Lognormal	Constant
Z_x (m³)	2.130×10^{-3}	0.05	Lognormal	Lognormal	Constant
D (kN/m)	45.959	0.10	Lognormal	Constant	Constant
L (kN/m)	16.05	0.25	Type I	Constant	Constant

TABLE 2. Reliability and Sensitivity Index for Strength Limit State (Nonlinear)

Case		Case 1		Case 2		Case 3	
Location		Beam at d	Column at d	Beam at d	Column at d	Beam at d	Column at d
β		3.31	3.08	4.392	4.374	4.788	4.658
γ	E	-0.2909	-0.3485	-0.3395	-0.4497	-0.3874	-0.5182
	F_y	-0.6921	-0.5743	-0.8084	-0.7414	-0.9212	-0.8522
	A	-0.0170	-0.0363	-0.0193	-0.0455	-	-
	I_x	-0.2342	-0.2720	-0.2735	-0.3517	-	-
	Z_x	-0.3381	-0.2699	-0.3951	-0.3491	-	-
	D	0.3410	0.4021	-	-	-	-
	L	0.3881	0.4888	-	-	-	-

cases: (i) all basic variables are random, (ii) all variables except the load-related basic variables are random, and (iii) only the Young's modulus E and yield stress F_y are random and all other variables are constants. The results for these three cases are shown in Table 2.

Several interesting observations can be made based on the results. For ease of discussion, all the variables shown in Table 1 can be grouped into three categories: material properties, sectional properties, and loads. When all the variables are considered to be random, the reliability index is found to be about 3.3, a value generally accepted in the profession. The sensitivity indexes for the variables are relatively large, except area A.

When the loads are considered to be constants in Case 2, the reliability and sensitivity indexes went up. This is expected, since a major source of uncertainty that comes from the unpredictability of the loads is removed from the problem, and the relative importance of the uncertainties in the material properties, namely the Young's modulus and the yield stress, becomes more significant. However, as in Case 1, the sensitivity index for A still remains very small. The results indicate that the cross-sectional areas can be treated as constants without compromising the accuracy of the results.

In Case 3, where only the material properties are considered to be random variables, both indexes went up. This is also expected, since more sources of uncertainty have been removed, and the material uncertainties have become the dominant source. But in all three cases, the sensitivity indexes for the material properties are relatively large, indicating that they can not be ignored in any realistic evaluation of risk of a structural systems. This simple example also indicates that the SFEM formulation discussed here can be used to estimate the relaibility considering the nonlinear behavior of structural systems.

6. Conclusions

The modeling of material uncertainties in terms of the method to be used for probabilistic analysis and the stochastic description of material properties are discussed. Using FORM, a nonlinear stochastic finite element method is proposed by tracking the stochastic variation of the response at every step of a deterministic analysis. In the context of SFEM, the random material properties are represented by continuous and discontinous random fields. Commonly used methods are discussed. To implement the alogorithm economically and efficiently, the concept of sensitivity index is introduced. The selection of the optimal random field element mesh size is briefly discussed. With the help of a small example, it is shown that the consideration of nonlinear material behavior and the uncertainty associated with it are essential in the estimation of the reliability of a structural system. The stochastic finite element algorithm proposed here can be used for this purpose.

7. Acknowledgment

This paper is based upon work partly supported by the National Science Foundation under Grants No. MSM-8896267 and CMS-9526809. Any opinions, findings, conclusions, or recommendations expressed in this publication are those of the authors and do not necessarily reflect the views of the sponsor.

8. References

1. Deodatis, G. and Graham, L. (1997) The weighted integral method and the variability response function as part of a SFEM formulation, *Uncertainty Modeling in Finite Element, Fatigue, and Stability of Systems*, A. Haldar, A. Guran and B.M. Ayyub (eds.), World Scientific Publishing Co., New Jersey, pp. 71-116.

2. Der Kiureghian, A. (1987) Multivariate distribution models for structural reliability, *Proc. 9th Int. Conf. Struct. Mech. in Reactor Technology*, Lausanne, Switzerland.

3. Der Kiureghian, A. and Ke, J.-B. (1987) The stochastic finite element method in structural reliability, *Lecture Notes in Engineering (31): Stochastic Structural Mechanics*, Y.K. Lin and G.I. Schueller (eds.), Springer, New York, pp. 66-83.

4. Elishakoff, I. and Ren, Y. (1977) Finite element method for stochastic structures based on inverse of stiffness matrix, *Uncertainty Modeling in Finite Element, Fatigue, and Stability of Systems*, A. Haldar, A. Guran, and B.M. Ayyub (eds.), World Scientific Publishing Co., New Jersey, pp. 51-70.

5. Gao, L. and Haldar, A. (1995) Safety evaluation of frames with PR connections, *J. of Structural Engineering*, *ASCE*, **121**(7), 1101-1109.

6. Haldar, A. (1984) Statistical site characterization, *Proc. 4th Australia - New Zealand Conference*, Perth, Australia, 530-534.

7. Haldar, A. and Gao, L. (1997) Reliability evaluation of structures using nonlinear SFEM, *Uncertainty Modeling in Finite Element, Fatigue, and Stability of Systems*, A. Haldar, A. Guran and B.M. Ayyub (eds.), World Scientific Publishing Co., New Jersey, pp. 23-50.

8. Haldar, A. and Mahadevan, S. (1993) First-order/second-order reliability methods (FORM/SORM), *Probabilistic Structural Mechanics Handbook*, C. Sundararajan (ed.), Chapman & Hall, New York, pp. 27-52.

9. Hisada, T. and Nakagiri, S. (1985) Role of the stochastic finite element method in structural safety and reliability, *Proc. 4th Int. Conf. Structural Safety and Reliability, ICOSSAR'85*, Kobe, Japan, **1**, 385-394.

10. Liu, W.K., Belytschko, T. and Mani, A. (1986) Random field finite elements, *Int. J. Numer. Meth. in Eng.*, **23**, 1831-1845.

11. Mahadevan, S. and Haldar, A. (1991) Practical random field discretization in stochastic finite element analysis, *J. of Structural Safety*, **9**, 283-304.

12. Spanos, P.D. and Ghanem, R.G. (1989) Stochastic finite element expansion for random media, *J. Eng. Mech., ASCE,* **115(5)**, 1035-1053.

13. Vanmarcke, E.H. (1983) *Random Fields: Analysis and Synthesis,* MIT Press, Cambridge, MA.

14. Vanmarcke, E.H. and Grigoriu, M. (1983) Stochastic finite element analysis of simple beams, *J. Eng. Mech., ASCE,* **109(5)**, 1203-1214.

15. Yamazaki, F., Shinozuka, M. and Dasgupta, G. (1988) Neumann expansion for stochastic finite element analysis, *J. Eng. Mech., ASCE,* **114(8)**, 1335-1354.

16. Zhang, J. and Ellingwood, B. (1995) Effects of uncertain material properties on structural stability, *J. of Structural Engineering, ASCE,* **121(4)**, 705-716.

STRUCTURAL APPROACH IN CONTINUUM MODELING OF DAMAGEABLE PARTICULATE COMPOSITES

V.V. MOSHEV, L. A. GOLOTINA, O. K.GARISHIN,
L. L. KOZHEVNIKOVA
Institute of Continuous Media Mechanics
of the Russian Academy of Sciences
Korolev Str. 1, 614013, Perm, Russia

1. Introduction

An experimental evaluation of mechanical behavior of damageable particulate composites is usually aimed at establishing relaxation, creep and constant strain-rate curves with subsequent deriving appropriate phenomenology more or less fitting the experimental data. The essence of internal structural mechanisms giving rise to the observed, often very complicated, effective behavior usually does not attract the particular attention of experimentalists, which remain quite satisfied with the constitutive relations obtained from experience.

However, an elucidation of the structural mechanisms providing the observed effective behavior of composites, undoubtedly, merits a serious examination, two reasons being put forth for the support of this opinion. First, an understanding of structural mechanisms, controlling effective mechanical behavior, is a background for a purposeful constructing constitutive relations. Second, the understanding of structural mechanisms is a good aid in development of new composites with a specified performance.

This paper is an attempt to create the bridge between the structure and the phenomenology in the domain of particulate polymeric composites.

2. Theoretical background

2.1. STRUCTURAL CELL CHARACTERIZATION

In our recent paper [1] a cylindrical isometric unite cell was offered as an element within a closely packed ensemble shown in Fig. 1. A rigid spherical inclusion (filler particle) is

G. N. Frantziskonis (ed.), PROBAMAT – 21st Century: Probabilities and Materials, 311–316.
© 1998 *Kluwer Academic Publishers.*

312

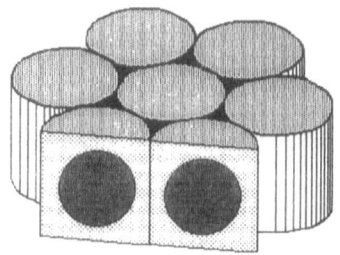

Figure 1. Packing of structural cells.

placed at the center of the polymeric (matrix) cylinder. In extension along the cylindrical axes, the restrictions, laid on every cell by surrounding, make its ends remain plane and its lateral surface remain cylindrical. This geometry is characterized by a remarkable feature. When the radius of the inclusion touches the lateral boundary and the ends of the cell, the overall solid fraction within the ensemble of Fig. 1, gaps between the cylindrical cells being taken into account, reaches 0.607. This value is close to that characteristic of ultimate packing of random structures composed of the uniformly sized spheres [2-4]. The calculation of the cell's effective modulus as a function of the filler volume fraction for cells keeping the bonded state demonstrated that the calculated modulus-concentration dependence [1] was close to the experimental observations [4] over a wide range of filler concentrations (Fig. 2) with the initial slope of the curve equal to 2.5 that coincides with the well-known Einstein's coefficient in the formula for the viscosity of suspensions.

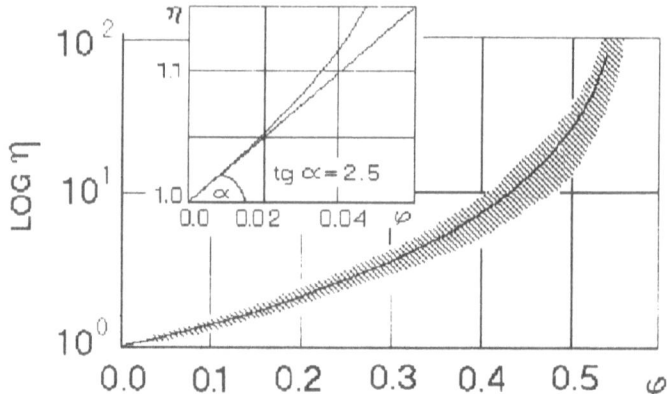

Figure.2 Designed and experimental modulus - concentration curves.

These results demonstrated a rather high predictive ability of cylindrical cells as constitutive elements controlling the effective behavior of particulate polymeric composites. Hence, a further examination of the cell behavior including a separation of the matrix from the inclusion and a final failure of the cell was undertaken aiming at obtaining other basic structural features that might be used in the derivation of the well-justified continuum representation.

All the calculations were carried out under assumption of a purely elastic matrix with no accounting for time-dependent processes. The matrix was represented by an elastomer following the neo-Hookean behavior with a low-strain shear modulus equal to 0.1 MPa. The inclusion was taken as perfectly rigid body. Griffith's approach was used in describing matrix separation process as it is done in other publications [5,6]. The finite element method described elsewhere [7] was used for calculations in the domain of large deformations.

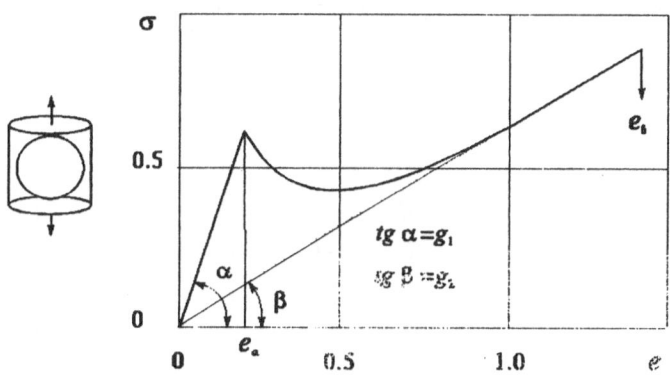

Figure 3 Tensile stress - strain relation of the cell.

The general appearance of the tensile curve for a cell under consideration is shown in Fig. 3. Its evolution is represented by three stages. The initial stage specifies the resistance of the bonded cell. The middle one describes the evolution of the matrix separation from the inclusion accompanied by the stress softening. The last stage presents the resistance of the completely debonded cell until its failure. It follows from Fig. 3 that describing such tensile curves requires establishing at least four basic controlling parameters: the initial rigidity, g_1 ; the strain, e_a , at which the matrix adhesive debonds begins; the final rigidity, g_2 ; and the breaking strain, e_b , of the cell. The middle (transition) part of the curve, also, should be somehow defined.

A solid volume fraction, φ, and matrix properties (elastic potential characteristic, G; energy of the debond, T_d ; breaking strain, ε_{mb}) are the basic structural features controlling the shape of the tensile curve of cells.

A number of boundary value problems have been solved for various combinations of φ, G, T_d and ε_b to establish the corresponding magnitudes for g_1 , e_a, g_2 and e_b and transform the results obtained into continuous constitutive relations having the form

$$\sigma_{ij} = 2 G (\varepsilon_1 , \sigma_0) (\varepsilon_{ij} - \theta / 3) + \delta_{ij} \sigma_0 ,$$
$$\theta = f(\varepsilon_1 , \sigma_0),$$

where the shear modulus, G , and volume compressibility, θ , are the functions of current magnitudes of the maximum main strain, ε_1 , and the mean stress, σ_0 .

This approach describes the complete life-cycle of the cell.

314

2.2. COMPOSITE STRUCTURAL INHOMOGENEITY

There exist at least two main sources of structural inhomogeneity: the geometrical local nonuniformity. contrasting with the ideal honeycomb structure of Fig. 1, and physical-chemical nonuniformity revealing in the debond, e_a, and breaking, ε_{mb}, strains scatter.
    To get insight into the first source of nonhomogeneity, the synthesis of the random geometrical structures consisting of identical spherical particles with the imposed mean filler volume fraction has been developed [8]. The structures generated are characterized by the coordinates of the sphere centers and
the magnitudes of their radii. The local matrix volumes were estimated by equating the mean gap between a given sphere and adjacent neighbors to the gap in the equivalent model cell.. This approach allows to get the distributions of the local solid volume fractions at various imposed mean solid volume fraction (Fig. 4).

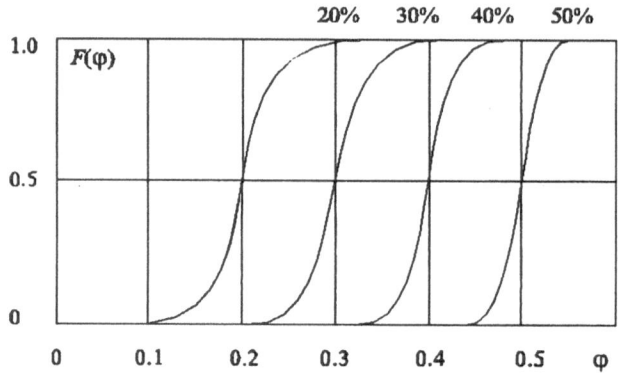

Figure 4. Distribution of the local solid volume fraction at different filler concentrations, shown near curves.

An analytical representation of these distributions has the general form
$$F(\varphi) = 1 - exp(-\alpha \, \varphi^m),$$
where $F(\varphi)$ is the probability to meet φ smaller that the indicated one. The random character of φ values induces the randomization of the appropriate cell parameters g_1, e_{a1}, g_2 and e_b.

2.3. COMPUTATION PROCEDURE

Replacing the discrete representation of cells behavior by the continuous one and accounting for the local cell nonhomogeneity allows us to examine particulate composites as piecewise continuous systems. The finite element potentialities seem to be most appropriate for computation of mechanical behavior of such bodies.
    Evidently, it is impossible to compose structure, where one finite elements represents the behavior of one structural cell. It would required accounting for thousands, possibly millions, finite elements that is beyond the possibilities of modern machinery.
    An implementation of the intermediate averaging procedure seems to be inevitable. This one can be performed through increasing the number of cells that

should to be enclosed in one finite element volume. So the number of the finite elements within the volume of the body can be reduced to a reasonable level. Obviously the mechanical variability of the finite elements becomes less than that of individual cells.

The calculations have shown that, in extension, the structural nonuniformity of the system provokes an appearance and progressive evolution of a large-scale nonuniformity leading to the loss of the elastic stability and macrocrack origination in the most compliant part of the body. In this approach, the description of the failure does not need to be referred to the so called strength criteria.

3. The influence of some structural parameters on macroscopic behavior of plane rectangular specimens

To illustrate the capabilities of the approach under consideration, a composite with a matrix having solid particles of 36 μm in diameter, Young's modulus of 3.0 MPa and breaking strain of 3.0 has been examined. The choice of these parameters allowed the comparison of calculated results with the available experimental data.

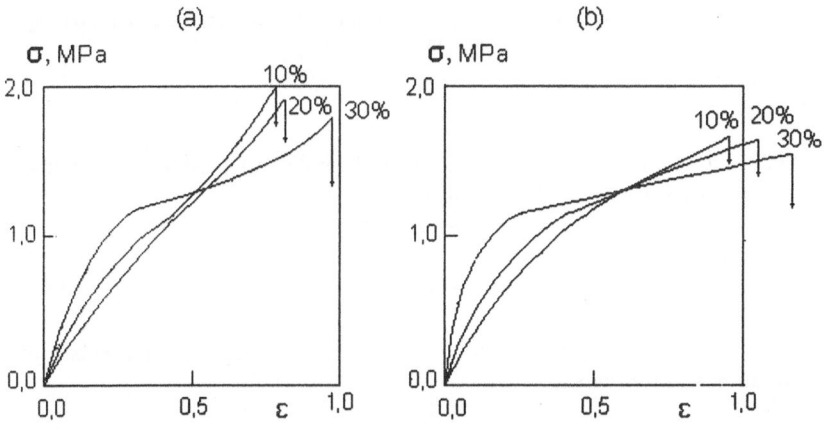

Figure 5. Stress - strain relations at various filler concentrations shown near the curves: (a) design; (b) experimental.

The calculations were performed for solid volume fractions of 0.1, 0.2 and 0.3. Fig. 5 (a) demonstrates the calculated stress-strain curves while Fig. 5 (b) represents experimental data from [9]. The similarity between the both picture is evident. The designed breaking strains and stresses of composite specimens are about 3 and 5 times smaller than those of the pure matrix.

The influence of particle sizes has been examined for particle of 36, 98 and 255 μm in diameter at 30% solid volume content (Fig.6). The designed and experimental [9] data are in a qualitative agreement. The diminution of the particle sizes, in itels, leading to higher uniformity with respect to the size of the specimen, provides higher ultimate properties.

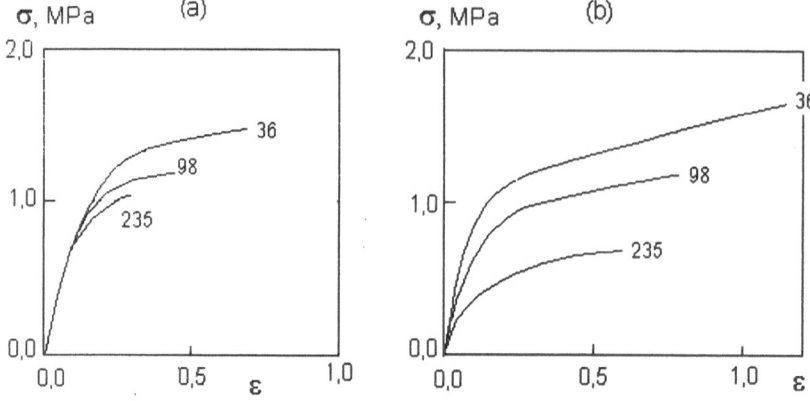

Figure 6. Stress - strain relations at various sizes of the filler particles shown near the curves: (a) design; (b) experiment.

4. Conclusions

1. The method of transforming of discrete structural models of particulate composites into adequate continuous representations is offered.

2. Three dimensional levels of structural damage are regarded those of cells, finite elements and specimens.

3. The designed and experimental data seem to be regarded as in good qualitative agreement.

4. The approach offered might be regarded as a breadboard model for subsequent detailed exploration and refinement.

References

1. Moshev, V.V and Kozhevnikova, L.L. (1997) Highly predictive structural cell for particulate polymeric composites, *J. Adhesion* **62**, 169-186.
2. Bernal, J. D. and Mason, G. (1960) Computations of dense random packings of hard spheres, *Nature* No. 4754, 910-911.
3. Farris, R. J. (1968) Prediction of the viscosity of multimodal suspensions from unimodal viscosity data, *Trans. Soc. Rheology* **12**, 281-301.
4. Chong, J. S., Christiansen, E. B. and Baer, A. (1971) Rheology of concentrated suspensions, *J. Appl. Polym. Sci.* **15**, 2007-2021.
5. Gent, A. N. (1980) Detachment of an elastic matrix from a rigid spherical inclusion, *J. Mater. Sci.* **15**, 2884-2888.
6. Gent, A. N. and Tobias, R. H. (1982) Threshold tear strength of elastomers, *J. Polym. Sci.:Polym. Phys. Ed.* **20**, 2051-2058.
7. Kozhevnikova, L. L., Moshev, V. V. and Rogovoy, A. A. (1993) A continuum model for finite void growth around spherical inclusion, *Int. J. Solids Structures* **30**, 237-248.
8. Moshev, V. V. and Garishin, O. K. (1993) Physical discretization approach to evaluation of elastic moduli of highly filled granular composites, *Int. J. Solids Structures* **30**, 2347-2355.
9. Schwarzl, F. R., Bree, H. B. and Nederveen, C. J. (1965) Mechanical properties of highly filled elastomers. I. Relationship between filler characteristics, shear moduli, and tensile properties, *Proc. 40th Intern. Congr. Rheology.* **3**, 241-263, Interscience/Wiley, NY.

FINITE ELEMENT AND RELIABILITY: COMBINED METHODS BY RESPONSE SURFACE

M. LEMAIRE

*LaRAMA – Institut Français de Mécanique Avancée
et Université Blaise Pascal
BP 265 – F-63175 AUBIERE – France*

1. Introduction

The analysis of reliability for materials and structures requires both a relevant model of the mechanical behavior (scenario of failure, performance function or limit state function) around crisis situations, and an efficient evaluation model of the probability of failure.

Based on relationships of the continuum mechanics and on the theory of structures, the Finite Element Method (*FEM*) is currently the most efficient numerical solution, under very large hypotheses, for taking into account degraded mechanical behavior (plasticity, damage, rupture, buckling).

Its use in sensitivity and reliability analyses considers as random some variables of the stiffness matrix and of the action vector [11]. It is then called Stochastic Finite Element Methods (*SFEM*). These terms gather a large range of methods according to the aimed analysis type and according to whether equations are written on a continuous set or on a discretized one.

This paper deals with the reliability analysis and shows how a response surface method allows us to combine a finite element code with an evaluation of the probability of failure by *FORM /SORM* methods (*First / Second Order Reliability Methods*) [6].

First, we shall recall three combined methods and then discuss the response surface method, by understressing possibilities to control approximations. It is illustrated on a bench example: a thick sphere under pressure.

G. N. Frantziskonis (ed.), PROBAMAT – 21st Century: Probabilities and Materials, 317–331.

2. Combined Finite Element and Reliability Methods

2.1. NOTATIONS

Let X_i be the random variable vector and $G(X_i)$ the performance function:

- $G(X_i) > 0$ is a success realization, $X_i \in \mathcal{D}_s$, safety domain,
- $G(X_i) \leq 0$ is a failure realization, $X_i \in \mathcal{D}_f$, failure domain,
- $G(X_i) = 0$ is the limit state function.

An isoprobabilistic transformation T associates with the vector X_i of any random variables in the physical space, the vector U_j of normally uncorrelated and standardized Gaussian variables $\mathcal{N}(0,1)$. The T transformation must be able to include all possible cases (independence, truncation, correlation and composition), and physical and standardized space dimensions n and m can be different. Nataf's transformation [5], that necessitates only limited information, generally available, on variables X_i (marginal densities and correlation), constitutes an efficient solution. It is thus:

$$U_i = T_i(X_j) \qquad G(X_j) = G(T_i^{-1}(U_i)) \equiv H(U_i) \qquad (1)$$

The Hasofer-Lind reliability index β is then calculated by solving the optimization problem:

$$\beta = \min \left(\sqrt{\sum_{i=1}^{m} U_i^2} \right) \quad \text{under the constraint} \quad H(U_i) \leq 0 \qquad (2)$$

The solution gives the value of β and the coordinates u_i^* of the design point P^* as well as the director cosines α_i (sensitivity factors) of the direction P^*O in the standardized space.

2.2. COMBINED METHODS

Three combined methods for coupling *FEM* and reliability are proposed:

1. **direct coupling**: the algorithm of optimization (for example Rackwitz-Fiessler [20] or Abdo-Rackwitz [1]) calls directly the *FEM* code each time a realization of the performance function $G(X_i)$ must be evaluated, including the calculation of the gradients. It is a robust method, but it leads to a too large number of calls. It will be truly efficient only when the *FEM* code will produce the calculation of the gradients directly. It has been implemented by our laboratory in the software *RYFES* [18] (*Reliability with Your Finite Element Software*).
2. **coupling by optimization**: the reliability problem is put as an optimization problem:

– in the physical space, by noticing that the design point P^* is the point where the joint density of probability $f_{X_i}(x_i)$ is at its maximum [14]. A direct research of its physical coordinates x_i^* is then processed, in principle, by:

$$x_i^* = \max_{x_i \in \mathcal{D}_f} f_{X_i}(x_i)$$

– or in the standardized space by changing the problem 2 in the following form:

$$\beta = \min \left(\sqrt{\sum_{i=1}^{m} U_i^2} \right) \qquad (3)$$

$$\text{under constraints} \quad G(X_i) \leq 0$$
$$\text{and} \quad U_i = T_i(X_j)$$

This approach is described in [13]. It necessitates that the *FEM* code have an optimization module.

3. **coupling by response surface**: simple approached explicit functions $\tilde{G}$ or $\tilde{H}$ replace the functions $G(X_i)$ or $H(U_i)$. The main advantages are:

  - to have then an analytic expression so that the solution of the reliability problem can be refined as much as desired by *FORM/SORM* methods, by direct or conditional Monte-Carlo simulations, or by an integration as *RGMR* [15].
  - to dissociate the mechanical calculation from the reliability calculation. However, it is necessary to verify the quality of $\tilde{G}$ or $\tilde{H}$ approximations. We shall now discuss this third combined mechanical and reliability method.

3. Coupling by response surface

3.1. INTRODUCTION

The utilization of response surfaces in reliability problems is not recent, but not achieved and new contributions are always proposed. Works have defined concepts [9],[8],[19], built solutions in the physical space [2], and proposed methods of evolution [7][4][10]. Generally, the reliability model is *FORM/SORM* but the response surface can be equally combined with simulations [12].

Proposed solutions are dissociated by the choice of the working space (physical or standardized) and by the evolution method towards the design point.

3.2. PRINCIPLE OF THE METHOD

The reliability calculation necessitates, by whatever method is used, a still important number of computations of realizations of the performance function $G(X_i)$. This number is a good measure of the efficiency of the method. Each realization can be obtained as a result from a complex model associated with robust numerical solution. It is indispensable to limit this number, and to do so, one of the possible methods is to construct a simple analytic representation close to the design point. The name *Response Surface (RS)* is given to such a representation. Its interest is in a simple expression on which the reliability calculation is simplified with a quasi-null cost; its difficulty is then to justify the chosen approximation and its validity domain.

To do so, two problems are to be solved:

1. to choose a basis of development of the *RS* and identify the unknown coefficients,
2. to build the *RS* around the design point P^*.

3.3. DEVELOPMENT OF THE *RS*

Three cases are to be considered, according to available information:

1. the mechanical model has to a known form and the *RS* depends on l coefficients. It is sufficient to identify the l coefficients with l calculations, checking that the equation system is not ill-conditioned.
2. a theoretical study gives indications on the *RS* form. It is the case, for example, when an expansion of the response is known round the design point (Taylor's or asymptotic expansion). It is also the case when simplified theories exist (approximate structural theory or elastic *FEM* calculation).
3. in case of a lack of information, a "blind" *RS* has to be constructed. An a priori development is chosen and coefficients are identified by interpolation or regression.

In this last case, a development built on a polynomial expansion is the most efficient. It is chosen by the majority of authors. The degree 2 (Quadratic Response Surface, *QRS*) it the best solution since it includes a possible calculation of curvatures and it avoids possible oscillations with a higher degree. With n random variables, the number of coefficients of the complete development is $l = (n + 2)(n + 1)/2$. For an approximation of $H(u)$, the development is then:

$$\tilde{H}(u) = \begin{bmatrix} \cdots & \left(u_i^p u_j^q\right)_k & \cdots \end{bmatrix} \begin{bmatrix} \vdots \\ a_k \\ \vdots \end{bmatrix} \tag{4}$$

with $p + q = 0, 1, 2 \, ; \, i, j = 1, m \, ; \, k = 1, l$

However, the designer disposes generally of information which allows him to suppress some interactions, and some coefficients a_k can be chosen as null or fixed. The dimension of the problem is reduced to $l^r \leq l$. The determination of coefficients is then undertaken by regression from n_e numerical experiments.

3.4. *RS* BUILDING: ORIENTED DESIGN OF EXPERIMENTS

After the choice of the development, it is now necessary to build it around the design point P^* whose position is unknown. The proposed method consists in researching a design point series $P^{*(s)}$ associated with a response surface series $RS^{(s)}$ whose domain of definition contains, at last, $P^{*(s)}$.

A numerical design of experiments (*DOE*) defines realizations to calculate. The minimum number is l^r for an interpolation but a higher number is necessary to guarantee the quality of the approximation.

The *DOE* theory [3] proposes classical plans:

– centered *DOE*:

- star-*DOE*, for example with 2 variables: $1, u, v, u^2, v^2...$, without interaction between factors,
- factorial-*DOE* of dimension 2^n: $1, u, v, uv...$,
- combined-*DOE* star + factorial $1, u, v, u^2, uv, v^2...$

– standardized tabulated *DOE*: the use of tables, such as Taguchi's, requires information on interactions between factors (variables).

Such *DOE* have been tested with bench examples and have validated methods, showing their aptitude to converge to the exact solution after few steps only [16]. However, this first approach does not take into account what the designer knows about the running of the studied structure. Indeed, for most random variables, the resistance or solicitation character is known, and the position of the gap from the average to the design point is also known a priori. Such *DOE*, that include all the expert knowledge of the designer, are called "oriented-*DOE*" and they must be favored during the first step while centered-DOE are to be built around the design point when its position is confirmed.

The choice of a first DOE allows us to construct a first RS and therefore to obtain a first design point $P^{*(1)}$. The analysis of results introduces new experiments around this first point to repeat the process by translation, homothetie [17] or projection [10].

3.5. CHOICE OF THE WORK SPACE

Let us consider the resolution of the reliability problem. Two possibilities are proposed to construct a RS: in the physical space or in the standardized space.

3.5.1. *In the physical space*

The physical space presents the advantage for directly physical solutions to be built, according to the designer's experience. However, the choice of physical points can lead to standardized points situated very far from of the origin in U-space if the standard variation is weak or, on the contrary, to almost merged points. More, the designer knows the average running area well, but can have difficulties for realizations around the failure to be identified. Finally, the quality of the reliability solution depends on the development of the limit state function obtained in normalized variables, in this case, by application of the T transformation (eq: 1) on the approximated expression $\tilde{G}(X_i)$. The physical space is more favorable for the sensitivity around the average than the reliability to be studied.

3.5.2. *In the standardized space*

To work directly in the standardized space guarantees a choice of realizations at a well controlled standard deviations number. The validity domain of the RS is then well known. However, the original physical point must be physically admissible. Aberrant physical solutions must be eliminated and the realistic character - or non - of physical points of calculation must be verified.

An a priori study of admissible domains of the physical variables allows us to obtain, by application of the T transformation, the standardized domain in which to place the DOE.

More, rules of correlation, composition and truncation of variables can lead to T transformations that are not unique. A realization $u_i^{(r)}$ always has an original physical realization $x_i^{(r)}$ but the inverse is not true.

Finally, with normalized variables, indexes are a-dimensional which facilitates their interpretation.

3.5.3. *Conclusion*

A response surface must be built in an acceptable physical domain and include sufficiently distributed experiments in the normalized space. A "round trip" process between physical points and normalized points is essential.

This process is independent of the mechanical model and can be largely undertaken with a quasi-null cost.

3.6. METHODOLOGY

The mechanical model response around P^* is evaluated by a process including successive convergent steps:

- some successive RS are built by assuming an evolution of the RS center that converges to a minimum. This minimum must belong to the definition domain of the DOE, it is, in general, possible to show that it is an absolute minimum on the RS, but not necessarily for the original problem.
- mechanical convergence: the mechanical model is continuously refined the further the process is advanced. It is useless to use the most refined solutions since the first realizations.
- reliability convergence: besides the reliability index and the $FORM$ probability obtained at each step, complementary data are calculated ($SORM$, conditional simulations).

This methodology is completed by creating a database that keeps all the calculation results (see 5).

No sensitive variables can be eliminated by fixing them to their average values when they appear as insignificant from the stochastic point of view.

The resolution of the reliability problem is then undertaken on the RS. Do not forget that the RS is valid only in a restricted domain around the obtained design point P^*.

4. How to validate?

For each set s of realizations r of a design of experiments leading to the construction of a response surface, tests give measures on the quality of obtained approximations. They can be classified according to necessary information for their evaluation.

4.1. PHYSICALLY ADMISSIBLE REALIZATIONS

The choice of a realization $u_i^{(r)}$ implies a physical realization $x_i^{(r)} = T_i^{-1}(u_j)$ that must be physically admissible and correspond to a compatible situation with the mechanical model hypotheses. If not, an aberrant realization is excluded from the *DOE*.

4.2. EXPERIMENT MATRIX

Each experiment (r) defines a row of the experiment matrix $[Z]$:

$$\left\langle z^{(r)} \right\rangle = \left\langle 1, ...u_i^{(r)}, ...u_i^{(r)} u_j^{(r)}... \right\rangle \quad \text{with} \quad [Z] = \begin{bmatrix} \vdots \\ \left\langle z^{(r)} \right\rangle \\ \vdots \end{bmatrix} \quad \text{(dim: } n_e \times l\text{)}$$

The theory of *DOE* suggests then three measures:

1. criterion of the *D*-optimal design: the value of the determinant of $[[Z]^t[Z]]$ is the greatest possible,
2. criterion of the *A*-optimal design: the trace of the matrix $[[Z]^t[Z]]^{-1}$ is the weakest possible,
3. criterion of the *G*-optimal design: the value of the greatest element on the diagonal of $[[Z]^t[Z]]^{-1}$ is the weakest possible.

Let us notice that these measures can be built without mechanical calculations and are therefore easy to evaluate. They can bring an answer to the question: if I have n numerical experiments, how do I choose the $n+1$ to improve measures?

4.3. QUALITY OF THE REGRESSION

It is measured after calculation of realizations $H(u_i^{(r)})$. The measure is the correlation R^2 between evaluations:

$$R^2 = 1 - \frac{\sum_{r=1}^{n_e} \left(H(u_i^{(r)}) - \tilde{H}(u_i^{(r)}) \right)^2}{\sum_{r=1}^{n_e} \left(E(\tilde{H}(u_i^{(r)})) - \tilde{H}(u_i^{(r)}) \right)^2} \longrightarrow 1$$

in which $E(.)$ is the average. This measure is equal to 1 for an interpolation and must be used only to control a regression where a value superior to 0.9 is expected.

4.4. VALUE OF THE LIMIT STATE FUNCTION

After reliability calculation, the design point P^* is obtained. Two measures are proposed:

— the first tests the belonging of P^* to the DOE:

$$I_{app} = 1 - \left(\frac{1}{n_e} \sum_{t=1}^{j} \frac{N_t}{t!} \right)^{-1} \longrightarrow 0$$

where j is the chosen limit for the development and N_t is the number of the DOE points to be less than t standard deviations of P^* according to a given norm:

$$N_t = card \left(P^{(r)} \mid \max_{i=1,m} (u_i^{(r)} - \tilde{u}_i^*) < t \right)$$

$$\text{or } N_t = card \left(P^{(r)} \mid \sqrt{\sum_{i=1}^{m} (u_i^{(r)} - \tilde{u}_i^*)^2} < t \right)$$

This measure tends to zero when the response surface surrounds the design point P^*.

— the second verifies that the design point P^* belongs to the limit state function. It requires a supplementary mechanical calculation:

$$I_a = \frac{H(\tilde{u}_i^*)}{H(E(u_i))} \longrightarrow 0$$

However, it does not prove that it is a minimum on the limit state function.

4.5. SENSITIVITY OF THE RS

Reliability software easily evaluates the sensitivity of the index β to RS parameters. This additional information illustrates the robustness (or non) of the RS. A weak sensitivity shows that approximations have only limited influence on reliability results.

4.6. RECAPITULATION

The figure 1 recapitulates the control tests and their overlapping.

5. Database

The construction of a database allows us to gather information issued from mechanical and reliability calculations. The database is constituted of three sets of data.

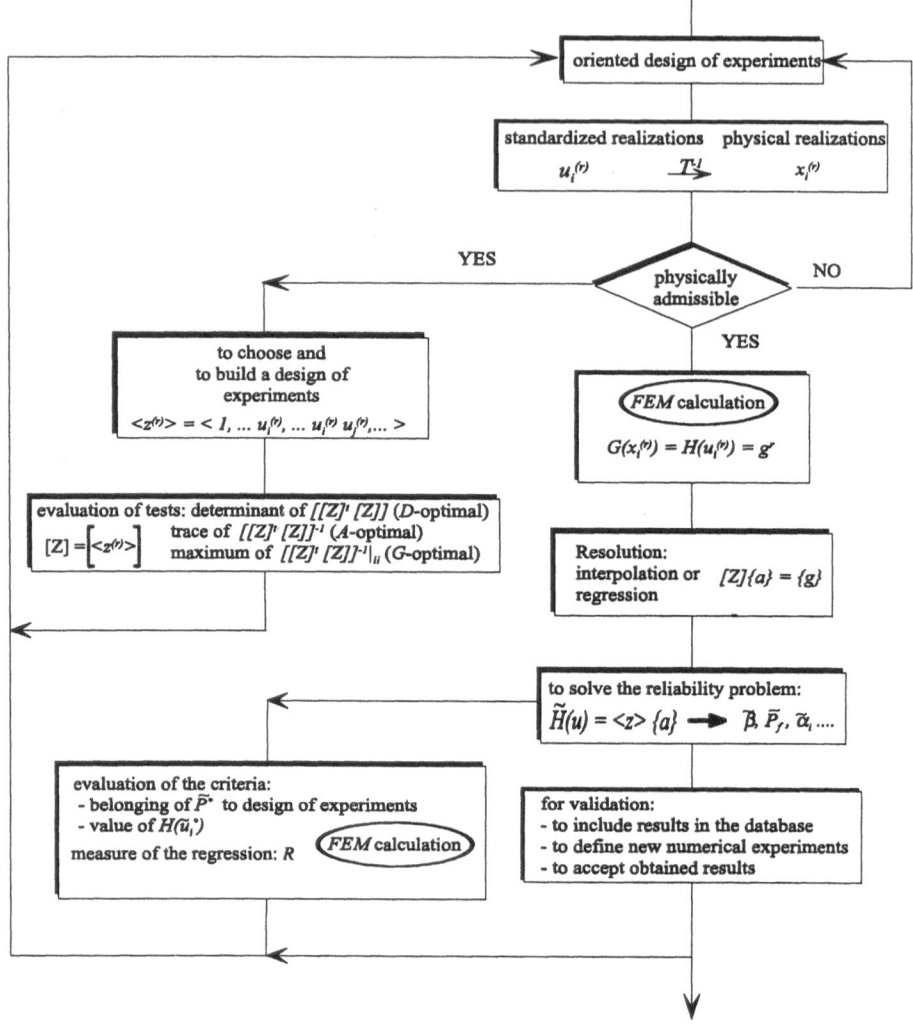

Figure 1. application of the response surface method.

1. relative data from a numerical experiment r: $u_i^{(r)}$; $x_i^{(r)}$; $\langle z^{(r)} \rangle$; $H(u_i^{(r)})$,
2. relative data from a design of experiments s: $[Z^{(s)}]$; $\tilde{H}(u_i)$; $a_k^{(s)}$; list of experiments, measures,
3. results of the reliability calculation: β; $u_i^{*(s)}$; $\alpha_i^{(s)}$; sensitivities of the $a_k^{(s)}$.

6. Example: thick sphere under pressure

First, the proposed method was the subject of applications for validation [16] and, then, for industrial projects. This example is a bench for an illustration to give.

6.1. MECHANICAL MODEL

This study deals with the reliability of a sphere under pressure p_0. The material is considered as homogeneous, without spacial variability. Several scenarios of failure are foreseeable, and we consider only the reach of the yielding stress f_e by the von Mises' stress σ_{eq}. Its interest is to dispose of an analytic solution giving results to obtain:

$$\sigma_{eq}(r = r_0) = \sqrt{\sigma_r^2 + \sigma_\theta^2 - 2\sigma_r\,\sigma_\theta} = |\sigma_r - \sigma_\theta| = \frac{3\,p_0}{2}\frac{r_1^3}{r_1^3 - r_0^3}$$

a relationship in which variables are defined figure 2.

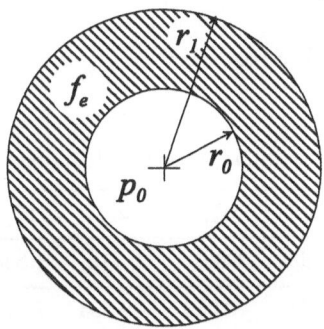

Figure 2. sphere under pressure.

6.2. ELASTOPLASTIC FAILURE

The limit state function represents the elastoplastic failure when σ_{eq} reaches the yielding limit f_e at any point of the sphere:

$$f_e - \sigma_{eq} = f_e - \frac{3\,p_0}{2}\frac{r_1^3}{r_1^3 - r_0^3} \tag{5}$$

4 variables are considered random. They are represented by log-normal densities. Data are gathered table 1.

The obtained *FORM* results (*COMREL* [21]) are given table 2.

TABLE 1. sphere under presure: data

variable X	m_X	σ_X	dens. func.
pressure p_0	$130\,MPa$	$8\,MPa$	log-N
inter. radius r_0	$50\,mm$	$2.5\,mm$	log-N
exter. radius r_1	$100\,mm$	$5\,mm$	log-N
yielding stress f_e	$300\,MPa$	$20\,MPa$	log-N

TABLE 2. reliability analysis ($FORM$): results

β	$\Phi(-\beta)$	$u_{p_0}^*$	$u_{r_0}^*$	$u_{r_1}^*$	$u_{f_e}^*$
3.055	$1.127 \cdot 10^{-3}$	1.884	0.899	-0.899	-2.041

Corrections $SORM$ and *Conditional Simulations* confirm obtained values, respectively $\beta = 3.003$ and $\beta = 3.004$.

6.3. RESPONSE SURFACE SOLUTION

TABLE 3. interpolation with 10 experiments

	β	$u_{p_0}^*$	$u_{r_0}^*$	$u_{r_1}^*$	$u_{f_e}^*$
results	3.196	1.851	0.715	1.880	-1.661
error	0.141	-0.033	-0.184	2.779	0.380
$I_a = 31.59\%$		$I_{app} = -12.33$		$R^2 = 100\%$	

According to the equation 5, the resistance and the solicitation parts are dissociated, and a RS is constructed for the second member only.

There are three random variables and a calculation is undertaken first by interpolation with 10 experiments. Table 3 gathers results. Absolute errors are calculated by comparison with $FORM$ results (table 2). The results are not satisfactory, especially for the variable r_1 and measures confirm this fact, except for R^2, without meaning in the case of an interpolation. 2 experiments are added for the I_a measure to be improved (table: 4). Important distances between experiment points do not modify I_{app} but I_a show that the new design point belongs to the limit state. A last regression is undertaken by bringing the experiment points closer for I_{app} to be improved

TABLE 4. regression with 12 experiments

	β	$u_{p_0}^*$	$u_{r_0}^*$	$u_{r_1}^*$	$u_{f_e}^*$
results	3.113	1.923	0.826	−0.661	−2.208
error	0.058	0.039	−0.073	0.238	−0.167

$$I_a = -1.56\% \qquad I_{app} = -12.33 \qquad R^2 = 99.9\%$$

TABLE 5. regression with 12 P^*-centered experiments

	β	$u_{p_0}^*$	$u_{r_0}^*$	$u_{r_1}^*$	$u_{f_e}^*$
results	3.053	1.884	0.895	−0.897	−2.040
error	−0.002	0.000	−0.004	0.002	0.001

$$I_a = 0.07\% \qquad I_{app} = 0.0 \qquad R^2 = 100\%$$

(all points are chosen to be less than one standard deviation). Table 5 shows final results.

7. Conclusion

The response surface method leads to a reliability problem whose resolution is significantly simplified. Its use brings interesting information with limited software time if the number of random variables is limited. Indeed, a complete identification requires at least $(n+1)(n+2)/2$ numerical experiments without the possibility to validate the results.

The proposed approach, developed in this paper, aims at organizing the method after having undertaken some methodological choices:

— utilization of Nataf's transformation,
— construction of the design of experiments in the standardized space,
— proposition of measures for the quality of the solution to evaluate.

Rather than hoping in an automatic method that will always be very heavy, it is judicious to call on the designer's expertise for the successive numerical experiments to choose. In the future, the database will be coupled with an expert system for series of DOE to be built.

Presently, the application of the proposed method allows us to process all reliability problems whose mechanical models require a FEM calculation. No adaptation of the FEM is necessary.

If another combined method is used, it implies, in all cases, the calculation of specific realizations of the limit state function. They can be gathered in the database and then used for a response surface to be built, thus bringing additional information on the quality of the obtained results.

Acknowledgement

The author wishes to acknowledge Electricité de France (EDF/SEPTEN) for the financial support of this research and M. Meynet and M. Pendola, graduate students at Institut Français de Mécanique Avancée, for their participation to the numerical calculation.

References

1. T. Abdo and R. Rackwitz. A new beta-point algorithm for large time-invariant and time variant reliability problems. In *Reliability and Optimization of Structures*. 3th WG 7.5 IFIP conference, Berkeley, 1990.
2. C.G. Bucher and U. Bourgund. A fast and efficient response surface approach for structural reliability problems. *Structural Safety*, (7):57–66, 1990.
3. D. Benoist, Y. Tourbier, and S. Germain. *Plans d'expériences : construction et analyse*. Lavoisier, 1994.
4. N. Devictor. *Fiabilité et mécanique : méthodes FORM/SORM et couplages avec des codes d'éléments finis par des surfaces de réponse adaptatives*. PhD thesis, Université Blaise Pascal – Clermont II, décembre 1996.
5. A. Der Kiureghian and P.L. Liu. Structural reliability under incomplete probability information. *J. Eng. Mechanics*, (112(1)), 1986.
6. O. Ditlevsen and H.O. Madsen. *Structural Reliability Methods*. John Wiley & Sons, 1996.
7. I. Enevoldsen, M.H. Faber, and J.D. Srensen. Adaptative response surface techniques in reliability estimation. In Schueller, Shinozuka, and Yao, editors, *Structural Safety and Reliability*, pages 1257–1264, 1994.
8. K. El-Tawil, J.P. Muzeau, and M. Lemaire. Reliability method to solve mechanical problems with implicit limit state. In R. Rackwitz and P. Thoft-Christensen, editors, *Reliability and Optimization of Structures, Proc of the 4th IFIP WG 7.5 conference*, pages 181–190. Springer-Verlag, 1991.
9. L. Faravelli. Response surface approach for reliability analysis. *Journal of Engineering Mechanics*, 115(12), 1989.
10. S.H. Kim and S.W. Na. Response surface method using vector projected sampling points. *Structural Safety*, 19(1):3–19, 1997.
11. M. Lemaire. Reliability and mechanical design. *Reliability engineering & system safety*, 55(2):163–170, February 1997.
12. Y.W. Liu and F. Moses. A sequential response surface method and its application in the reliability analysis of aircraft structural systems. *Structural Safety*, 16(1+2):39–46, 1994.
13. M. Lemaire, A. Mohamed, and O. Florès-Macias. The use of finite element codes for the reliability of structural systems. In D.M. Frangopol, R.B. Corotis, and R. Rackwitz, editors, *Reliability and Optimization of Structural Systems'96, 7th IFIP WG 7.5 Working Conference, Boulder, 2-4 April*. Elsevier (Pergamon), 1997.
14. G. Maymon. Direct computation of the design point of a stochastic structure using a finite element code. *Structural Safety*, 14(3):185–202, 1994.
15. J. C. Mitteau. Error estimates for FORM and SORM computations of failure probability. In *Probabilistic Mechanics and Structural Reliability, Worcester, August 7-9*. 7th ASCE speciality conference, 1996.

16. M. Meynet and M. Lemaire. Utilisation de plans d'expériences dans l'espace normé pour la construction de surface de réponse. Technical report, LaRAMA – Institut Français de Mécanique Avancée et Université Blaise Pascal, Clermont-Fd, France, 1996.

17. J.P. Muzeau, M. Lemaire, and K. El-Tawil. Méthode fiabiliste des surfaces de réponse quadratiques (srq) et évaluation des règlements. *Construction Métallique*, (3):41–52, septembre 1992.

18. A. Mohamed, F. Suau, and M. Lemaire. A new tool for reliability based design with ANSYS FEA. In *ANSYS Conference & Exhibition, Houston, USA*, pages 3.13–3.23. ANSYS, Inc, 1996.

19. M.R. Rajashekhar and B.R. Ellingwood. A new look at the response surface approach for reliability analysis. *Structural Safety*, 12(3):205–220, 1993.

20. R. Rackwitz and B. Fiessler. Structural reliability under combined random load sequences. *Computers and Structures*, (vol 9):489–494, 1979.

21. GmbH RCP and ApS RCP. *A Structural Reliability Analysis Program: STRUREL*. RCP GmbH, München, Germany, and RCP ApS, Mariager, Denmark, 1993.

PROBABILISTIC DESIGN OF PERFORMANCE IN GLUED LAMINATED TIMBER

H. Mihashi and N. Itagaki

Dept. of Architecture and Building Engineering, Tohoku University,
Sendai 980-77, Japan

Abstract

Because of the environmental adaptability and high strength–to-weight ratio, wood may be the most sustainable building material in the 21st century, though the large scatter of mechanical properties need to be reduced.

This study presents a design method for performance in glulam (glued laminated) beams of timber. An analytical model was made for predicting the modulus of rupture (MOR) of glulam beams and Monte-Carlo simulation was carried out for predicting the variability. Characteristic points of the present model are to consider the effects of reinforcement due to gluing laminae and of the plastic deformation in the compressive zone. The predicted results by means of the present model agreed well with the experimental results.

1. Introduction

Wood is one of the most traditional building materials which have been widely used. Then why is the wood still one of the most attractive and important building materials in the 21st century?

Now we know the remained amount of resources is limited and the earth has been polluted by CO_2 gas and various types of waste.   Especially we are strongly required to reduce the discharged quantity of CO_2 gas towards the 21st century. Therefore roles of environmental preservation and sustainable resource management   need to be seriously considered.

Modern building materials such as steel, concrete and plastic only consume the resources but can't be regenerated, require a great deal of energy to be produced, and discharge a tremendous   volume of waste. It is only wood that is a renewable natural

333

G. N. Frantziskonis (ed.), PROBAMAT – 21st Century: Probabilities and Materials, 333–345.

resource which can be planted and cultivated. Additionally wood requires only limited amounts of energy in preparation for end of use and it can be made more durable by various treatment. During the growth of wood, forest can consume CO_2 gas. Furthermore wood is easily recycled and naturally returned to the earth after the service life.

Wood has generally high strength-to-weight ratio in the longitudinal direction. The most serious weak point of wood as the modern building materials, is the scatter of the mechanical performance such as strength and stiffness. Since wood is a naturally produced material, the mechanical performances widely vary. For reducing such variability, glulam (that is, glued laminated) timber has been often used, which can be produced on the basis of designed arrangement of laminae whose properties are evaluated by E-rating or proof loading. Glulam is available in a wide variety of sizes and shapes, which can be suitable for all types of structures.

For the limit state design based on a reliability concept (for example, [1]), it is desired to produce timber with less variability in mechanical properties and to predict accurately the variability.

In order to calculate accurately the probability of failure, complete descriptions of the variability of the quantities involved must be available. Quantification of the variability is essential in order to assess the failure probabilities. There are various methods by which the variabilities can be described.

For reliability-based design, huge number of tests are usually carried out to build up a reliable database for the probability analysis. For predicting the probabilistic distribution of strength of glulam timber, however, conducting such a large number of tests is not realistic because of the cost and time consuming. To overcome such problem, simulation methods have been developed to predict the distribution on the basis of the properties of laminae.

In general, the flexural or tensile strength of a glulam beam is dependent upon: 1) the particular combination of lamina grades in the beam lay-up; 2) strength-reducing characteristics such as knots and slope-of-grain allowed in each lamination grade; 3) the E-rating or proof-loading of the individual laminae, or both; 4) strength and location of end-joints; 5) thickness of the laminae; 6) the size of the beam and the particular distribution of stresses [2].

A method widely used to predict the flexural strength of beams is based on a reduction of the flexural strength of clear wood by means of a factor that takes into account the particular lamina grades in the lay-up and the knots within each lamina [3]. However, this method doesn't allow the influence of end-joints nor for taking into account the results of E-rating or proof-loading. Foschi et al. [2] developed a more flexible approach to predict the flexural strength of glulam timbers, taking into consideration all the factors that influence the stiffness and the strength.

In this approach, however, the failure criterion for the lamina is based on the tensile strength and the nonlinearity in the compressive zone is not taken into consideration. Moreover the influence of the reinforcement due to gluing laminae is taken into account by the assumption that the specimen is fully restrained against lateral movement, though the failure load may be recorded before the lateral deflection starts.

Hayashi [4] developed a probabilistic model for predicting the flexural strength of glulam composed of visually-graded laminae. Hirashima and co-workers [5] also developed a probabilistic model for predicting the modulus of rupture (MOR) of glulam beams in which the criteria for rupture of a beam was critically discussed.

The present study is aimed to predict the variability in flexural strength of glulam timber for the reliability-based design. In this paper, an analytical model is presented for predicting MOR of glulam beams, though the influence of end-joints is not taken into account [6]. The procedure is shown as the flowchart in Figure 1. Monte-Carlo simulation was carried out for developing the strength database. Previous models for flexural failure of timber were improved by taking into consideration the influence of reinforcement due to gluing laminae and of the plastic deformation in the compressive zone. The predicted results by means of the present model agreed well with the experimental results.

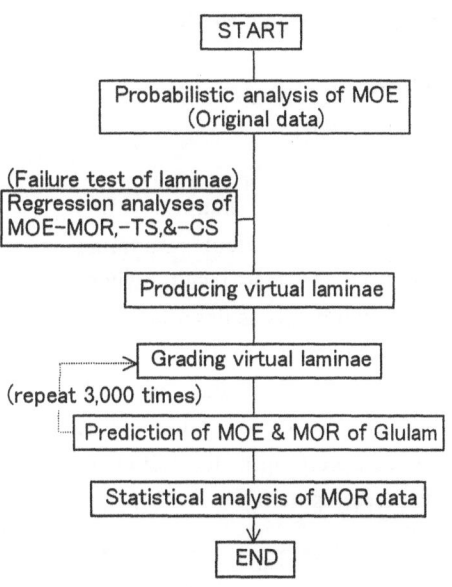

Figure .1 Flowchart to predict MOR

2. Material properties database

Since wood has generally large variability in its material properties as a building material, extensive testing has to be undertaken in order to get a meaningful distribution representing the strength of a particular configuration of glulam beams. In order to achieve such testing, a simulation program is developed which creates the database of lamina's material properties using Monte-Carlo simulation. In general, strength and modulus of elasticity (MOE) of wood are correlated each other. Since non-destructive E-rating tests are usually conducted on laminae, strength properties of laminae can be evaluated on the basis of the database of MOE.

The major factors affecting the beam strength are MOE and the ultimate strength of laminae that is strongly influenced by knots existing in the lamina. Since wood is a naturally grown material, these factors can be considered as random variables. If the distributions of the parameters such as MOE, ultimate strength and data about knots of a fairly large population of samples are known, beam strength values can be calculated by picking up the values of these parameters at random, using the rational model. The data of material properties can be approximated to known types of probabilistic distributions (normal, Weibull, log-normal, etc.) to facilitate Monte-Carlo simulations.

3. Probabilistic Model to Simulate MOR of Glulam Beams

3.1 FABRICATION OF VIRTUAL LAMINAE BY MEANS OF MONTE-CARLO SIMULATION

MOE of virtual laminae was determined by random sampling according to the test results whose probabilistic property is described by a probabilistic distribution function.

Then the mean values of each strength of the virtual lamina was determined according to the following equation (see Figure 2.):

$$f_i' = A\ MOE + B \qquad (1)$$

Specific values of the each strength were obtained by considering the deviation U_i from the mean value (Figure 2.).

In the present study, the deviation was evaluated by the assumption that the strength distributes around the regression line with the standard deviation of Se that is the standard error.

Thus the strength of a virtual

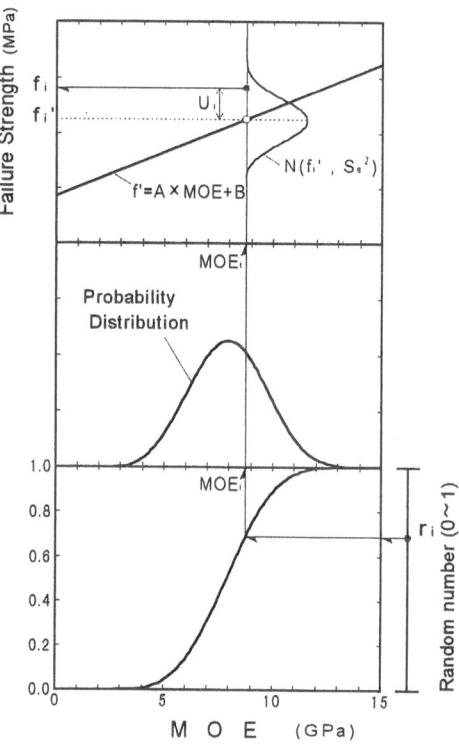

Figure 2. Determination of mechanical properties of virtual laminae.

lamina was obtained as follows:

$$f_i = f_i' + U_i \qquad (2)$$

3.2. FABRICATION OF VIRTUAL GLULAM WITH VIRTUAL LAMINAE

After grading all virtual laminae into three groups as shown in Figure 3 by means of MOE, virtual glulam beams were fabricated. MOE of the glulam is given by eq.(3).

$$E I = \sum_{i=1}^{n} E_i \; I_{i\text{-}nn'} \qquad (3)$$

where E: averaged Young's modulus of glulam beam, I: moment of second order of glulam beam, E_i : Young's modulus of the *i-th* lamina, $I_{i\text{-}nn'}$: moment of second order of the *i-th* lamina regarding the neutral axis of the glulam beam.

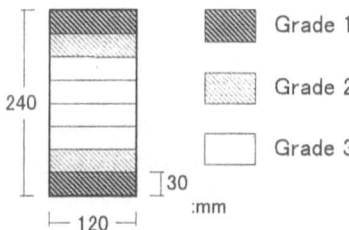

Figure 3. Composition of laminae in section of glulam.

4. Analysis of MOR

4.1. FAILURE CRITERION

In the previous study (Mihashi et al. [7]), the failure stress criterion was examined. When the failure of a glulam beam starts from laminae under tensile stress (Figure 4), the following three failure criteria are assumed.

[Model 1]: Failure occurs when the *i-th* fiber stress reaches MOR of the *i-th* lamina.

[Model 2]: Failure occurs when the *i-th* fiber stress reaches the tensile strength of the *i-th* lamina.

[Model 3]: Failure occurs when the *i-th* fiber stress reaches the following combined stress condition.

$$\left(\frac{\sigma_{i\text{-}b}}{f_{i\text{-}b}} \right)^2 + \left(\frac{\sigma_{i\text{-}t}}{f_{i\text{-}t}} \right)^2 = 1 \qquad (4)$$

where $\sigma_{i\text{-}b}$: flexural stress component of the *i-th* lamina, $\sigma_{i\text{-}t}$: tensile stress component of the *i-th* lamina, $f_{i\text{-}b}$: modulus of rupture of the *i-th* lamina, $f_{i\text{-}t}$: tensile strength of the *i-th* lamina.

338

Stress distribution of section of glulam beam

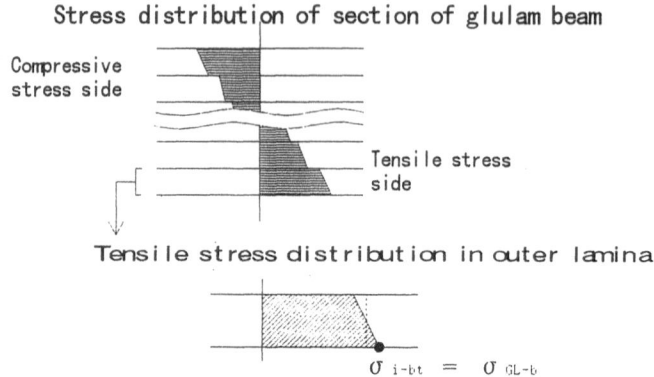

Figure 4. Stress condition of laminae in glulam beams.

Among these three criteria, simulated results obtained with Model 3 gave the best agreement with the corresponding experimental results, though a certain discrepancy was still observed. Then the reinforcing effect due to lamination and the nonlinearity of the lamina under compressive stress were taken into consideration. These two additional treatments finally gave a very good agreement with the experimental results (Mihashi et al. [7]).

4.2. REINFORCING EFFECT DUE TO GLUING LAMINAE

Database of material properties which are used to estimate the failure of the *i-th* lamina is based on the experimental results tested for single laminae. Lamina in the glulam beam, however, is no longer a single lamina but glued with the neighboring lamina. Especially the tensile strength of the lamina is strongly influenced by the existence of material defects such as knots. Once the laminae are glued each other, the weak points due to the material defects are reinforced and strength may be increased.

Mihashi [8] developed a stochastic model to describe the variability of strength in concrete in which the influence of material defects was taken into consideration. In the stochastic model, the non-failure probability and the expected strength of a specimen composed of m elements are described by eqs.(5) and (6), respectively.

$$P(\sigma) = exp\left\{-\frac{mL}{(\beta+1)\dot{\sigma}}\sigma^{\beta+1}\right\} \qquad (5)$$

$$\sigma = \left[\frac{\beta \dot{\sigma}}{m \, L} \right]^{\frac{1}{\beta+1}} \qquad (6)$$

where $\dot{\sigma}$ is the rate of loading, L is a material parameter considering the material defects, β is a material constant.

Now a lamina is assumed to be under a uniaxial tensile stress in which m' dominant material defects are included. The number m of the material defects may be reduced by gluing with other laminae from m to m'. Then the tensile strength of glued laminae may be increased as described by eq.(7).

$$\frac{\overline{\sigma'}}{\overline{\sigma}} = \left[\frac{m}{m'} \right]^{\frac{1}{\beta+1}} \qquad (7)$$

4.3. NONLINEARITY IN COMPRESSIVE ZONE OF GLULAM BEAMS

Figure 5. shows typical load versus deflection curves of glulam beams of Japanese cedar. The proportional limit loads are much lower than the maximum loads. Consequently it is obvious that the nonlinearity in the compressive zone needs to be taken into consideration. In the present study, a bi-linear stress strain relation (Figure. 6) was introduced for this purpose.

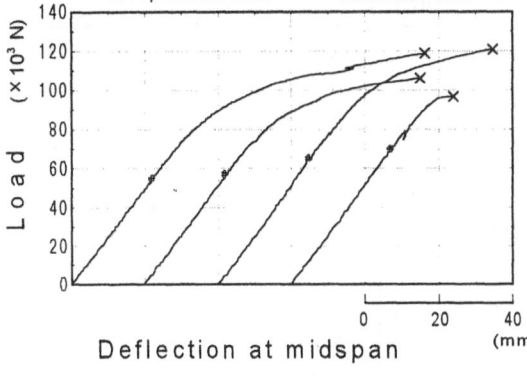

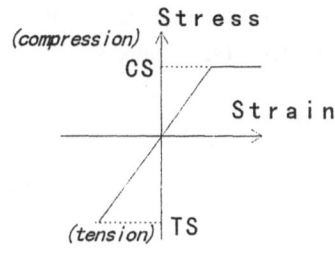

Figure 6. Stress-strain curve of lamina in the present model.

Figure 5. Load-deflection curves in bend tests of glulam beams of Japanese cedar.

5. Results and Discussion

5.1. PROBABILITY DISTRIBUTION OF MOE

MOE of laminae made of Japanese cedar was measured by non-destructive testing. Total number of the tested laminae was 126. The width. thickness and length of the lamina were 120mm, 30mm and 4,000mm, respectively. Main data of the MOE test are shown in Table 1. The obtained MOE data were approximated to well-known probabilistic types of normal, Weibull and log-normal distributions. The agreement to each distribution was approved by Kolmogorov-Smirnov statistical test, and the best fitness was obtained in case of Weibull distribution. The frequency distribution of MOE of laminae and the probabilistic density function are shown in Figure 7.

On the bases of the distribution, the population of the laminae is subdivided into three grades as shown in Table 2. For the purpose of optimizing the number of the laminae for producing glulam beams, the ratio between these three grades was made to be 1:2:2 (see Figure 3).

Table 1. Properties of laminae of used Japanese cedar

Total number of specimens: 126

	Specific Gravity (kg/m³)	Average Annual Ring Width (mm)	MOE (Gpa)
Minimum	300	1. 9	4. 02
Maximum	460	8. 5	11. 06
Mean	370	4. 0	7. 80
S.D.	30	1. 4	1. 59
C.V.(%)	8. 19	35. 0	20. 44

Table 2 Grade limits of laminae

Grade 1		MOE $\geq$ 8.97	Gpa
Grade 2	8.97 >	MOE $\geq$ 7.94	Gpa
Grade 3	7.94 >	MOE	Gpa

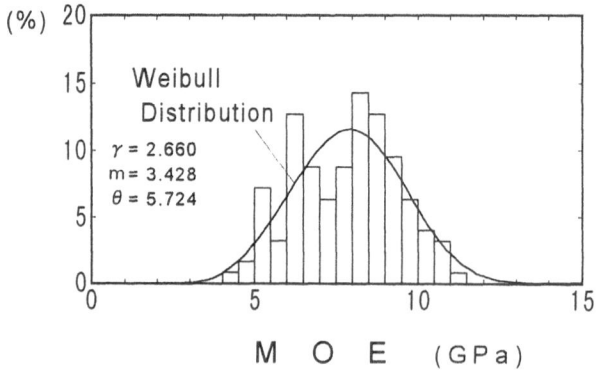

Figure 7. Frequency of MOE of laminae and probability distribution.

5.2 STRENGTH PROPERTIES OF LAMINAE

From each group of three grades for MOE, totally 118 pieces of lamina were randomly chosen   whose size was 120x30x2,000(mm). All of these pieces were used for tension test. After the tension test, unbroken parts of these specimens were cut into two pieces which were 120x30x690(mm) for bend test, and 120x30x180(mm) for compression test. In selecting specimens for these tests, existence of neither knots nor other defects was taken into consideration.

Distributions of strength properties of a fairly large population of samples are obtained. Bend   tests were carried out according to Bend Test C in JAS (Japanese Agricultural Standard) for Structural Heavy Timber of Glulam in which the flexural strength was evaluated by three point bending. Span length was 525mm in this case. The tensile strength (TS) was determined by direct tensile test and the compressive strength (CS) was determined by uniaxial compression tests.   MOE was measured by strain gauges on each specimen. Distributions of these three strength properties are shown in Figures 8, 9 and 10, where r is correlation coefficient and Se is standard error.

5.3   FABRICATING GLULAM BEAMS AND BEND TESTS

Using 126 laminae, fifteen glulam beams of 8 ply were fabricated according to laying-up shown in Figure 3. The size of lamina pieces used for the fabrication of glulam beams was of 120x30x4,000mm. Only MOE of each lamina was used as the parameter to subdivided laminae into three grades but the knot ratio was not taken into consideration in this case. Those beams were bent by a third-point loading whose span length was 3,600mm. MOE was calculated from the flexure   in the part of 960mm under the constant moment. For the tested glulam beams, the mean and standard deviation values of MOR are 43.7 and 3.57 MPa, respectively.

5.4. FABRICATION OF VIRTUAL LAMINA

The virtual MOE was expressed by a uniform random number: r as the inverse function of three-parameter Weibull distribution function (Figure 7).

$$MOE = \theta \times \{-ln(1-r)\}^{1/m} + \gamma \qquad (8)$$

where   m: shape parameter, θ : scale parameter, γ : position parameter.

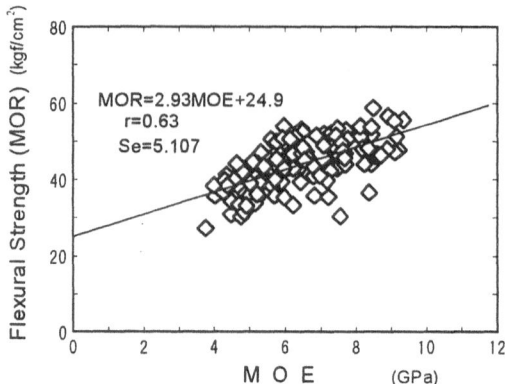

Figure 8. Relationship between MOE and flexural strength of laminae

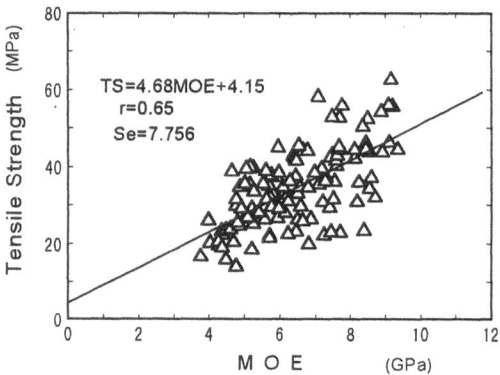

Figure 9. Relationship between MOE and tensile strength of laminae

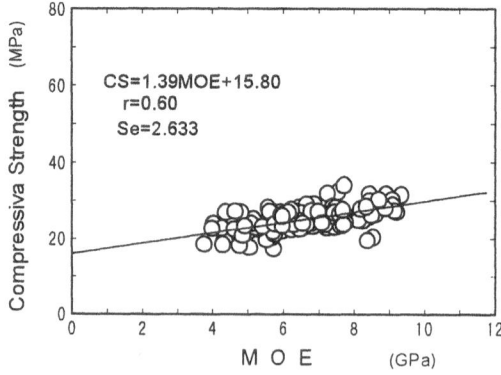

Figure 10. Relationship between MOE and compressive strength of laminae

5.5. DETERMINATION OF MATERIAL CONSTANT β

The relation between the non-failure probability $P(\sigma)$ and tensile strength of laminae is shown in Figure 11. From the slope of the linear regression driven from eq.(5), the material constant β can be determined as follows:

$$ln\{-ln\ P(\sigma)\} = (\beta + 1)\ ln\ \sigma\ +\ const. \tag{9}$$

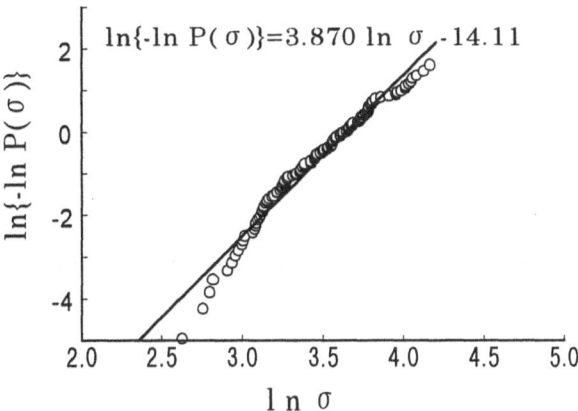

Figure 11. Relation between non-failure probability and tensile strength of laminae.

If size and location of knots in each lamina are measured, the strength properties may be adjusted according to the knot ratio by using rather simple calculations. Since the occurrence of knots in lumber pieces is considered to be at random, the influence of knots were not taken into consideration in the present study.

5.6. ANALYTICAL PREDICTION OF VARIABILITY IN MOR

According to the procedure shown in chapter 3, totally 914 glulam beams for simulation were fabricated. Then MOR of all these beams was analyzed.

Comparison of simulated probability function of MOR with experimental results of MOR distribution of glulam beams is shown in Figure 12. Very good agreement between these two results is observed. It means the present probabilistic model can predict MOR of glulam beams of Japanese cedar including the variability.

344

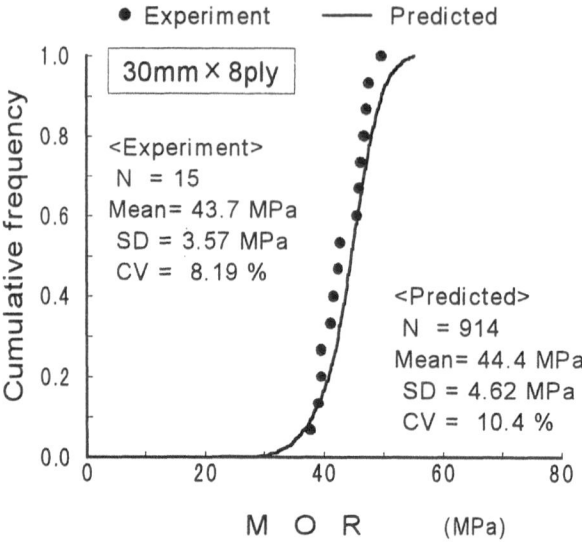

Figure 12. Comparison of predicted distribution curve (solid line) with experimental results.

6. Conclusions

While wood may be the most sustainable building material in the 21st century, probabilistic models are essential to utilize it as an advanced building material available for the reliability-based design because of the large variability of mechanical properties. The present paper proposed a procedure based on a probabilistic model to design the performance of glulam timber.

From the comparison between experiments and simulated results, it can be concluded that the proposed probabilistic model does sufficiently predict the main features required to predict the   strength of glulam beams including the variability.

Although the model has been applied to Japanese cedar, it is completely general and applicable to other kinds as long as the basic database of material properties are available.

Acknowledgment

The present study was performed as one of the collaboration projects between Building Materials Science Laboratory, Tohoku University and Forestry Testing Laboratory of Miyagi Prefecture. Authors would express their thanks for the collaboration.

References

[1] FOSCHI, R.O., FOLZ, B.R. and YAO, F.Z.: Reliability-Based Design of Wood Structures, *Structural Research Series*, Dept. of Civil Engineering, University of British Columbia, Vancouver, Canada, 1989, Report 34.

[2] FOSCHI, R.O. and BARRETT, J.D.: Glued-Laminated Beam Strength: A Model, *ASCE*, J. of Structural Division, 1980, 106 (8), 1725-1754.

[3] FREAS, A.D. and SELBO, M.L.: Fabrication and Design of Glued-Laminated Wood Structural Members, U.S. Dept. of Agriculture, Washington, D.C., 1954, *Technical Bulletin 1069*.

[4] HAYASHI, T.: Performance Prediction of Wood Laminates by a Probabilistic Model IV, *Journal of the Japan Wood Research Society*, 1990, 36 (10), 812-818.

[5] HIRASHIMA, Y., YAMAMOTO, Y. and SUZUKI, S.: Modeling for the Strength of Glulam Beams and for Their Probabilistic Distributions, *Journal of the Japan Wood Research Society*, 1994 40 (11), 1172-1179.

[6] ITAGAKI, N. and MIHASHI, H.: Prediction of Distribution of MOR in Glulam Beams by Means of A Probabilistic Model, *ICCOSSAR '97*, (in press)

[7] MIHASHI, H., ITAGAKI, N., ITO, Y. and SUZUKI, N.: Analytical Model for the Design of Performance in Glulam Beams of Japanese Cedar I, *Journal of the Japan Wood Research Society*, 1996, 42 (2), 122-129.

[8] MIHASHI, H.: A Stochastic Theory for Fracture of Concrete, *Fracture Mechanics of Concrete*, F.H. Wittmann (ed.), Elsevier Science Publishers B.V., Amsterdam, 1983, 301-339.

SCALE EFFECT FOR CONCRETE SPECIMENS: A NUMERICAL MODEL

Numerical modelling of concrete elements of increasing size

F. BONTEMPI
Deparment of Structural Engineering, Polytechnic of Milan,
P.zza L. da Vinci 32, 20133-Milan, Italy.

AND

F. CASCIATI
Department of Structural Mechanics, Univesity of Pavia,
Via Ferrata 1 - 27100 Pavia - Italy

Abstract

A numerical model for concrete specimens, holding at a micro-, meso- and macro-level is pursued. The constraints are: simplicity (the basic constitutive law must be described by two or three parameters), objectivity (the results are insensitive to the mesh size) and consistency (the results must agree with laboratory measurements: size effect, softening, crack localization and tension-compression ratio). A model was recently proposed by the authors to catch the essential of the phenomenon. In this paper, it is improved to satisfy some formal requirements and a better accuracy is achieved. This is emphasized through a detailed check of the constraints.

Keywords: concrete, discrete model, material chaos, numerical modelling, stochastic methods, softening.

1. Introduction

The size scale effect *(whole-structure vs. elementary-material volume)* for concrete specimens is the result of the interaction of several nonlinear and stochastic aspects. The size measure of the elementary material volume depends on a representative size of the aggregate; it can be regarded as the size of the maximum aggregate. The whole structure size is related with the full scale dimensions.

In synthesis, concrete is a material which shows:

- size-effect [1]: to increase the size of the structure results in a decrease of the nominal ultimate strength and of the overall ductility;

G. N. Frantziskonis (ed.), PROBAMAT – 21st Century: Probabilities and Materials, 347–366.

- a significant softening behaviour in both tension and compression. Unstable computations may occur and mesh refining may converge to physically inconsistent results [2] when numerical discretizations are made of strain-softening materials, The solution may depend on the size, shape and orientation of the mesh and the model could give rise to a *non-objective* representation of the structural problem.
- microstructure changes, which occur during the fracturing process [3]. Standard stress-strain relationships are derived from the force-displacement curves obtained from testing devices: the forces are divided by the initial load-carrying area and the changes in length by the original length of the specimen. Such a policy catches the essential of the physical behaviour in the hardening phase, but fails when tension and/or compression softening branches are present [4].

As a consequence, the use of standard continuum mechanics models results *incosistent* with the observed behaviour. This suggested the introduction of several regularization techniques, each showing its own analysis approach and computational technique [5]. In this way one can produce formulations for localized structural problems, like the interaction between reinforcement arrangement and concrete body, or formulations for global structural problems, like the analysis of a whole dam. The drawback is that, more or less explicitly, each of these regularization techniques focuses attention on a fixed structural size level and the results cannot be transferred to a different size.

Indeed each engineering application is focused on one of the following three size levels (see Figure 1):

- elementary material volume;
- structural element;
- whole structure.

At the first level, the interactions of aggregate, mortar and reinforcement (if any) are relatively simple, due to the well separated nature of its components. At the upper level, many of the nonlinear and stochastic aspects are averaged by a sort of integration on the large volume of the whole structure. These two situations, however, differs for how the crack alters the configuration (see Figure 2). At the elementary level, the cracks develop at the interface between aggregate and mortar and follow the border between the two homogeneous regions. The application of classical fracture theory methods is prevented due to sliding between aggregates and mortar. By contrast these methods are applicable and give quite realistic results when applied at the whole structural level.

The heterogeneity of the specimens is magnified at the intermediate level, where the interactions between the two phases (aggregate and mortar) must be carefully modelled even in view of a simple response analysis. Moreover, the disorder of the aggregate location plays a fundamental role in the development of the nonlinear aspects, initiating the crack patterns and deciding its evolution at the increase of the external load. Further considerations are required in the presence of reinforcement devices.

A simple model was recently proposed by the authors [6] to catch the essential of the phenomenon independently of the size level. In this paper, it is improved to

a

b

c

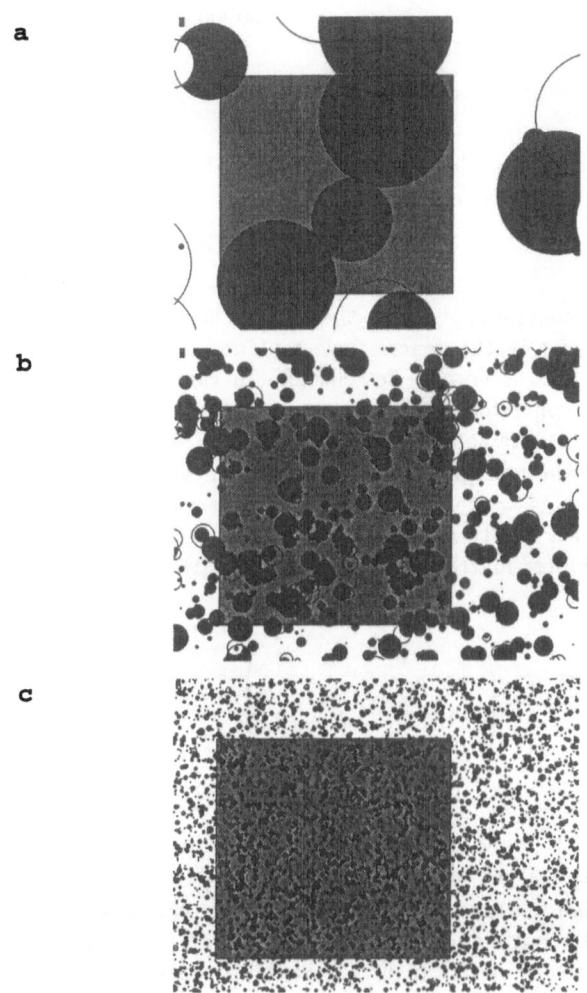

Figure 1. Geometrical aspects of specimens of the size a) of the maximum diameter of the aggregate; b) of a reference structural element; c) of the whole structure.

satisfy some formal requirements. A better accuracy is also achieved as emphasized by a detailed check of all the problem constraints.

2. Governing relations

The study is restricted to structural elements characterized by two dimensions larger than the third one (plane stress states).

2.1. AN EARLY MODEL

For the discretization of the concrete specimen, the authors suggested [6] to introduce a random distribution of nodes in the concrete element and to perform a Delauney triangulation [7] associated with a Voronoi tessellation. In this way one samples the spatial heterogeneity of the concrete through the randomness of the location of the nodes. From each triangle one obtains 3 elements, whose area A depends on the area A_T, on the perimeter P_T of the triangle itself and on the thickness d of the concrete element, through the relation $A = c \cdot \frac{A_T \cdot d}{P_T}$, c being a constant. The basic element connects two nodes at distance L with a straight element characterized by area A and material stiffness E. The stiffness matrix $\underline{\underline{K}}$ can be written with reference to the local coordinates system [8] as

$$\underline{\underline{K}} = \frac{E \cdot A}{L} \cdot \begin{pmatrix} 1 & 0 & -1 & 0 \\ 0 & f & 0 & -f \\ -1 & 0 & 1 & 0 \\ 0 & -f & 0 & f \end{pmatrix} \tag{1}$$

where the degrees of freedom are $\underline{q} = [q_1 \ q_2 \ q_3 \ q_4]^T$: of course $q_3 - q_1$ is the elongation along the axis of the element, while $q_4 - q_2$ is the difference of the displacements normal to that axis. It means that, in addition to the axial stiffness $k_a = \frac{EA}{L}$, there is a tangential stiffness represented as a fraction f of that axial one: $k_t = f \cdot k_a = f \cdot \frac{EA}{L}$. The nodes are supposed unable to rotate and hence the element end rotations are imposed to be zero. The element stiffness matrix of Eq.(1) is then transformed in the global reference system and assembled to form the global stiffness matrix $\underline{\underline{K}}_{global}$. The total Lagrangian formulation is used to take into account large displacements. To match the elastic behaviour with 0.15 as Poisson ratio, one puts $f = 1/4$ and $c = 25/8$. The global system is governed by the equation

$$\underline{\underline{K}}_{global}(\underline{q}_{global}) \cdot \underline{q}_{global} = \underline{P}_{global} \tag{2}$$

where $\underline{q}_{global}$ and $\underline{P}_{global}$ are respectively the displacements vector for the whole structural element and the applied load vector. Eq.(2) is solved using a secant approach [9], for given displacements. In this way, one can also follow the structural response in the post-critical branch. Possible snap-backs can be detected but cannot be followed.

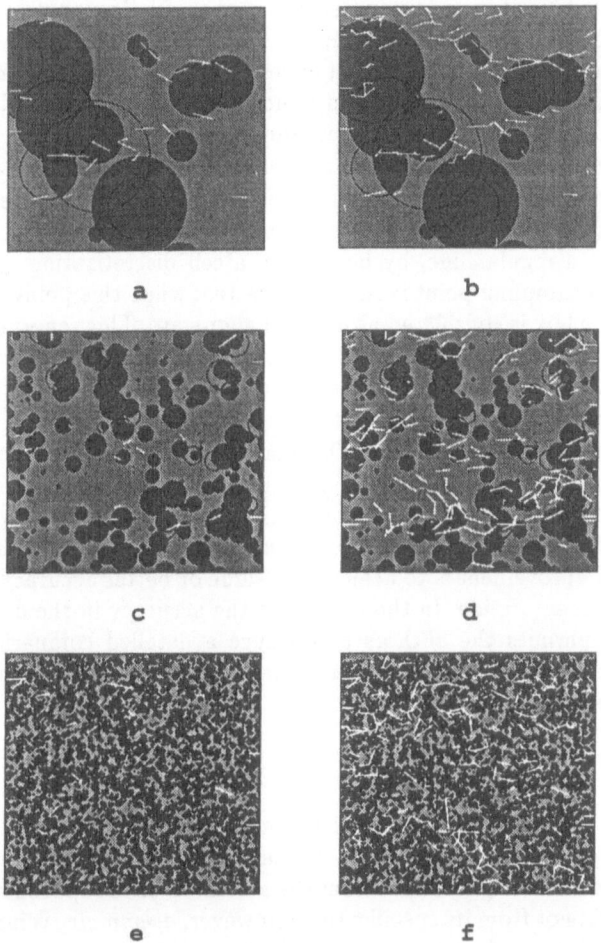

Figure 2. Crack pattern in a specimen of the size a)-b) of the maximum diameter of the aggregate; c)-d) of a reference structural element; e-f) of the whole structure. The left column refers to the initial load level; the right column to a damaged situation.

A Mohr-Coulomb failure criterion is adopted with given friction angle and given ratio between tension and compression strength; the strength depends on the element nature. Each element can result of three different types: *aggregate, concrete mortar* and *interface bond*. The words aggregate, mortar and bond, have to be interpreted as three different materials with decreasing values of resistance and stiffness. They partially reflect the real characteristics of aggregate, concrete matrix and interface bond.

A basic aspect of the proposed scheme is the renormalization [10] of the aggregate distribution. At a microscale level one encounters two materials (aggregate and concrete matrix) interfaced, along a regular line, by bonds. By a cell discretization, boxes are introduced. In each box a sampling point is randomly tested: when this point turns out to be aggregate, the whole box is considered as made of aggregate. This renormalization technique, that only acts upon the geometrical aspects of the aggregate distribution, is repeated until the size of the boxes is the same of the size of the basic element considered in the discretization.

Reference [6] provides an investigation of the material properties one can derive from such a model. Its peculiar aspect was the presence of a shear component in the stiffness matrix (without its bending moment companion which would be very low due to the bar length) smoothing some indesiderable features in the softening branch. The objective of the present improvement is to achieve the same or better accuracy without forcing the presence of a shear action. In the meantime the accuracy in the description of the softening branch permits the authors to produce a detailed comparison with current practice and the simplicity of the model makes it prone to be implemented in structural analysis computer codes.

2.2. SOME REMARKS

Material properties are expected as a result of the global model, not as an obvious consequence of suitable premises. For instance, the behaviour of the basic elements is assumed to be linear elastic until (brittle) failure: the softening of the specimen results from the model behaviour, not from its specification. However, *asymmetry* is introduced as a preliminary specification in the sense discusses below.

A correct modelling should include three different material components which are or are not detectable (in terms of image pixels) according to the size level:

— an aggregate phase;
— a mortar phase;
— an interface bond.

Standard concrete shows aggregates much stronger than mortar, while the interface is the weakest component. All these three components exhibit *asymmetry* between the behaviour in traction or compression. This comes from the over-microscopic coarse grain that has been selected.

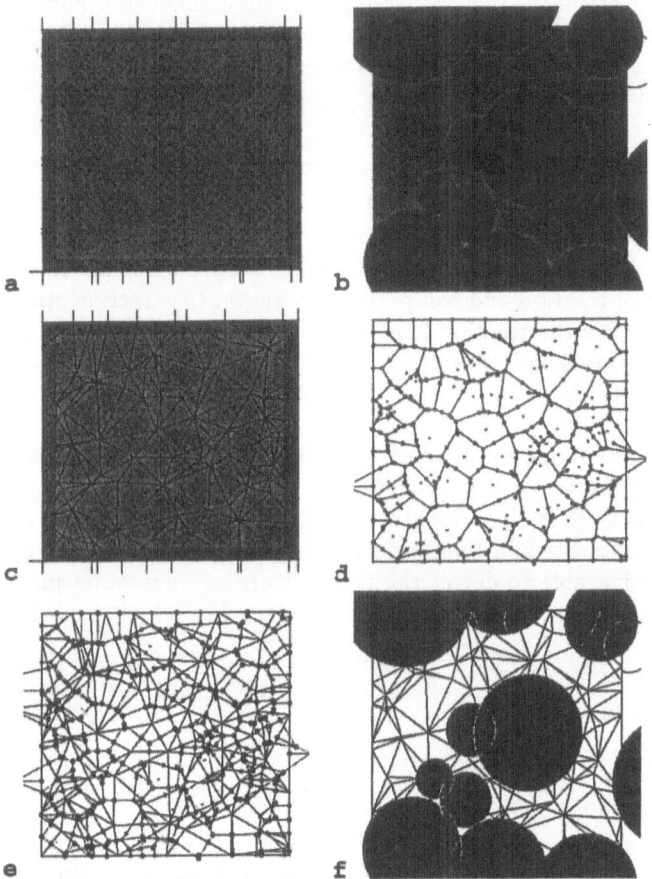

Figure 3. a) Specimen as given region; b) Generation of the aggregates; c) First Delauney triangulation; d) Associated Voronoi tesselation; e) Superposition of the two Delauney triangulations; f) Classification of the superimposed discrete bars.

In [6] holonomy was assumed, thus respecting the requirement of a poor input for a rich output. In the model of next section, however, it was decided to include a non-holonomous law for the basic element as more respectful of mechanical evidence. One has four simple alternatives:

- holonomous law;
- non-holonomous behaviour along the branch leading to failure;
- non-holonomous behaviour along both the branches after failure;
- a second type description for traction and a third type for compression.

The fourth model was selected thus implementing: (i) a perfect linear behaviour inside the limit values of traction and compression strength; (ii) traction stiffness and strength going to zero after the traction limit has been exceeded; in case of recover a normal behaviour in compression is preserved; (iii) a full crushing after the compression limit strenght has been exceeded.

In the early model, randomness only affects the distribution of the different components, i.e. of the aggregates inside the concrete matrix. Differently from other models [11], randomness did not affect the strength. The main drawbacks of this choice resulted in losing some ductility and mainly in putting the computation algorithm in the condition of not being able to detect the true pattern at bifurcation points. This suggested to have both strength and stiffness of the basic element as two independent random variables.

3. Updated numerical model

3.1. BUILDING THE GEOMETRY

The following steps are required to build the model geometry.

1. One considers plane stress situations (i.e. slabs); any out of plane aspect is neglected and the associated instability phenomena are disregarded. The specimen is a given area in the plane, delimited by one or more closed curves as represented in Fig. 3a). The thickness is constant.
2. Inside this area, one randomly assigns a given number of nodes. Indeed vertices are always nodes and a fraction of their total number is reserved to discretize the boundary (to include support and/or loading points). Too dense clusters are rejected.
3. On a larger area (to avoid any edge effect) an aggregate distribution is generated in agreement with a granular curve. Here, the aggregate arrangement follows the Fuller curve $p = 100\sqrt{D/D_{max}}$ in which p denotes the percent of aggregate of diameter lower than D, where D_{max} is the diameter of the largest aggregate particle (see [12] among others, for a discussion about this approximation). The shape of the aggregates is assumed to be circular or combination of circles (Fig. 3b)).
4. The nodes are connected through a double Delauney triangulation, in order to subdivide the admissble area in triangles, the most simple and the most general way

to partition an area. The Delauney configuration has the property of minimizing the acute angles in each triangle, and so it provides a sort of mesh optimization. The first triangulation connects all the nodes; the second one only serves one half of the nodes. The present way of considering two distinct layers of bars lies on the assumption that one half of the nodes of a single layer coincide with the ones of the other layer. This is the way the two layers interact each other. (A more sophisticated way would consider links between selected nodes of the two layer. Again these further connections lead to improoovements regarding the softening behaviour, because they simulate the real three dimensional structural behaviour.) The reason for developing two different superposed triangulations relies in the wish of simulating the tridimensional aspect of the crack development (in particular the bridge effect (see [13])). Moreover, doubling the bar number makes much more ductile the overall structural behaviour. There are many ways for obtaining such a 3D approximation: the one presented in this paper is poor (one of the two layers shows a limited number of nodes which are rigidly connected with the ones which substains the mesh in the other layer) but is a first implementation. Comparisons between richer models are presently in progress.

Associated with the Delauney triangulation (see Fig.3.c) the corresponding Voronoi tesselation [14] can be introduced as a tool for a conventional way to depict the cracks development. Indeed associated with each side of the Delauney triangulaton there is a perpendicularly disposed edge of the Voronoi tesselation (Fig.3.d). If the bar along the side of the Delauney triangulation breaks, one plots a crack along the associated edge in the Voronoi tesselation.

5. For each triangle, one conceives three bars, along the sides hinged in the three vertices. The characteristics of each element are determined on the basis of geometrical properties on the underlying aggregate distribution. One randomly takes two test points along the bar and the renormalization technique in [6] is used. Each of the resulting three classes of bars has a similar constitutive behaviour, but now the stiffness E_m and the traction strength $\sigma_{T,m}$, (being $m = 1, 2, 3$) show a random character (the experimental results suggest that $\sigma_{C,m} = -10 \times \sigma_{T,m}$). The two parameters of each bar are independent normally distributed random variables; they are also independent of the values in the neighbour bars. (This assumption is the simplest way to account for local fluctuations inside a single aggregate or a single block of matrix; alternatively one should simulate local average values and fluctuations around them.) The coefficient of variation is equal to 0.8: in this way one has a relatively large spread of the strength and stiffness values. Both strength and stiffness are restricted to be positive.

3.2. THE MODEL EQUATION

The discretization method described in the previous subsection belongs to the class of schemes which makes explicit reference to Hrennikof paper [15]. It is intrinsic in finite element models a link with continuity. This suggests to adopt truss schemes which allows one to represent cracks and crushing, i.e. displacement discontinuities [16, 17, 18].

Since linear plane stress can only be described by a truss scheme with isosceles triangles, the Voronoi irregular discretization must be coupled with a rule of area assignment to the trusses which results from a regressive scheme (see Appendix).

The stiffness matrix $\underline{\underline{K}}$ can be written with reference to the local coordinates system [8] as

$$\underline{\underline{K}} = \frac{E \cdot A}{L} \cdot \begin{pmatrix} 1 & -1 \\ -1 & 1 \end{pmatrix} \tag{3}$$

where the degrees of freedom are $\underline{q} = [q_1 \; q_2]^T$: of course $q_2 - q_1$ is the elongation along the axis of the element. The element stiffness matrix of Eq.(3) is then transformed in the global reference system and assembled to form the global stiffness matrix $\underline{\underline{K}}_{global}$. The strains along the bars are measured following the Total Lagrangian Formulation, in order to take into account the large displacements that develop as the cracks progress. The global system is governed by the equation

$$\underline{\underline{K}}_{global}(\underline{q}_{global}) \cdot \underline{q}_{global} = \underline{P}_{global} \tag{4}$$

where $\underline{q}_{global}$ and $\underline{P}_{global}$ are respectively the displacements vector for the whole structural element and the applied load vector. Eq.(4) is solved using a secant approach, for given displacements. In this way, one can also follow the structural response in the post-critical branch.

3.3. COMPUTATIONAL ASPECTS

The first computational aspect to be underlined is the need to avoid spurious singularities in the whole structural stiffness matrix after the failure of a bar. For this purpose, the bar is supposed to still show a bit of stiffness, with the new stiffness modulus down to $E_m = 0.0001 \times E_m$. This numerical trick avoids the adoption of very sophisticated techniques to handle the equilibrium of isolated points. Indeed, the equilibrium conditions of these particles of material that are cutting any link with the specimen body can only be conducted by a dynamical time integration scheme.

In structures made of softening materials, a local region may soften or crack, while the surrounding material unloads elastically [19, 2]. The strain localization may induce dynamic jumps to a new displacement state at a fixed load level (snap-through) or dynamic jumps to a new load level under a fixed displacement state (snap-back). To catch these effects, the adoption of special pseudo-static solution techniques is required.

Let the load vector be given by the product of a scalar factor λ by a constant reference force vector $\underline{y}$. Moreover, let $\underline{r}$ the restoring vector force due to the stress state corresponding to the displacement vector $\underline{x}$. A possible graphical representation shows a continuous line, with λ as ordinate, representing the equilibrium path $(\underline{x}, \lambda \underline{y})$.

For each equilibrium state, the vector of the residual forces is zero. The equilibrium path may show critical points.

One also introduces the variable t. It is a fictitious time playing the role of a reference searching parameter. By varying t, the representing point P moves along the equilibrium path. For non-linear elastic materials, the point P can also be identified, along the path by a curvilinear abscissa s rather than by t. When the symmetric tangent stiffness matrix is non singular, the displacement increment, corresponding to a load increment, identifies a regular point on the equilibrium path. When at least one eigenvalue of the stiffness matrix is zero, for a load variation the uniqueness of the displacement increment is lost. This occurrence identifies a singular point, which correspond to either a bifurcation point or a limit point. Also, if one has k zero eigenvalues, a non-simple bifurcation point with multiplicity k has been detected: it has k distinct branched off paths. At least, the ultimate point is defined on the basis of maximum allowable strains without material failure.

Structural imperfections or dynamic effects may shift the actual structural response from this purely ideal representation of the equilibrium state. From a mathematical point of view, this corresponds to a projection from the ideal space to another and in this projection operation the uniqueness of the solution is lost. However, when random imperfections occur (i.e. the equilibrium path is perturbed), the singular limit points tend to remain unchanged. Indeed, when imperfections are present, the concept itself of bifurcation point is lost. Since real structures are always affected by imperfections, only limit points actually exist. From a theoretical point of view, structural imperfections are accounted for to assess the structural sensibility to initial conditions [20]. In the case of reinforced concrete structures, these structural imperfections come either from non homogeneous strength or stiffness of the concrete or from dimensional non homogeneity of the structural elements: they are dominant aspects for a numerical modelling [21] These effects lighten the computational procedures, since they tend to ignore the presence of bifurcation points, but they also make the equilibrium paths extremely irregular, due to the softening behaviour as well as to some discontinuous mechanical aspects, like cracking and section partialization. As a consequence, the appearence of the computed load-displacement curve is in general too rough. It is necessary then to adopt some smoothing tecnique, but it is also a must to explain how this procedure works.

One considers the sequence of the computed points $({}^k\underline{x}, {}^k\lambda\underline{y})$, $k = 1, ..., K$ along the equilibrium path. One replaces the values ${}^k\lambda, k = 1, ..., K$ of the load multiplier by the smoothed set ${}^k\overline{\lambda}, k = 1, ..., K$, where

$$
{}^k\overline{\lambda} = \frac{{}^{k+1}\lambda + {}^{k-1}\lambda}{2}, \quad k = 2, ..., K - 1 \tag{5}
$$

with ${}^1\overline{\lambda} = {}^1\lambda$ and ${}^K\overline{\lambda} = {}^K\lambda$.

358

4. Numerical results

All the specimens that were numerically analyzed are squares of different side lengths L, varying from $5cm$ to $100cm$. The thickness $d = 10cm$ is constant in all cases. The maximum aggregate size is $1.5cm$.

All specimen shows: (i) restrained vertical displacements for the nodes along the bottom side; (ii) restrained horizontal displacement for the node in the bottom left corner to prevent rigid body motions. The equilibrium path is computed by the secant technique imposing increasing (in traction) or decreasing (in compression) vertical displacements for the top side nodes. The analysis continues until a final displacement equal to $L/500$ ($\epsilon_{max} = 0.002$) is reached in the case of a traction test, and until $-L/250$ ($\epsilon_{min} = -0.004$) in a compression test (compression stress and strain are conventionally regarded as negative).

A reference concrete strength of $2500N/cm^2$ is considered; the stiffness modulus is, as usual, equal to $2850000N/cm^2$. The following values are adopted for the basic element parameters:

- aggregate: $\sigma_{T,1} = 8000N/cm^2$, $E_1 = 7000000N/cm^2$;
- matrix: $\sigma_{T,2} = 4000N/cm^2$, $E_2 = 2500000N/cm^2$;
- bond: $\sigma_{T,3} = 1024N/cm^2$, $E_3 = 2500000N/cm^2$;

In Fig. 4, a specimen of size $L = 20cm$ is assigned. From the top to the bottom, three different meshes (with 100, 400 and 900 nodes) and the corresponding numerically simulated load-displacement diagrams (light gray curves with dots) are illustrated. Each diagram shows its simulated curve and the previous ones in order to make comparisons easy. A further curve of interest is drawn in solid line: the response that the specimen would follow if the classical Saenz law held. Finally, the straight line tangent in the origin represents the linear elastic behaviour. Some remarks follow from these numerical analyses: a) the initial stiffness the set of equivalent bars shows agrees with the linear elastic one; b) the overall appearence of the response is catched, even if there is a little overestimation of the traction strength: c) the softening branch is fully reached in agreement with Saenz's parabola; d) the results appear objective, the curves being almost identical (in a statistical sense).

The objectivity aspect is further investigated in Fig. 5. Three different specimen sizes (micro-meso-macro) are considered: $L_{micro} = 4cm$, $L_{meso} = 20cm$ and $L_{macro} = 100cm$. For each size specimen, 2 series of analysis are conducted, in both compression and traction, with K^2 ($K = 4, \ldots, 30$) nodes. For these three sizes, the results of all the simulations are plotted in Fig. 5.a, 5.c and 5.e, respectively, in terms of nominal stress and strain (the loads are divided by the nominal area and the displacements by the side length). The Saenz law and the linear elastic line are also drawn as a reference. It is worth noting that increasing the size of the square specimen, the Saenz parabola overestimates more and more the ultimate strength, the size-effect being neglected.

On the right column, three log-log plots are drawn: the logarithm of K (i.e. the number of mesh nodes) serves as abscissa, the logarithm of the (absolute) value of the

359

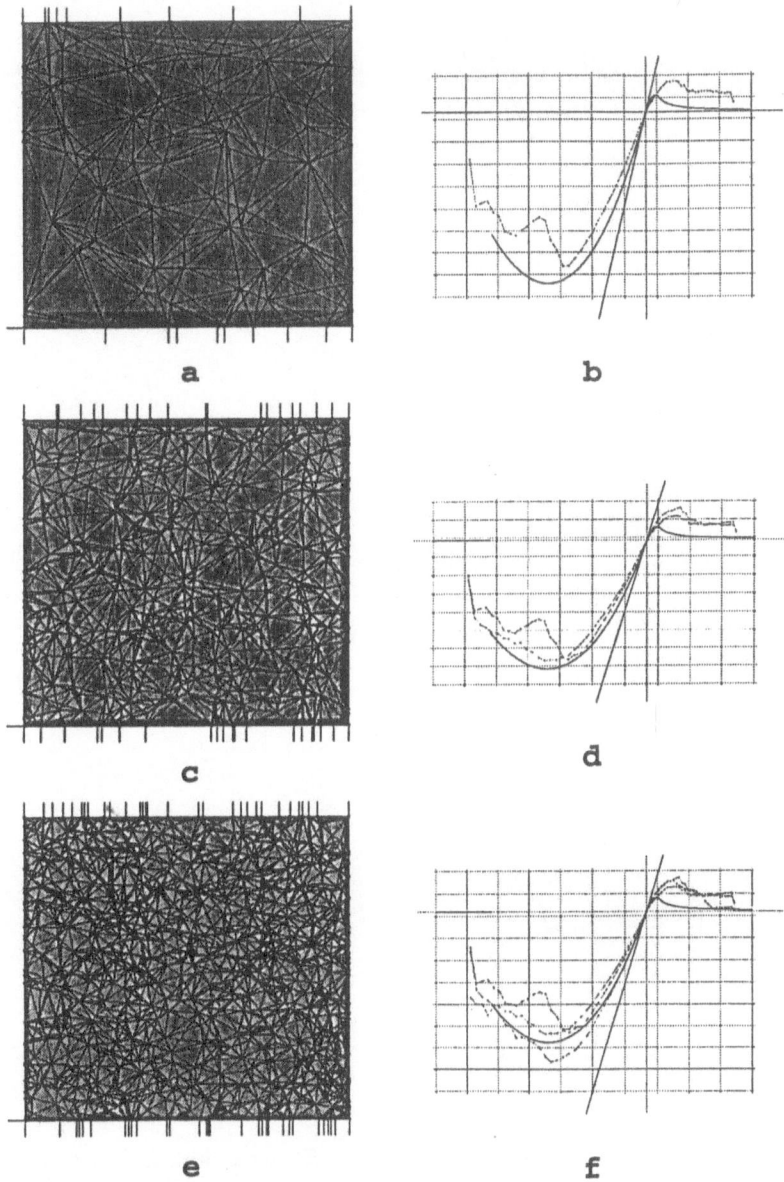

Figure 4. The square panel (20cm × 20cm with 10cm thickness) analyzed by three different meshes with a)-b) 100 nodes, c)-d) 400 nodes and e)-f) 900 nodes. On the left side the mesh and on the right side the associated compression-traction load-displacement diagram. (Abscissa range (-0.096, 0.048) cm; ordinate range (-53.848, 15.486) kN. The response curve in b) is also drawn in d) and f); the response curve in d) is also drawn in f).)

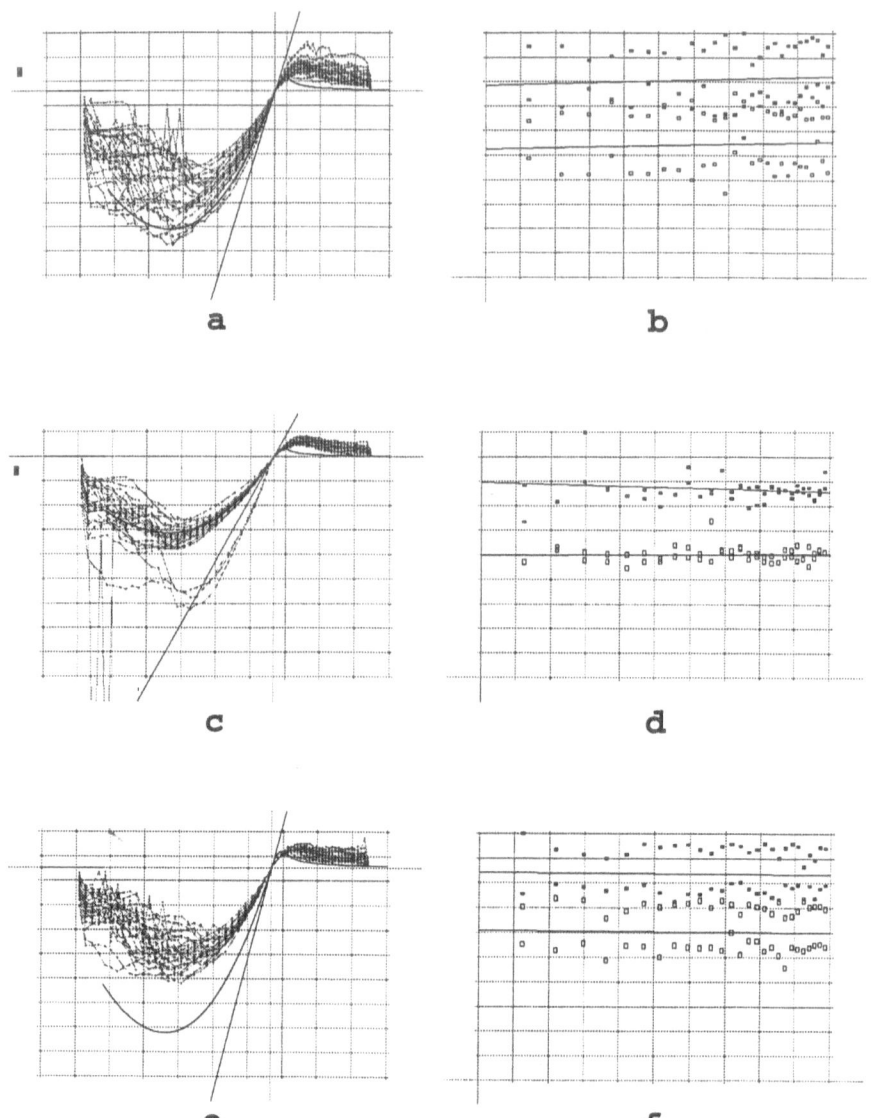

Figure 5. Objectivity of the analysis for meshes with K^2 nodes, with $K = 4, 5, ..30$. Three size levels are considered: micro, $L = 4cm$; meso, $L = 20cm$, macro, $L = 100cm$. Left column: nominal stress-strain diagrams. Abscissa ranges: a) (-0.0192, 0.0096) cm, c) (-0.096, 0.048) cm , e) (-0.48, 0.24) cm. Right column: logarithm of the nominal ultimate strength (compression: filled square; traction: unfilled square) as function of the logarithm of K (b), d), f), abscissa range: (ln 3, ln 31)). The least-square straigth-line interpolation is given.

ultimate strength as ordinate. Indeed, in each plot (*micro-meso-macro* situations), the logarithm of the maximum value computed along the equilibrium path is reported. The values obtained from the compression simulations are marked as filled squares; the ones obtained from traction analyses as unfilled squares. Despite some statistical scatter, the angular coefficients of the least square interpolating straigth lines are negligibles. In particular, let D the regression coefficient in a log-log plane:

- at the microscopic size one reads $D_c = 0.0857$ for compression and $D_t = 0.0650$ for traction;
- at the mesoscopic size one reads $D_c = -0.0986$ for compression and $D_t = 0.0105$ for traction;
- at the macroscopic size one reads $D_c = -0.0236$ for compression and $D_t = -0.0196$ for traction.

Objectivity can be regarded as fulfilled.

In Fig. 6.a one has the sizes (light gray squares) of the specimen analyzed: the side length is $L = 2 \times K\,cm$ ($K = 4, \ldots, 30$). The analysis was conducted in both compression and traction, the goal being the study of the sensitivity of the ultimate strength to variations of the size of the specimen itself. The sizes of the specimen that were adopted for showing objectivity ($L = 4, 20, 100\,cm$) are also shown by solid lines.

In Fig. 6.b the log-log plot of the ultimate strength versus K is given. For each side length, i.e. for given K, one analyzes 5 specimens. The angular coefficients resulting for the least square interpolation are:

- $D_c = -0.1055$ in compression;
- $Dt = -0.1498$ in traction.

The size effect is evident both in traction and in compression. The possibility to gather a more detailed size effect relationship, as the one pointed out in [1] would require to consider a broader specimen-size range. In the present study attention was focused on recognizing the feature of the central part of a size effect diagram (which shows *no size effect - size effect - no size effect* when moving from very small to medium, and from medium to very large structures).

5. Conclusions

This paper shows how the same numerical model fits well the behaviour of concrete specimens at a micro-, meso- and macro-level. The main characteristics of the behaviour of concrete specimens are reproduced through their discretization into a set of basic elastic brittle elements. Within the proposed statistical mechanics approach, the different global behaviours results from the interaction of these many, very simple, basic elements. A renormalization technique is applied to determine their characteristics (stiffness and strength).

The ability of the model to follow the actual behaviour of the material is emphasized in Fig.7, where the load-displacement diagram is illustrated by the progressive crack

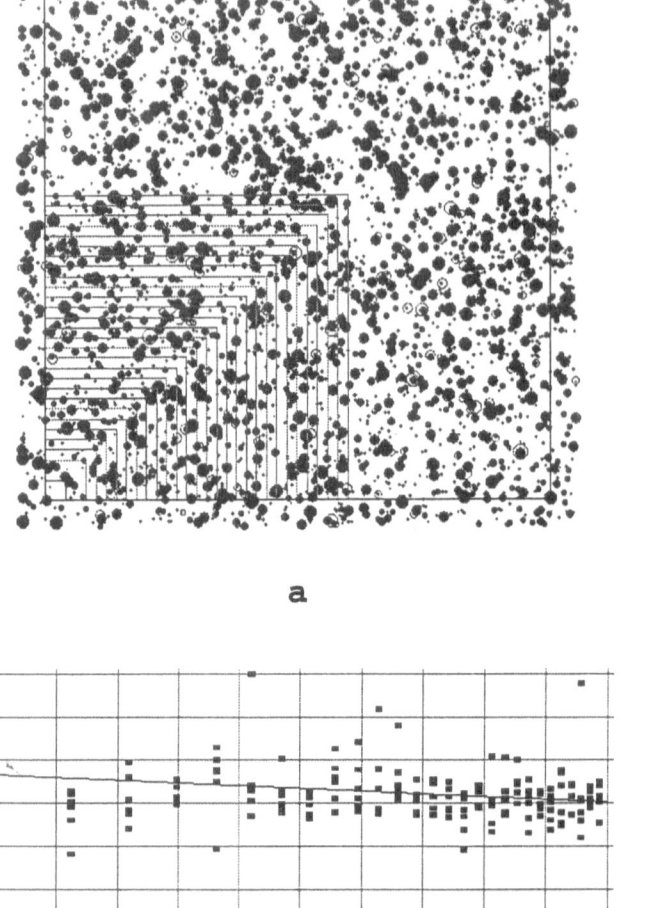

a

b

Figure 6. Size effect: a) size of the square specimens analyzed; b) logarithm of the ultimate strength (filled black square: compression; unfilled square: traction) wersus the logarithm of the specimen size. Abscissa range (ln 15, ln 155) cm; regression coefficients: $D_c = $ -0.11; $D_t = $ -0.15.

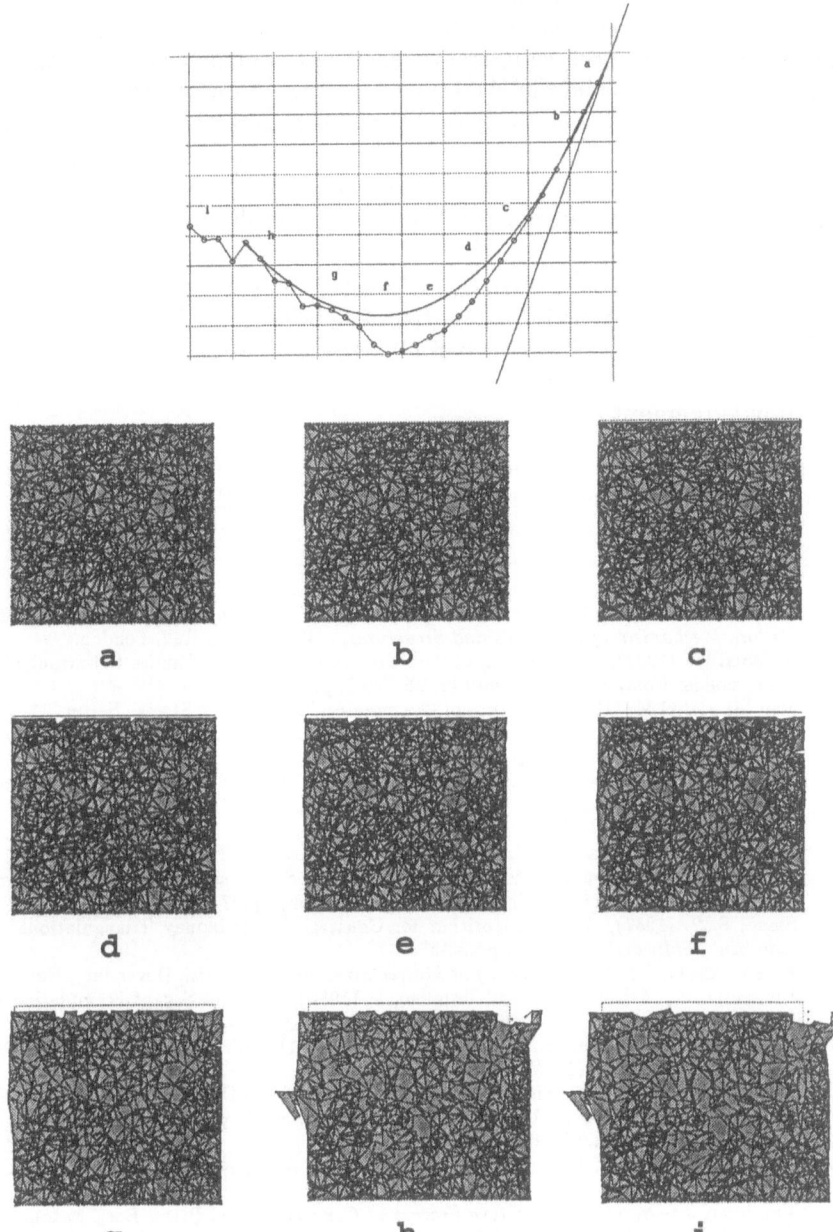

Figure 7. Compression test of a specimen of size $L = 20cm$: load-displacement diagram, compared with the Saenz law and the linear elastic straigth line (abscissa range (-0.08, 0.0) cm; ordinate range (-57.466, 0.0) kN). Progressive crack pattern development.

pattern development. After the peak load (i.e. label (f)), the volumetric deformation changes sign: it becomes positive due to the crack dilatancy. The model captures this feature as the large positive lateral strain shows. Moreover, at the final stage (i), the large displacements point out the explosive nature of the failure.

The constraints that the numerical model is required to satisfy are: simplicity (the basic constitutive law must be described by two or three parameters), objectivity (the results are insensitive to the mesh size), consistency (the results must agree with laboratory measurement: size effect, softening, crack localization and tension-compression ultimate strength ratio).

The model follows the principle of "a poor input for a rich output": the resulting simplicity makes it suitable for an implementation in structural analysis computer codes. An extension to cyclic loading processes is in progress.

Acknowledgement

This work has been supported by grants from the Italian Ministry of the University and of the Scientific and Technological Research (MURST).

References

1. Carpinteri, A. (ed.) (1996), *Proceedings of the IUTAM Symposium on Size-Scale Effects in the Failure Mechanism of Materials and Structures*, , Chapman & Hall, London.
2. de Borst, R. (1987), Computation of Post-Bifurcation and Post-Failure Behaviour of Strain- Softening Solids, *Computers & Structures*, **25**, No.2, pp. 211-224.
3. Van Mier, J.G.M. (1986), Fracture of Concrete under Complex Stress. *Heron*, **31**, no.3.
4. Ottosen, N.S. (1986), Thermodynamic Consequences of Strain Softening in Tension, *J. of Eng. Mech., ASCE*, **112**, No.11, pp.1152-1164.
5. Bontempi, F. and Malerba, P.G. (1996), The role of Tension Stiffening in the Numerical Analysis of R.C. Frame Structures, *Proceedings of Third Asian-Pacific Conference on Computational Mechanics (APCOM)*, Seoul, September 16-18, 1996.
6. Bontempi, F. and Cascati, F. (1996), A Renormalization Technique in the Study of Concrete Elements of Increasing Size, in Carpinteri A. (ed.), *Size-Scale Effects in the Failure Mechanism of Material and Structures*, Chapman & Hall, London, pp.27-42.
7. Sloan, S.W. (1987), A Fast Algorithm for Constructing Delaunay Triangulations in the Plane, *Adv. Eng. Software*, **9**, no. 1-2, pp.34-55.
8. Przemieniecki, J.S. (1969) *Theory of Matrix Structural Analysis.*, Dover Inc., New York.
9. Bontempi, F., Malerba, P.G. and Romano, L. (1995), A Direct Secant Formulation for the Reinforced and Prestressed Concrete Frame Analysis (in Italian), *Studi e Ricerche*, **16**.
10. Creswick, R.J., Farach, H.A. and Poole, C.P. (1991), *Introduction to Renormalization Group Methods in Physics.*, John Wiley & Sons, Inc.
11. Rossi, P. (1996), A Probabilistic Discrete Cracking Model of Concrete in Tension and Compression, *Size-Scale Effects in the Failure Mechanism of Materials and Structures*, A. Carpinteri (ed.), Chapman & Hall, pp.200-214.
12. Schlangen, E. (1993), Experimental and Numerical Analysis of Fracture Processes in Concrete. *Heron*, **38**, no.2.
13. Van Mier, J.G.M. (1997), *Fracture Process of Concrete.*, CRC Press, Boca Raton.
14. Sloan, S.W. and Houlsby, G.T. (1984), An Implementation of Watson's Algorithm for Computing 2-dimensional Delauney Triangulations., *Adv. Eng. Software*, **6**, 4, pp.192-197.
15. Hrennikoff, A. (1941), Solution of Problems in Elasticity by the Framework Method. *Journal of Applied Mechanics.*

16. Burt, N.J. and Dougill, J.W. (1977), Progressive Failure in a Model Heterogenous Medium. *Journal of the Engineering Mechanics Division, ASCE,* **103**, EM3, pp.365-376.
17. Charmet, J.C., Roux, S. and Guyon, E. (ed.) (1989), Disorder and Fracture. *Proceedings of a NATO Advanced Study Institute on Disorder and Fracture,* May 29 - June 9, Cargese (France).
18. Herrmann, H.J. and Roux, S. (ed.) (1990), *Statistical Models for the Fracture of Disordered Media,* Elsevier Science Publishers B.V., Amsterdam.
19. Crisfield, M.A., (1982), Local Instabilities in the Non-linear Analysis of Reinforced Concrete Beams and Slabs, *Proc. Instn. Civ. Engrs.,* Part 2, **73**, pp.135-145.
20. Seydel, R. (1988), *From Equilibrium to Chaos: Practical Bifurcation and Stability Analysis,* Elsevier Science Publishing Co. Inc., N.Y.
21. Carmeliet, J. and de Borst, R. (1995), Stochastic Approaches for Damage Evolution in Standard and Non-Standard Continua, *Int. J. of Solids and Structures,* **32**, No. 8/9, pp. 1149-1160.

Appendix – Assessing the correction factor

To implement a rule of area assignment to the discretization trusses, one consider the two following stiffness matrices:

1. as for the common costant strain triangle (CST) finite element of area A, one writes:

$$\underline{\underline{K}}_{CST} = \frac{1}{4A} \cdot \underline{\underline{B}}^T \cdot \underline{\underline{D}} \cdot \underline{\underline{B}} \tag{6}$$

where $\underline{\underline{B}}$ and $\underline{\underline{D}}$ are:

$$\underline{\epsilon} = \begin{pmatrix} \epsilon_x \\ \epsilon_y \\ \gamma_{xy} \end{pmatrix} = \frac{1}{2A} \begin{pmatrix} b_i & 0 & b_j & 0 & b_k & 0 \\ 0 & c_i & 0 & c_j & 0 & c_k \\ c_i & b_i & c_j & b_j & c_k & b_k \end{pmatrix} \cdot \underline{q} = \underline{\underline{B}} \cdot \underline{q} \tag{7}$$

with $b_i = y_j - y_k$, $b_j = y_k - y_i$, $b_k = y_i - y_j$, $c_i = x_j - x_k$, $c_j = x_k - x_i$, $c_k = x_i - x_j$ and

$$\underline{\underline{D}} = \frac{E}{1 - \nu^2} \begin{pmatrix} 1 & \nu & 0 \\ \nu & \nu & 0 \\ 0 & 0 & \frac{1-\nu}{2} \end{pmatrix} \tag{8}$$

respectively. Note that $\underline{\epsilon}$ is the vector of the strain components and $\underline{q}$ is the one of the nodal displacements.

2. one also writes, by assembling the matrices of the three bars which form the elementary triangle:

$$\underline{\underline{D}}_{bars} = A_B \cdot E \cdot \begin{pmatrix} \underline{\underline{k}}_{12} + \underline{\underline{k}}_{31} & -\underline{\underline{k}}_{12} & -\underline{\underline{k}}_{31} \\ -\underline{\underline{k}}_{12} & \underline{\underline{k}}_{12} + \underline{\underline{k}}_{23} & -\underline{\underline{k}}_{23} \\ -\underline{\underline{k}}_{31} & -\underline{\underline{k}}_{23} & \underline{\underline{k}}_{23} + \underline{\underline{k}}_{31} \end{pmatrix} \tag{9}$$

where

$$\underline{\underline{k}}_{ij} = \frac{1}{l_{ij}} \begin{pmatrix} c_{ij}^2 & c_{ij}s_{ij} \\ c_{ij}s_{ij} & s_{ij}^2 \end{pmatrix} \tag{10}$$

Considering the same material parameter, the first matrix $\underline{\underline{K}}_{CST}$ depends on the parameters E and ν, while the second matrix $\underline{\underline{K}}_{bars}$ only depends on E. So, anyway it is impossible in general to find a perfect equivalence between the two ways of analyzing the slab. Moreover, from the geometrical point of view, the first matrix depends on the area of the triangle A_T, while the second depends on the assumed area of the cross section of the bars A_B.

One introduces the following relationship for the cross sectional area of the bar:

$$A_B = \omega \times \frac{A_T \cdot d}{P} \tag{11}$$

where d and P are the thickness of the slab and the perimeter of the triangle, respectively. The parameter ω is a dimensionless correction factor that must be determined. Note that the area of the bar is proportional to the volume of the triangular portion of the slab considered divided by the thickness of the slab itself.

The exact value of the correction factor ω is determined imposing, in the least square sense, that the defomation energy produced by the CST stiffness matrix is equal to the same energy stored in the three-bars set for any arbitrary displacement vector:

$$\underline{q}^T \underline{\underline{K}}_{CST} \underline{q} = \underline{q}^T \underline{\underline{K}}_{bars} \underline{q} \tag{12}$$

This approach gives, for a Poisson ratio ν between 0 and 0.5, a suitable value of $\omega = 6$.

Relation (12) is only verified in an average sense. Therefore, one introduces here a sort of additional random perturbation in the discretized structure.

MULTISCALE MATERIAL CHARACTERIZATION AND APPLICATIONS

GEORGE N. FRANTZISKONIS
Department of Civil Engineering and Engineering Mechanics
University of Arizona, Tucson, AZ 85721-0072 USA

AND

MARK P. BLODGETT
WL/MLLP, Wright-Patterson AFB, 2230 Tenth St. Ste 1
Bldg. 655, Area B, Ohio 45433-7817

Abstract

Given the hierarchical structure of engineering materials, it is natural to seek multiscale characterization tools. This study explores wavelet analysis, a recently developed mathematical tool, for this purpose. In particular, multiscale microstructure characterization of certain alloys is studied on an exploratory basis. Ultrasonic B-scan data are used for multiscale characterization of various Ti 6-4 microstructures. Two such microstructures are studied in detail for the purpose of quantitative characterization as well as for extraction of microstructural features from the B-scan data .

1. Introduction

The objective of this research is to develop ultrasonic techniques to allow the characterization of microstructural features consistent with those found in conventional titanium alloys like Ti-6Al-4V, which is the workhorse alloy for the aerospace industry. Titanium alloys are used in many aerospace applications for airframe and engine structures. In aircraft applications, the titanium components are often subjected to cyclic stress and stress reversals which can lead to the initiation of surface connected fatigue cracks. Prior to the formation of a dominant crack, fatigue cracks may initiate in a multitude of locations. Initially, the fatigue damage will develop at the nanoscale, but, given continued cyclic loading, some of the damage will evolve into microscopic cracks and eventually into a visibly present crack or network of cracks.

Titanium alloys of interest to the aircraft industry are polycrystalline and significant effort is given to the homogenizing fabrication processes to remove any lingering as-cast microstructure, minimize micro-texturing, and promote the

G. N. Frantziskonis (ed.), PROBAMAT – 21st Century: Probabilities and Materials, 367–378.
© 1998 *Kluwer Academic Publishers.*

368

development of microstructural uniformity. Although significant effort is given to homogenization, often microstructural irregularities persist beyond the inhomogeneity associated with the random nature of the grain structure. Moreover, certain structural applications can benefit from the presence of macroscopic inhomogeneities of the elastic properties. Nondestructive evaluation techniques can be used to identify grossly anomalous microstructural flaws and elastic property variations associated with macroscopic inhomogeneities due to texturing; however, more research is needed to allow comprehensive nondestructive materials characterization.

The condition of the microstructure is of significant concern to the aircraft design community. Generally, aircraft structural applications require materials to have low density, high strength, high elasticity and toughness, corrosion resistance, and ease of processing. Beyond these general requirements, it is essential that the materials be resistant to fatigue crack initiation and propagation, and resistant to creep. While titanium alloy's are well suited to meet a great number of critical aircraft applications, it is extremely difficult to implement materials that can comfortably accommodate all of the environmental severity's encountered during flight, especially in military aircraft. One approach to ease this design constraint is to match specific applications with materials specially designed to meet those requirements. Materials are currently designed with a view towards developing microstructural conditions that more effectively allow the material to meet application specific structural requirements.

Certain microstructures, such as that of the beta annealed 64Ti, is known to perform well in terms of fatigue crack propagation resistance, owing to branching of the propagating crack tip, but perform poorly in terms of fatigue crack initiation, due to the presence of large colonies of similarly oriented material. The direct opposite is true for a recrystallization annealed 64-Ti microstructure, comprised of equiaxed a and intergranular b, with a narrow size distribution of fine grains. Knowledge of the microstructural make-up of a component is important to proper alloy performance capabilities and it would be satisfying to know that nondestructive evaluation techniques have been established to allow for in-depth microstructure characterization; unfortunately, this is not the case, at least not for materials with microstructural complexities consistent with commonly used titanium alloys.

With the important role of microstructure in mind, materials characterization is a vitally important aspect of nondestructive evaluation. Although not inherently nondestructive due to required surface preparation, Scanning Acoustic Microscopy and Electron Back-Scatter Diffraction are two approaches developed specifically for visualizing microstructural features for materials characterization. In polycrystalline alloys, grain to grain differences in the crystallographic orientation, and the presence of grain boundaries provide a source for the scattering for ultrasonic energy. In polycrystalline polyphase materials the scattering can be further complicated by the presence of variations in elastic modulus, density, and crystal lattice structure of the grains. This incoherent scattering of ultrasonic energy is generally considered as "grain noise" and can seriously hinder the detection of ultrasonic signals from defects or damage. On the other hand, the microstructure scatter can be used as a means of ultrasonically characterizing the grain structure of the material. Elastic anisotropy of single crystals plays an important role in ultrasonic materials characterization of polycrystalline materials. Microscopically homogeneous, but randomly oriented individual grains make up a macroscopically isotropic, but inhomogeneous medium, giving rise to incoherent grain scatter. While acoustic grain noise has an obvious adverse affect on flaw detection [Rose, 1992, Margetan et al 1994], it can be exploited for ultrasonic characterization of the grain structure [Goebbles, 1980, Hecht et al, 1981, Willems, and Goebbels, 1981, Guo et al, 1985].

Ultrasonic microstructure scatter data has been collected with a conventional C-scan data acquisition system. The system has a positioning gantry which can be set up to allow for scanning in a number of different configurations. The pulse-echo configuration was used for this work with simple normal incidence ultrasound entry from which the entire waveform was collected for each transducer position. This scanning resulted in the generation of two dimensional data files, which were composed of both spatial and temporal data. For each file the 10 Mhz transducer scanned a line over the titanium-64 samples. Each file was composed of 256 waveforms sampled at 100 MSPS. Three scan conditions were examined using 0.005", 0.020" and 0.050" step sizes, all of which maintained the same sampling rate. All other scan parameters were the same for the two specimens scanned. The tests were performed on 25.4 mm thick specimens (the depth being the direction of the propagated wave). The two specimens each had dramatically different microstructural features due to the processing history. One of the samples had the billet microstructure, which consisted of large colonies of similarly oriented a platelets as shown in figure 1a, the other specimen was recrystallization annealed, which consisted of equiaxed a and intergranular b. For convenience, herein we term the former as "noisy" and the latter as "quiet" microstructure.

2. Wavelet Analysis

Wavelet analysis has been used very recently to rationalize experimental data examined at various scales of observation, in several branches of physical sciences ranging from particle physics and biology to electrical engineering and fluid mechanics. However, this powerful technique has not been applied extensively to materials related problems, even though several open problems and critical questions in this field can be greatly benefited from it. This work addresses this issue.

The analysis of phenomena at multiple scales has received an ever increasing level of attention in the past few years. While it is not possible to provide a detailed commentary on the contributions of the theory of wavelets (harmonic analysis, phase space analysis and renormalization, stochastic and self-similar processes, computer vision, speech, etc.), it is worthwhile to refer to an example from fluid mechanics in order to illustrate the context in which this project is undertaken and in which it should be viewed. In turbulence, wavelet analysis has shown how during the flow evolution,

370

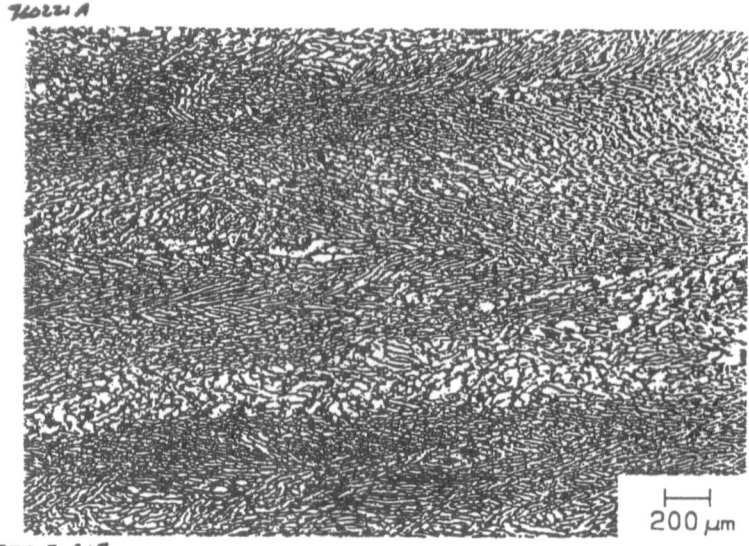

Figure 1a Typical micrograph of the "noisy" microstructure.

Figure 1b Typical micrograph of the "quiet" microstructure.

starting from an initial random distribution of vorticity, the smallest scales of the flow become more and more localized and concentrated in the centers of the so-called coherent structures [Farge, 1992, and references cited therein]. This led to conjecture that, contrary to generally accepted ideas, dissipation also acts at the center of coherent structures. Thus, wavelets are capable of capturing and analyzing the existence of fluid structures and patterns for which other techniques have proven inadequate, and the same is expected to be shown for solid microstructures, as described herein (on an exploratory basis).

Let us present an illustrative example of use of the wavelet transform, for the purpose of further understanding and facilitating the subsequent description of the present work. Figure 2a shows a "snowflake," where the "map" depicts the spatial position of snow particles for a specific value of the magnification. Let us consider that this "map" is at the finest detail (largest magnification) possible; thus, either due to physical/experimental limitations, or due to lack of interest, information at higher scales is not available. Figures 2b,c,d show the wavelet transforms of the snowflake at three different scales (only three scales are shown for simplicity). Figure 3 shows, again, the wavelet transform of the snowflake using 32 shades from white to black. Clearly, the wavelet representation provides spectacular evidence of its capability to describe spatial "patterns" at different scales realized as magnifications. In other words, each map (Fig. 3, Fig. 2b,c,d) illustrates the spatial pattern of the snowflake for a given value of the magnification. Note that the figures show the wavelet transform rather than the wavelet representation at different scales. In the present work, instead of a snowflake, different physical quantities are analyzed, i.e. ultrasonic B-scan data.

A more general goal of wavelet analysis is to provide representations of graphs (signals) as superposition of elementary functions. The corresponding representation is then used for different purposes, i.e. data compression, feature extraction, pattern recognition, etc. There are several publications on this rather new subject and applications can be found in a wide variety of scientific/engineering. avelet transforms provide both scale and location information about a given function. A wavelet $\psi(x)$ (with real values in our case) transforms a function $f(x)$ according to

$$W_f(a,b) = \int_{-\infty}^{\infty} f(x)\psi_{a,b}(x)dx \qquad (1)$$

The two-parameter family of functions, $\psi_{a,b}(x) = (1/\sqrt{a})\psi(\frac{x-b}{a})$ is obtained from a single one, ψ, called the mother wavelet, through dilatations by the factor a^{-1} and translations by the factor b. The factor $1/\sqrt{a}$ is included for normalization purposes and, with it, all the wavelets have the same energy. The scale parameter a can take any value on the positive real axis. The scalars defined in (1) measure, in a certain sense, the fluctuations of $f(x)$ around point b at the scale a.

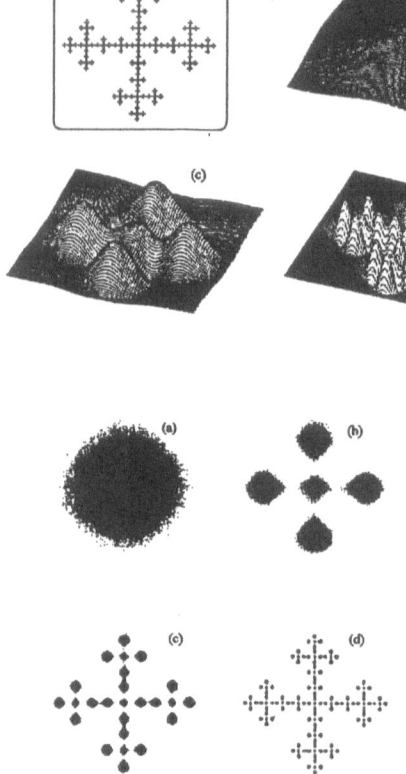

Figure 2. Wavelet transforms of the one-scale (fixed magnification) snowflake shown in (a). The analyzing wavelet is the second Gaussian derivative (Mexican hat, discussed in the text). The scale in (b) is three times larger than the scale in (c), and the scale in (c) is three times larger than that in (d). Figures adopted from Arneodo et al. (1992).

Figure 3. Wavelet transforms of the snowflake shown in Fig. 2. It is coded using 32 shades from white to black. Each scale is three times smaller than the previous one, i.e. the scale at (d) is three times smaller than the scale at (c). Figure adopted from Arneodo et al. (1992).

A wavelet analysis can either be continuous or discrete. The second one, based on an orthogonal decomposition of a signal, is convenient for data in the sense that no redundant information is present and an extensive "library" of wavelets is available. In the following, we concentrate on the discrete wavelet analysis and its physical relevance to the problems addressed herein. Due to, in general, non symmetry in the discrete transform (using wavelets with full orthogonality and compact support) physical quantities may often be "misrepresented," thus for the case at hand we use symmetric bi-orthogonal wavelets.

Not every function ψ can qualify as a wavelet. In particular, its Fourier transform,

$$\hat{\psi}(k) = \int_{-\infty}^{\infty} \psi(x) e^{-2i\pi kx} dx \qquad (2)$$

should ensure that the coefficient c_ψ (given below) is bounded [Chui, 1992]

$$c_\psi = \int_0^{\infty} \frac{|\hat{\psi}(k)|^2}{|k|} dk = \int_0^{\infty} \frac{|\hat{\psi}(-k)|^2}{|k|} dk < \infty \qquad (3)$$

which implies, in particular, that the moment of zero-order vanish, i.e. $\int_{-\infty}^{\infty} \psi(x) dx = 0$.

Given the wavelet coefficients $W_f(a,b)$ associated to a function f, it is possible to reconstruct f in real space $f = f(x)$ through the inversion formula

$$f(x) = \frac{1}{c_\psi} \int_0^{\infty} \int_{-\infty}^{\infty} W_f(a,b) \psi_{a,b}(x) \, db \frac{da}{a^2} \qquad (4)$$

The presence of the coefficient c_ψ in this inversion formula motivates the boundedness condition expressed by (3). At any given scale $a > 0$, f is decomposed into the summation of a "trend" at scale s and of a "fluctuation" around this trend. The trend is the contribution from all scales $s > a$ in (4), and the fluctuation is given by the scales $s < a$.

Within the context of continuous wavelet transform, wavelets commonly used are the so-called Gaussian derivatives of order n, $n = 1,2,3...$,

$$\psi_n(x) = (-1)^n \frac{d^n}{dx^n} (e^{-x^2/2}) \qquad (5)$$

For $n = 2$ we have the so-called "Mexican hat." There are several discrete wavelet constructions present in the literature, and as mentioned above certain bi-orthogonal wavelets are advantageous for our purpose. A good source for details on bi-orthogonal and other wavelets is the manual for the program used in this study, c.f. next section.

3. Multiscale Characterization of Microstructure from Ultrasonic Tests

As explained in the introduction the microstructure is crucial for the properties of the material and for its performance under service and it is important to be able to characterize the microstructure, preferably in a nondestructive fashion.

It is important to quantitatively assess the repeatability of the tests. Since the volume of the specimen represented in the images is large compared to the length (volume) characteristic of the microstructure, the tests should be repeatable, statistically, to say the least. In other words, all images (from the same microstructure) should have some similarity to each other, in a statistical sense. Since the volume of the specimen tested is large, a statistical description may not be necessary. This is indeed the case. Figure 4 shows the wavelet coefficients from 4 tests (different B-scan locations) on "quiet" microstructure and 4 tests on the "noisy" one. The top four curves in each plot are from the "noisy" microstructure and the bottom ones from the "quiet" one. Clearly, the fact that each group of (four) curves are closely together indicates the repeatability of the tests. In other words the microstructure is spatially uniform in a statistical sense, and wavelets are able to capture this similarity. Of course, there should be a minimum volume below which this similarity is not repeatable for single measurements (realizations in a statistical sense).

The wavelet coefficients represent (are related to) the local regularity of the image. In turn, this local regularity is related to the scattering of the wave and thus to the microstructure. It is feasible to track such a relation, i.e. wavelet coefficients to microstructure, analytically and/or numerically. This task (not attempted herein) involves understanding (modeling) the scattering process through the inhomogeneous material., and is presently being worked out.

From the metallurgical as well as the material application point of view, it is important to know whether the desired microstructure is produced consistently, whether there are regions with "undesired" microstructure, and whether there are defects. The wavelet analysis shows significant potential to address this since: (a) wavelets provide local information thus spatial irregularities are detectable; (b) the indication of repeatability of the microstructure signature provides a robust tool for discriminating one microstructure from another; (c) wavelet coefficients are sensitive enough so even small irregularities within the scanned material volume produce deviations from the signature. Importantly, these deviations have local information, e.g. they can pinpoint the region(s) where irregularities are present as well as information on the nature of the irregularities.

One important parameter is the volume of the material scanned. If this volume is small, compared to the length (volume) characteristic of the microstructure, the scan data will not be reproducible from single realizations (a large number of small sample volume data would have to be averaged. Thus it is important to identify the minimum volume of material for consistent repeatability of results.

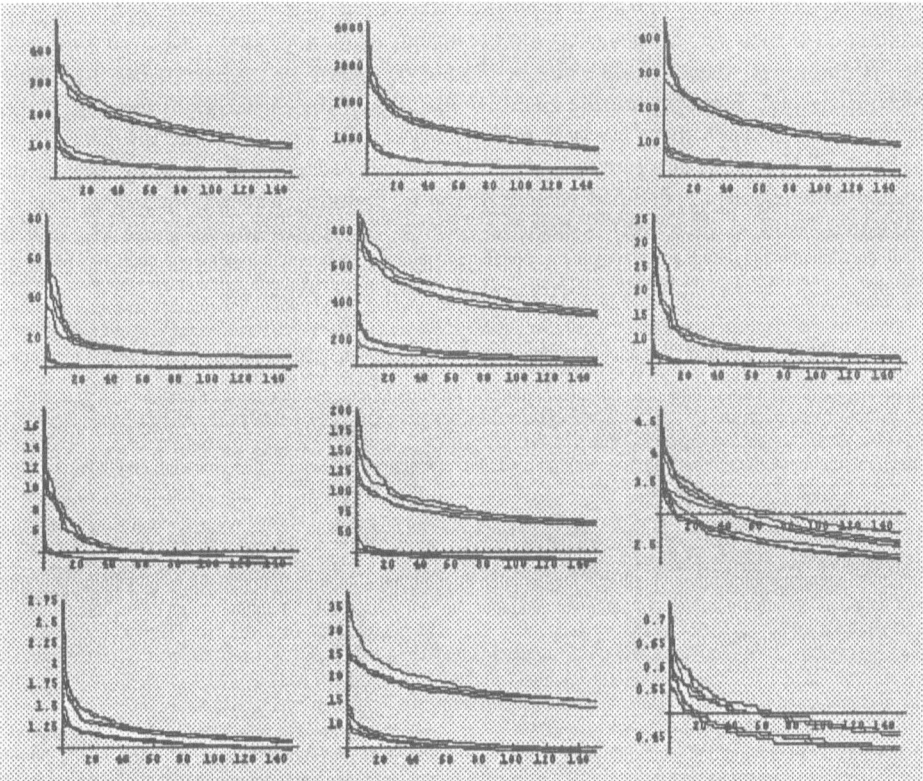

Figure 4 Plot of wavelet coefficients (150 largest ones for each scale) obtained from four images of "noisy" microstructure (top four curves in each plot) and from four images of "quiet" microstructure (bottom four curves in each plot). The first (left) column of plots show the coefficients from the wavelet transform in the y-direction, the second one from the x, and the third one from the x-y. Scale decreases from top to bottom.

The following question is addressed : is it possible to extract visual information about the microstructure, e.g. grain structure, position of grain boundaries, etc. ? As discussed before the wavelet coefficients have information on the microstructure (they represent the local regularity of the image, which is in turn related to the way the ultrasonic wave is scattered, the last being related to the microstructure). It may be possible to extract relevant information from the wavelet coefficients. For example, we can use the fact that local maxima of the wavelet coefficients represent sharp transitions in the image. Figure

5 shows typical features extracted from the wavelet coefficient maxima. After the local maxima (and their spatial position) are evaluated, contour plots of these maxima yield the features. A clear difference in "features" can be seen (noisy vs. quiet microstructures). Furthermore the essential features of the microstructures, Fig. 1, are reproduced. Although these are only preliminary results, they illustrate the potential of the method. Several issues remain to be examined, i.e. whether the spacing used in the B-scans is adequate for detailed representation of features, whether the digitization of the signal (vertical direction) is detailed enough, etc. Furthermore, there are several ways to extract features from wavelet coefficients, and a large number of publications on the subject can be found. Here a very simple one has been explored.

Noisy   Quiet

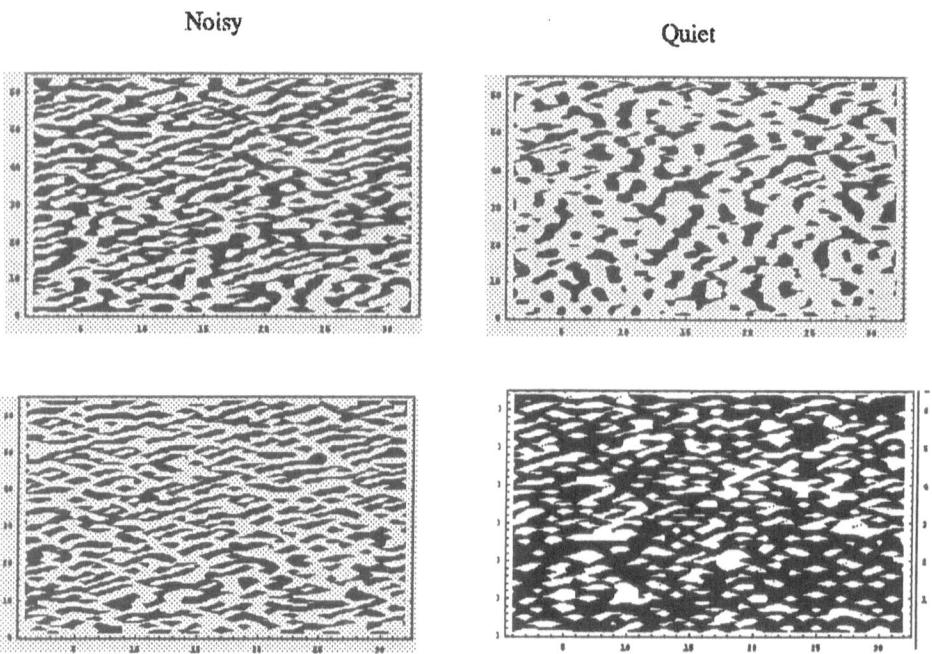

Figure 5 Features extracted form the wavelet coefficient maxima for the "noisy" (first column) and the "quiet" Note that 256 units correspond to 32.5 mm. The first row of images is the contour plot of the wavelet maxima at the corresponding scale. The second one was produced as follows: the local wavelet coefficients were averaged over a radius equal to the support of the wavelet at the corresponding scale and then the local maxima were plotted.

Conclusions

A "fresh" approach to material characterization with a solid mathematical background has been examined on an exploratory basis, and the potential of the approach has been demonstrated. the proposed approach should have important consequences on material characterization and relevant properties (e.g. toughness, remaining material/structure life), nondestructive evaluation, statistically based reliability, material modeling, etc. Similarly to the success of wavelets in several fields of physical science, wavelets show to be an effective tool for investigating materials related properties at various scales.

Acknowledgments

The experiments reported herein were performed at Wright-Patterson AFB by the second author; the support from personnel is fully acknowledged. The first author was supported under grand No. F49620-93-C-0063 from the US Air Force.

References

ARNEODO A., ARGOUL F., BARCY E., ELEZGARAY J., FREYSZ E., GRASSEAU G., MUZY J.F. & POULIGNY B., 1992, I. From the transition to chaos to fully developed turbulence; II. Optical wavelet transform of fractal growth phenomena. In Wavelets and applications, Y. Meyer (ed.), Masson, Springer-Verlag, Berlin.

BASDEVANT C., PERRIER V., PHILIPOVITCH T. & DO KHAC M., 1993, Local spectral analysis of turbulent flows using wavelet transforms. In Vortex flows and related numerical methods, J.T. Beale et al. (eds.), Kluwer Academic Publishers, The Netherlands, 1-26.

BENEDETO J.J. & FRAZIER M.W., (Eds.), 1994, Wavelets, Mathematics and Applications, CRC Press, Boca Raton, Florida.

CHUI C.K., et al, 1992, An introduction to wavelets, Volume 1, Academic Press, San Diego, Also volumes 2,3,4 & 5, 1992-1995.

DAUBECHIES I., 1993, Two recent results on wavelets: wavelet basis for the interval, and biorthogonal wavelets diagonalizing the derivative operator, In Recent Advances in Wavelet Analysis, L.L. Schumaker & G. Webb (eds.) Academic press, pp 237-257.

FARGE M., 1992, The continuous wavelet transform of two-dimensional turbulent flows. In Wavelets and their applications, M. Ruskai et al. (ed.), Jones and Bartlett, Boston.

FRANTZISKONIS G., 1995, Heterogeneity and implicated surface effects - statistical, fractal formulation and relevant analytical solution, Acta Mechanica, 108, 157-178.

FRANTZISKONIS G. & LORET B., 1995, Scale dependent constitutive relations - information from wavelet analysis and application to localization problems, A Eur. J.

Mechanics A/Solids, 14, 873-892, 1995.

GOEBBLES K., 1980, Structure analysis by ultrasonic radiation. Research Techniques in Nondestructive Testing. New York, Academic: 85-157.

GUO C. B. et al., 1985, Scattering of ultrasonic waves in anisotropic polycrystalline metals, Acoustica Vol. , no. 59: 112-120

HECHT, A. et al., 1981, Nondestructive determination of grain size in austenitic sheet by ultrasonic backscattering," Mat. Eval. Vol. , no. 39: 934-938.

HOUDRE H., Wavelets, probability, and statistics: some bridges, in Wavelets, J.J. Benedetto & M.W. Frazier Editors, CRC press, Boca Raton, 1994.

MARGETAN, F. J. et al., 1994, Backscattered microstructural noise in toneburst inspections, J. Nondestr. Eval. Vol. , no. 13.

MEYER Y., ed, 1991, Wavelets and applications, Masson, Springer Verlag, Paris.

MEYER Y. & ROQUES S., eds, 1993, Wavelet analysis and applications, Editions Frontier Press, Singapore.

NEUHAUSER H., 1988, The dynamics of slip band formation in single crystals, Res Mechanica, 23, 113-135.

ROSE J. H., 1992, Ultrasonic backscatter from microstructure, in Rev. Progr. Quant. Nondestr. Eval.11B, Plenum, 1677-1684.

RUSKAI M.B. et al (eds.), 1992, Wavelets and their applications, Jones and Bartlett Publ., Boston.

WILLIAMS J.R. & AMARATUNGA K., 1995, A discrete wavelet transform without edge effects, IESL (MIT) technical report No. 95-02.

WILLEMS, H. and K. GOEBBELS, 1981, Characterization of microstructure by backscattered ultrasonic waves, Met. Sci. Vol. , no. 15 (1981): 549-553.

PROBABILISTIC ASPECTS IN SHAPE MEMORY ALLOY MODELLING

Constitutive law for energy dissipators

L. FARAVELLI AND L. PETRINI
Department of Structural Mechanics, Univesity of Pavia
Via Ferrata 1 - 27100 Pavia - Italy

Abstract

Shape-memory alloys (SMA) have potential for use in dissipation devices that are employed in the passive control of structural systems subjected to seismically induced vibration. The present paper extends to SMA an hysteretic constitutive law developed originally by Mròz for steel. After the formulation of the multisurface hardening model, the way the parameters are identified by testing is dicussed. Their randomness is taken into account by suitable probabilistic models.

Keywords: constitutive law, energy dissipation, hardening, hysteretic cycle, yielding surface.

1. Introduction

SMA materials can recover plastic strain by suitable thermal cycles [1, 2]. This is the result of a change in the crystal structure of the alloy interpreted as martensitic phase transformation . Martensite forms on cooling from the high temperature phase called austenite. The transformation from austenite to martensite and the reverse transformation from martensite to austenite is not at the same temperature. The complete thermal cycle is characterized by four temperatures: austenite start temperature A_s, austenite finish temperature A_f, martensite start temperature M_s, martensite finish temperature M_f. Austenite and martensite show distinct differences in their stress/strain curve. Different models of the behaviour of these materials have recently been implemented [3, 4, 5]. The material response of interest for the purposes of this research is a typical stress-strain curve for TiNi at temperatures below M_f as given in Fig. 1. When the alloy is deformed, the elastic deformation (1) is followed by yielding in which the stress remains approximately constant. The deformation occuring in (2) is due to twinning deformation of martensite. Unloading in (2), the strain due to the plastic deformation remains (3). This residual strain reverts to the original state upon the application of

G. N. Frantziskonis (ed.), PROBAMAT – 21st Century: Probabilities and Materials, 379–391.

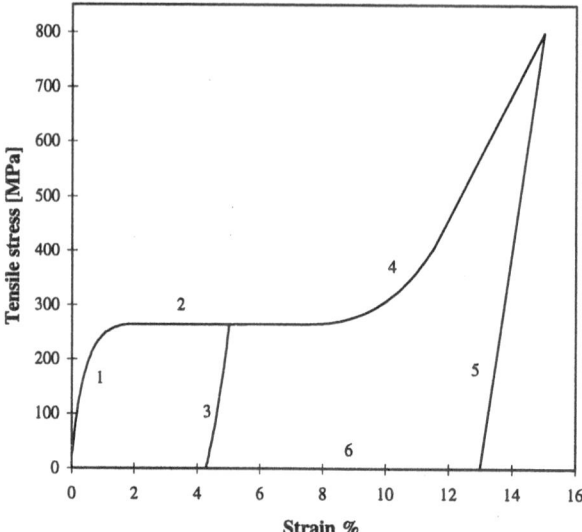

Figure 1. The stress-strain relationship of interest for TiNi alloy.

heat, thus giving rise to the shape-memory effect (SME). If the deformation strain increases, the hardening begins (4) and during the unloading stress the SME can be incomplete. In order to avoid this, the strain should be kept below a fixed quantity. The previous behaviour shows two aspects which together form a basic requirement in the realization of energy dissipators: a significant yielding stress and a suitable ductility. When combined with the possibility of removing the plastic strain by a thermal process, such a material becomes a formidable candidate for conceiving devices with recoverable deformation [6, 7, 8, 9]. For this purpose one produces in SMA short bars the analysis of which requires 3-dimensional finite element discretizations. In the following a 3D constitutive model able to reproduce the described shape-memory alloy behaviour, will be discussed. The results of experimental tests are shown and discussed toward the evaluation of the parameters of the proposed model. Finally the parameters affected with uncertainty are identified and probabilistic models are developed.

2. Basic aspects of the plasticity law modelling

The plasticity aspects of shape-memory alloy are the specific goal of thermo-mechanic studies developed during the last decade [10]. In this study, however, the thermal transformation follows the plasticity progress and, hence, classical mechanical models can be adopted in the energy dissipation analysis. The endochronic approach [11] is under examination, but within this paper attention is focused on classical plasticity. In particular, one tries to extend to SMA the Mròz's multisurface hardening model [12].

Consider the stress-strain relationship obtained in a simple axial test. In order to facilitate the extension to the 3-dimension case, describe the strain (ϵ) and the stress (σ) in terms of the components of their deviators e and s, respectively. Assume that the stress-strain relationship be piece-wise linear. Let k be the number of branches and let (E_j, R_j), $j = 1, \ldots, k$, be the coordinates of the discontinuity points in the (e^p, s) plane, e^p being the plastic strain. Instead of E_j, one can make use of the hardening parameter sequence:

$$H_1 = \frac{R_2 - R_1}{E_2 - E_1}, \quad \ldots \quad H_{k-1} = \frac{R_k - R_{k-1}}{E_k - E_{k-1}} \tag{1}$$

For the virgin curve, one has $E_1 = 0$, but after a loading-unloading cycle a different value can be detected for E_1 and hence a further value $H_0 = \frac{R_1}{E_1}$ is required. The stress-strain relation of a material showing a symmetric response is initially centered around the axis $s = 0$. But this is no longer true as plasticity progresses: the moving center of symmetry of the stress deviator will be denoted as back-stress and indicated by α.

In a 3-dimension situation, stress and back stress become the vectors s and α, respectively. To enter a new branch, say the j-th one, means to have as active loading surface:

$$\|s - \alpha_j\| - R_j = 0 \tag{2}$$

with associated flow rule:

$$de^p = \frac{1}{H_j} n(s \cdot n) \tag{3}$$

n being the normalized $((n \cdot n)^{\frac{1}{2}} = 1)$ stress gradient of the loading surface. Note that Eq. (2) denotes the lateral surface of a cylinder of circular section, radius R_j and center α_j: both of them require the definition of an evolution law. In the absence of isotropic hardening R_j remains constant and the evolution law for the active surface is simply:

$$d\alpha_j = ds_j \tag{4}$$

In the general case the evolution law of α_j also involves the evolution law of α_{j+1}:

$$d\alpha_j = d\alpha_{j+1} + (dR_{j+1} - dR_j)m + \frac{ds_j \cdot m - d\alpha_{j+1} \cdot m - dR_{j+1}}{(s_{j+1} - s_j) \cdot m}(s_{j+1} - s_j) \tag{5}$$

m being the unit vector along $s_j - \alpha_j$. One still has to define the term s_{j+1} and the evolution law of $d\alpha_j$ for $j < l \leq k$, k being the number of loading surfaces. Let s_j any point on the previous surface; on the nested surface one can define an associated point:

$$s_{j+1} = \alpha_{j+1} + \frac{R_{j+1}}{R_j}(s_j - \alpha_j) = \alpha_{j+1} + R_{j+1}m \tag{6}$$

This associated point has just the characteristic that the same normal vector s_j shows. Let $\eta_h = s_{j+1} - s_j$ be the so-called hardening translation vector and $\eta_r = \alpha_{j+1} - \alpha_j$ be the so-called recovery translation vector. The Mròz-Rodzik model [12] assumes that the more external circle (the k-th one) is fixed and that the translation rule mixes hardening and recovery contributions with weights χ_h and χ_r to be calibrated:

$$da_j = da_{j+1} + \chi_h \eta_{hj} + \chi_r \eta_{rj} \qquad (7)$$

with

$$dx_h = A(de^p \cdot de^p)^{\frac{1}{2}}(B(\xi_a)^L - \chi_h) \qquad (8)$$
$$dx_r = C(de^p \cdot de^p)^{\frac{1}{2}}(D(\xi_m)^M - \chi_r) \qquad (9)$$

In the previous equation A, B, C, D, L and M are parameters to be calibrated while ξ_a and ξ_m are the measures of amplitude and maximal prestress, respectively. They are simply computed as the ratio, with R_1, of the current index at the instant of unloading and the maximum index achieved during the loading evolution, respectively. Initially all these quantities show a zero value.

The isotropic hardening only develops for the non-proportional loading history and has the form:

$$R_j = (1 + \beta_j \psi) R_{0j} \qquad (10)$$

with β_j a scaling factor of expression $\frac{R_{0j} - R_{01}}{R_{0k} - R_{01}}$ and k the index of the boundary surface. The material function ψ has the incremental expression:

$$d\psi = P[1 - (\frac{da_j \cdot \alpha_j}{||da_j||)^2] \, ||\alpha_j||}(Y(\frac{R_{0h}}{R_{01}})^N - \psi) \qquad (11)$$

thus adding the further three parameters N, P and Y. The index h makes explicit the dependence on either the maximal loading surface reached during the non-proportional loading or the actual amplitude surface during the non-proportional unloading.

3. Testing versus modelling

It is very important to consider the manner in which the experimantal data have been collected. The design of an experiment must be consistent with the requirements of the problem. There are two important sources of experimental error:

1. the variability of the material itself;
2. the use of poor experiment techniques.

For this reason it is necessary to identify the parameters of the model to which the result of the structural analysis could show a significant sensitivity, the final goal being the capability of reliable numerical simulations. In this context, some parameters will always require the introduction of a probabilistic model, some of them will not need it at all and some others should be carefully checked before a decision is achieved.

3.1. GENERAL FEATURES

The first item to be considered is the degree of heterogeneity of the material, mainly due to its production technique. A plasma technology offers a great homogeneity, while standard production processes give rise to the situation in Fig. 2. This figure shows an image obtained from analyses with the microanalyzer (JEOL JXA-840A electron probe microanalyzer test). They pointed out a certain level of heteregeneity within the bar. It is justified by the precipitation of Titanium during the process of creation of the alloy. The dark areas indicates the presence of zone with Ti and Ni in percentage different from the one of reference. But it does not seem to have influence in the results of other tests. Moreover, the dimension of these impurities is much smaller than the size of any finite element one can conceive for his structural analyses. As a consequence, the material can be regarded as homogeneous, provided a preliminary check at a microscopic level has been done. Of course this material homogeneity does not involve necessarily a homogeneous spatial distribution of the material parameter values.

The second item to be studied is the sequence of the transformation temperatures. The differential scanning calorimeter (DSC) test estimates the temperatures of phase transition. A suitable sample of weight 20 mg is prepared, treated by chemical etching, to remove superficial martensite, and then subjected to a variation of temperature; the goal is to measure the heat flow. Two different Ti_{51}-Ni_{49} bars have been analysed. For them the results of Fig. 3 were obtained. Since one requires a martensitic phase at the working temperature, the only parameter of interest is M_f. The DSC test will generally give a positive answer (i.e. M_f is higher than the working temperature), but Figure 3 shows the different reliability that different bars can offer. Sharper peaks guarantee a greater likelihood that the thermal transformation be successful in recovering the mechanically induced plastic strain. The dependency of curve smoothness on the bar diameter is a topic which require further investigation. From the DSC curve of bar # 1 the following values were calculated: $M_f = 54.9°$, $M_s = 80.8°$, $A_s = 84.2°$, $A_f = 112.5°$.

3.2. CONSTITUTIVE LAW PARAMETERS

At the present stage of the research attention is focused on the definition of the parameters which characterize the multisurface constitutive law discussed in the previous section. One distinguishes the parameters of mechanical meaning from those introduced to fit experiment data.

With reference to this second set, the parameters A, B, C, D, L, M, N, P and Y, one makes use of a regression model. In the space of these nine parameters, each point generates a special (e^p, s) line under fixed strain (or stress) amplitude histories. The goal is to find the set of values which minimizes the error from an experimental set of data.

Coming to the two sets of correlated vectors R_j and H_j with direct mechanical meaning, statistical inference must be used. The principle of simplicity [13] should

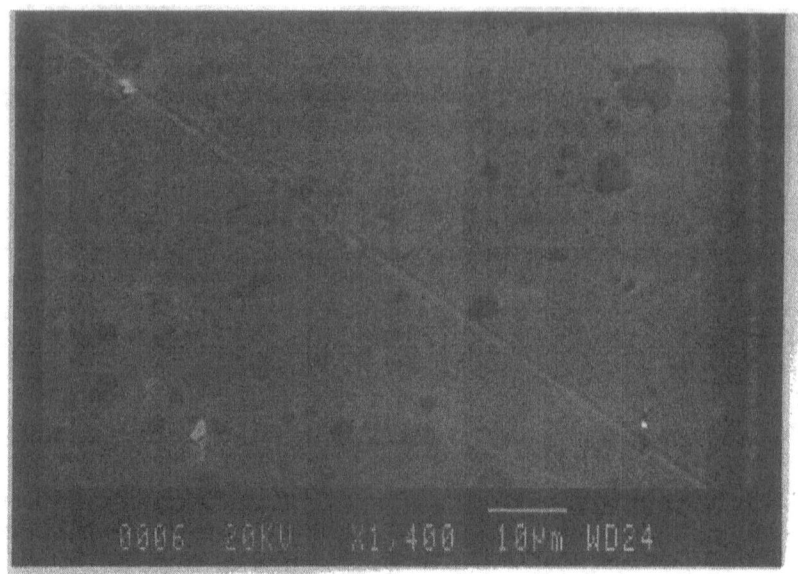

Figure 2. Secondary Electron Image obteined for reference bar with the electron probe microanalyzer.

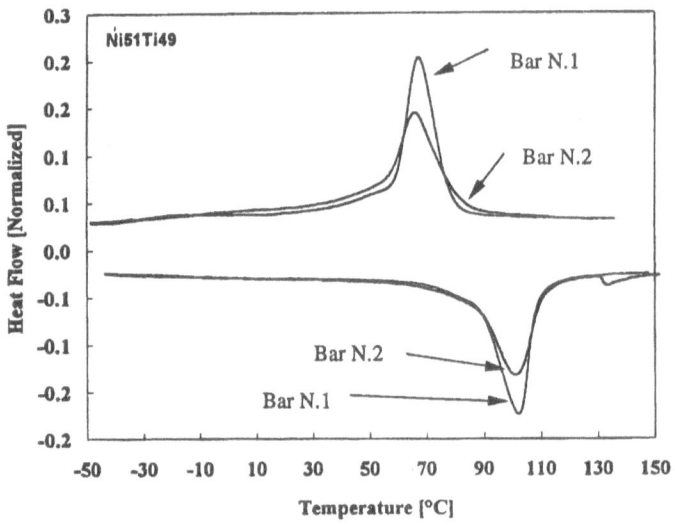

Figure 3. DSC curves: heating and cooling cycles for two different bars.

drive in choosing the values of the parameters $\theta_1, ..., \theta_q$ that fix the probability measure. Instead of the deterministic principle of least variability it is reasonable to choose $\theta_1, ..., \theta_q$ so that the probability density has its maximum at the actual set of interpolation functions. This is the well-known principle of maximum likelihood estimation in the theory of mathematical statistics. From an operational point of view the most attractive probability measure to choose is the Gaussian measure.

3.3. SPATIAL VARIABILITY

Mainly the latter set of parameters (the ones with direct mechanical meaning) could show spatial variability, so that one must introduce a probabilistic model for a multivariate random field. The numerical approach most widely used in mechanics is based on the descritization of the continuum in finite elements. As a consequence any random field describing material properties has to be discretized ("stochastic finite element"). The stochastic field is replaced by a random vector which takes into account the interelement correlations. There are three levels of modelling sophistication:

1. Material parameters random but spatially homogeneous;
2. Material parameters random and spatially varying but constant over a finite element;
3. Materials parameters random within a finite element.

The field discretization methods were developed at level 2 in order to perform structural analysis with practicable computational effort. A list of these methods is described in [13]. In particular the field can be replaced by:

- the values at a central point of each element (Midpoint Method);
- the average field values over each element (Integral Average Method);
- an interpolation function by shape functions within each element (Shape - function Method);
- linear regression methods;
- truncated series expansion methods.

For a probabilistic structural analysis which makes use of the extended response surface model [14], the random field is replaced by the product or the sum of a scalar central value and the deviation vector from it.

4. Response probabilistic model

An experiment design for numerical simulations can be planned on the basis of the probabilistic model of the material discussed in the previous section. Let n_e be the number of available experiments. and n_c the number of parameters in the model. Each random vector, as the ones of components R_j and H_j, is modelled by a single random parameter and a set of deviations from a reference realization associated scaled by the value of the previous associated parameter [14]. The same idealization is easily extended to the case of actual spatial variability. One considers then, as a model, a quadratic function of the observed variables

$$y = A\vartheta + \epsilon_L + \epsilon. \tag{12}$$

where A is the observation matrix of the regressor variables including the constant, i.e. an $n_e \times n_c$. and ϑ is an n_c-dimensional vector of the regression coefficients composed of ϑ_0 (the constant term), ϑ_1 (the coefficients of the linear part) and Θ_2 (the coefficients of the quadratic part), which has to be estimated. Further $y = (y^{(1)}, \ldots, y^{(n_e)})^T$ is the vector of the observed responses. The components of $\epsilon_L = (\epsilon_L^{(1)}, \ldots, \epsilon_L^{(n_e)})^T$ are the systematic errors due to lack of fit and the components of $\epsilon = (\epsilon_1, \ldots, \epsilon_{n_e})^T$ the random errors associated with the vector deviations. It will be assumed that the random errors have mean zero, are uncorrelated and have the same variance σ^2. This gives for the matrix of the variances of the random errors:

$$\mathbb{E}[\epsilon\epsilon^T] = \sigma^2 I_{n_e}. \tag{13}$$

The estimated vector of the regression coefficients is then

$$\hat{\vartheta} = (A^T A)^{-1} A^T y. \tag{14}$$

The estimated value of the response vector y results

$$\hat{y} = A(A^T A)^{-1} A^T y = P_A y. \tag{15}$$

This is the projection of the vector y on the subspace spanned by the columns of A, which has dimension n_c, and P_A is the projection matrix of this subspace (see [14]). The residual sum of squares, is

$$SS_R = (y - \hat{y})^T (y - \hat{y}). \tag{16}$$

This gives then, using equation (15),

$$SS_R = ((I_{n_e} - P_A)y)^T ((I_{n_e} - P_A)y) \tag{17}$$

One can also write

$$|y|^2 = |\hat{y}|^2 + |P(\epsilon_L + \epsilon)|^2. \tag{18}$$

The sum of the squared y's is decomposed into the sum of the $\hat{y}$'s, the projection on the space spanned by the rows of A and the remaining error terms, which are orthogonal to this subspace. From the model above one also estimates the error variance of the model by repeating m times the experiment at the central point. We assume that these are the first m observations. Since there is no variation of the regressor variables, the value $\epsilon_L^{(1)} = \ldots = \epsilon_L^{(m)}$ of the systematic error is always the same. One has for the pure error sum of squares the form

$$SS_E = \sum_{j=1}^{m} (y^{(j)} - \bar{y})^2 = \sum_{j=1}^{m} (y^{(j)})^2 - m\bar{y}^2 \tag{19}$$

with $\bar{y} = m^{-1} \sum_{j=1}^{m} y^{(j)}$. This can be written as

$$SS_E = \epsilon_0^T A_E \epsilon_0 \tag{20}$$

with $\epsilon_0 = (\epsilon^{(1)}, \ldots, \epsilon^{(m)})^T$. The matrix A_E is defined by $A_E = I_m - E_m$, where $E_m = m^{-1}[(1)_{i,j=1,\ldots,m}]$.

One considers then the lack of fit sum of squares SS_L. It is defined as the residual sum of squares minus the pure error sum of squares SS_E

$$SS_R = SS_L + SS_E. \tag{21}$$

A possible F-test is to compare the error variance with the pure random error variance to find out if there is a systematic error component. In this case we check if the value of $|P(\epsilon_L + \epsilon)|^2/\sigma^2$ is too large to come from a central χ^2-distribution with $n_e - n_c$ degrees of freedom. It is possible to show [14] that the squared length of the random vector obtained by projecting this random vector $\epsilon + \epsilon_L$ on a subspace with dimension $n_e - n_c$ is a non-central χ^2-square distribution with non-centrality parameter defined by

$$\Delta = \sigma^{-2}|P\epsilon_L|^2. \tag{22}$$

In the case that there is no lack of fit, i.e. $\Delta = 0$, the SS_L and SS_E sum of squares are independent and both have a central χ^2- distribution, with $\nu_L = n_e - n_c - m - 1$ and $\nu_E = m - 1$ degrees of freedom, respectively. Then the ratio

$$\lambda^2 = \frac{SS_L \cdot \nu_E}{\nu_L \cdot SS_E} \tag{23}$$

has an F-distribution with ν_L and ν_E degrees of freedom. In [14] the use of the statistic of λ is used to find the best model among the non-rejected ones. The expected value of λ is, for all models, larger than or equal to unity. Only for the correct model it is approximately equal to unity. In conclusion, in order to fit a quadratic regression model, one estimates the regression coefficients and the λ statistic. The latter one should be used in two ways. Firstly one checks if there is a systematic error by performing an F-test and then one identifies the best model among them which were not rejected by the test.

5. Laboratory results

A number of tests have been devoted to know the mechanical behaviour and the shape memory effect of a SMA bar subjected to stress loading/unloading at room temperature. The TiNi bar used in the tests has circular cross section with diameter 10.0 mm. It has been submitted to tension tests by loading-unloading sequences, up to different levels of plasticity. After each test the bar was subjected to heating in a oven to investigate its recovery capacity.

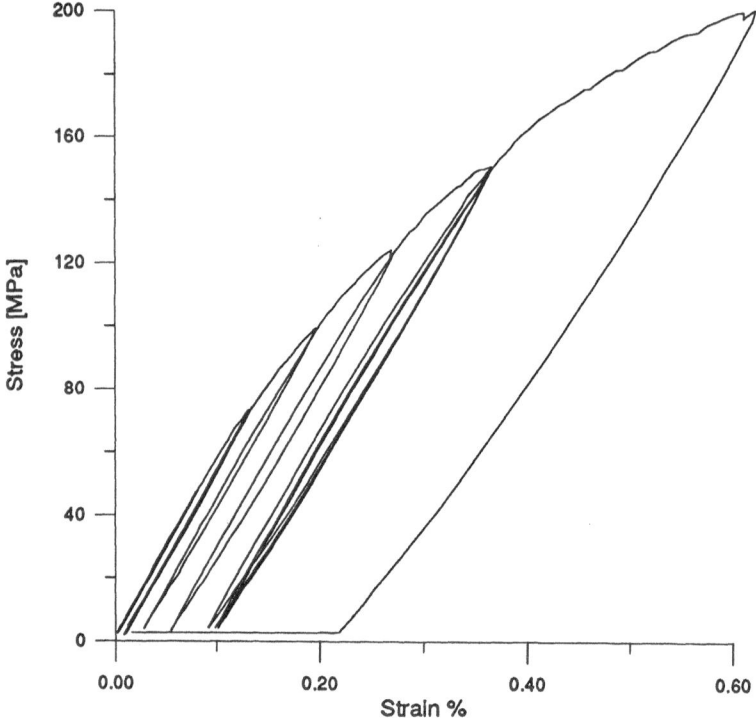

Figure 4. The laboratory stress-strain relationship for a TiNi alloy bar of diameter 10.0 mm.

The first test, Fig. 4, corresponds to small values of the plastic strain. It provided the standard elastic mechanical parameters: Young's modulus E=57000 MPa, Poisson ratio ν=0.35.

The second test, Fig. 5, at the end of which the plastic deformation achieves 2%, made possible to define eight discontinuity points and the relevant yielding stresses and hardening slopes. The following values of the yielding surfaces radii and of the internal plastic hardening moduli were identified:

$R^1 = 78.8 MPa$
$R^2 = 118 MPa$ $H^1 = 49.8 GPa$
$R^3 = 150 MPa$ $H^2 = 20.3 GPa$
$R^4 = 160 MPa$ $H^3 = 9.34 GPa$
$R^5 = 166 MPa$ $H^4 = 6.46 GPa$
$R^6 = 173 MPa$ $H^5 = 3.56 GPa$
$R^7 = 176 MPa$ $H^6 = 1.66 GPa$

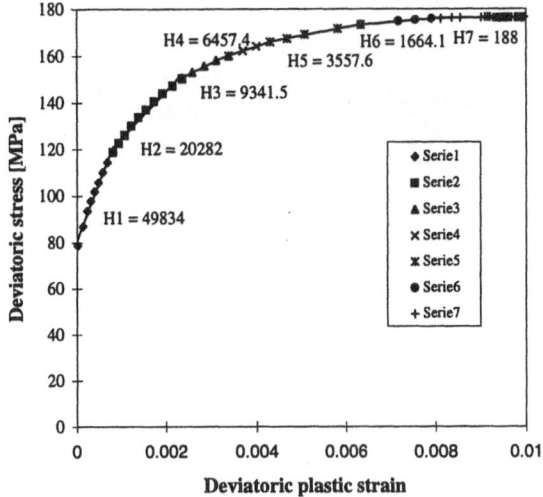

Figure 5. Identification of the hardening slopes from a laboratory test on the TiNi alloy bar.

$$R^8 = 178MPa \qquad\qquad H^7 = 0.19GPa$$

In Fig. 4, the strain recovering after heating is also evident: after the last cycle of loading-unloading, the bar is left at the constant temperature of 150°C until the recovering phase is stable. The recovering seems to be satisfactory also after several cycles in plasticity.

Fig. 6 shows the variation of the mechanical behaviour of the SMA bar at two different strain-gauges. These parameters should be modelled as a random field in the probabilistic model, in order to incorporate such an evident spatial variability.

6. Conclusions

This paper presents some preliminary results of a long-term research program toward the exploitation of SMA bars in energy dissipation devices. In particular, the way the material parameters for the 3-D constitutive law are identified from experimental tests is discussed. A probabilistic model is also proposed making the numerical simulation analyses adequately accurate.

390

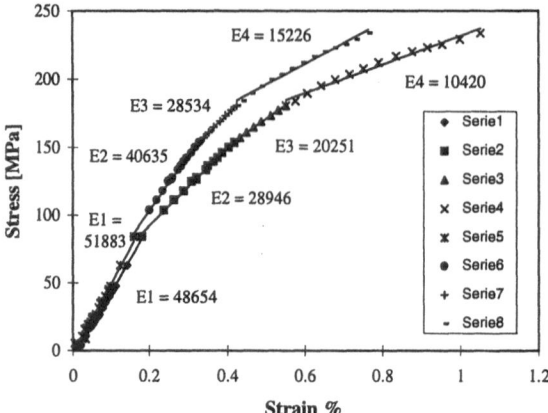

Figure 6. Comparision between the slopes E of the diagrams $\sigma - \epsilon$ in a tensil test at two different strain-guages.

7. Acknowledgement

This work has been supported by grants from the National Research Council, Strategic Project on Large System Design, with Professor E. Benvenuto of the University of Genova as coordinator (N.96.05415.ST74).

References

1. Funakubo, H. (1987), *Shape Memory Alloys*, Gordon and Breach Science Publishers.
2. Hodgson, D.E. and Biermann, R.J. (1992), Shape Memory Alloys, *Metals Handbook*, Vol. 2.
3. Liang, C. and Rogers, C.A. (1990), One-dimensional Thermomechanical Constitutive Relations for Shape-Memory Materials, *J. Intell. Mater. Sys. Structures*, 1, 207-234.
4. Graesser, E.J. and Cozzarelli, F.A. (1991), Aseismic Isolation, *Journal of Engineering Mechanics*, 117, No. 11, 2590-2608.
5. Auricchio, F. and Taylor, R.L. (1997), Shape-memory Alloys: Modelling and Numerical Simulations of the Finite-strain Superelastic Behavior, *Comput. Methods Appl. Mech. Engrg.*, 143, 175-194.
6. Ciampi, V. and Marioni, A. (1991), New Types of Energy Dissipating Devices for Seismic Protection of Bridges, *Proc. of Third World Congress on Joint Sealing and Bearing Systems for Concrete Structures*, Toronto.
7. Attanasio, M., Battaini, M. and Faravelli, L. (1996), SMA in Energy Dissipation Devices, *Proceedings of the First European Conference on Structural Control*, Barcelona; A. Baratta and J. Rodellar (eds.), World Scientific, Singapore, 24-31
8. Attanasio, M., Battaini, M., Casciati F. and Marioni, A. (1996), Use of Shape Memory Alloy

for Seismic Isolation Devices, *Proceedings of the 7th International Conference on Adaptive and Technologies*, Roma; Technomic publ. co., Lancaster, 267-276

9. Attanasio, M., Faravelli L. and Marioni A. (1996), Exploiting SMA Bars in Energy Dissipators, *Proceedings of the 2nd International Workshop on Structural Control*, Hong Kong; HKUST, 41-50

10. Achenbach, M. and Müller, I. (1985), Shape Memory as Thermally Activated Process, in *Plasticity Today*, A. Sawczuk and G. Bianchi (eds.), Elsevier, 515-534

11. Valanis, K.C. and Fan, J. (1985), Experimental Verification of Endochronic Plasticity in Spatially Varying Strain Fields, in *Plasticity Today*, A. Sawczuk and G. Bianchi (eds.), Elsevier, 153-174

12. Mròz, Z. and Rodzik, P. (1995), The Modified Multisurface Hardening Model and its Application to Cyclic Loading Problems, in *Inelastic Behaviour of Structures under Variable Loads*, Z. Mroz et al. (eds.), Kluwer Academic Publishers, 3-18.

13. Ditlevsen, O. (1996), Dimension Reduction and Discretization in Stochasic Problems by Refression Method, in *Mathematical Models for Structural Reliability Analysis*, F. Casciati and B. Roberts (eds.), CRC, Boca Raton, 51-132.

14. Breitung, K. and Faravelli, L. (1996), Response Surface Methods and Asymptotic Approximations, in *Mathematical Models for Structural Reliability Analysis*, F. Casciati and B. Roberts (eds.), CRC, Boca Raton, 227-285.

HOMOGENIZATION OF RANDOM GRANULAR MEDIA

K. SAB
CERAM-ENPC
6 & 8, avenue Blaise Pascal - Cité Descartes - Champs-sur-Marne - 77455 Marne-la-Vallée CEDEX 2. France.

Abstract

A constitutive law for granular material is derived based on a new micro-mechanics approach taking into account the mechanisms of sliding and separation of particles. In this approach, the representative volume element is divided into tetrahedral elements where each node is occupied by a particle's centre. Stress and strain tensors are defined on each tetrahedron and the stress-strain relationship is derived based on the contact law between particles. Then, using the same localization process as the one used by Zaoui [8], the macroscopic constitutive law is derived and compared with experimental results.

1. Introduction

This paper is concerned with micro-mechanical modelling of rate-independent plasticity of random granular materials undergoing small deformations.

Following Zaoui [8], it can roughly be said that a micro-mechanics model is a combination of three kinds of elements :
a) a characterization of the constituent elementary phases
b) an averaging procedure which defines overall variables from local (or microscopic) ones
c) a localization modelling which refers to the derivation of local variables from overall ones (e.g. Voigt and Reuss localizations, self-consistent models, ...).

G. N. Frantziskonis (ed.), PROBAMAT – 21st Century: Probabilities and Materials, 393–403.

The main difficulty in the homogenization of a granular material (a soil, for example) undergoing small deformations is the discrete nature of the modelling at the microscopic scale.

Indeed, at this scale, the material is perceived as a random assembly of rigid particles which interact at contact points. A contact law between two particles A_c and B_c in contact at point $\underline{c}$ relates $\underline{f}_c$, the force exerted by A_c on B_c at $\underline{c}$, to $\underline{d}_c$, the velocity of A_c relatively to B_c at point $\underline{c}$.

$$\underline{d}_c = \underline{u}^A - \underline{u}^B + \underline{\omega}^A \wedge (\underline{c} - \underline{a}) + \underline{\omega}^B \wedge (\underline{b} - \underline{c}) \tag{1}$$

Here $\underline{a}, \underline{b}$ are the centres of A and B, $\underline{\omega}^A, \underline{\omega}^B$ their rotation rates and $\underline{u}^A, \underline{u}^B$ are the velocities at their centres, respectively.

Except in where the elementary constituent is formed by a particle and the set of particles which are in contact with the considered one, in almost all micro-mechanics modelling the contact point is considered as the elementary constituent.

The overall stress tensor, $\underline{\underline{\Sigma}}$, in a Representative Volume Element (RVE), V, is expressed in terms of contact forces according to the following well-known relationship [4],[5]:

$$\underline{\underline{\Sigma}} = \frac{1}{|V|} \sum_{c \in V} \underline{f}_c \otimes^s \underline{l}_c \tag{2}$$

where $\underline{l}_c = \underline{a} - \underline{b}$ is the branch vector connecting the centre of B_c to the centre of A_c, and $\otimes^s$ is symmetric tensor product.

The difficulty in these models is to express, $\underline{\underline{\mathring{E}}}$, the overall strain rate tensor, in terms of relative velocities, $\underline{d}_c$, at contact points in V. Thanks to the mathematical theory of homogenization, it has been rigorously proved in [7] that this is possible only under very restrictive conditions.

In [7], spherical particles have been considered and it has been assumed that the branch vectors in V form a tessellation, $\mathcal{T}_V$, made of tetrahedrons, $T \in \mathcal{T}_V$ (the assembly is 'dense'). Notice that the nodes of this triangulation are the centres of particles. Then, the following equation has been established:

$$\underset{=}{\dot{E}} = \left\langle \underset{=}{\dot{\varepsilon}} \right\rangle = \frac{1}{|V|} \sum_{T \in \mathcal{T}_v} |T| \, \underset{=T}{\dot{\varepsilon}} \tag{3}$$

where, $\langle \ \rangle$ is the volume average on V, and $\underset{=}{\dot{\varepsilon}}$ is the strain rate field computed using a piecewise-linear interpolation of the velocity at the nodes of the above described triangulation. The gradient of the interpolated velocity field is piecewise-constant. It defines a strain rate tensor, $\underset{=T}{\dot{\varepsilon}}$, and a rotation rate vector, $\underline{\omega}_T$, on each $T \in \mathcal{T}_v$.

If the particles are spherical, then $\underset{=T}{\dot{\varepsilon}}$ can be expressed in terms of the relative velocities $\underline{d}_c$ of contact points lying on the boundary of T [7].

If the assembly is not dense, one can always add to the set of 'real' branch vectors a set of 'fictitious' branch vectors which join the centres of particles which are not in contact in such manner that the set of all real and fictitious branch vectors form a triangulation of V, also noted $\mathcal{T}_v$. Such triangulation always exists (but is not unique) if the particles have smooth convex shapes as it has been shown in [1]. Then, the fields $\underset{=}{\dot{\varepsilon}}$ and $\underline{\omega}$ can be defined similarly to the case of a dense assembly, and it can be shown that relationship (3) still holds true. In this case, $\underset{=}{\dot{E}}$ does not express in terms of $\underline{d}_c$ any more.

Our purpose is to provide a micro-mechanical modelling where the tetrahedrons, $T \in \mathcal{T}_v$, are considered as the elementary constituents. The strain rate tensor, $\underset{=T}{\dot{\varepsilon}}$, and the rotation rate vector, $\underline{\omega}_T$, of T have already been defined. We define a symmetric stress tensor, $\underset{=T}{\sigma}$, and an asymmetric stress vector, $\underline{\tau}_T$, of T as follows :

$$\underset{=T}{\sigma} = \sum_{\underline{c} \in \partial T} \frac{1}{|D_c|} \underline{f}_c \otimes^s \underline{l}_c \qquad \underline{\tau}_T = \sum_{\underline{c} \in \partial T} \frac{1}{|D_c|} \underline{l}_c \wedge \underline{f}_c \tag{4}$$

where the summation is extended to contact points lying on ∂T, the boundary of T; D_c is the domain formed by the union of all the tetrahedrons that have contact point $\underline{c}$ in common.

Then, it is easy to check that (2) equivalently writes:

$$\underline{\underline{\Sigma}} = \langle \underline{\underline{\sigma}} \rangle = \frac{1}{|V|} \sum_{T \in \mathcal{T}_v} |T| \, \underline{\underline{\sigma}}_T \tag{5}$$

Besides, the overall momentum equilibrium of contact forces in V writes:

$$\langle \underline{\underline{\tau}} \rangle = \frac{1}{|V|} \sum_{T \in \mathcal{T}_v} |T| \, \underline{\underline{\tau}}_T = \frac{1}{|V|} \sum_{c \in V} \underline{l}_c \wedge \underline{f}_c = 0 \tag{6}$$

2. Local constitutive law

A straightforward computation shows that (1) equivalently writes :

$$\underline{c} \in \partial T \Rightarrow \underline{d}_c = \underline{\underline{\dot{\varepsilon}}}_T \cdot \underline{l}_c + \left(\underline{\omega}_T - \underline{\omega}^A \right) \wedge \underline{l}_c + \left(\underline{\omega}^A - \underline{\omega}^B \right) \wedge \left(\underline{c} - \underline{b} \right) \tag{7}$$

Following Chang [3], we neglect the effect due to the last term in (7) ($\underline{\omega}^A \approx \underline{\omega}^B$); We define on each T a relative rotation rate vector, $\underline{\omega}_T^r$, such that (8) holds true:

$$\underline{c} \in \partial T \Rightarrow \underline{d}_c = \underline{\underline{\dot{\varepsilon}}}_T \cdot \underline{l}_c + \underline{\omega}_T^r \wedge \underline{l}_c \tag{8}$$

We shall adopt the following unilateral elasto-plastic contact law with Coulomb's friction, where $k^n > 0$, $k^t \geq 0$ and $C > 0$ are material constants.

Local basis $\left(\underline{n}_c, \underline{t}_c, \underline{s}_c \right)$:

$$\underline{n}_c = \frac{\underline{l}_c}{|\underline{l}_c|}, \quad f_c^n = \underline{f}_c \cdot \underline{n}_c, \quad \underline{f}_c^t = \underline{f}_c - f_c^n \cdot \underline{n}_c, \quad \underline{t}_c = \frac{\underline{f}_c^t}{|\underline{f}_c^t|} \tag{9.1}$$

Criterion:

$$F_c = \left|\underline{f}_c^t\right| + C f_c^n \leq 0 \qquad (9.2)$$

Unilateral contact:

$$\underline{f}_c = 0, \, d_c^n \geq 0 \Rightarrow \dot{\underline{f}}_c = 0 \qquad (9.3)$$

Elasticity:

$$F_c < 0 \text{ or } F_c = 0, \dot{F}_c < 0 \Rightarrow \dot{\underline{f}}_c = k^n d_c^n \cdot \underline{n}_c + k^t \left(d_c^t \cdot \underline{t}_c + d_c^s \cdot \underline{s}_c \right) \qquad (9.4)$$

Otherwise, plasticity:

$$\dot{\underline{f}}_c = k^n d_c^n \cdot \underline{n}_c - C k^n d_c^n \cdot \underline{t}_c + k^t d_c^s \cdot \underline{s}_c \qquad (9.5)$$

This contact law joined with (4) and (7) allows us to write the formal rate-independent behaviour of T:

$$\begin{pmatrix} \dot{\underline{\underline{\sigma}}}_T \\ \dot{\underline{\tau}}_T \end{pmatrix} = \mathbf{A}_T \begin{pmatrix} \dot{\underline{\underline{\varepsilon}}}_T \\ \dot{\underline{\omega}}_T^r \end{pmatrix} \qquad (10)$$

where $\mathbf{A}_T$ is the multi-branched fourth order tensor of incremental moduli of T.

3. Localization

We shall adapt to granular materials the localization modelling proposed by Zaoui [8] for polycrystals.

Notice that if contact forces in V, $\{\underline{f}_c\}$, are statically and plastically compatible with overall stress tensor, $\underline{\underline{\Sigma}}$, i.e. (5), (6) and (9.2) hold true, then, contact forces $\{\lambda \underline{f}_c\}$ with $\lambda > 0$ will be statically and plastically compatible with $\lambda \underline{\underline{\Sigma}}$.

So, let us introduce the following normalised stress variables:

$$P \equiv -\frac{1}{3}\mathrm{Tr}\left(\underline{\underline{\Sigma}}\right) > 0, \; \underline{\underline{\sigma}}_{T,n} \equiv \underline{\underline{\sigma}}_T/P \;, \; \underline{\tau}_{T,n} \equiv \underline{\tau}_T/P, \; \underline{\underline{\Sigma}}_n \equiv \underline{\underline{\Sigma}}/P = \left\langle \underline{\underline{\sigma}}_{T,n} \right\rangle \quad (11)$$

The normalised strain rate variables are defined *via* the behaviour (10) by:

$$\begin{pmatrix} \overset{\circ}{\underline{\underline{\varepsilon}}}_{T,n} \\ \overset{\circ}{\underline{\omega}}^r_{T,n} \end{pmatrix} \equiv A_T^{-1} \begin{pmatrix} \overset{\circ}{\underline{\underline{\sigma}}}_{T,n} \\ \overset{\circ}{\underline{\tau}}_{T,n} \end{pmatrix} \qquad \begin{pmatrix} \overset{\circ}{\underline{\underline{E}}}_n \\ \overset{\circ}{\underline{\Omega}}^r_n \end{pmatrix} = \begin{pmatrix} \left\langle \overset{\circ}{\underline{\underline{\varepsilon}}}_{T,n} \right\rangle \\ \left\langle \overset{\circ}{\underline{\omega}}^r_{T,n} \right\rangle \end{pmatrix} \qquad (12)$$

The localization modelling writes:

$$\overset{\circ}{\underline{\underline{\Sigma}}}_n = 0 \quad \Rightarrow \quad \begin{pmatrix} \overset{\circ}{\underline{\underline{\sigma}}}_{T,n} \\ \overset{\circ}{\underline{\tau}}_{T,n} \end{pmatrix} = 0 \qquad (13)$$

$$\overset{\circ}{\underline{\underline{\Sigma}}}_n \neq 0 \quad \Rightarrow \quad \begin{pmatrix} \overset{\circ}{\underline{\underline{\varepsilon}}}_{T,n} \\ \overset{\circ}{\underline{\omega}}^r_{T,n} \end{pmatrix} - \begin{pmatrix} \overset{\circ}{\underline{\underline{E}}}_n \\ \overset{\circ}{\underline{\Omega}}^r_n \end{pmatrix} = -\alpha \frac{\left|\overset{\circ}{\underline{\underline{E}}}_n\right|}{\left|\overset{\circ}{\underline{\underline{\Sigma}}}_n\right|} \left[\begin{pmatrix} \overset{\circ}{\underline{\underline{\sigma}}}_{T,n} \\ \beta \overset{\circ}{\underline{\tau}}_{T,n} \end{pmatrix} - \begin{pmatrix} \overset{\circ}{\underline{\underline{\Sigma}}}_n \\ 0 \end{pmatrix} \right] \qquad (14)$$

where α and β are positive constants, and $\left|\underline{\underline{X}}\right|^2 = \underline{\underline{X}}:\underline{\underline{X}} = \sum_{i,j} X_{ij}^2$.

For $\beta = 0$, $\underline{\omega}^r_{T,n}$ is uniform in all the tetrahedrons; for $\alpha = 0$, (14) is the Voigt localization of the normalised strain rates ; and, ; for $\alpha = +\infty$, (14) is the Reuss localization of the normalised stress rates.
Of course, the proposed localization modelling makes sense if P is strictly positive. In particular, it does not apply if the initial state is not pre-stressed.

We consider in the following samples which are pre-stressed by an isotropic compression. The contact forces resulting from this initial compression are computed using the Voigt localization model ($\overset{\circ}{\underline{\underline{\varepsilon}}}_T = \overset{\circ}{\underline{\underline{E}}}$, $\underline{\omega}^r_T = 0$, $\forall T$).

4. Random geometry

We present, in this section, a first simple modelling of the randomness of the triangulation $\mathcal{T}_v$.

We assume that all the branch vectors have the same length, $\bar{l}$. We are aware that this assumption is geometrically unacceptable because we know that there is no triangulation of the space with identical regular tetrahedrons. However, we believe that it could be *statistically* reasonable to neglect the effect of the dispersal of branch vector length.

Consequently, a tetrahedron is characterised by its random orientation which is uniformly distributed in the case of initially isotropic assemblies. Besides, we assume that an edge of a tetrahedron is occupied by a real branch vector with a probability of p, and by a fictitious one with a probability of $1-p$, independently from the other edges.

In order to connect probability the p to the (initial) void ratio, e, we shall compute p in terms of c, the co-ordination number of the assembly (i.e. the average number of particles which are in contact with a given particle). Indeed, it is well-known that the co-ordination number is a function of the void ratio, and that it is independent of the grain size distribution as it has been experimentally shown by Oda [6].

Dividing the solid angle at the vertex of a tetrahedron by the total solid angle, 4π, we find that a node of the triangulation (i.e. a particle) belongs to $\approx 22{,}79$ tetrahedrons on average. Similarly, dividing by 2π the angle between two faces having in common an edge of a tetrahedron, we find that an edge belongs to $\approx 5{,}10$ tetrahedrons on average. Therefore, the following equations are straightforward :

$$\left|D_c\right| = 5{,}10 \times |T| = 5{,}10 \times \frac{\sqrt{2}}{12}\bar{l}^3 \tag{15}$$

$$c = 2 \times p \times \frac{6}{5{,}10} \times \frac{22{,}79}{4} \approx 13{,}41 \times p \tag{16}$$

The proposed micro-mechanics modelling depends on the following parameters : $k^n/\bar{l}$ (in Pa) and $k^t/k^n \geq 0$, $C \geq 0$, $\alpha \geq 0$, $1 \geq p \geq 0$.
Practically, the orientation, O, of a tetrahedron is described by the three Euler's angles

$$O = (\theta, \psi, \phi) \in \mathcal{E} = [0, \pi[\times [0, 2\pi[\times [0, 2\pi[\qquad (17)$$

between the local basis $\left(\underline{e}_1', \underline{e}_2', \underline{e}_3'\right)$ of the tetrahedron and the reference basis, $\left(\underline{e}_1, \underline{e}_2, \underline{e}_3\right)$.

Actually, the orientation of a regular tetrahedron depends only on the *set* of the three *directions* defined by vectors $\left\{\underline{e}_1', \underline{e}_2', \underline{e}_3'\right\}$. One can easily check that the same tetrahedron can be defined with 24 values of O, each one corresponding to a choice of the order and the sign of vectors $\left\{\underline{e}_1', \underline{e}_2', \underline{e}_3'\right\}$. So, the set of orientations is in one-to-one correspondence with the following subset $\mathcal{E}' \subset \mathcal{E}$ defined as follows:

$$\mathcal{E}' =$$
$$\left\{ O \in \mathcal{E}, \quad 0 \le \min \left\{ \underline{e}_i' \cdot \underline{e}_3, \ i = 1,2,3 \right\} < \max \left\{ \underline{e}_i' \cdot \underline{e}_3, \ i = 1,2,3 \right\} = \underline{e}_3' \cdot \underline{e}_3 \right\} \qquad (18)$$

Hence, a tetrahedron is described by $\omega = \left(O, i_1, \ldots, i_6\right) \in \Omega = \mathcal{E}' \times \{0,1\}^6$ where $i_k = 1$ means that the *k*th edge is occupied by a real branch vector , and $i_k = 0$, otherwise.

The set Ω is endowed with the probability (19) which corresponds to a uniform distribution of the orientations.

$$d\mathbb{P} = p^{\Sigma i_k} (1-p)^{6-\Sigma i_k} \times \frac{24}{8\pi} \sin(\theta) d\theta d\phi d\psi \qquad (19)$$

Under these assumptions, a volume-average is transformed into an *ensemble*-average on Ω endowed with (19). For example,

$$\underline{\underline{E}} = \left\langle \overset{\circ}{\underline{\underline{\varepsilon}}} \right\rangle = \frac{1}{|V|} \sum_{T \in \mathcal{T}_V} |T| \overset{\circ}{\underline{\underline{\varepsilon}}}_T = \overline{\overset{\circ}{\underline{\underline{\varepsilon}}}_T} \equiv \int_{\omega \in \Omega} \overset{\circ}{\underline{\underline{\varepsilon}}}(\omega) \, d\mathbb{P} \qquad (20)$$

5. The numerical problem

We have to solve the following problem : given the overall strain increment, $\Delta \underline{\underline{E}} = \overset{\circ}{\underline{\underline{E}}} \times \Delta t$, find the random variable

$$\omega \to \left(\Delta \underline{\underline{\sigma}}(\omega), \Delta \underline{\tau}(\omega), \Delta \underline{\underline{\varepsilon}}(\omega), \Delta \underline{\omega}^r(\omega) \right) = \left(\overset{\cdot}{\underline{\underline{\sigma}}}(\omega), \overset{\cdot}{\underline{\tau}}(\omega), \overset{\cdot}{\underline{\underline{\varepsilon}}}(\omega), \underline{\omega}^r(\omega) \right) \times \Delta t \quad (21)$$

solution of (3), (5), (6), (10), (13) and (14) where the *ensemble*-average (19), (20) is substituted for volume-average on V.

The numerical computations are carried out by discretizing $\mathcal{E}'$ into 727 weighted orientations. For each orientation, we simulate, once for all, a real or a fictitious branch vector on each edge. So, we get 727 tetrahedrons corresponding to 727 elements ω^k of Ω which are representative of the microstructure.

The discrete problem is solved according to the following iterative scheme : $\Delta \underline{\underline{\varepsilon}}(\omega^k), \Delta \underline{\omega}^r(\omega^k)$ being given, $\Delta \underline{\underline{\sigma}}(\omega^k), \Delta \underline{\tau}(\omega^k)$ are computed with (10) ; then, the overall variables and the error in (14) are computed; finally, each $\Delta \underline{\underline{\varepsilon}}(\omega^k), \Delta \underline{\omega}^r(\omega^k)$ is conveniently modified and the procedure is repeated until convergence occurs.

6. Example of predictions compared with experiments

All the simulations have been carried out with $\beta = 0$ which means that the effect of relative rotations has been neglected. The stress-strain behaviour of the idealised material with $k^n/\bar{l} = 451\,kPa$, $k^t/k^n = 1$, $C = 0.32$ has been predicted for two loading conditions : two-dimensional test and cubical triaxial loading.

The path load in two-dimensional test is:

$$\underline{\underline{\Sigma}}(t) = -S\underline{\underline{1}} - Q(t)\underline{e}_1 \otimes \underline{e}_1 \quad Q(t){\uparrow}, \quad Q(0) = 0, \quad S > 0 \quad (22)$$

The predicted results are compared with the experimental results of Canou and Dupla [2] on a sand (Hostun fin) with two initial void ratios. For each void ratio, the probability p is determined as described above using Oda's results [6] and (16); Then, α is fit for each p. The predictions give good agreement with the experimental behaviour in the ascending branch of the curve. Figure 1.

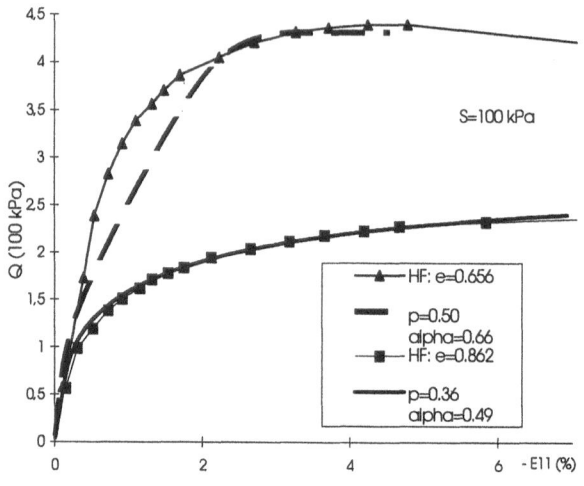

Figure 1: predicted and measured two-dimensional compression behaviour.

The cubical tiaxial loading has constant mean stress and depends on parameter $0 \le b \le 1$.

$$\underline{\underline{\Sigma}}(t) = -S\underline{\underline{1}} - Q(t)\underline{\underline{b}}$$
$$\underline{\underline{b}} = (2-b)\underline{e}_1 \otimes \underline{e}_1 + (2b-1)\underline{e}_2 \otimes \underline{e}_2 - (b+1)\underline{e}_3 \otimes \underline{e}_3 \qquad (23)$$
$$Q(t)\uparrow, \quad Q(0) = 0, \quad S > 0$$

Our purpose is to compute the failure surface of the idealised material and to compare it with Coulomb's surface. Remember that the friction angle, ϕ, defined by $\sin(\phi) = (\Sigma_{III} - \Sigma_I)/(\Sigma_{III} + \Sigma_I)$ does not depend on loading parameter b for Coulomb's failure surface. Figure 2 shows that the friction angle of the idealised material varies very slightly with b. So, the failure surface is very close to a Coulomb's failure surface.

Considering tetrahedrons as the elementary constituents of the micro-mechanics modelling, we have shown that it was possible to take into account the randomness of the geometry of a granular assembly.

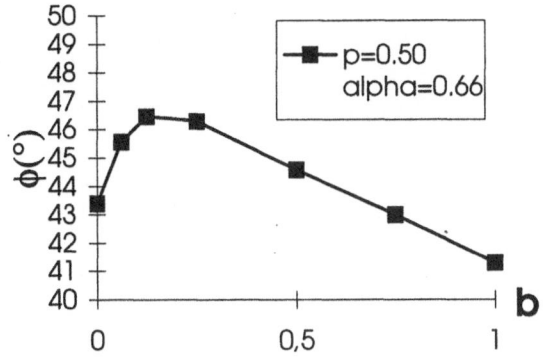

Figure 2: friction angle versus b

7. References

1. Bagi, K. (1996) Stress and strain in granular assemblies, *Mech. Mater.*, **22**, n° 3, 165-177.
2. Canou, J. (1989) Contribution à l'étude et à l'évaluation des propriétés de liquéfaction d'un sable, *P.H.D. thesis E.N.P.C.*, Paris.
3. Chang, C.S. (1993) Micro-mechanical modelling of deformation and failure for granulates with frictional contacts, *Mech. Mater.*, **16**, n° 1- 2, 13-24.
4. Christoffersen, J., Mehrabadi, M.M., and Nemat Nasser, S., (1981) A micro-mechanical description of granular material behaviour, *J. Appl. Mech.*, **48**, 339-344.
5. Love, A.E.H. (1929) *A treatise of the mathematical theory of elasticity.*
6. Oda, M. (1977) Co-ordination number and its relation to shear strength of granular material, *Soils and Foundations*, **17**, n°2, 29-42.
7. Sab, K. (1996) Déformations microscopiques et macroscopiques dans un assemblage dense de particules rigides, *C. R. Acad. Sci. Paris*, **322**, Série II*b*, 715-721.
8. Zaoui, A. (1986) Plastic behaviour of polycrystals. in Gittus J. and Zarka J. (eds), *Modelling small deformations of polycrystals*, Elsevier Applied Science Publishers, London and New York, 187-225.

TOWARDS QUANTITATIVELY CORRECT MICROMECHANICS MODELS

J.G.M. VAN MIER and A.VERVUURT[*]
Delft University of Technology, Stevin Laboratory,
P.O. Box 5048, 2600 GA Delft, The Netherlands

Abstract
The application of micromechanics models for simulating mechanical and physical properties of composites has become widespread in the past years. Often qualitatively correct results are obtained, whereas the quantitative predictions of load-displacement curves is not as straightforward as a result of inherent difficulties in direct measurement of the model parameters. In a recently developed lattice model for simulating brittle fracture of concrete and sandstone, the interface between aggregate particles and the matrix plays a decisive role. A new splitting test set-up was developed for studying the interfacial fracture behaviour between the aggregate and matrix material. A statistical analysis was used for identifying the significant interface parameters. The inverse analysis is but an isolated step in the parameter identification process. Combined experimental and numerical research of fracture of the aforementioned composites in different geometries and under a wide variety of loading situations is equally important for developing a quantitatively sound model. In this the situation is not different from macro-models, except that the physics of the discrete fracture process are captured in much more detail in the micromechanics approach.

1. Introduction

Following the reductionist approach, behaviour at a certain scale of observation can be explained from phenomena and mechanisms occurring at a lower scale of observation. For material science in general, the three levels denoted as macro, meso and micro-level are of particular importance. For example, for engineering purposes, the behaviour of a material at the macro (continuum) level must be know. No internal material structure is recognised at this level, i.e. the properties are smeared over a representative volume, which should be larger than - at least - five times the largest structural dimension of the material at the lower level of observation. For some materials, like ceramics, which are built from grains with dimensions at the μm scale, the representative volume can thus be very small indeed. For other materials like some rocks and concretes, the grains can be as large as a few centimetres, and the representative

---

[*] present address: TNO Building and Construction Research, Rijswijk, The Netherlands

G. N. Frantziskonis (ed.), PROBAMAT – 21st Century: Probabilities and Materials, 405–417.

volume is then much larger. The mechanical (stiffness, strength and toughness) and physical properties (conductivity, permeability) at the macro-level can be explained - at least in part - by considering the material structure at the meso-level. Again, the meso-level can have quite different absolute dimensions for different types of materials. The next step brings us to the micro-level where explanations must be sought for the meso-level behaviour of the materials. For cement and concrete, all three levels are of importance to describe the properties (i.e. behaviour laws) at the macro-level. At the meso-level, a composite like, for example, concrete consists of aggregate particles that are glued together by means of cement paste. The thickness of the cement paste layers is usually very small, i.e. not exceeding 100 μm. One should therefore be careful in schematising the meso-level structure of concrete as particles swimming in a sea of cement paste. Interactions between the inclusions cannot be neglected at such short distances. In particular for compressive loading, an approach where the material is schematised as a stack of particles separated by means of a narrow brittle-elastic layer seems to make much more sense. The properties of the cementitious material between the particles depend on the micro-structure of cement paste, which is very complex, e.g. Scrivener [1]. The hardened cement paste usually contains numerous pores: a porosity above 25 % is not exceptional. The pores are of importance for strength and ductility [2], and have a large effect on the permeability of the material as well [3]. Note that the particle stack in concrete, rock and ceramics is generally not very regular. Of course, in some sandstones and ceramics rather regular particle distributions can be found, but engineered materials like concrete rely heavily on the density of the particle structure. The shape of the particles can vary widely as well.

The brittle disordered materials are all important structural materials. Concrete is the most widely used building material, ceramics are nowadays applied more frequently in, for example, refractory applications, and structural properties of rock must be known in the mining and oil industry. For engineering applications, macro-level models are usually preferred, but when it comes to fracture, the meso-level and micro-level structure of these materials seems to play an eminent role [4]. Boundary condition and size/scale effects have a significant effect on the fracture response, and direct measurement of the macroscopic fracture properties (for example a critical stress intensity factor or a softening diagram) is simply impossible. In addition to that, when continuum based models are used, spurious mesh sensitivity affects the results, and higher order continuum models must be used, [5]. The disadvantage of these higher order continua is however that an additional parameter must be introduced, namely an internal length scale, which is a representation of the meso-structure of the material. For example for gradient plasticity, the constitutive law for the material is written as,

$$\sigma = f(\varepsilon) - c\frac{d^2\varepsilon}{dx^2} \qquad (1),$$

where $f(\varepsilon)$ is the non-linear softening function and c is the length scale, [6]. At present it is not clear how this length scale should be measured directly, and it seems quite likely that inverse techniques must be used, or alternatively, meso-levels models should be applied to derive the length scale on a theoretical basis. Looking at the various stages in the fracture process of the aforementioned brittle disordered materials, it seems quite likely that the internal length scale is not a constant at all, but its value changes (and decreases) probably when the fracture process progresses [4,7]. In addition to that, due to randomness in microcrack distributions, subsequent growth and mutual interactions, it is not certain whether a smooth continuous analytical macroscopic curve gives a correct description of reality, [14]. This means that the macroscopic material description must be underscored by meso- and micro-level models where it is tried to estimate the correct energy dissipation.

One of the problems of the continuum models is of course that the discrete nature of the fracture process are denied. Other approaches, like numerical models based on Linear Elastic Fracture Mechanics principles, try to avoid the mesh sensitivity by means of a remeshing technique, e.g. [8]. The essential material parameter in such models is either a critical stress intensity factor, or a softening function. The softening function is necessary when the model is used at a scale where the energy dissipation in a process zone near the crack tip must be taken into account. For concrete this is true for almost every existing structures (with the exception of perhaps the largest dams); for rocks and ceramics in many cases the LEFM approach seems applicable.

As mentioned before, the macroscopic fracture parameters can be determined through an inverse analysis only, [4]. Applying the reductionist approach, one might try to derive the relevant macroscopic parameters from meso-level analyses. The application of numerical tools is quite convenient in such an approach as virtually no geometrical constraints exist. The essential difference between macro- and meso-level models is that the meso-material structure is projected on top of the finite element mesh. Different properties are then assigned to the different phases in the material. In applying numerical meso-mechanics models to concrete, rock and ceramics, usually three material phases are distinguished, namely the aggregate, the cementing matrix and the interface between these two major constituents. In addition, pre-existing flaws and pores may be included in the model as well. As a basis for the analysis, conventional finite element methods may be used, but recently, lattice type models in which the material is discretised in a regular or random network of beam elements have become quite popular. In Delft we have gained quite some experience with a simple lattice model [9], which is based on developments in theoretical physics [10]. A simple elastic-brittle strength-based fracture law is used in the model. Because the parameters in the model are single valued (both strength and stiffness of the lattice elements must be specified), a simple statistical technique might be used for their determination. The determination of the elastic parameters in the lattice is no problem, and has been described in detail in earlier papers, [4], [9], [11]. The definition of the fracture strengths of the lattice elements is more problematic, in particular the strength of the interface between the aggregate and cementing phase. In this paper, we will focus on an inverse technique that may be adopted for estimating the properties of the interfacial transition zone between aggregate and matrix in concrete. Because in many places around the world, high quality aggregates are becoming scarce, new alternative aggregate materials are looked for. Using the lattice model, and the method of deriving the interfacial properties, new cement composites with alternative aggregates may be developed having improved mechanical properties for specific applications. Needless to say that the same approach might be used for other properties of the composites as well, for example the permeability of the material.

2. Parameters in the discrete lattice model

In a lattice model the material is discretised in a network of brittle breaking truss or beam elements. The breaking of the linear elements is done using a simple effective stress criterion, for example,

$$\sigma = \beta \left\{ \frac{F}{A} \pm \alpha \frac{\left| M_{i,j} \right|_{max}}{W} \right\} \leq f_t \tag{2}$$

where A and W are the cross-sectional area and the section modulus of the lattice elements respectively, and $M_{i,j}$ are the bending moments in the nodes (when beam elements are used). The factors α and β are two empirical constants. The constant α determines how much bending is taken into account in the fracture law (when beam elements are used of course). This affects directly the length of the tail of the softening diagram. The factor β is used to scale the peak of a tensile softening diagram. For the beam strengths, strength ratio's are defined as will be discussed in section 2.2. In the remainder of this paper α and β are kept constant and will not be discussed further. For more information, see for example in [11].

As mentioned in the introduction, two sets of parameters must be determined when the lattice model is applied, namely the elastic properties of the lattice elements and their fracture strengths.

2.1. ELASTIC PARAMETERS

When the material is discretised as a lattice, the elastic properties of the truss or beam elements must be determined. This is done by comparing the macroscopic Young's modulus and Poisson's ratio to experimental values. In the process, the cross-sectional shape of the lattice elements must be determined, their dimensions, and the Young's moduli of the lattice elements falling in the different constituents of the composite (i.e. E_a, E_m and E_b for aggregate, matrix and interface respectively must be set). In the procedure, the absolute values of the Young's moduli are not important, but rather the ratios between the different moduli for aggregate, matrix and bond zones. For the procedure to be followed, the reader is referred to earlier papers, e.g. [4], [11].

2.2. FRACTURE PARAMETERS

Which fracture parameters must be determined depends completely on the type of fracture law that is used. In the present paper the fracture law (eq. (2)) was used for a lattice of beam elements. When different material phases are distinguished, different failure strengths must be specified. The application of such a simple lattice model implies a revival of classical strength based analysis. When the effective stress computed following eq. (2) exceeds the strength assigned to a particular beam element, the beam is removed from the lattice, thus creating a micro-fracture. Note that in the model no friction in the cracks is taken into account. At present, steps are undertaken to allow for frictional energy dissipation in the lattice model too. Under subsequent beam removals, the damage patterns evolve to more complex patterns, until collapse in a single macro fracture occurs. Although the fractal dimension of the total microcrack pattern keeps increasing during the process [12], the structure of the *active* cracks becomes increasingly more simple. Note that the majority of the microcracks is shielded by the larger cracks, and may extend only when the loading on the structure is seriously changed.

For concrete at the meso-level, three phases are distinguished, namely the aggregate, matrix and bond zones. For each of these three phases a critical strength must be set. These will be denoted as f_{tA}, f_{tM} and f_{tB} for the aggregate, matrix and bond zones respectively. Also here the ratio's between the different strengths is important. In general the same critical value is used for all the lattice elements falling in a certain phase of the material. One might also choose to select a given statistical distribution of strengths for the beams falling in the same material phase [12]. An example where a normal distribution for the strength of the beams is used is included in this paper.

3. Statistical approach to estimate interface parameters

For evaluation of the interface strength, no direct test can be devised. The normal approach in measuring the interface strength is by casting a block of rock against hardened cement paste, and pulling the two materials apart until failure occurs. The bond strength is then defined as the quotient between pull-off force and net area. A different approach was followed by Vervuurt [11], who proposed to apply a splitting load to a thin notched plate of hardened cement paste or mortar containing a single (off-centred) cylindrical aggregate particle of 20 mm diameter. Aggregates of different materials were used, namely a porous open sandstone (Bentheimer), a dense granite (Polar White granite) and a residue from an industrial process (phosphorous slag). For a detailed description of the test set-up and results, the reader is referred to [11]. It will be obvious that in such a test no uniform state of stress is applied to the interface (but this is so for tensile tests as well). This means that only through inverse analysis the interface properties can be estimated. A strong assumption is made in this approach, namely, *it is assumed that the model used is a correct representation of reality.*

Because the strength and stiffness parameters, or rather the ratio's between fracture strengths and Young's moduli (because the ratio's determine the disorder in the behaviour of the material), are single valued parameters, a simple inverse technique may be used to estimate their value. First, however, the significant parameters must be found. For this purpose, a systematic variation of a number of model parameters was carried out, and the statistical theory of linear models was applied to identify the significant parameters. The aforementioned splitting test was used in the simulations. Because the splitting experiments were performed on very thin specimens containing cylindrical aggregates, a plane stress analysis was carried out.

The theory of linear models was applied to estimate the model parameters. Following this approach the numerical results are described by means of a model which is linear in the parameters β_i, and which contains k variables,

$$y = \beta_0 + \beta_1 x_1 + \beta_2 x_2 + \ldots\ldots + \beta_k x_k + e \qquad (3),$$

where y is a vector of size n containing the numerical results. The results from the numerical simulations should preferably be expressed as single values, like the maximum strength, the work of fracture, or the fractal dimension of the crack pattern at a characteristic stage of the simulation. The vector e contains all random errors, which are assumed to be mutually independent, and have a normal distribution with expectation $E(e) = 0$ and variance $var(e) = \sigma_0^2$. The estimate b of the variables β_i can be determined through

$$b = (X'X)^{-1} Xy \qquad (4),$$

where X is the design matrix that contains all the information regarding the parameter settings in the numerical simulations. The purpose of the simulations is to estimate the variables β_i by means of the method of least squares. The vector b has size $(k+1)$ and expectation $E(b) = \beta$. Normally one should strive to an orthogonal experimental design. In that case the matrix $X'X$ reduces to a diagonal matrix. The random error is then determined through

$$\sigma_0^2 = \frac{KS_r}{n-k-1} \qquad (5),$$

where KS_r is the sum of squares of deviations of the simulation results. By means of a student-t test, hypotheses regarding the individual β_i can be tested. For hypotheses regarding linear combinations of β_i an F-distribution must be used. For example the hypothesis H_0 that k-g variables are equal to zero,

$$H_0: \beta_{g+1} = \beta_{g+2} = \ldots\ldots = \beta_k = 0 \quad (g < k) \qquad (6),$$

may be checked using an F-distribution with $(k-g)$ and $n-(k+1)$ degrees of freedom according to

$$F = \frac{(KS_{r0} - KS_r)/(k-g)}{KS_r/(n-k-1)} \tag{7},$$

where KS_{r0} is the sum of squares of deviations of the simulations for the reduced model with $(k-g)$ values of β_i equal to zero, and KS_r is the sum of squares of the original model. When the hypothesis cannot be rejected, the model can be reduced by setting the relevant β_i to zero. Confidence limits may be determined as well. The above type of linear model is excellently suited for testing the significance of the various model parameters on the outcome of the numerical simulations. Using the reduced model, comparisons to experimental results can be made, and the value of a certain model parameter can be estimated. Quite important in this approach is to check whether the simulated and experimental fracture mechanisms correspond to one another. Moreover, the method will work only when the random error is sufficiently small. This can be guaranteed by having a large number of simulations, i.e. by making the value of $(n-k-1)$ in eq. (5) as large as possible. As mentioned, before an attempt is made to estimate the various model parameters using the above statistical approach, one should ensure that a number of basic failure mechanisms can indeed be simulated qualitatively by means of the model. For the Delft lattice model, the failure mechanisms under various loading regimes have been examined exhaustively, see for example [9], [11], [13].

For the splitting test mentioned in the beginning of this section, two sets of simulations were carried out to get some insight in the sensitivity of the model to the value of the stiffness ratio's E_M/E_A and E_M/E_B, as well as for different values of f_{tM}/f_{tB}. In the first set of simulations, a high strength (granite) aggregate with stiffness ratio's $E_M/E_A = 1.0$ and 0.33, $f_{tM}/f_{tA} = 0.5$ was simulated. The parameters E_M/E_B, and f_{tM}/f_{tB}. were varied at four ($E_M/E_B = 10, 8, 4$ and 2) and three levels ($f_{tM}/f_{tB}. = 8, 4$ and 2) respectively.

In a second series of simulations, representing the situation of a weak (sandstone) aggregate with stiffness ratio's $E_M/E_A = 6$ and 3, and relative strength $f_{tM}/f_{tA} = 1.0$, the parameters E_M/E_B, and f_{tM}/f_{tB}. were varied at four ($E_M/E_B = 6, 3, 1.5$ and 1) and five levels ($f_{tM}/f_{tB}. = 0.75, 1.0, 1.25, 1.0^*)$ and 1.0^{**}) respectively. For the case with a single asterisk, a normal strength distribution with a standard deviation of 1.3 MPa for the aggregates and 0.3 MPa for the matrix and bond zones was assumed, whereas for the analyses indicated with a double asterisk the standard deviations were 0.7 MPa and 0.3 MPa respectively for aggregate and matrix/bond beams.

In all the simulations it was assumed that the matrix was made of a mortar. The value for β in the fracture law equation (2), i.e. the parameter for scaling the peak strength, was found to be equal to 1.41 for this material. The parameter α was held constant at 0.005.

412

4. High strength aggregate in a mortar matrix

The linear model for this case contains a total of seven main factors and two- and three factor interactions, as follows,

$$y = \beta_0 + \beta_1 x_1 + \beta_2 x_2 + \beta_3 x_3 +$$
$$+ \beta_4 x_1 x_2 + \beta_5 x_1 x_3 + \beta_6 x_2 x_3 + \beta_7 x_1 x_2 x_3 + e \qquad (8),$$

where β_0 represents the weighted average, $x_1 = E_M /E_A$, $x_2 = f_{tM}/f_{tB}$, and $x_3 = E_M /E_B$. The load versus crack mouth opening displacement (CMOD) diagram is characterised by a single peak. The work of fracture is defined as the area under the load-CMOD curve. Both the computed peak loads and the work of fracture were selected as response variables in eq. (8), i.e. the values for P (or G_f) form the elements of the vector y. From the statistical analysis it was found that the model can be reduced to

$$y = \beta_0 + \beta_2 x_2 \qquad (9).$$

It should be mentioned however that the factor β_2 was not significant according the F-test, but the F-value was rather close to the critical value that it has been included here. Thus, only the interface strength has a significant effect on the peak strength and work of fracture. Looking at the influence of the interfacial stiffness (Figure 1), reveals that the effect can be neglected indeed. A more precise conclusion can probably be drawn when the number of simulations is increased. In that case, the number of degrees of freedom available for estimating the random error becomes larger, and the various F-tests can be carried out with higher accuracy. Another option might be to increase the fineness of the lattice used. Figure 2 shows that this was rather fine already, and a further refinement would lead to a huge increase in computational effort.

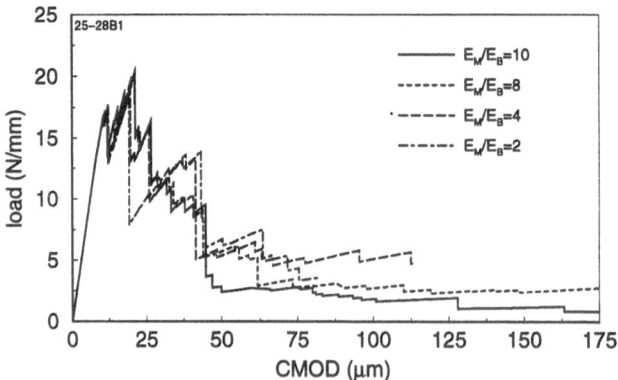

Figure 1. Effect of the interfacial stiffness on the load-CMOD curve for the case with a high strength aggregate.

Because all simulations (in total over 60 simulations were preformed) already took about half a year, it is not realistic to explore these options at the moment.

Next to the strength and work of fracture, also the fracture patterns should match the experimental observations. It was found that in all simulations, the crack ran along the interface, and never penetrated the aggregates. Two examples of crack patterns are included in Figure 2. There the influence of the interfacial stiffness at the maximum load P and at the stage with 300 beams removed are shown. The figures at the left are for an interfacial stiffness $E_M/E_B = 10$, and those at the right for $E_M/E_B = 2$. For both analyses, f_{iM}/f_{iB}, = 4, and $E_M/E_A = 1.0$. The crack patterns are representative for all simulations with a high strength aggregate. For a higher interfacial stiffness, more microcracking was observed (i.e Figure 2b) as compared to the case where a lower interfacial stiffness was used. In the final crack patterns, however, hardly any differences were observed (compare Figure 2c and 2d).

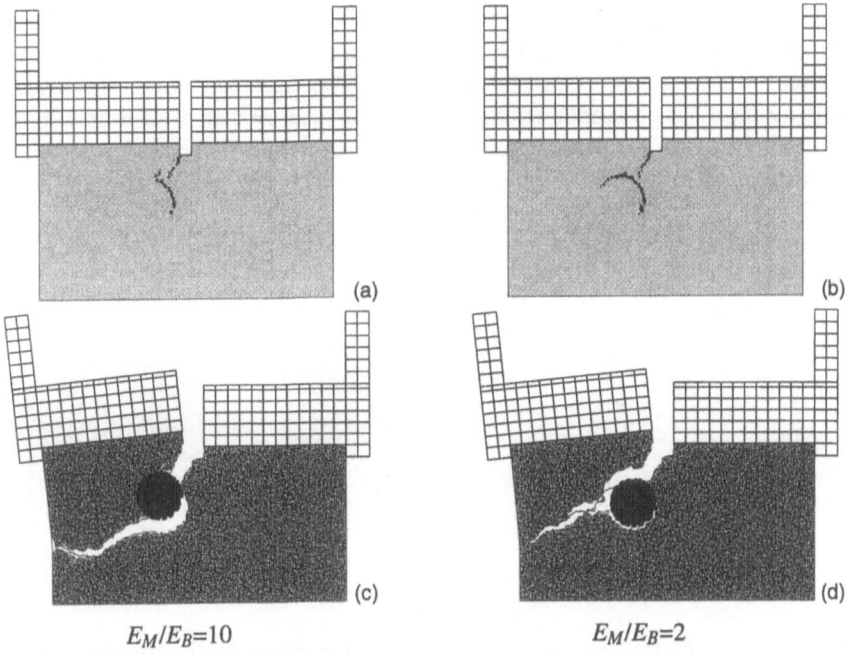

$E_M/E_B=10$ $E_M/E_B=2$

Figure 2. Influence of varying the interfacial stiffness for simulations with a high strength aggregate. Crack patterns at peak load (a,b), and at 300 beams removed (c,d). The top two figures are shown in inverted form because the cracks are hardly visible in the normal representation of figures (c,d), after [11].

The crack patterns are in agreement with experimental observations, [11]. The crack growth in the pre-peak regime and at peak was very difficult to detect in the

414

experiments, in spite of the fact that a long distance microscope was used with a resolution of about 1 μm. Further experiments are needed to better estimate the amount of interface cracking. Moreover, because of the extreme sensitivity of cement to temperature and hygral gradients during the hydration process, it is to be expected that many flaws are present along the interface before any mechanical loading has been applied. Future investigations should show the significance of these initial flaws, for example with respect to strength.

5. Sandstone aggregate in a mortar matrix

In the second example, the same linear model of eq. (8) was used. The results from the analysis showed that only the factor β_0 had to be included in the model when the maximum load in the load-CMOD diagram was considered. However, looking at the crack patterns, and the appearance of a so-called second peak in the load-CMOD diagrams, there was a definite influence of the interface strength f_{tM}/f_{tB}. The statistical analysis was however also net very discriminating as far as this second peak is concerned. In Figure 3, the load-CMOD curves and the final crack patterns for two simulations with a weak aggregate are shown. The difference is in the interfacial strength. With a low interface strength, a pronounced second peak is found, which resembles the experimental result for a sandstone aggregate embedded in a cement or mortar matrix.

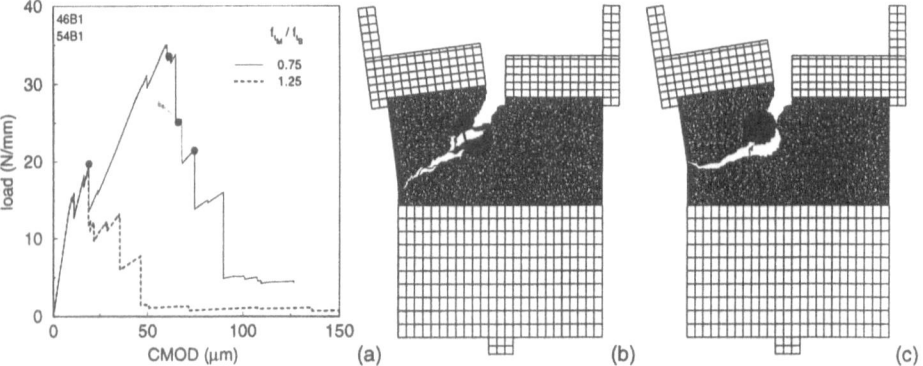

Figure 3. Load-CMOD curves (a) and final crack patterns for the simulations with $f_{tM}/f_{tB}, = 0.75$ (b) and $f_{tM}/f_{tB}, = 1.25$ (c), after [11].

After first growth of the crack through the rather homogeneous matrix material, the crack is stopped by the aggregate particle. This is shown in Figure 4a. At this stage a distinct first peak is found in the analysis, i.e. the first black dot along the load-CMOD curve in Figure 3a. Some cracks appear around the aggregate, as shown in Figure 4b. Next the crack propagates at the lower end of the aggregate particle, and only later the aggregate fractures.

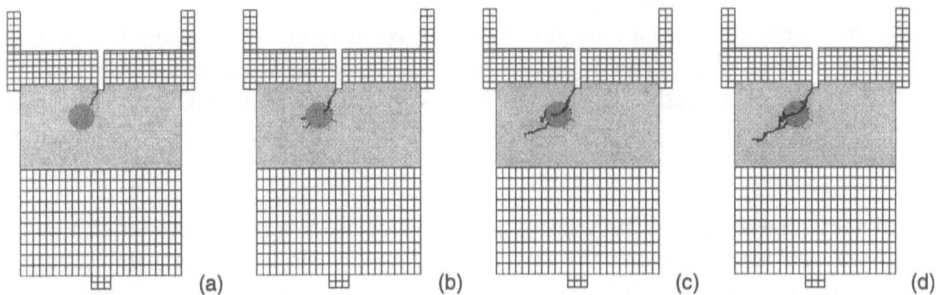

Figure 4. Crack history for the simulation of a weak sandstone aggregate with a low interface strength f_{tM}/f_{tB}, = 0.75. Crack patterns are shown at 25, 75, 150 and 225 beams removed, after [11].

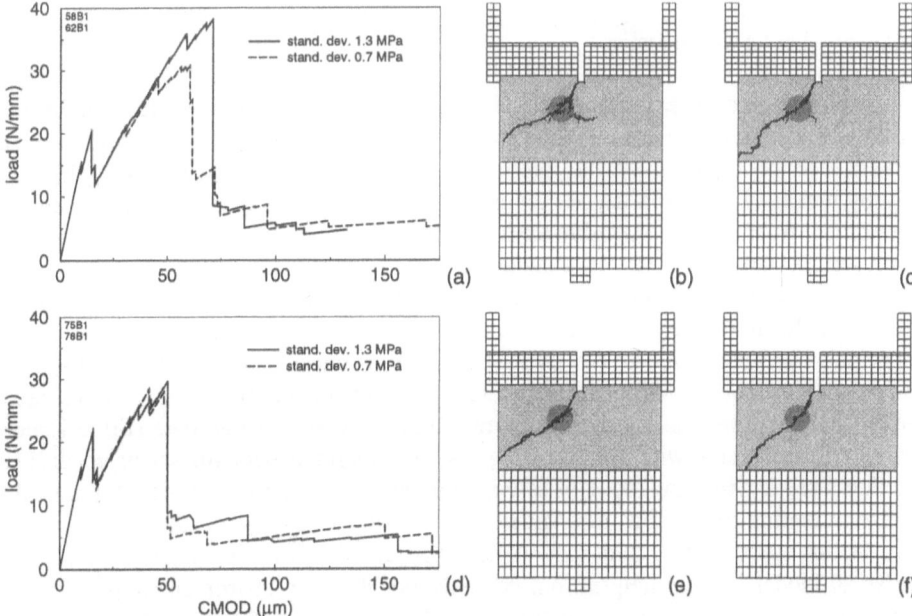

Figure 5. Effect of varying the standard deviation in the normal distribution of the aggregate strength on the load-CMOD curves (a,d) and on the final crack patterns (b,c and e,f). The difference between the two sets of analyses is in the ratio E_M/E_A, which is 6.0 for Figures a-c and 3.0 for Figures d-f, after [11].

A number of additional analyses was carried out in which a normal strength distribution was assigned to the beams in the aggregate, matrix and bond zones. This means that randomness is not only caused by varying the length of the lattice beams, but also from the variation in beam strength. At this stage it is not clear whether a normal distribution is the correct choice, but for simplicity this one has been selected first. In Figure 5 some of the results are shown. They concern the load-CMOD curves for an analysis with $E_M/E_A = 6.0$ (Figures 5a-c), and for the case with $E_M/E_A = 3.0$ (Figures 5d-f).

From these figures it can be seen that the second peak increases when either the aggregate stiffness decreases (i.e. an increase of E_M/E_A), or when the standard deviation of the normal distribution of the aggregate strength increases. Note that the average beam strength of all phases was set at 5 MPa. Form the figures it can be seen that the first peak is not affected by the aggregate: the value is identical irrespective of the strength variation in the aggregates. When a larger standard variation for the matrix and bond zones is selected, more distributed cracking will appear in the pre-peak regime. In that case a larger variation in the first peak will be observed.

6. Discussion and conclusion

Quantifying lattice type micro-mechanics models for simulating fracture in brittle disordered materials like concrete and rock is not a straightforward process. The direct measurement of, for example the interface strength and stiffness is not possible, and only by means of a back-analysis, the most likely values for such parameters can be determined. By means of extensive experimentation and numerical simulations with an elastic-brittle lattice model, it is attempted to quantify the interfacial properties between aggregate and cement paste in concrete. Idealised, single aggregate specimens were used. In this paper only the results of a number of numerical simulations are presented. The application of the statistical theory of linear models may be helpful in determining the significant interface parameters in the model. So far the analyses have shown that the interface strength is more important than the interface stiffness. For example in concretes with weak aggregates, the interface strength seems to decide whether a crack will grow along or through an aggregate. The lattice model is capable of discriminating the various mechanisms. Moreover the aggregate in the plate works as a stress concentrator, and may deflect the crack growth to some extent. The statistical model is quite helpful, but certainly not the only instrument of developing quantitatively correct micromechanics models. Many simulations are needed on extremely fine lattices. At present a full parameter study along these lines is rather time consuming, and thus not very appealing. Improving the statistical model to better discriminate between the various parameters seems essential. Next to that it should be mentioned that the single particle geometry which has been studied in this paper is by no means representative for real concretes. Further work is needed to investigate whether the interfacial parameters derived from the present study can be used in simulations of real concretes as well.

7. References

1. Scrivener, K. (1989) The microstructure of concrete, *in Materials Science of Concrete*, Skalny, J.P. ed., American Ceramic Society, Westerville, OH, 127.
2. Arslan, A., Schlangen, E. and Van Mier, J.G.M. (1995), Effect of model fracture law and porosity on tensile softening of concrete, *in Proceedings FraMCoS-2*, Wittmann, F.H., ed., AEDIFICATIO Publishers, Freiburg, 45.
3. Visser, J.H.M. (1997) *Tensile hydraulic fracture of concrete*, PhD thesis, Delft University of Technology, Delft, The Netherlands.
4. Van Mier, J.G.M. (1997) *Fracture processes of concrete - Assessment of material parameters for fracture models*, CRC Press, Boca Raton, FL.
5. De Borst, R. and Mühlhaus, H.-B. (1991), Continuum models for discontinuous media, in *Fracture Processes in Concrete, Rock and Ceramics*, Van Mier, J.G.M., Rots, J.G. and Bakker, A. eds., E&FN Spon, London/New York, 601.
6. Pamin, J. (1994), *Gradient-dependent plasticity in numerical simulation of localization phenomena*, PhD thesis, Delft University of Technology, Delft, The Netherlands.
7. Van Mier, J.G.M. (1992), Scaling in tensile and compressive fracture of concrete, in *Applications of Fracture Mechanics to Reinforced Concrete*, Carpinteri, A. ed., Elsevier Applied Science, London/New York, 95.
8. Ingraffea, A.R. and Saouma, V. (1984), Numerical modelling of discrete crack propagation in reinforced and plain concrete, in *Application of Fracture Mechanics to Concrete Structures*, Sih, G.C. and Di Tomasso, A. eds., Martinus Nijhoff Publishers, Dordrecht, Chap. 4.
9. Schlangen, E. and Van Mier, J.G.M. (1992), Experimental and numerical analysis of micromechanisms of fracture of cement-based composites, *Cement & Conc. Comp.*, **14**, 105.
10. Herrmann, H.J., Hansen, H. and Roux, S. (1989), Fracture of disordered elastic lattices in two dimensions, *Phys. Rev. B.*, **39**, 637.
11. Vervuurt, A. (1997) *Interface fracture in concrete*, PhD thesis, Delft University of Technology, Delft, The Netherlands
12. Van Mier, J.G.M., Chiaia, B.M. and Vervuurt, A. (1997), Numerical simulation of chaotic and self-organising damage in brittle disordered materials, *Comput. Methods Appl. Mech. Engrg.*, **142**, 189.
13. Schlangen, E. (1995), Computational aspects of fracture simulations with lattice models, *in Proceedings FraMCoS-2*, Wittmann, F.H., ed., AEDIFICATIO Publishers, Freiburg, 913.
14. Krajcinovic, D. (1997), Essential structure of the damage mechanics theories, in *Proc. 19th ICTAM, Kyoto*, August 1996, Tatsumi,T., Watanabe, E. and Kambe, T., eds., Elsevier Science, Amsterdam, 411.

MULTISCALE PERMEABILITY AND DISPERSION
IN RANDOMLY HETEROGENEOUS GEOLOGIC MEDIA

V. DI FEDERICO AND S.P. NEUMAN[1]
D.I.S.T.A.R.T., Università di Bologna
Viale Risorgimento 2, 40136 Bologna, Italy
[1]*Department of Hydrology and Water Resources, The University of Arizona*
Tucson, Arizona 85721, U.S.A.

1. Introduction

Flow and transport in natural soils and rocks have been traditionally described by means of partial differential equations (pde's). These equations are generally taken to represent basic physical principles (conservation and constitutive laws) that operate on some macroscopic scale (theoretical support volume) at which the geologic medium may be viewed as a continuum. The precise nature of this theoretical macroscopic support scale remains generally unclear though some derive comfort from associating it in the abstract with a "representative elementary volume" (REV). Unfortunately, the concept of an REV is equally difficult to define without ambiguity or to apply in practice. Flow and transport pde's are local in the sense that all quantities (parameters; forcing functions including initial, boundary and source terms; dependent variables) which enter into them are defined at a single point (x, t) in space-time. Parameters such as permeability, porosity, and dispersivity are generally regarded as macroscopic medium properties that are well-defined, and can thus be determined (at least in principle) experimentally and more-or-less uniquely, at any point x in the flow domain.

In reality, geologic media are heterogeneous and exhibit both discrete and continuous spatial variations on a multiplicity of scales. One may therefore anticipate that the flow and transport properties of these media would exhibit similar variations. Indeed, one manifestation of such variations is the observed, and by now well-documented, dependence of permeabilities and dispersivities on their scale of measurement (support volume). Figure 1 summarizes laboratory and field permeability data from crystalline rocks at a variety of sites and suggests that these data vary with the scale of measurement. Figure 2 shows longitudinal dispersivities, derived by means of traditional advection-dispersion models from a variety of laboratory and field tracer studies world-wide, and suggests that these parameters also vary with the scale of observation. In other words, permeabilities and dispersivities seem not to be unique properties of the geologic medium but to depend additionally on support and observation scales.

G. N. Frantziskonis (ed.), PROBAMAT – 21st Century: Probabilities and Materials, 419–453.
© 1998 *Kluwer Academic Publishers.*

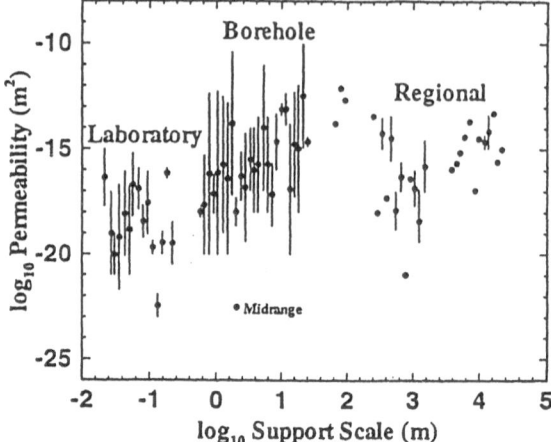

Figure 1. Permeability of crystalline rocks as a function of experimental scale (after *Clauser*, 1992, and *Neuman*, 1994).

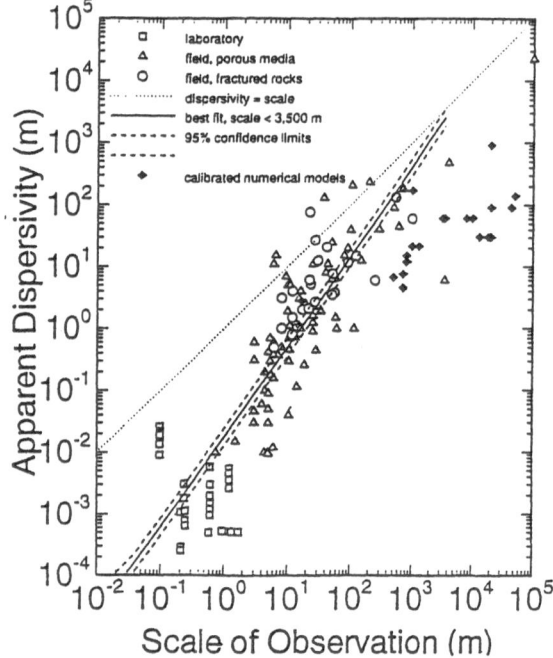

Figure 2. Apparent longitudinal dispersivities versus scale of observation (after *Neuman*, 1995).

Internal consistency of flow and transport pde's requires that all quantities which enter into them (parameters, forcing terms, dependent variables) be defined on one unique support scale; consistency of these pde's with data (without which the equations are not operational) require that the latter be defined and measured on this same scale. To the extent that measurements and their statistics exhibit dependence on support scale, it is necessary in theory and advisable (though not always possible) in practice to work with one consistent support scale. The selection of a working ω depends not on the abstract question whether or not it constitutes an REV, but on the pragmatic question whether or not it allows in principle, and renders technically feasible as well as otherwise desirable, the definition and measurement of all relevant quantities on this scale throughout the domain if interest.

One key question of practical and theoretical interest is whether it might be possible to relate quantitatively flow and transport parameters and associated statistics, determined on one scale of measurement, to those determined on another scale. Transforming such information from a small to a larger support scale constitutes upscaling, while transforming data and statistics from a large to a smaller scale constitutes downscaling. Only on rare occasions can upscaling or downscaling be properly accomplished by linear convolution (weighted averaging) or deconvolution of small- or large-scale data; one notable case where this may be possible is that of permeabilities in perfectly stratified porous media. For most other cases, there is a need to develop alternative ways of relating measurements and their statistics across disparate support scales. One approach that has led to useful insight into the scale-dependence of subsurface flow and transport phenomena is their analysis within geostatistical and stochastic frameworks. This approach has been motivated by the recognition that subsurface data and especially permeability, which often exhibit seemingly erratic spatial fluctuations, nevertheless tend to be spatially autocorrelated as well as mutually cross-correlated. The key to understanding scale-dependence in flow and transport seems to lie in an understanding of how permeabilities (and, to a lesser extent, other parameters) as well as their statistics (primarily spatial correlations), defined and measured on one scale, affect related parameters and their statistics on larger scales. We shall pursue this question on the premise that all relevant flow and transport quantities are defined and measured on one consistent support scale, ω.

Spatial fluctuations in permeability are due in part to random errors of measurement and errors arising from the indirect assessment of permeability based on measurements of pressure and flow rate. In large part, such fluctuations are caused by spatial variations in the makeup of ω-scale soil and rock volumes that comprise the geologic medium, *i.e.*, by medium heterogeneity on a resolution scale ω. Since permeabilities can be determined only at selected well locations and within discrete depth intervals, their values elsewhere in the subsurface remain unknown. As measured values tend to fluctuate erratically in space, they cannot be extended into untested portions of the subsurface with certainty. Hence the subsurface distribution of ω-scale permeabilities is at once random and uncertain. The same may be true about ω-scale forcing terms which may additionally fluctuate, in an uncertain fashion, with time. Upon introducing such random inputs into flow and transport pde's, the latter become stochastic and their ω-scale solutions (in terms of hydraulic heads, concentrations, fluxes, and velocities) are likewise random. Hence these solutions cannot be specified directly (other than as random samples, or realizations, of an infinite set, or ensemble, of equally likely, therefore nonunique, solutions) but only in terms of ω-scale ensemble statistics. Such theoretical, deterministic ensemble statistics represent the results of sampling and averaging ω-scale quantities, at a given point (x,t), across their infinite ensemble of equally likely values (realizations) in probability space (sampling and averaging across a random finite sample of these realizations yields sample statistics which are random variables). Constraining ensemble moments to be consistent with specific measurements renders them conditional on data; otherwise, if they are consistent with data statistics but not with specific data values, they are unconditional. A conditional ensemble mean solution represents an optimum (minimum variance) unbiased prediction of the unknown true ω-scale solution at any point (x,t); the corresponding conditional ensemble variance is a measure of the associated prediction error.

In general, ensemble mean quantities satisfy nonlocal (integro-differential) rather than local (differential) flow (*Neuman and Orr*, 1993; *Neuman et al.*, 1996; *Tartakovsky and Neuman*, 1997a) and transport (*Neuman*, 1993a; *Zhang and Neuman*, 1996) equations of the kind we review briefly in this paper. In these equations, ensemble mean (predicted) fluxes at any point (x,t) depend not only on mean gradients at this point but also on additional quantities at other points in space-time. In other words, the mean fluxes are nonlocal and therefore non-Darcian (in the case of flow) and non-Fickian (in the case of transport). It follows that the standard notions of permeability and dispersivity, as material properties (coefficients in Darcy's and Fick's laws) that are well defined at each point in space, generally lose meaning when one deals with ensemble averaged quantities under uncertainty. This is especially true under conditioning where the local (depending on one point) and nonlocal (depending on more than one point) parameters that control ensemble mean flow and transport are additionally functions of the quantity and quality (information content) of ω-scale data. For standard notions of permeability and dispersivity to apply in the ensemble mean, one must restrict consideration to special situations without conditioning. The most common among these special situations is that of mean uniform steady state flow in a statistically homogeneous permeability field. If the flow domain is additionally infinite (so large that, for practical purposes, its boundaries can be assumed to lie at infinite distance from the area of interest), one may achieve ergodicity whereby the now constant ensemble mean flux and hydraulic gradient also represent spatial averages of their random ω-scale counterparts over a sufficiently large domain in one or more realizations. In this case, the effective hydraulic conductivity, which by definition relates the ensemble mean flux to the ensemble mean gradient, is equal to the equivalent hydraulic conductivity which, by definition, relates the spatially averaged flux to the spatially averaged gradient; all of these quantities are deterministic. If the flow domain is finite, ergodic conditions may not be achieved and spatial averages of flux and gradient, over one or a few realizations, may remain random; in this case, the deterministic ensemble mean flux and gradient represent not spatial averages of these quantities but only the ensemble mean (*i.e.*, best estimates) of such spatial averages. By the same token, the effective hydraulic conductivity that relates ensemble mean flux to gradient represents not any actual value, but only the ensemble mean (best estimate), of random equivalent hydraulic conductivities associated with random spatial averages of flux and gradient over a domain of restricted size. Similar ideas apply in principle to the more complex case of transport.

One important aspect of geostatistical analysis is the inference from such data of a (semi)variogram (structure function) $\gamma(s)$ where s is a vector defining the spatial separation between any two points. Natural log hydraulic conductivity ($Y = ln$ K) and transmissivity ($Y = ln$ T) data often appear to fit variogram models associated with a constant sill (variance) σ^2 and integral (spatial) autocorrelation scale, λ. Figure 3 illustrates such behavior for 1 m-scale packer-test permeability data from fractured granites at the Stripa mine in Sweden. This is commonly taken to imply that the data are representative of a statistically homogeneous (stationary) random field $Y(x)$. Indeed, most stochastic analyses of subsurface flow and transport are based on the assumption that $Y(x)$ is statistically homogeneous. Such analyses show (among others) that, in an infinite domain subject to a uniform ensemble mean hydraulic gradient, the steady state ensemble mean flux is related to this gradient via Darcy's law with an effective hydraulic conductivity that depends primarily on the variance and integral scale of $Y(x)$. This effective hydraulic conductivity doubles as an equivalent conductivity provided that one averages the flux and gradient over a domain which is much larger than the integral scale of $Y(x)$; for an answer to the question "how much larger" refer to *Paleologos et al.* (1996) and *Tartakovsky and Neuman* (1997b). By the same token, the mean spread of ensemble averaged solute concentrations (sampled on the scale ω) of an inert solute, under these same conditions, is controlled by an effective dispersivity tensor whose longitudinal component first grows linearly with mean travel distance (or time) but later tends asymptotically to a constant (Fickian) value. The asymptotic Fickian regime is established when the mean travel distance becomes large in comparison to the integral scale of $Y(x)$; the corresponding effective longitudinal dispersivity depends on the variance and integral scale of $Y(x)$. Under ergodic conditions (for an answer to the question "when do such conditions develop" refer to *Dagan*, 1990, 1991, and *Zhang et al.*, 1996), the same dispersivity tensor describes the mean spread of an actual plume (in a single realization). In other words, a statistically homogeneous random $Y(x)$ field can, under the above conditions, be associated with effective as well as equivalent hydraulic conductivity and longitudinal dispersivity that tend to constant values as the scale of observation increases.

We have seen earlier that, in reality, permeabilities (Figure 1) and longitudinal dispersivities (Figure 2) appear to vary continuously over a broad range of experimental scales. Likewise, the integral scale of $Y(x)$ (Figure 4), and to a lesser extent its variance (this will become clear later), appear to increase consistently with the size of the domain under investigation. These findings suggests that statistical homogeneity, as is often suggested by standard variogram analyses, may not be a true property of $Y(x)$ but rather an artifact of the scale of investigation and method of inference. It further suggests a possible link between this apparent scale dependence of λ and σ^2 of $Y(x)$ on one hand, and that of larger-scale equivalent permeability and solute dispersivity on the other hand.

Indeed, when sample variograms are plotted on logarithmic rather than on arithmetic paper, the data often lie close to a straight line which is not consistent with the assumption of homogeneity. Such behavior is being observed at an increasing number of sites (*Neuman*, 1995) on distance scales ranging from a few meters (Figure 5) to 100 km (Figure 6). The

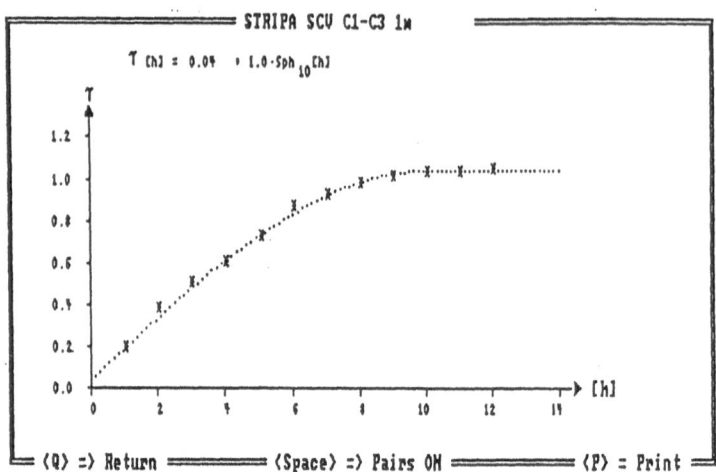

Figure 3. Variograms of natural log hydraulic conductivities of fractured granites at Stripa, Sweden, as determined in 1 m sections of three boreholes (after *Winberg*, 1991).

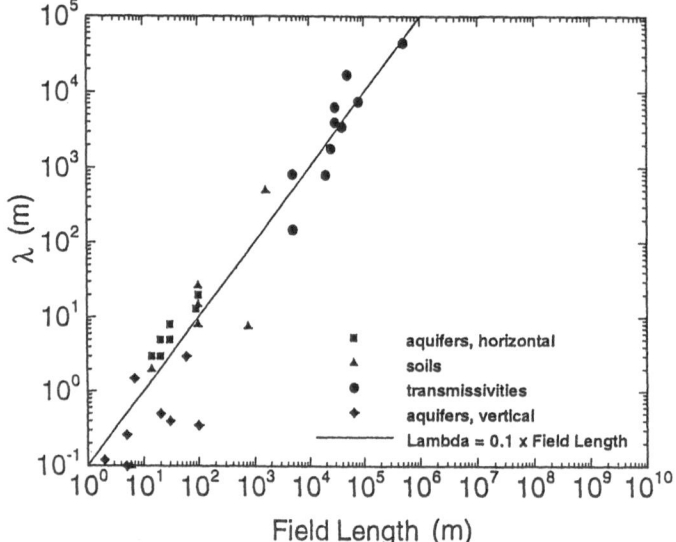

Figure 4. Correlation scales λ of natural log hydraulic conductivities and transmissivities at various sites versus field lengths (after *Neuman*, 1994; data from *Gelhar*, 1993, Table 6.1).

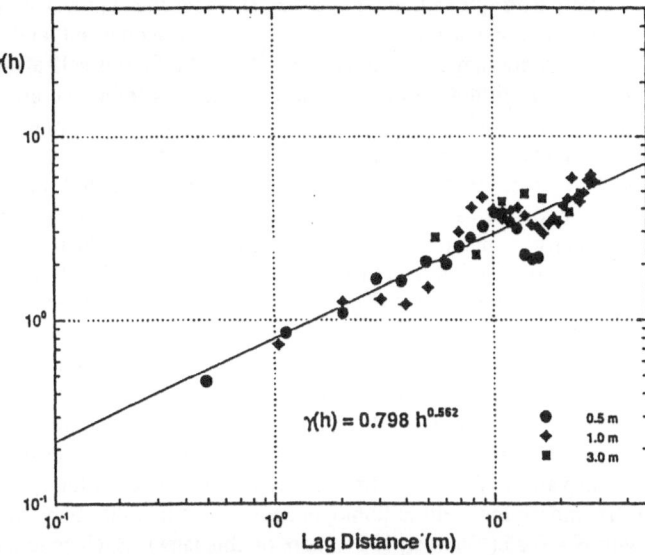

Figure 5. Log-log variogram of natural log air permeabilities of unsaturated fractures tuffs near Superior, Arizona, as determined in 1 m sections of several boreholes (after *Guzman and Neuman*, 1996).

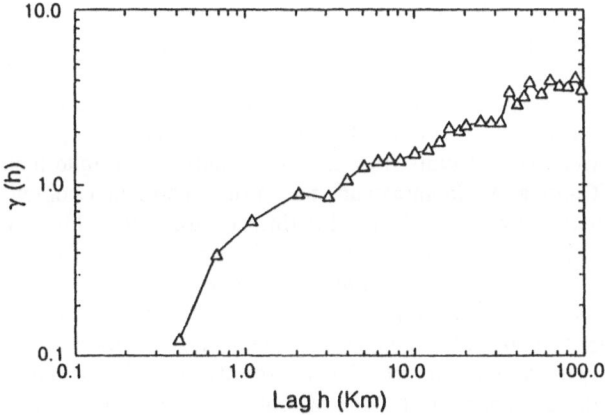

Figure 6. Log-log variogram of natural log transmissivities in a sandstone aquifer (after *Desbarats and Bachu*, 1994).

straight line is indicative of a nonstationary field with homogeneous spatial increments. If the field is statistically isotropic, a line with slope $2H$ represents a power variogram $\gamma(s) = as^{2H}$ where s is separation distance (the magnitude of s), a is a constant, and H is the Hurst coefficient. Since the variogram scales as $\gamma(rs) = r^{2H}\gamma(s)$ the field is self-affine and, within the range $0 < H < 1$, constitutes a random field that exhibits fractal geometry (a random fractal) with dimension $D = d + 1 - H$ where d is Euclidean (topologic) dimension (*Voss*, 1985). If the field is additionally Gaussian, it constitutes fractional Brownian motion (fBm; *Mandelbrot and Van Ness*, 1968). When $0.5 < H < 1$, random spatial increments in field values are positively correlated so that positive and negative deviations from the mean tend to persist over distance, a phenomenon known as persistence. When $0 < H < 0.5$, the increments are negatively correlated so that positive and negative devations from the mean tend to alternate rapidly, a phenomenon called antipersistence. When $H = 0.5$, the increments are uncorrelated and the field represents Brownian motion.

The Hurst coefficient of $Y(x)$ is not the same at each site though it has been found to lie near the midrange of $0 < H < 0.5$ in several recent studies quoted by *Neuman* (1995), and in the study of *Molz and Boman* (1995). Within this range the increments are negatively correlated and relatively noisy, exhibiting antipersistent behavior. Nevertheless, when one juxtaposes apparent values of σ^2 and λ from many different sites, inferred from Y data by assuming that the underlying field is homogeneous, one finds that they fit a generalized power model with $H \approx 0.25$ (*Neuman*, 1994; more on this later). Such generalized behavior has been deduced earlier by *Neuman* (1990) from the observed scale-dependence (Figure 2) of juxtaposed apparent dispersivities reported for a large number of tracer studies world-wide. This too will be discussed in more detail later.

Neuman(1990) has shown that any random field with homogeneous increments can be viewed as an infinite hierarchy of mutually uncorrelated homogeneous fields (modes) characterized by exponential autocovariance functions and variances which increase as a power of scale. He noted that accounting deterministically for large-scale spatial variability is equivalent to filtering out low-frequency modes from this hierarchy. *Di Federico and Neuman* (1997a) have extended these ideas by demonstrating that both the power variogram and associated spectra of a random field with homogeneous isotropic increments can be constructed as weighted integrals from zero to infinity (an infinite hierarchy) of either exponential or Gaussian variograms and spectra of uncorrelated homogeneous isotropic modes. They then analized the effect of filtering out (truncating) high and low frequency modes from this infinite hierarchy in the real and spectral domains. They showed that a low-frequency cutoff renders the truncated hierarchy statistically homogeneous with a spatial autocovariance function which varies monotonically with separation distance in a manner not too dissimilar from that of its constituent modes. The integral scales of the lowest and highest frequency modes (cutoffs) are related, respectively, to domain and sample (support) size. Taking this relationship to be one of proportionality renders their expressions for integral scale and variance dependent on domain size in a manner consistent with observations. We summarize some of their results in this paper and describe how their hierarchical theory allows bridging across scales. We also discuss flow and transport in isotropic multiscale, random log hydraulic conductivity fields formed by the weighted superposition of mutually uncorrelated exponential or Gaussian modes between a lower and an upper cutoff based on the work of *Di Federico and Neuman* (1997b-c).

2. Scaling of Random Fields by Means of Truncated Power Variograms and Associated Spectra

2.1. SUPERPOSITION OF MODES

Following *Di Federico and Neuman* (1997a) consider a power variogram of the form

$$\gamma(s) = \frac{1}{2} <[Y(x+s)-Y(x)]^2> = C_0 s^{2H} \quad (0<H<1) \tag{1}$$

where s is distance (lag), x is position in one-dimensional space, $<\cdot>$ indicates ensemble mean (expectation), C_0 is a constant, and H is Hurst coefficient. The associated one-dimensional random field $Y(x)$ is then self-affine (*Yaglom*, 1987) and devoid of a characteristic correlation scale. The random field $Y(x)$ is statistically nonhomogeneous but possesses homogeneous spatial increments. When these increments are multivariate Gaussian, the field constitutes fractional Brownian motion (fBm; *Mandelbrot and Van Ness*, 1968) that can be constructed as a moving average of white noise from $-\infty$ to x (*Bras and Rodriguez-Iturbe*, 1985). When $0 < H < \frac{1}{2}$, the increments are negatively autocorrelated and thus antipersistent; when $\frac{1}{2} < H < 1$, the increments are positively autocorrelated and persistent: when $H = \frac{1}{2}$, the increments are uncorrelated and $Y(x)$ degenerates into ordinary Brownian motion. In a d-dimensional Euclidean domain R^d, the location x and displacement s are vectors. Then $Y(x)$ is an isotropic self-affine random field provided that

$$\gamma(s) = \frac{1}{2} <[Y(x+s)-Y(x)]^2> = C_0 s^{2H} \quad (0<H<1) \quad s = |s|. \tag{2}$$

2.1.1. *Exponential Modes*
Consider an infinite hierarchy of mutually uncorrelated, statistically homogeneous and isotropic, random fields (modes) each of which is associated with an exponential variogram

$$\gamma(s,\lambda) = \sigma^2(\lambda)[1 - \exp(-s/\lambda)] \tag{3}$$

and dimensionless variance

$$\sigma^2(n) = \frac{C}{n^{2H}} \tag{4}$$

where λ is integral scale, $n = 1/\lambda$ is mode number, and C is a constant (dimensions $[L^{-2H}]$) so that the variance decreases as a power of the mode number. Integrating a continuous hierarchy of such modes over all possible scales, and weighting each contribution by a factor $1/n$, gives

$$\gamma(s) = \int_0^\infty \gamma(s,n) \frac{dn}{n}. \tag{5}$$

Substituting (3) and (4) into (5) and evaluating yields

$$\gamma(s) = C_0 s^{2H} \tag{6}$$

where

$$C_0 = C \frac{\Gamma(1-2H)}{2H} \qquad (0 < H < 1/2) \tag{7}$$

is a constant proportional to C and having the same dimensions. The weighted superposition of exponential modes is thus a power variogram of the form (1).

Consider now the case

$$\gamma(s,n_l) = \int_{n_l}^\infty \gamma(s,n) \frac{dn}{n} \tag{8}$$

where integration is performed with a lower cutoff $n_l = 1/\lambda_l$, so that all modes with integral scale larger than λ_l are filtered out (excluded). Then

$$\gamma(s,n_l) = \frac{C_0}{\Gamma(1-2H)n_l^{2H}} [1 - \exp(-n_l s) + (n_l s)^{2H} \Gamma(1-2H, n_l s)] \tag{9}$$

where $\Gamma(a,x)$ is the incomplete gamma function; in the limit as $n_l \to 0$, (9) reduces to (6). The variogram (9) defines a homogeneous field associated with a constant variance

$$\sigma^2(n_l) = \frac{C_0}{\Gamma(1-2H)n_l^{2H}} \tag{10}$$

an autocovariance $C(s,n_l) = \sigma^2(s,n_l) - \gamma(s,n_l)$ given by

$$C(s,n_l) = \frac{C_0}{\Gamma(1-2H)n_l^{2H}} [\exp(-n_l s) - (n_l s)^{2H} \Gamma(1-2H, n_l s)] \tag{11}$$

and a corresponding finite integral scale

$$I(n_l) = \frac{1}{\sigma^2(n_l)} \int_0^\infty C(s,n_l) ds = \frac{2H}{1+2H} \frac{1}{n_l} = \frac{2H}{1+2H} \lambda_l. \tag{12}$$

Since these results hold for $0 < H < \frac{1}{2}$, the integral scale $I(n_l)$ of the truncated multiscale random field is proportional to (and always smaller than) the integral scale λ_l of the lowest mode retained, the constant of proportionality increasing with H. Note also that for $n_l s \ll 1$, or equivalently $s \ll \lambda_l$, $\gamma(n_l, s) \approx C_0 s^{2H}$ still holds despite the truncation. This implies that the cutoff has no effect on lags much smaller than the integral scale of the smallest active mode.

If the hierarchy of modes is truncated both below a lower cutoff n_l and above an upper cutoff n_u according to

$$\gamma(s,n_l,n_u) = \int_{n_l}^{n_u} \gamma(s,n) \frac{dn}{n} \qquad (13)$$

then one obtains, in analogy to (9),

$$\gamma(s,n_l,n_u) = \gamma(s,n_l) - \gamma(s,n_u). \qquad (14)$$

When $n_l \to 0$ and $n_u \to \infty$, (14) reduces to (6). The variance, autocovariance and integral scale associated with (14) are, respectively,

$$\sigma^2(n_l,n_u) = \sigma^2(n_l) - \sigma^2(n_u) \qquad (15)$$

$$C(s,n_l,n_u) = C(s,n_l) - C(s,n_u) \qquad (16)$$

$$I(n_l,n_u) = \frac{2H}{1+2H} \frac{n_u^{1+2H} - n_l^{1+2H}}{n_l n_u (n_u^{2H} - n_l^{2H})} \qquad (17)$$

where $n_l = 1/\lambda_l$ and $n_u = 1/\lambda_u$, λ_l and λ_u being, respectively, the integral scales of the lowest and highest mode retained in the superposition. Note that (14) - (17) differ little from (9) - (12) when $n_u \gg n_l$; in particular, for given H, C_0, and n_l, the introduction of a higher cutoff n_u brings about a reduction in variance and an increase in integral scale.

2.1.2. Gaussian modes
Consider now modes having Gaussian variograms

$$\gamma(s,\lambda) = \sigma^2(\lambda)[1 - \exp(-\pi s^2/4\lambda^2)] \qquad (18)$$

where λ is again integral scale. Substituting (4) and (18) into (5) yields

$$\gamma(s) = C_0' s^{2H} \qquad (19)$$

where

$$C_0' = C \frac{\Gamma(1-H)}{2H} \left(\frac{\pi}{4}\right)^H \qquad (0 < H < 1) \qquad (20)$$

is a constant proportional to C and having the same dimensions. The weighted superposition of Gaussian modes is thus a power variogram of the same form as (6) but is valid over a wider range of H values. Superimposing Gaussian modes with a lower cutoff $n_l = 1/\lambda_l$ according to (8) gives

$$\gamma(s,n_l) = \frac{C_0'}{\Gamma(1-H)(\pi/4)^H n_l^{2H}} [1 - \exp(-\frac{\pi}{4} n_l^2 s^2) + (\frac{\pi}{4} n_l^2 s^2)^H \Gamma(1-H, \frac{\pi}{4} n_l^2 s^2)]. \qquad (21)$$

Since $\Gamma(1-H,0) = \Gamma(1-H)$, (21) reduces to (19) in the limit as $n_l \to 0$. The variance, autocovariance and integral scale associated with (21) are, respectively,

$$\sigma^2(n_l) = \frac{C_0'}{\Gamma(1-H)(\pi/4)^H n_l^{2H}}$$ (22)

$$C(s,n_l) = \frac{C_0'}{\Gamma(1-H)(\pi/4)^H n_l^{2H}} [\exp(-\frac{\pi}{4}n_l^2 s^2) - (\frac{\pi}{4}n_l^2 s^2)^H \Gamma(1-H, \frac{\pi}{4}n_l^2 s^2)]$$ (23)

$$I(n_l) = \frac{1}{\sigma^2(n_l)} \int_0^\infty C(s,n_l) \, ds = \frac{2H}{1+2H} \frac{1}{n_l} = \frac{2H}{1+2H} \lambda_{l'}$$ (24)

The latter is identical to (12) corresponding to exponential modes, even though (21) - (23) differ from (9) - (11). From (22) it follows that (21) remains approximately equal to (18) when $n_l s \ll 1$.

If the hierarchy of modes is truncated both below a lower cutoff n_l and above an upper cutoff n_u according to (13) then the variogram, variance, autocovariance, and integral scale are given respectively by (14) - (17).

The autocorrelation functions $\rho(s,n_l) = C(s,n_l)/\sigma^2(s,n_l)$ corresponding to exponential and Gaussian modes with a lower cutoff are plotted in Figures 7 and 8, respectively, as functions of dimensionless distance s/I at various values of H. Since the variance (10) and (22), and integral scale (12) and (24), depend on H, we vary C_0 or C_0' and n_l with H so as to maintain the values of σ^2 and I constant. Exponential and Gaussian models with the same variance and integral scale are included for comparison. As H increases from 0 to 0.5, the multiscale autocorrelation function in Figure 7 approaches the exponential model; as H increases from 0 to 1, the multiscale autocorrelation function in Figure 8 approaches the Gaussian model.

Figures 9 and 10 show the multiscale autocorrelation functions $\rho(s,n_l,n_u) = C(s,n_l,n_u)/\sigma^2(s,n_l,n_u)$ with both lower and upper cutoffs for exponential and Gaussian modes, respectively, versus s/I at $H = 0.25$ for various $\beta = n_l/n_u$, $0 \leq \beta \leq 1$; the case without upper cutoff corresponds to $\beta = 0$. As β increases, a wider range of high modes is excluded, and the resulting truncated autocorrelation functions approach the corresponding exponential or Gaussian models.

All of the above results in the real domain have a direct counterpart in the spectral domain (*Di Federico and Neuman*, 1997a).

2.2. VERIFICATION OF SCALING RELATIONS

To verify directly the above multiscale model, *Di Federico and Neuman* (1997a) consider values of spatial autocorrelation (integral) scale and variance, inferred by various researchers from natural log hydraulic conductivity and transmissivity data at diverse sites on the assumption that the corresponding fields are statistically homogeneous and isotropic, recently summarized by *Gelhar* (1993, Table 6.1). Gelhar's table also lists a characteristic length for each domain within which conductivity or transmissivity was sampled. Figure 4 is a plot of reported integral scales versus characteristic lengths of the various field sites (domains).

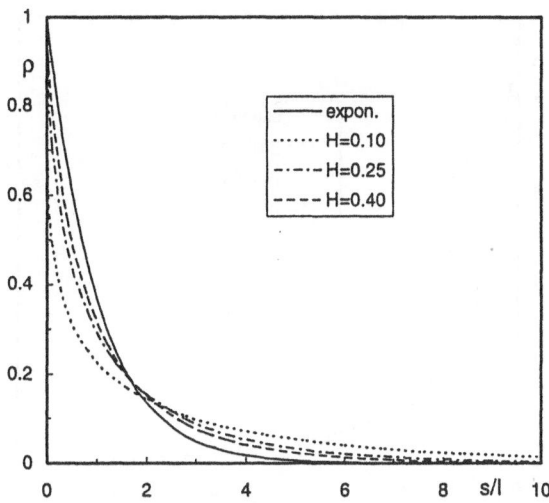

Figure 7. Autocorrelation function corresponding to exponential modes with lower cutoff versus dimensionless distance s/l for various values of H.

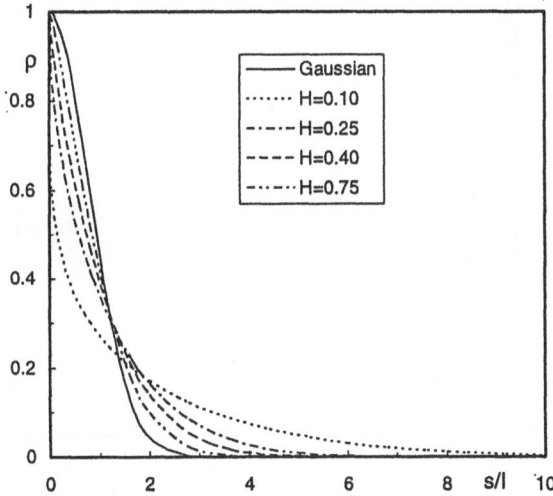

Figure 8. Autocorrelation function corresponding to Gaussian modes with lower cutoff versus dimensionless distance s/l for various values of H.

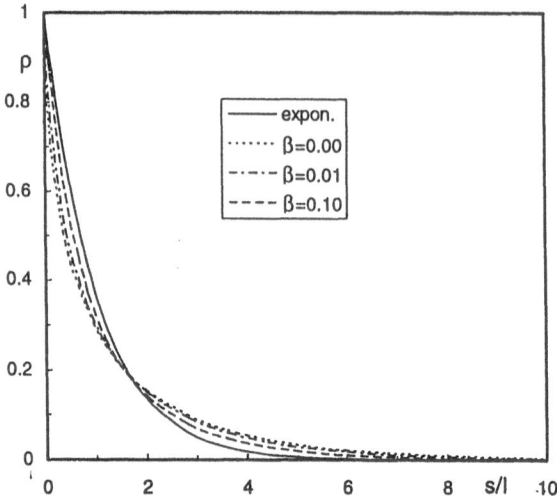

Figure 9. Autocorrelation function corresponding to exponential modes with lower and upper cutoffs versus dimensionless distance s/l for $H = 0.25$ and various $\beta = n_l/n_u$, $0 \le \beta \le 1$.

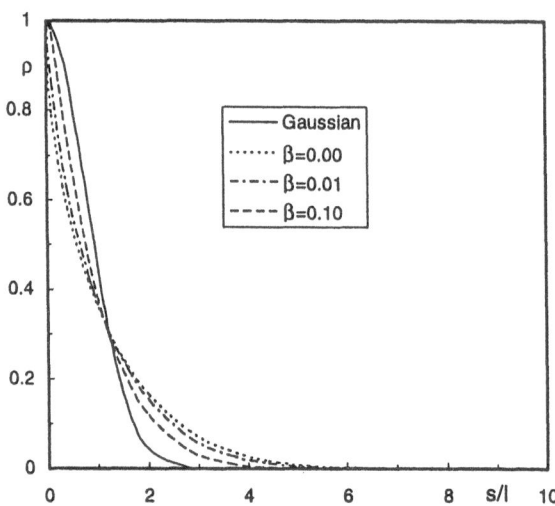

Figure 10. Autocorrelation function corresponding to Gaussian modes with lower and upper cutoffs versus dimensionless distance s/l for $H = 0.25$ and various $\beta = n_l/n_u$, $0 \le \beta \le 1$.

It suggests that integral scales tend to increase linearly with domain size, at a generalized rate of about 1/10, over domains that range in length scale from a few meters to several hundred kilometers. This is consistent with the above truncated hierarchical model provided only that one postulates a fixed ratio μ between the length scale of any sampling window superimposed on the infinite hierarchy, and the integral scale of the lowest frequency mode that is filtered out by this window. Thus, if ω is the length scale of the data support (sample size) and Ω is the length scale of the domain being sampled, then the sampling window has a length scale bounded from below by ω and from above by Ω, so as to introduce an upper frequency cutoff $n_u = 1/\lambda_u = 1/\mu\omega$ with associated integral scale $\lambda_u = \mu\omega$, and a lower frequency cutoff $n_l = 1/\lambda_l = 1/\mu\Omega$ with associated integral scale $\lambda_l = \mu\Omega$. In the common case where $\Omega \gg \omega$ and thus $n_u \gg n_l$, it is appropriate to work with the limiting case $n_u \to \infty$. Then, according to (12) or (24), the integral scale of the filtered field is given by

$$I(\lambda_l) = \alpha \lambda_l \qquad \alpha = \frac{2H}{1+2H} \tag{25}$$

where $0 < \alpha < \frac{1}{2}$ for exponential modes and $0 < \alpha < 2/3$ for Gaussian modes. Hence

$$I(\lambda_l) = \alpha \mu \Omega \tag{26}$$

which we saw is supported by the data in Figure 4.

We mentioned earlier that *Neuman* (1990, 1995) had derived, on the basis of juxtaposed apparent dispersivity data from tracer studies at many diverse sites, a generalized power variogram for natural log hydraulic conductivities having the simple form $\gamma(s) \approx cs^{1/2}$ where $c \approx 0.027$. His interpretation seems to hold (*Neuman*, 1993b, 1995) despite the fact that some of the field-derived dispersivities may not be entirely reliable (*Gelhar et al.*, 1992). The corresponding variogram is identical to (6) or (19) provided one adopts the generalized Hurst coefficient $H \approx 0.25$. If one takes the variogram to consist of exponential modes then, according to (7), $C_0 \approx 0.027$ and $C \approx 0.027/(2\pi^{1/2}) = 0.0076$. If one takes it to consist of Gaussian modes then, according to (20), $C_0 \approx 0.027$ and $C \approx 0.027(4/\pi)^{0.25}/[2\Gamma(3/4)] = 0.0117$. It follows that C has a generalized value close to 0.01. Figure 4 suggests that, for juxtaposed hydraulic conductivity and transmissivity data from many diverse sites, $\alpha\mu \approx 0.1$. According to (25), $H \approx 0.25$ corresponds to a generalized value of $\alpha \approx 1/3$ and so one obtains a generalized value of $\mu \approx 1/3$.

Equations (10) and (22), together with the postulate $\lambda_l = \mu\Omega$, imply that when $\Omega \gg \omega$ the variance of a truncated hierarchy increases as a power of domain size according to $\sigma^2 \propto \Omega^{2H}$ and so $\sigma^2\lambda \propto \Omega^{2H}$. For juxtaposed natural log hydraulic conductivities and transmissivities from many sites we expect $H \approx 0.25$ and hence $\sigma^2\lambda \propto \Omega^{1.5}$. Indeed, Figure 11 shows that when the hydraulic conductivity data (squares and triangles) from *Gelhar's* (1993) Table 6.1 are juxtaposed on a log-log plot of $\sigma^2\lambda$ versus λ (or equivalently Ω), they are scattered closely about a (dashed) line having the predicted slope of 1.5; the corresponding transmissivity data (circles) are scattered closely about another (solid) line having the same slope, and an intercept corresponding to *Neuman's* (1990) predicted value (on the basis of tracer study data) of $c \approx 0.027$ (see previous paragraph). The offset between the two lines was attributed by *Neuman* (1994) to a reduction in variance obtained when

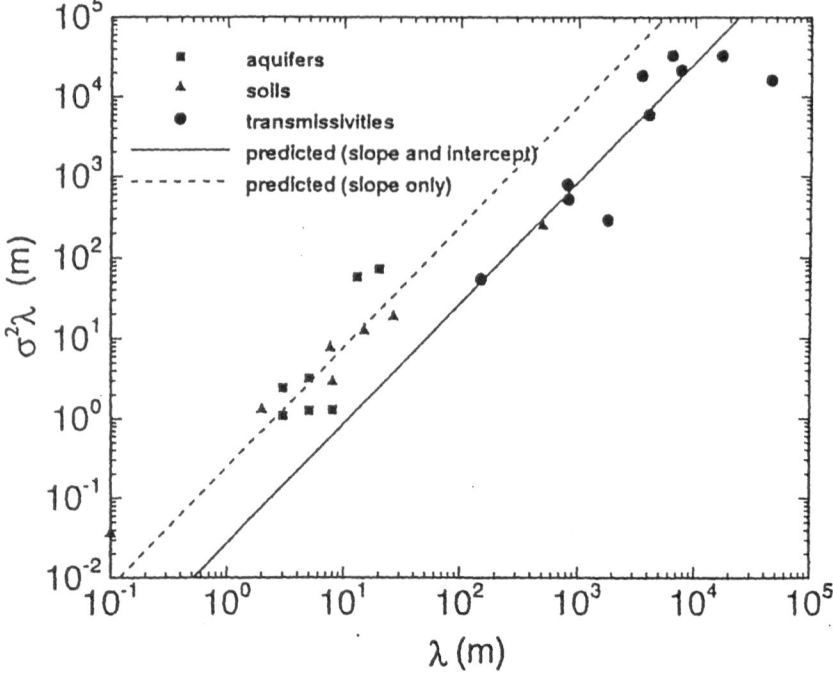

Figure 11. Product of variance σ^2 and horizontal correlation scale λ of natural log hydraulic conductivities and transmissivities at various sites versus λ (after *Neuman*, 1944; data from *Gelhar* (1993, Table 6.1). Predictions correspond to $\gamma(s) = c\sqrt{s}$).

hydraulic conductivities are averaged over the vertical to yield transmissivities. This explanation is in turn qualitatively consistent with (15) which, together with the postulate $\lambda_u = \mu\,\omega$, predict a reduction in variance with support scale according to

$$\sigma^2 \propto (\Omega^{2H} - \omega^{2H}) \tag{27}$$

2.3. BRIDGING ACROSS SCALES

The above hierarchical theory allows bridging across scales by predicting the effect of viewing a multiscale random field, defined and measured on a given support scale, through a larger window defined by the domain under investigation. More precisely, the theory predicts the effect of domain size on the integral scale, variance, and covariance of such a multiscale random field. At a given locale, one can in principle calibrate a truncated variogram model (estimate its parameters C, H, μ) to sample data observed on a given support length scale ω (which, with μ, defines the upper frequency cutoff n_u) in one domain of length scale Ω (which, with μ, defines the lower cutoff n_l), then predict the autocovariance structure of the corresponding field in domains that are either smaller ($< \Omega$) or larger ($> \Omega$). One may also venture (though with less predictive power) to transfer such predictions from one locale to another by introducing the generalized parameters $C \approx 0.0076$ for exponential modes, $C \approx 0.0117$ for Gaussian modes, $H \approx 0.25$, and $\mu \approx 1/3$ into the truncated variogram models. Selection between alternative models (exponential modes, Gaussian modes, with or without lower cutoffs, other) is also possible in principle by means of formal model discrimination criteria, coupled with methods such as maximum likelihood cross-validation, as proposed by *Samper and Neuman* (1989).

2.4. EFFECTIVE HYDRAULIC CONDUCTIVITY

It is common in the stochastic groundwater literature to treat log hydraulic conductivity as a statistically homogeneous Gaussian field. For an unbounded flow domain subject to a uniform mean hydraulic gradient, the effective conductivity corresponding to a statistically isotropic field has been conjectured by *Matheron* (1967) to be

$$\mathrm{K}_{eff} = \mathrm{K}_g \exp\left[\sigma^2\left(\tfrac{1}{2} - \tfrac{1}{d}\right)\right] \tag{28}$$

where K_g is the geometric mean conductivity. This result is rigorously valid for the one-dimensional case ($d = 1$) where K_{eff} is the harmonic mean $\mathrm{K}_h = \mathrm{K}_g \exp(-\sigma^2/2)$ and for the two-dimensional case ($d = 2$) where $\mathrm{K}_{eff} = \mathrm{K}_g$. Its validity in the three-dimensional case ($d = 3$) has been demonstrated numerically for σ^2 as large as 7 in *Neuman and Orr* (1993), and for somewhat smaller values of σ^2 by *Ababou* (1988), *Desbarats* (1992), and *Dykaar and Kitanidis* (1992).

The effective conductivity for a multiscale isotropic field consisting of exponential modes is obtained by substituting the variance (15), coupled with (10), into (28). The results is (*Di Federico and Neuman*, 1997b)

436

$$K_{eff} = K_g \exp\left[\frac{C_o}{2\Gamma(1-2H)}\left(\frac{1}{2}-\frac{1}{d}\right)\left(\frac{1}{n_l^{2H}}-\frac{1}{n_u^{2H}}\right)\right] \quad (29)$$

The same can be done with Gaussian modes by using (22) instead of (10).

Figure 1 shows only ranges of measured permeability which are insufficient to validate quantitatively the theoretical scaling relation (29). However, the figure does provide qualitative support for this relation. The laboratory and borehole permeability data appear to exhibit a systematic increase with the reported scale of measurement. No such trend is indicated by the regional data. This could be due to a combination of factors: The available sample is too small to indicate a trend; some of the data correspond to near two-dimensional flow regimes; and some of the data are derived from calibrated numerical models which account explicitly for medium heterogeneity and thus filter out large-scale, low-frequency variations from the hierarchy.

2.5. FIRST-ORDER THEORY OF FLOW AND ADVECTIVE TRANSPORT

Di Federico and Neuman developed a first-order theory of flow (1997b) and advective transport (1997c) in isotropic multiscale, random log conductivity fields formed by the weighted superposition of mutually uncorrelated exponential or Gaussian modes with lower and upper cutoffs in two and three dimensions. We present below some of their two-dimensional results.

Consider steady state flow satisfying the continuity equation and Darcy's law
$$\nabla \cdot q(x) = 0 \qquad q(x) = -K(x)\nabla h(x) \quad (30)$$
where q is Darcy flux and h is hydraulic head. The seepage velocity is given by $u(x) = q(x)/\phi$ where ϕ is porosity (assumed to be constant). The natural logarithm $Y(x) = ln\,K(x)$ of hydraulic conductivity is treated as a statistically homogeneous, multivariate Gaussian random field, uniquely defined by its constant ensemble mean (expected value) $\langle Y \rangle$, variance σ^2, and spatial covariance function $C(s)$ where $s = x - \chi$ is a separation vector. Furthermore, $Y(x)$ is assumed to result from a weighted superposition of exponential modes with a lower cutoff n_l so that σ^2, $C(s)$, and the associated finite integral scale are given, respectively, by (10) - (12). This allows deriving explicit expressions for the cross-covariance $C_{Yh}(x,\chi)$ between Y and h, the head variogram $\gamma_h(x,\chi)$, the cross-covariance $C_{uY}(x,\chi)$ between any component of u and Y, the cross-covariance $C_{uh}(x,\chi)$ between any component of u and h, and the cross- and auto-covariance $C_{uu}(x,\chi)$ between any two components of u (see *Di Federico and Neuman*, 1997b).

Following *Di Federico and Neuman* (1997c), consider an indivisible particle of inert solute which has finite mass but zero volume. At time $t = 0$ the particle is known to be located at the origin of the coordinates. It is then swept by the random groundwater velocity $u(x)$ to describe a time-dependent, random trajectory $X(t)$. Under steady state uniform mean flow, the mean displacement is $\langle X(t) \rangle = \langle u \rangle t$ where
$$\langle u \rangle = \frac{K_G}{\phi} J = U \quad (31)$$
and $J = -\nabla\langle h(x) \rangle$ is the constant mean hydraulic gradient. Write $X(t) = \langle X(t) \rangle + X'(t)$ and note that $X_{ij}(t) = \langle X_i'(t)X_j'(t) \rangle$ is the covariance of particle displacement. Explicit first-order

expressions for X_{11} and X_{22} are given, for the case of zero local dispersion, by *Di Federico and Neuman* (1997c). From these, they derive corresponding expressions for the macrodispersion coefficients

$$D_{ij}(t) = \tfrac{1}{2}\tfrac{dX_{ij}}{dt} \tag{32}$$

The longitudinal macrodispersion coefficient is

$$D_{11}(r) = U\frac{\sigma^2}{4(1+2H)(2+H)(1+H)n_l}[8H(2+H)(1+H)+12H(1+2H)(1+H)\tfrac{1}{r}$$
$$-12H(1+2H)(1+H)\tfrac{e^{-r}}{r^3}-12H(1+2H)(1+H)\tfrac{e^{-r}}{r^2}-6H(1+2H)(2+H)\tfrac{1}{r}$$
$$+6H(1+2H)\tfrac{e^{-r}}{r}-6He^{-r}+3re^{-r}-3r^{1+2H}\Gamma(1-2H,r)] \tag{33}$$

and the transverse coefficient is

$$D_{22}(r) = U\frac{\sigma^2}{4(2+H)(1+H)n_l}[-12H(1+H)\tfrac{1}{r}+12H(1+H)\tfrac{e^{-r}}{r^3}+12H(1+H)\tfrac{e^{-r}}{r^2}$$
$$+2H(2+H)\tfrac{1}{r}+2H(1+2H)\tfrac{e^{-r}}{r}-2He^{-r}+re^{-r}-r^{1+2H}\Gamma(1-2H,r)] \tag{34}$$

where U is the magnitude of **U** and $r = n_l s$, s being the magnitude of the displacement vector **s**. Asymptotically, as $r \to \infty$, all terms on the right-hand side of (33) vanish, except the first, and so

$$D_{11}(\infty) = U\sigma^2\frac{2H}{(1+2H)n_l} = U\frac{2HC_o}{(1+2H)\Gamma(1-2H)n_l^{1+2H}} \tag{35}$$

Note that this Fickian asymptote can be recast as $D_{11}(\infty) = U\sigma^2 I$ which yields the asymptotic (Fickian) longitudinal dispersivity

$$\alpha_{11}(\infty) = \tfrac{D_{11}(\infty)}{U} = \sigma^2 I \tag{36}$$

We saw earlier that $\sigma^2 I$ grows as a power $1+2H$ of distance, hence (36) implies that the same is true about $\alpha_{11}(\infty)$. Considering that the apparent longitudinal dispersivities in Figure 2 were obtained by fitting tracer experimental data to Fickian models, we now have a theoretical explanation (and prediction) of their observed supralinear increase with mean travel distance (or time). This is true about the open symbols in Figure 2; the solid symbols had been obtained from numerical models which accounted explicitly (deterministically) for grid-scale medium heterogeneity and thus fall outside the scope of the theory discussed here. Note that the above theoretical prediction as to how $\alpha_{11}(\infty)$ scales with distance corresponds exactly to, and thus confirms, the scaling relation derived semiempirically by *Neuman* (1990) from Figure 2.

The first-order transverse dispersion coefficient in (34) tends to zero at large mean travel distance (or time).

Figures 12 and 13 show how dimensionless D_{11} and D_{22} (normalized with respect to $U\sigma^2 I$ so that they also represent dimensionless dispersivities), respectively, vary with the dimensionless distance s/I as H takes on different values. Since the integral scale $I = (2H)/[(1+2H)n_l]$ and the variance $\sigma^2 = C_o/[\Gamma(1-2H)n_l^{2H}]$ depend on H, C_0 and n_l are made to vary with H so as to maintain the values of I and σ^2 constant. The longitudinal dispersivity for the multiscale model is seen to be always inferior to that for an exponential model with a similar integral scale. The difference increases as H decreases, and the rate at which D_{11} approaches its asymptote goes down. At a travel distance of 80 integral scales, dimensionless D_{11} for $H = 0.10$ is still 0.93, while the exponential model gives 0.98. The greatest relative difference between the two models occurs in the intermediate range of a few tens of integral scales. The transverse dispersivity shows lower rising limbs and peaks as H decreases; the peaks become lower and the descending limbs are offset to the right.

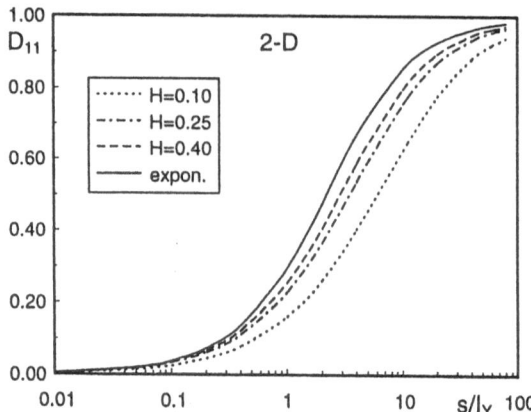

Figure 12. Dimensionless longitudinal dispersion versus dimensionless distance as function of H in the presence of a lower cutoff, compared with exponential model.

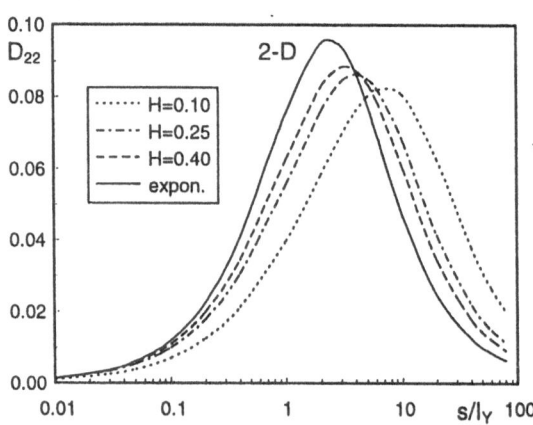

Figure 13. Dimensionless transverse dispersion versus dimensionless distance as function of H in the presence of a lower cutoff, compared with exponential model.

3. Effect of Conditioning on Effective Dispersivity

3.1. EFFECT ON APPARENT FICKIAN DISPERSIVITIES

The dispersivities designated by open symbols in Figure 2 were obtained by means of Fickian (mostly analytical) models which consider the permeability (or transmissivity) to be uniform in each tracer study. Dispersivites designated by solid symbols were obtained from the calibration of numerical models against hydraulic and concentration data corresponding to large-scale plumes. Such models account explicitly for intermediate-scale (larger than the support scale ω, smaller than the flow domain Ω) medium heterogeneity by allowing the permeability (or transmissivity) to vary spatially in a deterministic manner. In the language of stochastic theory, such models are conditioned (though not in a rigorous manner) on site-specific measurements. What effect does this have on the computed apparent Fickian dispersivities?

The apparent dispersivities from calibrated, explicitly heterogeneous models are seen to increase much more slowly with scale than is predicted by (36). Applying regression to these dispersivities yields a straight line with a slope of 0.54 (not shown in the figure). The null hypothesis that the slope is 1 or larger must be rejected at the 99% confidence level. Hence the recorded dispersivities from calibrated models are inconsistent with the generalized power variogram, and associated fBm model, which we saw were consistent with the rest of the data. The two data sets constitute samples from different populations and must be analyzed separately. This explains why data from calibrated (conditional) models were not included in the previous discussion of unconditional fields.

The calibrated dispersivities in Figure 2 grow approximately in proportion to $\sqrt{s}$, and the discrepancy between them and values predicted by (36) grow in proportion to s. *Neuman* (1990) attributed this reduction in scale partly to a nonlinear effect and partly to conditioning on (accounting explicitly and deterministically for) site-specific data. Such conditioning renders the model better able to resolve spatial variations in the advective velocity field. It introduces an upper bound (cutoff) on the correlation scales that affect dispersivities, causing the latter to diminish as information about the spatial variability of log permeabilities improves in quantity and quality. The diminution in apparent longitudinal dispersivity occurs because only those spatial variations in advective velocity which are not resolved explicitly have an effect on dispersion. As these variations are now limited to higher frequencies (and shorter wavelengths), so are the dispersivities reduced.

3.2. EFFECT ON PREASYMPTOTIC DISPERSION

3.2.1. *Preasymptotic Dispersion in Fractal Velocity Fields*
Following *Neuman* (1995), let us consider steady state advection in a random velocity vector field $\mathbf{v}(x)$. The field has homogeneous spatial increments and is characterized by a tensor

(dyadic) power variogram

$$\gamma(s) = b_0 \rho \left(\frac{s}{L_0} \right)^{2H} \qquad 0 < H < 1 \qquad (37)$$

where s is lag distance, L_0 is a suitable reference scale, b_0 is a constant (dimensions $L^2 T^{-2}$), ρ is a constant dyadic, and the scaling exponent H is the Hurst coefficient. As such, $v(x)$ is a self-affine random field with fractal dimension $D = d + 1 - H$, d being topologic dimension. Its ensemble mean and variance are generally undefined. We define them conditionally by assuming, without loss of generality, that the velocity is known (from hydraulic and/or tracer data) to be v_0 at some point x_0. Hence v_0 is the conditional ensemble mean of v. Our (37) is identical to a corresponding expression due to *Neuman* (1995) except that he did not include a reference length in his definition of γ so that b_0 acquired dimensions $L^{2-2H} T^{-2}$.

At time t_0, a mass M_0 of tracer is introduced into the stream at another point y_0. We associate this mass with an indivisible particle which advects with the random velocity $v(x)$ without diffusion or local dispersion. Let $\tau = t - t_0$ be the particle residence time at any $t > t_0$ so that its conditional mean displacement from the origin y_0 is $h = v_0 \tau$. Let β be the angle between h and the separation vector $r = y_0 - x_0$. Then the conditional mean distance s of the particle from x_0 is given by $s^2 = r^2 - 2rh \cos \beta + h^2$ where r and h are the respective magnitudes of r and h. The corresponding displacement vector is $s = r + h$. The velocity of the particle after a displacement s is $v(x_0 + s) = v_0 + \Delta v(s)$ where $\Delta v(s)$ is a velocity increment. As velocity increments are homogeneous, the conditional velocity cross-covariance (not auto-covariance!) is simply $V(v) = V(\Delta v) = \gamma(s)$. If $r = 0$ (the particle originates at the conditioning point), $s = h$ and the velocities experienced by the traveling particle have a cross-covariance dyadic that grows indefinitely as a simple power of the residence time τ,

$$V(\tau) = b_0 \rho \left(\frac{\tau}{\tau_0} \right)^{2H} \qquad (38)$$

where $\tau_0 = L_0 / v_0$ is a time scale. The same happens for $r > 0$ following a displacement h large enough to satisfy $h \gg r$. We restrict ourselves below to r values that are small enough, or (equivalently) τ values that are large enough, to satisfy this requirement.

It is most important to note that $v(x)$ is autocorrelated spatially over all distance scales; it has an infinite integral scale λ_∞. Hence the particle velocities are autocorrelated temporally at all times. In other words, the particle velocities are associated with an infinite (Lagrangian, conditional) correlation time $T = \lambda_\infty / v_0$.

Let $p(v)$ be the conditional univariate probability density function (pdf) of v. It is well established (*Batchelor*, 1952a, pp. 349-350; *Neuman*, 1993, Appendix A) that, in a velocity field with finite Lagrangian correlation scale $T < \infty$, the conditional ensemble mean concentration due to the above instantaneous point source is given at early time $\tau \ll T$ by

$$<c(y,\tau)> = \frac{M_0}{\phi \tau^k} p \left(\frac{y}{\tau} \right) \qquad (39)$$

where $y = x - y_0$ is a radius vector measured from the particle origin. Whereas in general $<c(y,\tau)>$ satisfies a nonlocal (space-time integro-differential) conditional mean transport equation (*Neuman*, 1993), the early time solution (39) satisfies a local (partial-differential) equation (*ibid*, Appendix A).

Since in a fractal velocity field T is infinite, it follows that $<c(y,\tau)>$ is local and given by (39) at all times $\tau < \infty$. This simple but far-reaching fact appears to have gone unnoticed in the literature. Our simple demonstration that it must be valid in a fractal velocity field implies that mean advective transport in such a field is inherently local rather than nonlocal as proposed by some authors. This is true provided that the actual (random) concentration, c, is taken to be governed by a local advective transport equation as considered by *Neuman* (1993) rather than by a nonlocal equation.

The local transport equation satisfied by $<c(y,\tau)>$ in (39) is not the standard advection-dispersion equation (e,g., *Neuman*, 1993, Appendix A). As such, it is neither Fickian nor quasi-Fickian and allows $<c(y,\tau)>$ to have a non-Gaussian spatial profile (note however that $<c(y,\tau)>$ is spatially Gaussian if v is univariate Gaussian, a point to which we will return later). It is nevertheless well known that one can describe its mean spread by a dispersion tensor (dyadic)

$$\mathbf{D}(\tau) = \frac{1}{2}\frac{d\Omega(\tau)}{d\tau} \tag{40}$$

where $\Omega(\tau)$ is the second spatial moment of $<c(y,\tau)>$ about its conditional center of mass,

$$\Omega(\tau) = \frac{\phi}{M_0}\int(y - v_0\tau)(y - v_0\tau)^T <c(y,\tau)> dy \tag{41}$$

Since $\phi<c(y,\tau)>/M_0$ is the probability density function of particle displacements (*Batchelor*, 1949, 1952b; *Dagan*, 1987), $\Omega(\tau)$ also represents the covariance of particle displacements y about its conditional mean displacement $h = v_0\tau$. Substitution of (39) into (41), followed by substitution of the result into (40) and the change of variables $z = y/\tau$, yields

$$\mathbf{D}(\tau) = \frac{1}{2}\frac{d}{d\tau}[\tau^2\mathbf{V}] \tag{42}$$

which is known in the literature to hold for $\tau \ll T$. Since in our case T is infinite, both (39) and (42) hold for all $\tau < \infty$. This also appears to have remained unnoticed in the literature.

In the special case where $v(x)$ is a homogeneous field with mean v_0, variance σ^2, and a finite Lagrangian correlation time $T < \infty$, $\mathbf{V}$ is a constant and (42) reduces to the classical early-time result (*Taylor*, 1921)

$$\mathbf{D}(\tau) = \mathbf{V}\tau \qquad \tau \ll T \tag{43}$$

The corresponding dispersivity tensor, defined as

$$\alpha = \frac{\mathbf{D}}{v_0} \tag{44}$$

can be expressed in the form

$$\alpha(h) = C_v^2\rho h \tag{45}$$

where h is mean displacement, $C_v = \sigma/v_0$ is a velocity coefficient of variation, and ρ is the cross-correlation dyadic of velocity components. Note that $\alpha(h)$ in such a homogeneous velocity field grows in proportion to h during the early preasymptotic regime (*Taylor*, 1921).

Only in the case of a finite T is an asymptotic regime ever reached. In our fractal case where T is infinite, the transport remains always preasymptotic as given by (39) and (42). Recalling that we consider only cases where r is small enough, and/or (equivalently) τ is large enough, so that $h \approx s$ and (38) holds, substituting the latter into (42) and evaluating the derivative yields a power-law time-growth for the dispersivity dyadic,

$$\mathbf{D}(\tau) = (1+H)\,b_0\rho\,\frac{L_0}{v_0}\left(\frac{\tau}{\tau_0}\right)^{1+2H}$$
(46)

By virtue of (44), the dispersivity then grows as a power of the mean travel distance according to

$$\alpha(h) = (1+H)\,\bar{C}_v^2\rho L_0\left(\frac{s}{L_0}\right)^{1+2H}$$
(47)

where $\bar{C}_v = \sqrt{b_0}/v_0$ is a normalized velocity coefficient of variation. Since H is theoretically confined to the range $0 \le H \le 1$, α in (47) grows with h at a rate faster than linear. We saw earlier that this predicted growth rate is strongly supported by the apparent longitudinal dispersivities measured in the field.

We saw earlier that any random field having a power variogram $\gamma(s) = c s^{2H}$ with $0 < H < 1/2$ can be viewed as the superposition of an infinite hierarchy of mutually uncorrelated, statistically homogeneous fields (modes) characterized by exponential correlation functions. By virtue of the central limit theorem (*c.f.*, *Christakos*, 1992, p.47), the univariate conditional pdf of the velocities encountered by a traveling particle at time τ should be Gaussian with cross covariance given by (38), and therefore (39) should read

$$<c(y,\tau)> = \frac{M_0}{\phi[\tau(2\pi)^{1/2}]^d\,|\mathbf{V}|^{1/2}}\exp\left[-\frac{1}{2}\left(\frac{y-v_0\tau}{\tau}\right)^T|\mathbf{V}|^{-1}\left(\frac{y-v_0\tau}{\tau}\right)\right]$$

$$= \frac{M_0}{\phi[(\tau/\tau_0)^{1+H}\bar{C}_vL_0(2\pi)^{1/2}]^d\,|\rho|^{1/2}}\exp\left[-\frac{(y-s)^T|\rho|^{-1}(y-s)}{2L_0^2\bar{C}_v^2(\tau/\tau_0)^{2(1+H)}}\right]$$
(48)

where the last term is written for mean velocity oriented parallel to the x_1 coordinate so that $s = (s,0,0)^T$, and d is dimension.

We now adopt the dimensionless coordinates

$$y' = \frac{y}{L_0} \qquad s' = \frac{s}{L_0} \qquad \tau' = \frac{\tau}{\tau_0} = \frac{\tau v_0}{L_0} \qquad <c'> = \frac{<c>}{c_0} = \frac{c}{\dfrac{M_0}{\phi}L_0^d}$$
(49)

and note that $|s'| = s' = \tau'$ (dimensionless time and dimensionless mean travel distance coincide). With (49), (48) can be recast in dimensionless form as

$$<c(y,\tau)> = \frac{1}{[(2\pi)^{1/2}\bar{C}_v\tau^{1+H}]^d|\rho|^{1/2}} \exp\left[-\frac{(y-s)^T|\rho|^{-1}(y-s)}{2\bar{C}_v^2\tau^{2(1+H)}}\right] \tag{50}$$

where primes are omitted for brevity.

When ρ is diagonal, (50) simplifies to

$$<c(y,\tau)> = \frac{1}{[(2\pi)^{1/2}\bar{C}_v\tau^{1+H}]^d(\rho_{11}\rho_{22}\rho_{33})^{1/2}} \exp\left\{-\frac{1}{2\bar{C}_v^2\tau^{2(1+H)}}\left[\frac{(y_1-\tau)^2}{\rho_{11}} + \frac{y_2^2}{\rho_{22}} + \frac{y_3^2}{\rho_{33}}\right]\right\} \tag{51}$$

in three dimensions where $d = 3$; corresponding results for $d = 2$ or 1 are obtained upon omitting terms with indeces 3 or 2 and 3.

3.2.2. Nonpoint Sources

Let us consider a nonpoint, instantaneous source of constant strength s_0 (dimensions ML^{-k}) in a box of sides $2\alpha_1,...2\alpha_d$. The resulting mean concentration is derived by integrating (51) over the box (c.f., *Zhang* 1992, 1993) to yield

$$<c(y,\tau)> = \frac{s_0}{\phi 2^d} \prod_{i=1}^{d} z_i(y_i,\tau)$$

$$z_i(y_i,\tau) = \text{erf}\left[\frac{y_i+\alpha_i-<v_i>\tau}{(2V_{ii})^{1/2}\tau}\right] - \text{erf}\left[\frac{y_i-\alpha_i-<v_i>\tau}{(2V_{ii})^{1/2}\tau}\right] \tag{52}$$

Substituting the expression of V_{ii} given by (38), considering that $<v_i> = \delta_{1i}v_0$, using (49) and defining

$$\alpha_i' = \frac{\alpha_i}{L_0} \qquad <c'> = \frac{<c>}{c_0} = \frac{c}{\dfrac{s_0}{\phi}} \tag{53}$$

yields in dimensionless form (primes omitted for brevity)

$$<c(y,\tau)> = \frac{1}{2^d} \prod_{i=1}^{d} z_i(y_i,\tau)$$

$$z_i(y_i,\tau) = \text{erf}\left[\frac{y_i+\alpha_i-\delta_{1i}\tau}{(2\rho_{ii})^{1/2}\bar{C}_v\tau^{1+H}}\right] - \text{erf}\left[\frac{y_i-\alpha_i-\delta_{1i}\tau}{(2\rho_{ii})^{1/2}\bar{C}_v\tau^{1+H}}\right] \tag{54}$$

Analogous solutions can be derived for i) a line source of length $2\alpha_2$ along x_2 in two dimensions; ii) a line source of length $2\alpha_3$ along x_3 in three dimensions; iii) a rectangular source, with sides having length $2\alpha_2$ and $2\alpha_3$ along x_2 and x_3 respectively, in three dimensions. Defining the dimensionless concentration as $<c'> = <c>/c_0 = c/[(s_0)/(L_0\phi)]$ in cases i) and iii), and as $<c'> = <c>/c_0 = c/[(s_0)/(L_0^2\phi)]$ in case ii), yields for (primes omitted for brevity)

i) Line source in two dimensions:

$$<c(y,\tau)> = \frac{1}{(2)^{3/2}(\pi)^{1/2}(\rho_{11})^{1/2}\,\bar{C}_v\,\tau^{1+H}}\exp\left[-\frac{(y_1-\tau)^2}{2\rho_{11}\bar{C}_v^2\,\tau^{\,2(1+H)}}\right]z_2(y_2,\tau) \tag{55}$$

ii) Line source in three dimensions:

$$<c(y,\tau)> = \frac{1}{4\pi(\rho_{11}\rho_{22})^{1/2}\,\bar{C}_v^2\,\tau^{2(1+H)}}\exp\left\{-\frac{1}{2\rho_{11}\bar{C}_v^2\,\tau^{\,2(1+H)}}\left[\frac{(y_1-\tau)^2}{\rho_{11}}+\frac{y_2^2}{\rho_{22}}\right]\right\}z_3(y_3,\tau) \tag{56}$$

iii) Rectangular source in three dimensions:

$$<c(y,\tau)> = \frac{1}{4\pi(\rho_{11}\rho_{22})^{1/2}\,\bar{C}_v^2\,\tau^{2(1+H)}}\exp\left\{-\frac{1}{2\rho_{11}\bar{C}_v^2\,\tau^{\,2(1+H)}}\left[\frac{(y_1-\tau)^2}{\rho_{11}}+\frac{y_2^2}{\rho_{22}}\right]\right\}z_2(y_2,\tau)\,z_3(y_3,\tau) \tag{57}$$

For a uniform instantaneous nonpoint source with constant concentration $c_0 = s_0/\phi$, the concentration variance can then be computed according to (*Dagan*, 1989, p. 285)

$$\sigma_c^2 = <c>(c_0 - <c>) \tag{58}$$

3.2.3. *On Dispersion in Fractal Log Permeability Fields*

Under what circumstances does the velocity field form a random fractal with a power-law cross-covariance such as in (38)? Consider a statistically homogeneous and isotropic Gaussian (natural) log permeability field with variance σ_Y^2 and an exponential autocovariance function. Then, for mean uniform flow parallel to the x_1 coordinate, the velocity cross covariance is, to first order in σ_Y^2, diagonal with nonzero components (*Dagan*, 1989, p.256-257)

$$V_{11} \approx \sigma_Y^2 \, K_g^2 \frac{J^2}{\phi^2} \approx \sigma_Y^2\, v_0^2 \tag{59}$$

in one dimension,

$$V_{11} \approx \frac{3}{8}\sigma_Y^2\, K_g^2 \frac{J^2}{\phi^2} \approx \frac{3}{8}\sigma_Y^2\, v_0^2 \qquad V_{22} \approx \frac{1}{8}\sigma_Y^2\, v_0^2 \tag{60}$$

in two dimensions, and

$$V_{11} \approx \frac{8}{15}\sigma_Y^2\, v_0^2 \qquad V_{22} = V_{33} \approx \frac{1}{15}\sigma_Y^2\, v_0^2 \tag{61}$$

in three dimensions. Here v_0 is the magnitude of the mean velocity, V_{11} is the variance of the longitudinal velocity component v_1, and V_{22}, V_{33} are the variances of the transverse components v_2, v_3, respectively.

Consider now a log permeability field with homogeneous increments characterized by a power variogram $\gamma(h) = c\,(s/L_0)^{2H}$ where c is a constant and $0 < H \le 1/2$. We recall that any such field can be viewed as the superposition of an infinite hierarchy of mutually uncorrelated, statistically homogeneous fields characterized by exponential autocorrelation functions. Hence, for $\gamma(h) < 1$, the variances of velocities encountered by a traveling particle are given to first order by

$$V_{11}(h) \approx c \, v_0^2 \left(\frac{h}{L_0} \right)^{2H} \tag{62}$$

in one dimension,

$$V_{11}(h) \approx \frac{3}{8} c \, v_0^2 \left(\frac{h}{L_0} \right)^{2H} \qquad V_{22}(h) \approx \frac{1}{8} c \, v_0^2 \left(\frac{h}{L_0} \right)^{2H} \tag{63}$$

in two dimensions, and

$$V_{11}(h) \approx \frac{8}{15} c \, v_0^2 \left(\frac{h}{L_0} \right)^{2H} \qquad V_{22}(h) \approx \frac{1}{15} c \, v_0^2 \left(\frac{h}{L_0} \right)^{2H} \tag{64}$$

in three dimensions. The velocity thus behaves approximately as a fractal for mean travel distances h which satisfy $\gamma(h) < 1$.

It follows from (38), (42), (44) and (62) - (64) that, for $\gamma(h) < 1$, conditional preasymptotic dispersivity is a diagonal tensor with scale-dependent components given to first order by

$$\alpha_{11}(h) \approx c(1 + H)L_0 \left(\frac{h}{L_0} \right)^{1 + 2H} \tag{65}$$

in one dimension,

$$\alpha_{11}(h) \approx \frac{3}{8} c(1 + H)L_0 \left(\frac{h}{L_0} \right)^{1 + 2H} \qquad \alpha_{11}(h) \approx \frac{1}{8} c(1 + H)L_0 \left(\frac{h}{L_0} \right)^{1 + 2H} \tag{66}$$

in two dimensions, and

$$\alpha_{11}(h) \approx \frac{8}{15} c(1 + H)L_0 \left(\frac{h}{L_0} \right)^{1 + 2H} \qquad \alpha_{22}(h) = \alpha_{33}(h) \approx \frac{1}{15} c(1 + H)L_0 \left(\frac{h}{L_0} \right)^{1 + 2H} \tag{67}$$

in three dimensions. Here α_{11} is the longitudinal dispersivity parallel to v_1, and α_{22}, α_{33} are transverse dispersivities parallel to v_2, v_3, respectively.

Equating (62) - (64) with (38) when the velocity dyadic is diagonal yields

$$b_0 = c \, v_0^2$$
$$\rho_{11} = 1 \qquad (1\text{-D})$$
$$\rho_{11} = \frac{3}{8}; \qquad \rho_{22} = \frac{1}{8} \qquad (2\text{-D}) \tag{68}$$
$$\rho_{11} = \frac{3}{8}; \qquad \rho_{22} = \rho_{33} = \frac{1}{15} \qquad (3\text{-D})$$

Note that the first expression in (68) implies $c = \bar{C}_v$.

446

3.2.4. *Computational Results*

We present below computational results in which $\bar{C}_v$ and H vary while v_0 and L_0 remain constant. Figure 14 depicts breakthrough curves of dimensionless mean concentration versus dimensionless time at a dimensionless distance $y_1 = 10$ downstream from the center of a line source of unit dimensionless lentgh in one dimension (top row), a square source of unit dimensionless area in two dimensions (middle row), and a cube source of unit dimensionless volume in three dimensions (bottom row) for $H = 0.10$ (left column) and $H = 0.25$ (right column). The concentration is seen to exhibit a peak which diminishes, and is shifted to the left, as $\bar{C}_v$ increases. This is so because an increase in $\bar{C}_v$ implies an increase in velocity variance and dispersion. As the dimensionality, d, of the domain increases from 1 to 3 the peak remains virtually unaffected when $H = 0.10$ but diminishes slightly when $H = 0.25$ (it decreases significantly in both cases for a point source, which we do not illustrate), the breakthrough curve narrows, and the effect of $\bar{C}_v$ becomes more pronounced. The peak diminishes and the breakthrough curve generally widens as H increases from 0.10 to 0.25. This again happens because variance and dispersion increase with H.

Though we don't show this, the effects of $\bar{C}_v$ and H become more pronounced as the size of the source diminishes and as one moves closer to it; an increase in source size causes the peak to diminish and the breakthrough curve to broaden. As one moves farther downstream from a source, its size has less and less effect on the breakthrough curve.

Figures 15 shows dimensionless concentration variances (normalized with respect to s_o/ϕ) corresponding to each mean concentration breakthrough in Figure 14. The two sets of curves are qualitatively very similar.

Figure 16 shows dimensionless mean concentration breakthroughs at a short dimensionless distance $y_1 = 0.1$ downstream from a point source where the early time Batchelor solution applies. The figure compares results in a fractal velocity field (f) with those in a statistically homogeneous velocity field (r) for $\bar{C}_v = 0.75$ and 1.50. The peak mean concentration is seen to be higher for the fractal velocity field than for the homogeneous field. Contrary to later time, at this early time the peak is higher when $H = 0.25$ than when $H = 0.10$, and increases with dimensionality. At both early and late times an increase in $\bar{C}_v$ tends to smear out the concentration peak.

4. Summary

We have shown that both the power (semi)variogram and associated spectra of a random field with homogeneous isotropic increments can be constructed as weighted integrals from zero to infinity (an infinite hierarchy) of either exponential or Gaussian variograms and spectra of uncorrelated homogeneous isotropic random fields (modes). We investigated mathematically the effect of filtering out (truncating) high and low frequency modes from this infinite hierarchy in one-, two-, and three-dimensional real and spectral domains. Our analytical results showed that a low-frequency cutoff renders the truncated hierarchy statistically homogeneous with a positive spatial autocovariance function that decays monotonically with separation distance in a manner not too dissimilar from that of its constituent (exponential or Gaussian) modes.

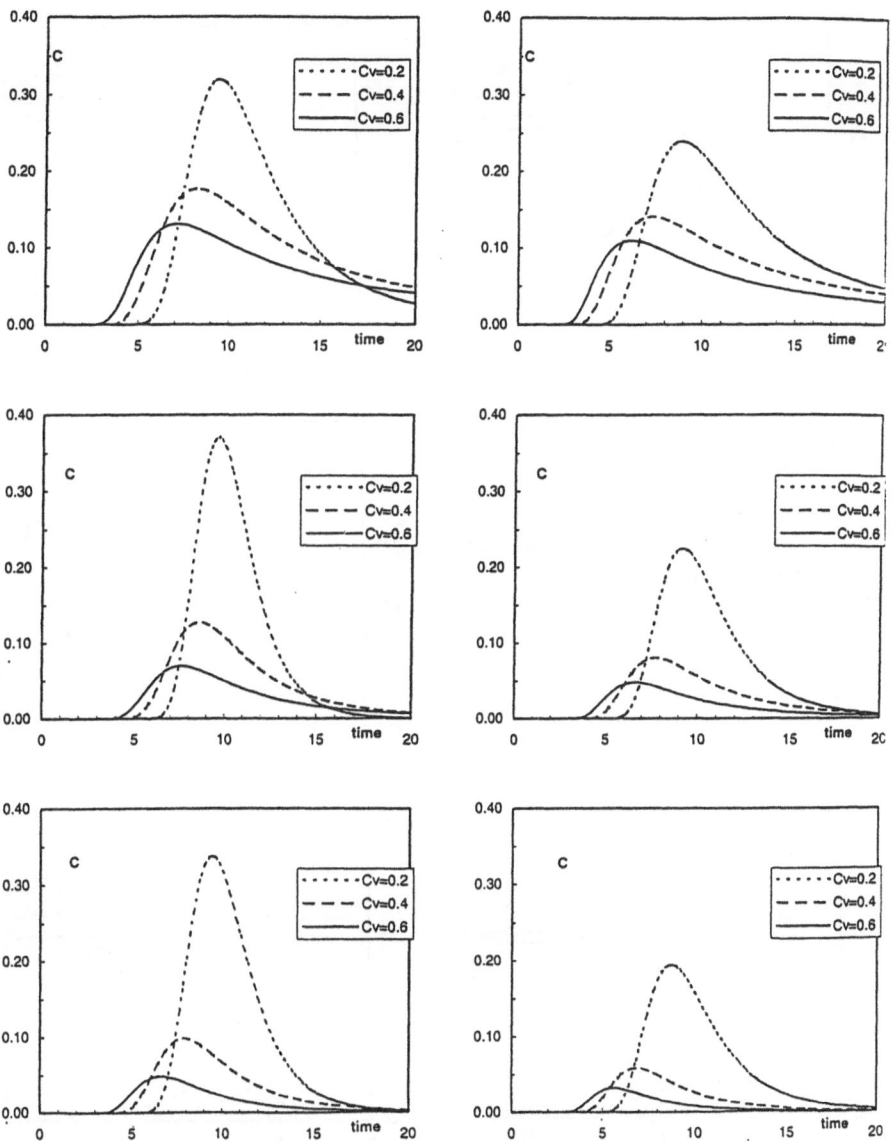

Figure 14. Dimensionless mean concentration versus dimensionless time at dimensionless distance $y_i = 10$ downstream from a nonpoint source of unit length (1-D, top row), area (2-D, middle row), and volume (3-D, bottom row) for $H = 0.10$ (left column) and $H = 0.25$ (right column).

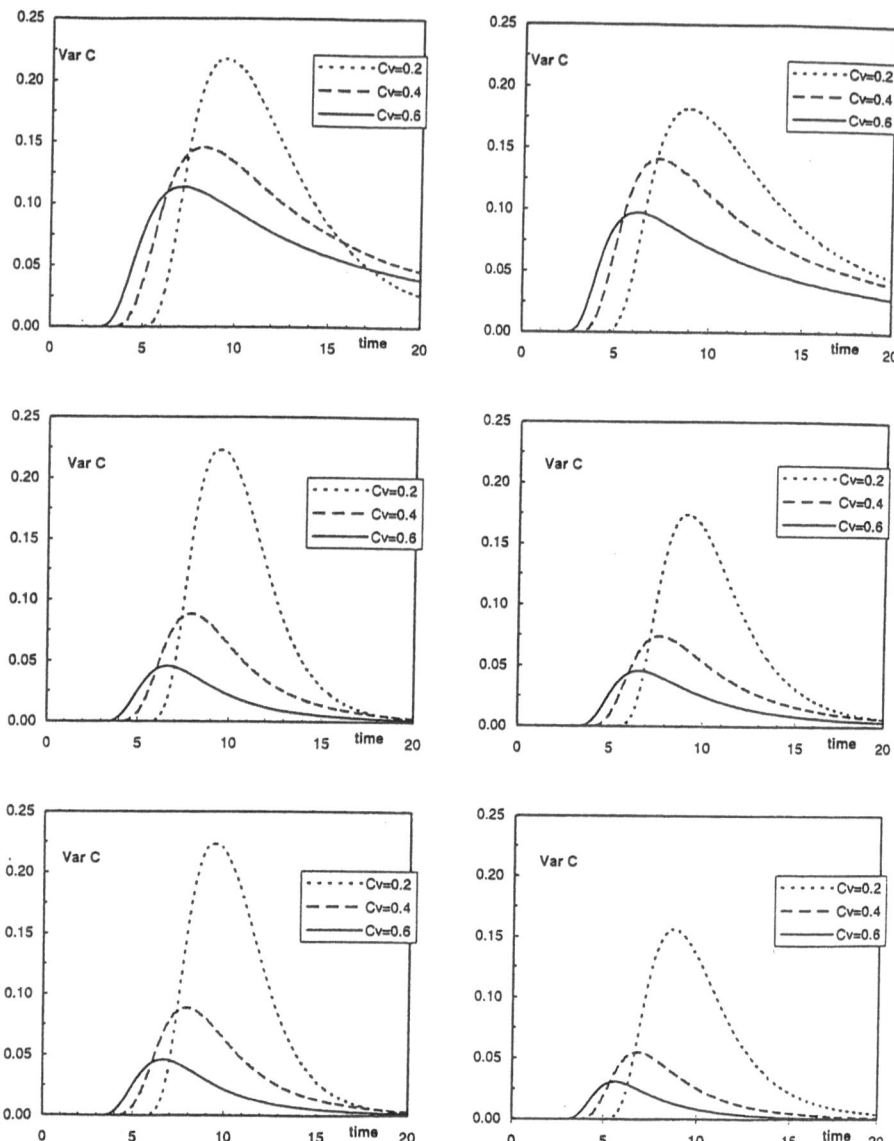

Figure 15. Dimensionless concentration variance versus dimensionless time at dimensionless distance $y_1 = 10$ downstream from a nonpoint source of unit length (1-D, top row), area (2-D, middle row), and volume (3-D, bottom row) for $H = 0.10$ (left column) and $H = 0.25$ (right column).

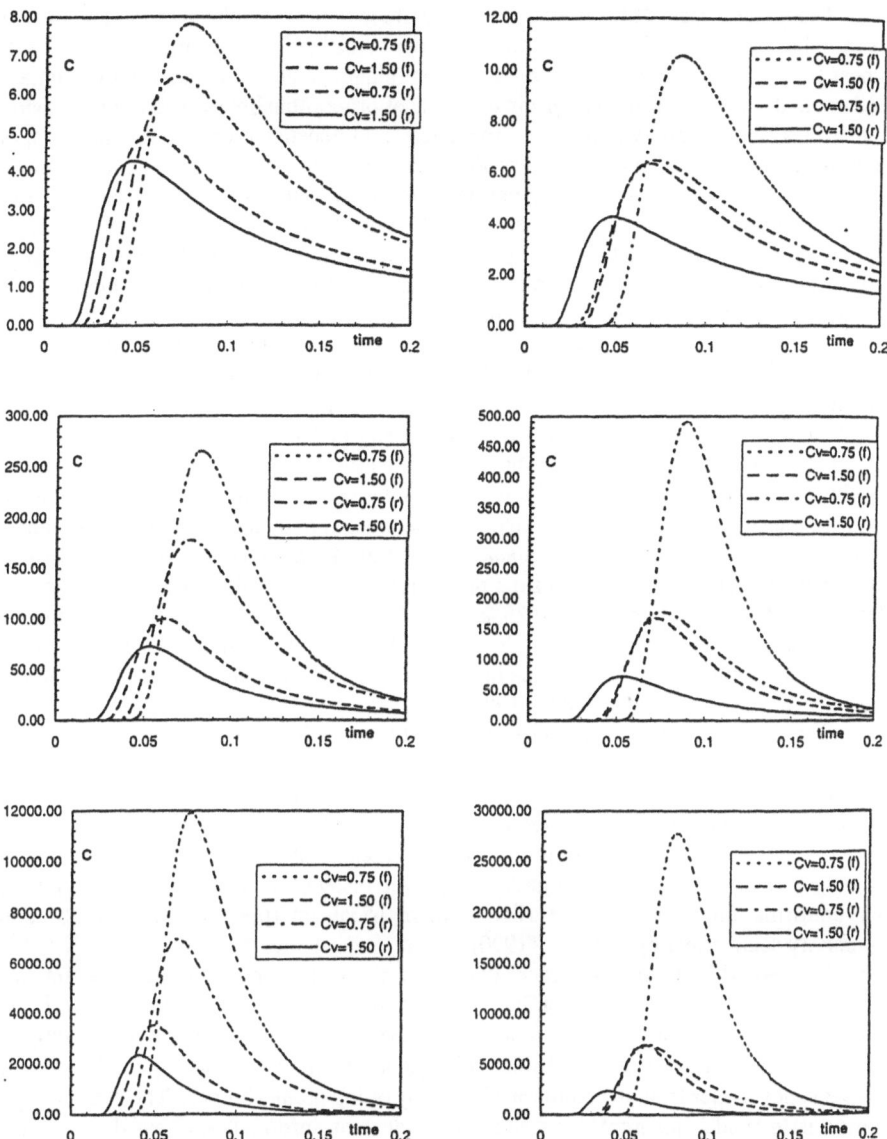

Figure 16. Dimensionless mean concentration versus dimensionless time at dimensionless distance $y_1 = 0.1$ downstream from a point source (1-D top row, 2-D middle row, 3-D bottom row) for homogeneous (r) and fractal (v) velocity fields with $H = 0.10$ (left column) and $H = 0.25$ (right column).

The integral scale of the lowest frequency mode (cutoff) is related to the length scale of a sampling window defined by the domain under investigation. The integral scale of the highest frequency mode (cutoff) is related to the length scale of data support (volume of measurement). Taking each relationship to be one of proportionality renders our expressions for the integral scale and variance of a truncated field dependent on window and support scales in a manner consistent with observations. In particular we predict, and confirm on the basis of field hydraulic conductivity and transmissivity data, that the integral scale of a truncated multiscale hierarchy increases linearly with the length scale of a superimposed window (domain of investigation) provided that the latter is much larger than the length scale of data support. We likewise predict and confirm that the variance of a hierarchy so truncated increases as a power $2H$ of window (domain) scale.

Our hierarchical theory allows bridging across scales by predicting the effect of viewing a multiscale random field, defined and measured on a given support scale, through a larger window defined by the domain under investigation. At a given locale, one can in principle calibrate a truncated variogram model to sample data observed on a given support scale in one domain, then predict the autocovariance structure of the corresponding field in domains that are either smaller or larger. One may also venture (with lesser predictive power) to transfer such predictions from one locale to another by introducing the generalized variogram parameters (derived by *Neuman* (1990, 1994, 1995) and ourselves on the basis of juxtaposed hydraulic and tracer data from many sites) $C \approx 0.0076$ for exponential modes, $C \approx 0.0117$ for Gaussian modes, $H \approx 0.25$, and $\mu \approx 1/3$ into our truncated variogram models. Selection between alternative models is possible in principle by means of formal model discrimination criteria, coupled with methods such as maximum likelihood cross-validation, as proposed by *Samper and Neuman* (1989).

Our hierarchical theory allows solving problems of flow and transport in multiscale random log permeability fields. We presented expressions for effective hydraulic conductivities in such fields under uniform mean flow, and first-order preasymptotic as well as asymptotic longitudinal and transverse macrodispersivities under advective transport. Our theory predicts that apparent longitudinal dispersivities, derived from tracer data by means of standard Fickian models, should grow as a power $1+2H$ of mean travel distance or time. This confirms theoretically an earlier semiempirical interpretation of the observed dispersivity scale effect by *Neuman* (1990).

Juxtaposed hydraulic and tracer data from many sites and geologic environments, world-wide, seem to exhibit generalized scaling behaviors that are captured well, and consistently, by our theory. There is evidence in the tracer data that accounting explicitly for site-specific medium heterogeneity by conditioning is tantamount to filtering out large-scale, low-frequency modes from the multiscale hierarchy of log permeabilities. The effect of such filtering is to render apparent dispersivity dependent on information about the advective velocity field. As the quantity and quality of this information increases, the rate at which apparent dispersivity increases with scale goes down.

A theoretical analysis was presented of preasymptotic advective transport in a steady state random velocity field with homogeneous increments. Expressions were derived for mean concentration in such a field due to instantaneous point and nonpoint sources, and both the mean concentration and its variance were illustrated graphically. As the mean and variance of the underlying fractal velocity field are undefined, the theory is conditioned on knowledge of the velocity at some point x_o. If a tracer is introduced at another point y_o, then its conditional mean dispersion is local at all times. Its conditional mean concentration and variance are given explicitly by well-established expressions. Once the conditional mean travel distance h of the tracer becomes large compared to the distance between y_o and x_o, the corresponding dispersion and dispersivity tensors grow in proportion to h^{1+2H} where $0 < H < 1$. This supralinear rate of growth is similar to that we have established by means of an unconditional theory of transport in a truncated fractal hydraulic conductivity field; both are consistent with juxtaposed apparent longitudinal dispersivities obtained by standard methods of interpretation from tracer behavior observed in a variety of geologic media under varied flow and transport regimes.

A self-affine natural log permeability field gives rise to a self affine velocity field while h is sufficiently small to insure that the variance of the log permeabilities, which grows as a power of h, remains nominally less than one.

5. References

Ababou, R. (1988) *Three-Dimensional Flow in Random Porous Media*, Ph.D. dissertation, Ralph Parsons Laboratory, MIT, Cambridge, Massachusetts.

Batchelor, G.K. (1949) Diffusion in a field of homogeneous turbulence, I, Eulerian analysis, *Aust. J. Sci. Res. A.* **2**(4), 437-450.

Batchelor, G.K. (1952a) Diffusion in a field of homogeneous turbulence, II, The relative motion of particles, *Proc. Cambridge Philos. Soc.* **48**, 345-363.

Batchelor, G.K. (1952b) The effect of homogeneous turbulence on material lines and surfaces, *Proc. R. Soc. London A* **213**, 349-366.

Bras, R. L., and Rodriguez-Iturbe, I. (1985) *Random functions and hydrology*, Addison-Wesley.

Christakos, G. (1992) *Random Field Models in Earth Sciences*, Academic Press, San Diego, 1992.

Clauser, C. (1992) Permeability of Crystalline Rocks, *Eos Trans. AGU* **73**(21), 233, 237-238.

Dagan, G. (1987) Theory of solute transport by groundwater, *Ann. Rev. Fluid Mech.* **19**, 183-215.

Dagan, G. (1989) *Flow and Transport in Porous Formations*, Springer-Verlag, Berlin.

Dagan, G. (1990) Transport in heterogeneous porous formations: spatial moments, ergodicity, and effective dispersion, *Water Resour. Res.*, **26**(6), 1281-1290.

Dagan, G. (1991) Dispersion of a passive solute in non-ergodic transport by steady velocity fields in heterogeneous formations, *Jour. Fluid Mech.* **233**, 197-210.

Desbarats, A.J. (1992) Spatial averaging of transmissivity in heterogeneous fields with flow toward a well, *Water Resour. Res.* **28**(3), 757-767.

Desbarats, A.J., and Bachu, S. (1994) Geostatistical analysis of aquifer heterogeneity from the core to the basin scale: A case study, *Water Resour. Res.* **30**(3), 673-684.

Di Federico, V., and Neuman, S.P. (1997a) Scaling of random fields by means of truncated power variograms and associated spectra, *Water Resour. Res.* **33**(5), 1075-1085.

Di Federico, V., and Neuman, S.P. (1997b) Flow in multiscale log conductivity fields with truncated power variograms, *Water Resour. Res.* (Under review).

Di Federico, V., and Neuman, S.P. (1997c) Transport in multiscale log conductivity fields with truncated power variograms, *Water Resour. Res.* (Under review).

Dykaar, B.B., and Kitanidis, P.K. (1992) Determination of the effective hydraulic conductivity for heterogeneous porous media using a numerical spectral approach. 2. Results, *Water Resour. Res.* **28**(4), 1167-1178.

Gelhar, L. W. (1993) *Stochastic Subsurface Hydrology*, Prentice Hall, N.Y.

Gelhar, L. W., Welty, C., and Rehfeldt, K.R. (1992) A critical review of field-scale dispersion in aquifers, *Water Resour. Res.* **28**(7), 1955-1974.

Guzman, A., and Neuman, S.P. (1996) Field air injection experiments, pp. 52 - 94 in Rasmussen, T.C., Rhodes, S.C., Guzman, A., and Neuman, S.P. 1996. *Apache Leap Tuff INTRAVAL Experiments, Results and Lessons Learned*, NUREG/CR-6096, U.S. Nuclear Regulatory Commission, Washington, D.C.

Mandelbrot, B. B., and Van Ness, J. W. (1968) Fractional Brownian motions, fractional noises and applications, *SIAM Rev.* **10**, 422-437.

Matheron, G. (1967) *Elements Pour une Theorie des Millieux Poreux*, 166 pp., Masson et Cie, Paris, France.

Molz, F. J., and Boman, G. K. (1995) Further evidence of fractal structure in hydraulic conductivity distribution, *Geophysical Research Letters* **22**(18), 2545-2548.

Neuman, S. P. (1990) Universal scaling of hydraulic conductivities and dispersivities in geologic media, *Water Resour. Res.* **26**(8), 1749-1758.

Neuman, S.P. (1993a) Eulerian-Lagrangian theory of transport in space-time nonstationary velocity fields: Exact nonlocal formalism by conditional moments and weak approximations, *Water Resour. Res.* **29**(3), 633-645.

Neuman, S.P. (1993b) Comment on "A Critical Review of Data on Field-Scale Dispersion in Aquifers" by L.W. Gelhar, C. Welty and K.R. Rehfeldt, *Water Resour. Res.* **29**(6), 1863-1865.

Neuman, S. P. (1994) Generalized scaling of permeabilities: validation and effect of support scale, *Geophysical Research Letters.* **21**(5), 349-352.

Neuman, S. P. (1995) On advective transport in fractal velocity and permeability fields, *Water Resour. Res.* **31**(6), 1455-1460.

Neuman, S.P., and Orr, S. (1993) Prediction of steady state flow in nonuniform geologic media by conditional moments: Exact nonlocal formalism, effective conductivities, and weak approximation, *Water Resour. Res.* **29**(2), 341-364.

Neuman, S.P., Tartakovsky, D., Wallstrom, T.C., and Winter, C.L. (1996) Correction to "Prediction of steady state flow in nonuniform geologic media by conditional moments: Exact nonlocal formalism, effective hydraulic conductivities, and weak approximation" by S.P. Neuman and S. Orr, *Water Resour. Res.* **32**(5), 1479-1480.

Paleologos, E.K., Neuman, S.P., and Tartakovsky, D. (1996)   Effective hydraulic conductivity of bounded, strongly heterogeneous porous media, *Water Resour. Res.* **32**(5), 1333-1341.

Samper, F. J., and Neuman, S. P. (1989) Estimation of spatial covariance structures by adjoint state maximum likelihood cross validation, 1, Theory, *Water Resour. Res.* **25**(3), 351-362.

Tartakovsky, D.M. and S.P. Neuman (1997a) Transient flow in bounded randomly heterogeneous domains: 1. Exact conditional moment equations and recursive approximations, *Water Resour. Res.* (in press).

Tartakovsky, D.M. and S.P. Neuman (1997b) Transient flow in bounded randomly heterogeneous domains: 2. Localization of conditional moment equations and temporal nonlocality effects, *Water Resour. Res.* (in press).

Tartakovsky, D.M. and S.P. Neuman (1997c) Transient effective hydraulic conductivities under slowly and rapidly varying mean gradients in bounded three-dimensional random media, *Water Resour. Res.* (in press).

Taylor, G.I. (1921) Diffusion by continuous movements, *Proc. London Math. Soc.* **2**(20), 196-214.

Voss, R. F. (1985) Random fractals: characterization and measurement, in *Scaling Phenomena in disordered systems*, edited by R. Pynn and A. Skjeltorp, NATO ASI Ser. **133**.

Winberg, A. (1991) *Analysis of Spatial Correlation of Hydraulic Conductivity Data from the Stripa Mine*, Stripa Project Tech. Rep. 91-28, Swedish Nuclear Fuel and Waste Management Co. SKB, Stockholm.

Yaglom, A. M. (1987) *Correlation Theory of Stationary and Related Random Functions I: Basic Results*, Springer Verlag, New York.

Zhang, D., and Neuman, S.P. (1996)   Effect of local dispersion on solute transport in randomly heterogeneous media, *Water Resour. Res.* **32**(9), 2715-2723.

Zhang, Y.-K., Zhang, D., and Lin., J. (1996)   Non-ergodic solute transport in three-dimensional heterogeneous isotropic aquifers, *Water Resour. Res.* **32**(9), 2955-2963.

Zhang, D. (1992) *Some aspects of stochastic flow and transport in complex geologic media*, M.S. Thesis, Department of Hydrology and Water Resources, The University of Arizona, Tucson.

Zhang, D. (1993) *Conditional stochastic analysis of solute transport in heterogeneous geologic media*, Ph.D. Dissertation, Department of Hydrology and Water Resources, The University of Arizona, Tucson.

6. Acknowledgments

This work was supported in part by USIA-CIES, the Fulbright Scholarship Program, and the U.S. Nuclear Regulatory Commission under Contract NRC-04-95-038.

PROBABILISTIC MODELING OF GRANULAR MEDIA ANISOTROPY

G. AUVINET
Professor
Instituto de Ingeniería, UNAM, Cd. Universitaria, Apdo. Postal 70-472,
Coyoacán, C.P. 04510, México, D.F.

Abstract

Mechanical properties of granular media are strongly influenced by their structure. Sand samples prepared by different techniques present different stress-strain behavior and different sensitivity to phenomena such as liquefaction. Anisotropic mechanical behavior is a consequence of the inherent or induced anisotropic arrangement of the grains constituting the medium. The nature of geometrical anisotropy in granular media structure can be studied by direct observations on sand or gravel samples, or on physical models of these materials, but also by simulating numerically assemblies of spherical and non-spherical granules. It can be shown that geometrical anisotropy is mainly related to the position of interparticle contacts and to preferred orientation of non-spherical grains. In this paper it is shown that probability theory can be used to model these random structural features. Structural probabilistic models can then be incorporated into micro-mechanical models that have proved useful to explain some basic aspects of the macroscopic behavior of granular materials.

1. Introduction

All granular materials present an inherent anisotropy due to the effect of gravity forces during the formation process of the medium (Arthur, 1977). In the case of discs or spherical particles, contact points tend to concentrate on specific zones of the grains (Konishi, 1982). On the other hand, in materials with oblong particles, the grains tend to deposit with their longer axis perpendicular to the sedimentation direction. Orientation of tangential planes at contact points present a preferred distribution (Mitchell, 1976). Anisotropy of loose materials is especially pronounced. Mullilis (1977) and Alberro (1992) showed that susceptibility to liquefaction of sand is affected by the particular anisotropy introduced by the preparation method of the samples. Application of stresses to granular materials deeply modifies their fabric and inherent anisotropy. Biarez (1961) showed that tangential planes at contacts tend to take a position normal to the largest principal stress. Konishi (1982), Rotherburg (1989) and others verified this on a variety

G. N. Frantziskonis (ed.), PROBAMAT – 21st Century: Probabilities and Materials, 455–464.

of natural and artificial materials. Probabilistic models can be used to represent the random structural features of granular media and in particular their anisotropy. These models can be based not only on physical observations but also on numerical simulations of granular media. Structural probabilistic models can then be incorporated into micro-mechanical models representing the behavior of granular materials.

2. Numerical simulations

2.1 Spherical granules

Random assemblies of spherical granules can be simulated numerically using a simple algorithm (Auvinet, 1975 and 1977). A container is defined and grains "falling" along a random vertical axis are placed one at a time in a position that guarantees static equilibrium until a structure with a sufficient number of grains is built. The fabric of such structures is totally defined. The exact position of the center of the particles and of contact points between grains is known. The assemblies can be inspected to assess structural features such as porosity, pore-size distribution, heterogeneity due to segregation and wall effect, and geometrical anisotropy (Auvinet, 1986). The analysis of such structures confirms that the existence of a preferred vertical axis corresponding to the gravity direction leads to a conspicuous structural anisotropy, even for spherical particles. This anisotropy affects in particular the orientation of the vector normal to tangential planes at contact points (Fig 1).

As can be seen on Fig 2, for loose uniform materials, anisotropy with axial symmetry is observed. Contacts tend to concentrate on grains at a "latitude" of about 50 degrees ($\delta \approx 40^0$). In the case of materials with a wider range of particle sizes, an apparent global isotropy is observed (Fig 3). However, a different picture is obtained when grains belonging to different size fractions are considered separately. In the case of the smallest particles, contacts are mainly concentrated on the lower part of the particles (where they are required for static equilibrium) while the largest particles are covered with a "rain" of small grains (Fig 3). This type of anisotropy should be considered typical for sand samples formed by dry pluviation or slow sedimentation in water.

2.2 Non-spherical granules

Direct simulation of structures made of non-spherical granules present serious difficulties. On the contrary, it is relatively easy to obtain certain particular structures made with this type of grains by applying a linear geometrical transformation to assemblies of spheres. To attain this objective, these assemblies are "compressed" or "stretched" along one or several orthogonal directions.

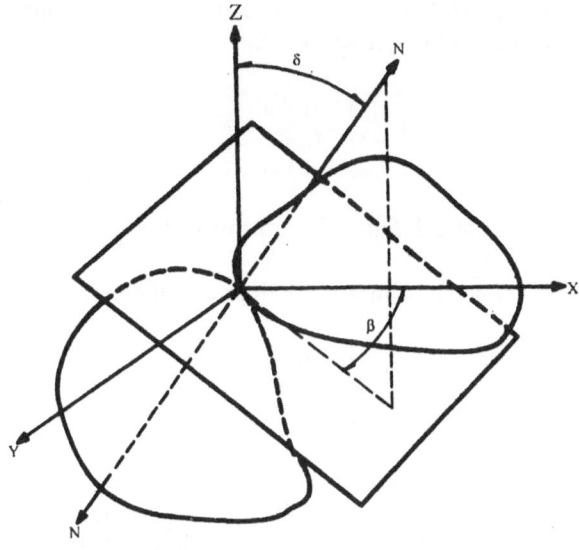

Figure 1 Tangential plane at contact

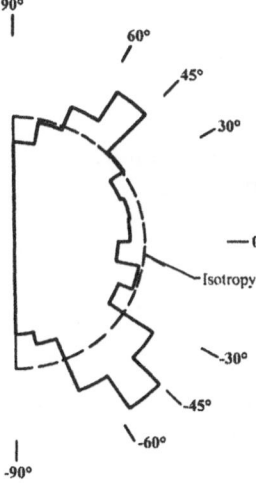

Figure 2 Anisotropic distribution of contacts on grains. Uniform material

In the reference system (1,2,3) defined by these orthogonal directions, the coordinates x_{10}, x_{20} and x_{30} of the centers of the spheres become:

$$
\begin{aligned}
x_1 &= r_1\, x_{10} \\
x_2 &= r_2\, x_{20} \\
x_3 &= r_3\, x_{30}
\end{aligned}
\tag{1}
$$

Where r_1, r_2 and r_3 are the factors defining the transformation.

Spheres with a diameter D become ellipsoids with principal diameters equal to:

$$
\begin{aligned}
D_1 &= r_1\, D \\
D_2 &= r_2\, D \\
D_3 &= r_3\, D
\end{aligned}
\tag{2}
$$

All particles are oriented along the same directions. This is obviously an extreme case of anisotropy. However, these assemblies can be expected to present strong similarities with the structure of natural granular materials such as sand and gravel of alluvial origin with flat or oblong and rounded grains.

3. Probabilistic modeling

Probabilistic models idealizing the frequency distributions of contacts observed empirically on natural or simulated materials can be used to describe the random anisotropic fabric of granular media. The distribution of orientation of vectors normal to tangential planes at contact points can be represented by a function $\Gamma(\delta, \beta)$, satisfying the condition:

$$
\int_{\Omega} \Gamma(\delta, \beta)\, d\omega = 1
\tag{3}
$$

Where
δ and β are the angles represented on Fig 1.
$d\omega$: solid elementary angle equal to $\sin\delta\, d\beta\, d\delta$
Ω : sphere with unitary radius

The quantity $\Gamma(\delta, \beta)\, d\omega$ represents the probability that the normal vector to a tangential plane at a particular interparticle contact point be located within elementary solid angle $d\omega$. For structure with axial symmetry, the distribution no longer depends on angle β and a similar function $\Gamma(\delta)$ can be introduced. This function will be such that:

$$\int_0^\pi \Gamma(\delta)\sin\delta\, d\delta = 1 \tag{4}$$

It can be easily verified that, if the distribution is uniform, then:

$$\Gamma(\delta,\beta) = \frac{1}{4\pi} \quad \text{and} \quad \Gamma(\delta) = \frac{1}{2} \tag{5}$$

Deviations from these values will mean that some anisotropy exists.

The type of probability distribution $\Gamma(\delta,\beta)$ associated to the structures made of ellipsoids described in 3.2 can be easily determined. After transformation, direction cosines u_{10}, u_{20} and u_{30} of the vector normal to the tangential plane at a given interparticle contact become:

$$u_1 = \frac{u_{10}}{r_1\sqrt{\left(\dfrac{u_{10}}{r_1}\right)^2 + \left(\dfrac{u_{20}}{r_2}\right)^2 + \left(\dfrac{u_{30}}{r_3}\right)^2}}$$

$$u_2 = \frac{u_{20}}{r_2\sqrt{\left(\dfrac{u_{10}}{r_1}\right)^2 + \left(\dfrac{u_{20}}{r_2}\right)^2 + \left(\dfrac{u_{30}}{r_3}\right)^2}} \tag{6}$$

$$u_3 = \frac{u_{30}}{r_3\sqrt{\left(\dfrac{u_{10}}{r_1}\right)^2 + \left(\dfrac{u_{20}}{r_2}\right)^2 + \left(\dfrac{u_{30}}{r_3}\right)^2}}$$

In the particular case of a transformation along one direction only, ($r_1 = 1$, $r_2 = 1$ and $r_3 \neq 1$), we have:

$$u_3 = \frac{u_{30}}{r_3\sqrt{(u_{10})^2 + (u_{20})^2 + \left(\dfrac{u_{30}}{r_3}\right)^2}} \tag{7}$$

i.e.:

$$\cos\delta = \frac{\cos\delta_0}{\sqrt{r_3\sin^2\delta_0 + \cos^2\delta_0}} = \frac{1}{\sqrt{r_3 tg^2\delta_0 + 1}} \tag{8}$$

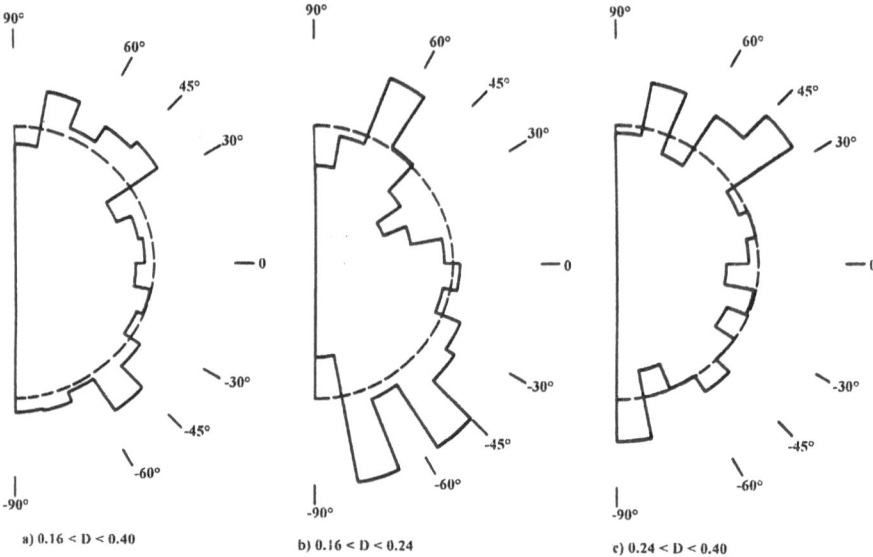

a) 0.16 < D < 0.40   b) 0.16 < D < 0.24   c) 0.24 < D < 0.40

Figure 3 Distribution of contacts on grains. Well graded material

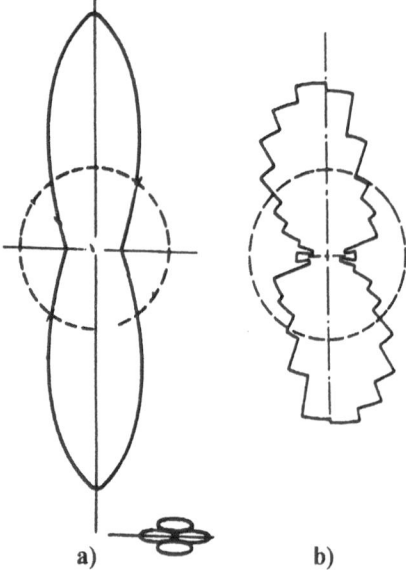

a)   b)

Figure 4 Anisotropy of non-spherical granules assemblies

Where:

δ_0 : initial angle of the normal vector to tangential plane at a given contact with the direction of the transformation

δ : same angle after deformation

The above equation allows evaluating the new distribution of the orientation of vectors normal to tangential planes at interparticle contacts. As an example, we show on Fig 4.a the modification suffered by a uniform distribution for a factor r_3 >1. In that case, a vertical cross section on the volume defined by $\Gamma(\delta,\beta)$ is presented.

The results presented are comparable to those obtained experimentally for loose materials by Mahmood (Mitchell, 1976) and shown on Fig 4.b. The distribution obtained has the same shape, including a characteristic contraction near the origin. It is worth noting that this distribution hardly fits an ellipsoid and that it cannot be represented, as was proposed by several authors, by a second order tensor. The type of anisotropy presented on Fig 4 should be considered typical for sands with non-spherical particles deposited by dry pluviation or slow sedimentation in water.

4. Micro-mechanical modeling

Probabilistic models representing geometrical anisotropy have been included in analytical micro-mechanical models (Auvinet and Cambou, 1987). Another approach consists of modeling the mechanical behavior of granular materials by simulating numerically the successive transformations suffered by their structural parameters, including distribution $\Gamma(\delta,\beta)$, as a consequence of progressive external loading. This is an attractive alternative to time-consuming explicit simulation of interactions between particles using discrete element method. A model of this type has been developed for 2D materials (discs; Auvinet, 1986) and is being extended to 3D granular media. The main features of the model in its present state are the following:

a) Probability distribution $\Gamma(\delta,\beta)$ of the orientation of the vector normal to tangential planes at contact points is assumed to be known. This distribution represents the inherent anisotropy of the material and can be assessed from the results presented in 3, or from those obtained experimentally by other authors.

b) Contacts forces vectors are assumed to form a random angle ρ with the vector normal to the tangential planes at interparticle contacts. The probability density of angle ρ is assumed to be Gaussian. The unknown standard deviation of this angle is a parameter of the model. The expected module and orientation of the contact forces corresponding to a given value of angle δ are estimated from continuum mechanics considerations.

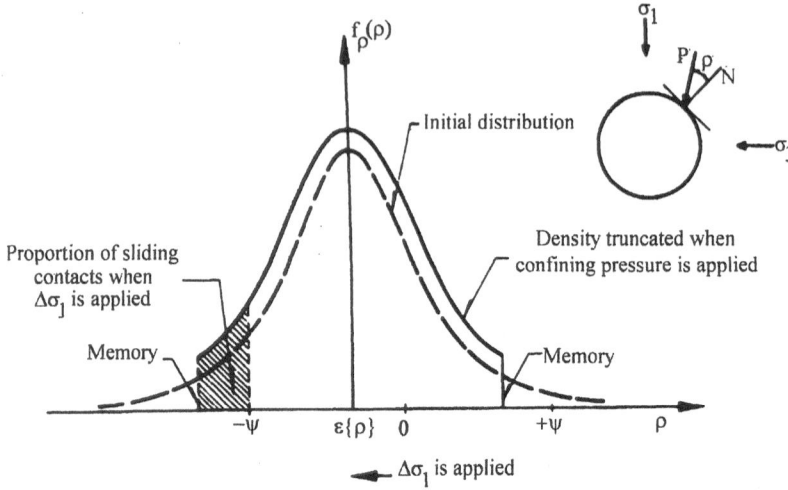

Figure 5 Probability density of orientation of contact forces.

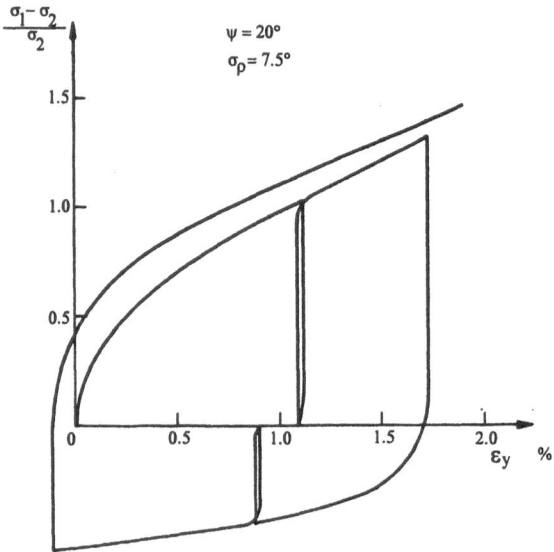

Figure 6 Numerical simulation of the mechanical behavior of 2D granular media.

c) For any given incremental isotropic or deviatoric loading, the probability of sliding due to the existence of contacts forces located out of Coulomb's friction cone is estimated for different values of angle δ, in intervals of one degree. It is admitted that sliding occurs along the tangential planes. The corresponding strains, assumed to be proportional to the probability of sliding, are calculated based on the "solid path" concept introduced by Horne (1968).

d) Probability distribution of contact forces is truncated in the direction of loading or unloading representing the forces getting out of Coulomb's friction cone. The truncation is progressively eliminated in the opposite direction, erasing the memory effect (Fig 6).

e) Simultaneously, distribution $\Gamma(\delta,\beta)$ is modified to take into account the elimination of unstable contacts and random generation of new ones. The evolution of distribution $\Gamma(\delta,\beta)$ (induced anisotropy) is thus the result of a process describing the "survival" of better-oriented contacts.

A typical stress-strain curve obtained with the model for a biaxial stress path including several loading and unloading cycles is presented on Fig 6. Stress hardening and Baushinger effect are represented satisfactorily. Parametric studies performed with this model have shown that anisotropic mechanical behavior is a consequence of both anisotropic structure and anisotropic orientation of contact forces. However, it was found that strain hardening behavior is mainly related to reorientation of contact forces. Reorientation of tangential planes only becomes significant for large deformations.

Influence of inherent structural anisotropy can be evaluated using the model by assuming different distributions $\Gamma(\delta,\beta)$ considered appropriate to represent the initial structural state of the material. The main effect of anisotropy with an unfavorable orientation is of course a reduction of initial stiffness. In the case of non-spherical grains with preferred orientation, the stress tensor leading to the largest deformability and lowest strength will be as indicated in Fig 7. This was confirmed by experimental results obtained by Konishi (1982).

An advantage of this kind of model is that it allows exploring the effects of any stress path, including those that are very difficult to follow in the laboratory.

5. Conclusion

Probabilistic modeling of granular media anisotropy can be used to describe and explain the macroscopic behavior of these materials using a micro-mechanical approach. Numerical simulation of the structure of particulate media and of their statistical behavior can be a useful complement to laboratory tests. It can be expected that the type of probabilistic simulation presented in this paper will become increasingly attractive to researchers and engineers in the future.

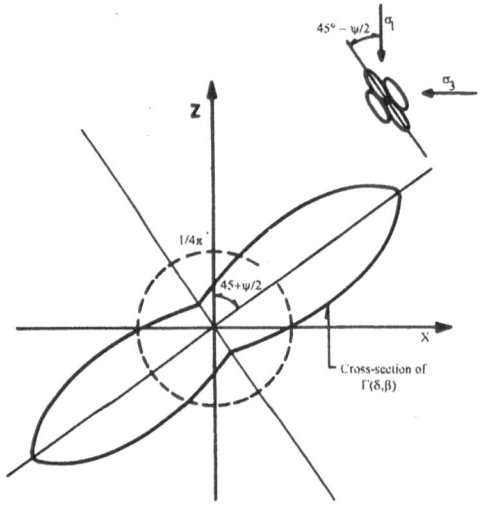

Figure 7 Critical orientation of grains in triaxial test.

6. References

1. Alberro, J. (1992) Liquefaction in sands, *R.J. Marsal Volume*, Sociedad Mexicana de Mecánica de Suelos, Mexico
2. Arthur, J.R.F. (1977) Induced anisotropy in sand, *Geotechnique*, 27
3. Auvinet, G. (1975) Generation of granular media by computer, *Proceedings, Vth Pan-American Conference on Soil Mechanics and Foundation Engineering*, vol 1, pp. 205-216, Buenos Aires, Argentina
4. Auvinet, G. (1977) Structure des milieux pulvérulents, *Proceedings, IXth International Conference on Soil Mechanics and Foundation Engineering*, Tokyo, Japan.
5. Auvinet, G. and Cambou (1987) Contribution à l'étude de la structure des milieux granulaires, *IVème Colloque Franco-Polonais de Mécanique des Sols appliquée*, 1, 1-15, Grenoble , France
6. Horne, M.R. (1965) The behaviour of an assembly of rotund rigid cohesionless particles, *Proc. Roy. Soc. A.*, Vol 286, 62-97, Great Britain
7. Konishi, J., Oda, M. and Nemat-Nasser, S. (1982) Inherent anisotropy and shear strength of assembly of oval cross sectional rods, *Proceedings, IUTAM Conference on Deformation and Failure of Granular Materials*, Delft, Holland, 403-412, Balkema.
8. Mitchell, J.K. (1976) Fundamentals of Soil behavior, Wiley, New York, USA
9. Mullilis, J.P. et al (1977) Effects of sand preparation on sand liquefaction. *Jour. Soil Mech. Found Div., ASCE*, feb., GT-2, 91-108
10. Rothenburg, L. and Bathurst, R.J. (1989) Analytical study of induced anisotropy in idealized granular materials, *Geotechnique*, 39, No 4, 601-614

A UNIQUE MODEL FOR THE SAND AND CLAY BEHAVIOR

FAVRE J. L.; BIAREZ J.; HACHI F.
Ecole Centrale Paris
Grande Voie des Vignes
92295 Châtenay-Malabry
France.

1. Introduction

Since more complex and ambitious works are built and since more powerful calculation tools are used for their dimensioning, finer behavior laws are required. The aim is to adjust the most representative models to the labor and in situ tests in order to evaluate a growing number of parameters.

But the model sophistication leads to a higher parameter definition imprecision and may give rise to serious errors on the structure behavior. Consequently, it is necessary to find the "right" models in order to minimize as much as possible the influence of the model's imprecision, or the parameter's imprecision on the structures behavior prevision. Finding a right model means to show the typical behaviors on models with few parameters on one hand, and to have the best possible tests (reference behavior or reference tests) in order to evaluate those parameters on the other hand. In this paper, we are dealing with this problem and presenting our first results.

After underlining the correlation between behavior and discontinuous medium parameters (grain distribution) and continuous medium (soil), we are showing the typical behavior of the assembly of grains (sand and clay) without connecting glue and the reference tests collected on those materials. We then describe a unique mathematical equation covering all the tests. The equation parameters are linked with the physical or soil identification parameters (discontinuous medium parameters). We are presenting as well a model error through a comparison between the reference tests and the mathematical model, and we give a definition of the systematic and hazardous measures error dues to several factors.

2. From The Discontinuous to the Continuous Medium

In order to understand the fictive continuous medium or soil behavior, it is necessary to look after the constituting bodies and their assembly. We often speak about the micro or macro behavior of materials, but for soil, we prefer to speak about continuous (MC) or discontinuous (MD) medium. We basing our self on the material grain scale (solid, liquid, and gaseous) composing the soil, and representing a discontinuous medium.

G. N. Frantziskonis (ed.), PROBAMAT – 21st Century: Probabilities and Materials, 465–479.
© 1998 *Kluwer Academic Publishers.*

The continuous medium parameters are soil rheological parameters (behavior model parameters), the discontinuous medium parameters are the behavior and boundaries conditions of their interface laws. The transition from the MD parameters to the MC parameters is done with the help of the mechanic of continuous medium method through the phase law integration in their particular boundary conditions. As for the MD we are not using behavior and boundaries conditions parameters but physical and soil identification parameters showing more or less directly and in detail his behavior. This leads us to a logical parameter classification and the observation of non-hazardous and coherent correlation among them. This thesis and the first research results have been already presented at the PROBAMAT event (Favre and al, 1994). You can synthesize this theory with the following equation :

Nature (MD) + Compacity (MD) = Rheology (MC)

For the dry or saturated media without glue (no capillary force-unsaturated soil, no cohesion –natural clays, natural clays, no natural cement formation , no artificial binding), in constraint area where there is no breaking of grains, we make the hypothesis that the grains are not loosing their shape within the assembly, that they are rather slipping than rolling one on the top of the others, and that the assembly is unshaped but without any important contact points number variation. Details have been presented during the previous PROBAMAT event (Favre and al., 1994). Those characteristics apply to sand and clay, it is therefore justified to treat them with the same logic and identical models.

The (MD) nature parameters best suited to explain the (MC) rheology are the following :
-   For clay : W_L and W_P, strongly correlated and representing only one parameter,
-   For sand : e_{min}, e_{max} and $U=d_{60}/d_{10}$, strongly related and representing only two parameters.

The compacity (MD) is taken into account under two aspects :
-   Its initial value if the material is overconsolidated, is deducted from the overconsolidation level,
-   Its instantaneous value deducted from the loading state independently from a normal or overconsolidated state.

Sands present an overconsolidated natural assembly. So, this notion must be extended to them. It is nevertheless possible to create naturally consolidated sand grain assemblies. Clays are usually consolidated or overconsolidated.

3. The typical behavior and reference tests

The reference tests (Biarez; 1993) have been obtained after the many analysis tests and the elimination of some tests or parts of tests having undergone experimental difficulties. Some simple formulations that seem enough satisfactory to give an order of magnitude for areas where there are no tests, have been added. These formulations put also in obviousness originalities of behavior laws of formed materials of elastic grains without " paste ". The proposed formulations have for goal to be typical behaviors for the usual test analysis of laboratory. It's from these typical behaviors that we can fit different models of behaviors (Biarez and al.; 1995).

3.1. NC AND OC BEHAVIOR

We can't make a choice of the typical behavior separately from that of the reference tests, because the thinking on the experimental quality has leads us to track behaviors which are not typical, and the comparison of the tests with typical behavior has permit to detect factors errors on tests, and thus to dismiss them.
First, we have choose to begin by presenting typical behavior deduced from experimental analysis.

3.1.1. NC behavior

Biarez (1995) proposes to represent the NC behavior by a reference behavior of Cam Clay type for the triaxial path for deformations $\varepsilon > 10^{-2}$, where properties depend more on the relative grain displacement, than the grain elasticity. We will retain a simplified Roscoe-Hyorsley formulation which connects void ratio e to the average pressure p ', and to $\eta = q/p$ with following parameters :

- M' or φ'_{pp} perfect plasticity friction angle
- Isotropic compression e(p')

The void ratio is a linear function of the logarithm of the pressure p' :
$$e = e_0 - C_c \, Log(p'/p'_0) \tag{1}$$
C_c(or λ_c, $C_c = 2.3 \, \lambda_c$) is the slope of this straight line e(Log p ').
A point $e_0(p_0)$ that could be :

$$\begin{cases} e_{W_L} = 2.7\dfrac{W_L}{100} \; pour \; p' = 7kPa \\[4mm] e_{W_P} = 2.7\dfrac{W_P}{100} \; pour \; p' = 1MPa \end{cases} \tag{2}$$

- Deviatoric compression e(q)→C_d or λ_d

C_d is the slope of the straight : $e = e_0 - C_d \, Log(1+\eta'/M')$.
C_d is often close to 1/3.
The deviatoric deformation is given by the Cam Clay plastic potential.

The three coefficients M', C_c, C_d are related to the grains nature: the limit liquidity W_L or the plastic limit W_P.

3.1.2. OC behavior

A not normally consolidated soil has to be characterized by the initial geometry of the arrangement of grains.

Since the isotropic hypothesis has been emitted, Biarez (1993) has suggested using the void ratio e or e_{oc}, as a unique variable of the isotropic geometry.

In practice, the overconsolidated behavior (OC) is generally defined by a scalar noted

$$OCR = \frac{P'_{ic}}{p'_i} = \frac{\text{isotropic consolidation constraint on NC path}}{\text{constraint after unloading on OC path}} \qquad (3)$$

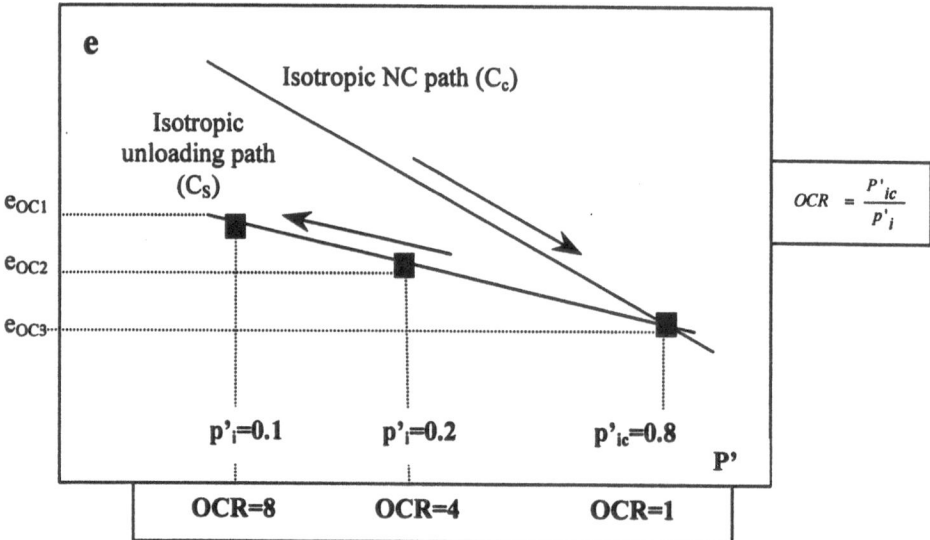

Figure 1. : Definition of isotropic overconsolidation (according to Saim; 1997)

The consolidation constraint for clays can be determined with precision, by continuing the loading to higher stresses, until the obtaining of the NC behavior.

On the other hand for usual densities of sands and aggregates, it is difficult to observe this NC behavior, because the increase of the isotropic stress, leads to the rupture of particles, with the result that we pass from a curve of consolidation to an other. Thus, we pass directly from the overconsolidated area to the particle rupture area (Biarez and Saim, 1997).

Thus to characterize the overconsolidated behavior of sands, Biarez (1993) defined a fictive consolidation pressure of the manner illustrated on the next figure:

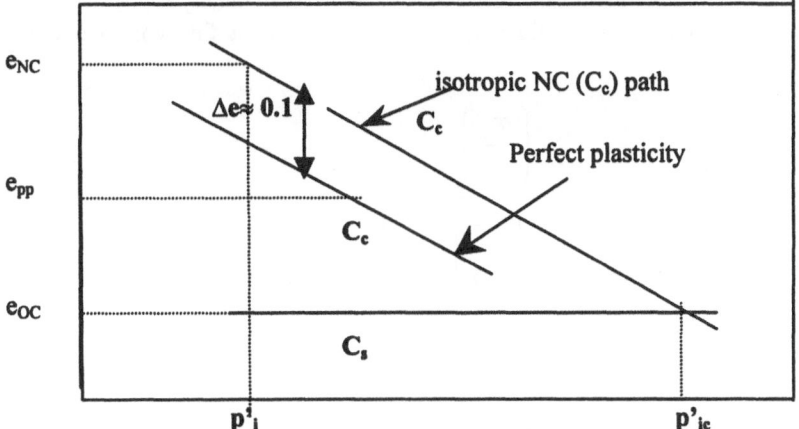

Figure 2. : Definition of fictive consolidation pressure

For a same average pressure, the perfect plasticity line is offsetted upwards of $\Delta e \approx 0.1$ to approach the curve of isotropic loading. Then from the initial state of the classic triaxial test, we extrapolate to higher stresses the isotropic path Cs, until to reach the NC path of slope Cc, the point of encounter of these two straight lines gives the pressure of consolidation p'_{ic}.

Biarez proposes to replace the overconsolidation told " classic ", expressed by the OCR and void ratio by the difference ($e_{NC}-e_{OC}$). This difference is only the distance that separates the overconsolidation state (OC) from the normally consolidated state (NC).

The relationship of Rowe

The Rowe's law applies to a pulverulent material without cohesion, the friction angle between two grains being constant and equal to φ_μ. This approach of the behavior of sands is based in particular on an energy balance analysis done during a monotonous material loading. In the case of the triaxial device this law is written as follows:

$$\frac{\sigma'_1}{\sigma'_3} = tg^2\left(\frac{\pi}{4}+\frac{\varphi_{pp}}{2}\right)\left(1-\frac{d\varepsilon_v}{d\varepsilon_1}\right) \qquad (4)$$

Where: φ'_{pp} is the perfect plasticity friction angle that corresponds to M' in the plan q (p'),

$\left(1-\dfrac{d\varepsilon_v}{d\varepsilon_1}\right)$ is the dilatancy speed generated by the grains slipping.

By making the report " peak - landing " (Biarez and Hicher, 1990), we observe that it depends on the maximal dilatancy and this, whatever is the type of overconsolidated soil:

$$\frac{\left(\dfrac{\sigma'_1}{\sigma'_3}\right)_{max}}{\left(\dfrac{\sigma'_1}{\sigma'_3}\right)_{pp}} = \left(1 - \frac{d\varepsilon_v}{d\varepsilon_1}\right)_{max} \qquad (4bis)$$

If we claim that the work consists of:

The work of the déviatoric deformation + the work of the isotropic deformation,

this leads to think that the corresponding supplementary work will give a superior mechanical resistance, we conceive therefore, that the former grows with the dilatancy, expressed here by $(d\varepsilon_v/d\varepsilon_1)_{max}$.

This dilatancy is linked to the difference $(e_{NC} - e_{OC})$ that can be retained as a coefficient characterizing similar degrees of overconsolidation for different materials, inducing curves comparable $e\ (\varepsilon_1)$ and $\varepsilon_v(\varepsilon_1)$. We will retain the next work hypothesis:

Same dilatancy in $\varepsilon_v(\varepsilon_1)$ gives same curves in $\eta'/M'\ (\varepsilon_1)$ with $\eta'/M'=q/p'$

The comparison of more of two hundred drained triaxial tests, has allowed to release reference tests (Biarez and al., 1993) used as base to judge the quality of a new test by superposing it to this set of tests, so as to reflect on possible anomalies that would present the test.

3.2. TESTS OF REFERENCE

The analysis of experimental results obtained on sands and clays, between them and to typical behaviors (NC, OC...) leads to extract the possible difference causes, that can come from :

Simplifications of typical behaviors:

- elastic   grains (bit deformable)
- friction between grains φ_μ ,
- grains without « glue » $c'<<\sigma_3$
- no role of deformation speed (time)
- initial state isotropic for the usual triaxial test

- monotonous compression paths with constant deformation speed ($\sigma_2 = \sigma_3$)
- deformation without simultaneous rotation
- constraints area $0.01 < p < 1$ Mpa

Experimental difficulties:

The hypothesis of homogeneity of constraint and deformation states is often emitted to establish laws of behavior. In general this homogeneity ceases to be valid in triaxial tests from a certain level of deformation because of:

- the restraining effect that exists at the sample extremities;
- the ratio $H/D \geq 2$ favors the appearance of kinematics discontinuities that prevent to reach the perfect plasticity, even with a H/D " small " that allows the rigid head to limit the relative displacement, one observes the birth of others types of locations;
- in general, measures of deformations in triaxial tests are not satisfactory for deformations $\varepsilon_1 < 10^{-2}$ because of :
   * the bad initial sample head with the piston,
   *sensors placed to the exterior of the device and not on the sample as in tests of precision.
- the initial void ratio e_0 is sometimes bad known, especially during the saturation of sand samples with very low density ($e \geq e_{max}$);
- no taken into account of the resistance of the membrane in tests realized with very weak lateral constraints σ_3;
- to respect the hypothesis of isotropy, it is necessary to pay attention to the mode of fabrication of the material and to detect for tests realized with weak low σ_3, if there is an initial anisotropy due to :
   * the one-dimension consolidation k_0 in a tube,
   * the dump of the sand and its compaction,
   *the secondary consolidation.

The test curves retained for the drained reference curves are:

- Tests of Flavigny and al. (1991) realized on the sand of Hostun RF ($e_{min} = 0.624$; $e_{max} = 0.96$; $\gamma_s = 25.97$ kN/m^3). A reliable system of the minimize the restraining effect for deformations $10^{-2} < \varepsilon_1 < 10^{-1}$ has been used during these tests (but this system does not allow to reach the perfect plasticity).
- Tests of Bouyard (1982) realized on the sand of Hostun ($d_{60}/d_{10} = 1.4$; $\gamma_s = 27$ kN/m^3 ; $e_{min} = 0.529$; $e_{max} = 0.813$). Bouvard has used a system allowing the homogeneous deformation development in great deformations, but can be harmful for the small deformations.
- Tests of Zervoyannis on the kaolinite ($W_L = 15$ %)
- Tests of Ladd on the simple clay ($W_L = 46\%$; $W_P = 24\%$)

- Tests of Ziani on the sand of Hostun RF ($d_{60}/d_{10}=2.6$)

The reference curves are represented on the figure 3.

Therefore, up till now, the comparison of a new test to the reference curves was made in visual manner, by superposing the test to the reference curves and in the case where it intersects the curves, this test has to be thoroughly analyzed in detail. If a curve of dilatancy $\varepsilon_V(\varepsilon_1)$ coincides approximately with a curve of the reference curves, it has to be close to the curve η'/M' (ε_1) corresponding to this dilatancy.

The aim of this paper is therefore to propose a mathematical model quantifying all these considerations. We are trying to give mathematical tools to enabling the evaluation of the quality of a test and the possible anomalies that it would present.

4. The mathematical model

First, we must establish a mathematical equation that would allow representing the reference curves. Frossard (1978) proposes a cubical function for the representations of curves $\varepsilon_V(\varepsilon_1)$. This function seems suitable to represent the curve for a certain level of deformation, but its extrapolation to the high deformations does not seem to give a perfect plasticity level.

Therefore, we are looking for an equation that reproduce in satisfactory manner the volumetric deformation curves, by respecting conditions imposed by these curves:

- A perfect plasticity level
- The contractance - dilatancy transition for the overconsolidated behavior

We suggest the following equation:

$$\varepsilon_v = \theta_0 + \theta_1.\exp(-\theta_2\,\varepsilon_1) - \theta_3.\exp(-\theta_4\varepsilon_1) \qquad (5)$$

Where $\theta_0, \theta_1, \theta_2, \theta_3, \theta_4$ are the coefficients of the model that we will specify thereafter.

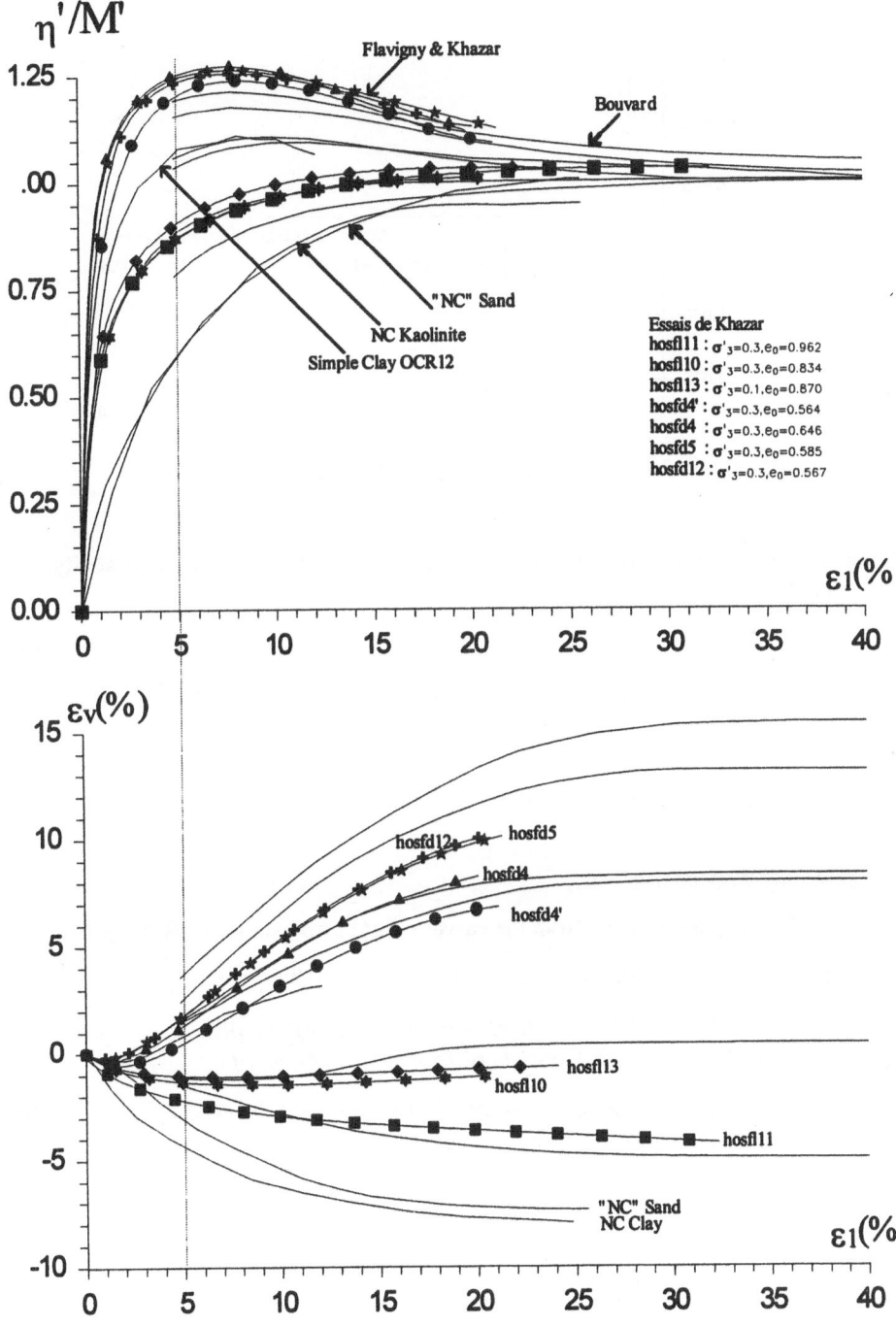

Figure 3. Reference curves (according to Biarez and al.; 1993)

474

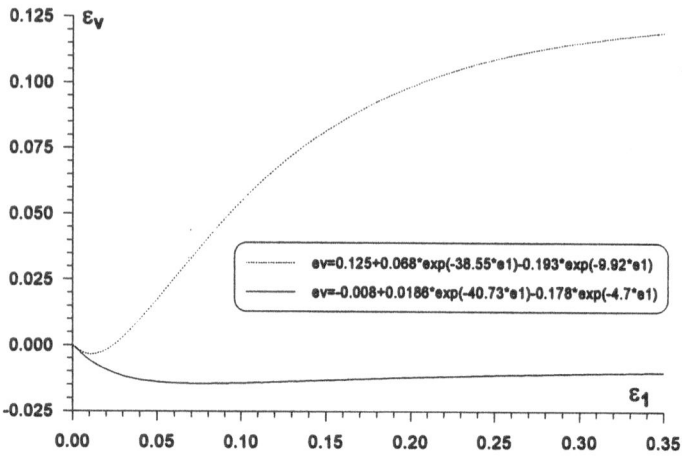

Figure 4. The dilatancy curve given by the mathematical model.

When $\varepsilon_1 \to + \infty$, $\varepsilon_v \to \varepsilon_{vpp}$ (volumetric deformation at the perfect plasticity) $\Rightarrow$ according to (5):

$$\theta_0 = \varepsilon_{vpp} \qquad (6)$$

When $\varepsilon_1 = 0$, $\varepsilon_v = 0 \Rightarrow$ according to (5): $\varepsilon_{vpp} + \theta_1 \theta_3 = 0 \Rightarrow \theta_3 = \varepsilon_{vpp} + \theta_1$ (7)

Thus the model becomes:

$$\varepsilon_v = \varepsilon_{vpp} + \theta_1 . \exp(-\theta_2 \, \varepsilon_1) - (\theta_1 + \varepsilon_{vpp}) . \exp(-\theta_4 \varepsilon_1) \qquad (8)$$

We have only three coefficients.

Curves η'/M' (ε_1) are deduced from the curves $\varepsilon_v(\varepsilon_1)$ by means of the Rowe's law.

The injection of the equation (8) in the law of Rowe gives:

$$\frac{\eta'}{M'} = \frac{3}{M'} \frac{3M' + (3 + 2M')(-\theta_1 . \theta_2 . \exp(-\theta_2 \, \varepsilon_1) + \theta_4 (\theta_1 + \varepsilon_{vpp}) . \exp(-\theta_4 \varepsilon_1))}{9 + (3 + 2M')(-\theta_1 . \theta_2 . \exp(-\theta_2 \, \varepsilon_1) + \theta_4 (\theta_1 + \varepsilon_{vpp}) . \exp(-\theta_4 \varepsilon_1))} \qquad (9)$$

Curves η'/M' (ε_1) corresponding to those of figure 4, and deduced by means of the equation (9) are on the next figure:

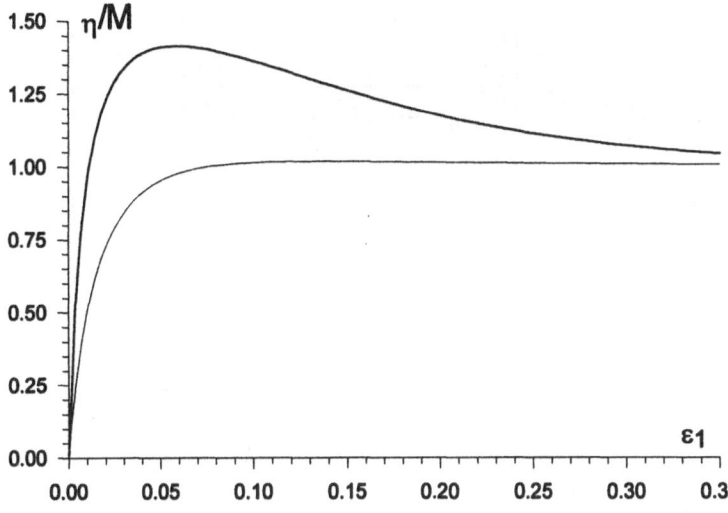

Figure 5. The deviatoric curve given by the mathematical model.

The curves η'/M' (ε_1) introduce a supplementary condition:

when $\varepsilon_1=0$, $\eta'/M'=0 \Rightarrow$ according to (9):

$$\theta_4 = \left(\theta_1\theta_2 - \frac{3M'}{3+2M'}\right)/\left(\varepsilon_{vpp} + \theta_1\right) \qquad (10)$$

and then we have only two coefficients.

The reference curves are parameterized in dilatancy, therefore the two coefficients θ_1, θ_2 would have to depend on the dilatancy.

Simulation of reference curves

So as to specify the model coefficients, we have begun by simulating tests of Flavigny because these tests are enough the most correct for the small deformations. The simulations were carried out with the help of the software SiDoLo developed at the E.N.S Cachan.

The fitting has been made on volumetric deformation curves first. Then the deviator curves have been deduced by means of the equation (9). The experience - calculation comparison, is represented on the figure (6):

These simulation leads to the coefficients showed on the table below:

Table 1. : Simulations results

Test	θ_1	θ_2	ε_{vpp}
hosfd12	0.068	38.55	0.125
hosfd4'C	0.18	23.64	0.0816
hosfd4C	0.049	43.15	0.11
hosfd5C	0.073	37.48	0.12
hosfl10C	0.0186	40.73	-0.008
hosfl11	0.0355	18.45	-0.0364
hosfl13	0.014	53.47	-0.0016

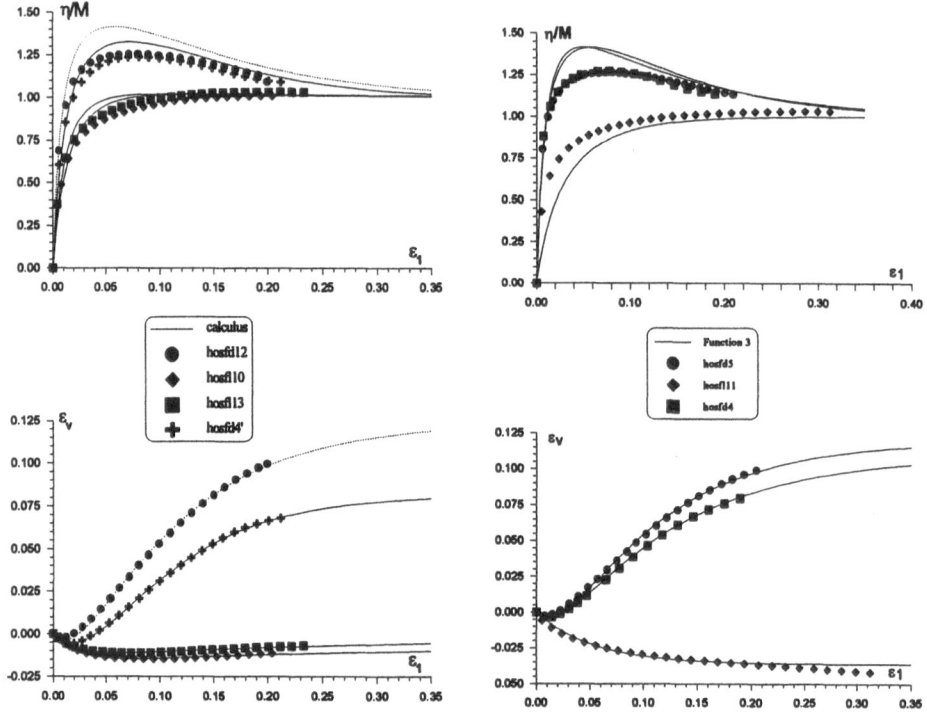

Figure 6. Simulation of Flavigny tests

According to these figures, we notice that the model is able to simulate the deformations curves fairly accuracy. On the other hand, curves η'/M' (ε_1) show peaks higher than those of tests. This is possibly due to the form of the Rowe's law that we have adopted. For the moment we keep this model formulation, set a base on which test quality controls criteria stand.

We tried to parameter our formula as a dilatancy function, i.e. to link coefficients θ_1, θ_2 as a function of dilatancy measure. This measure could be ε_{vpp} because we have already postulated that $e_{pp} - e_{OC}$ was a measure of the dilatancy.

The first calculations seems to indicate that θ_1, θ_2 are third degree polynomes in ε_{vpp}.

$$\theta_1 = 0.026 + 1.19\varepsilon_{vpp} + 27.0\varepsilon_{vpp}^2 - 283.0\varepsilon_{vpp}^3$$
$$\theta_2 = 48.0 + 251.7\varepsilon_{vpp} - 12389.1\varepsilon_{vpp}^2 + 50337,3\varepsilon_{vpp}^3$$

(11)

Therefore, we have been able to establish a model capable of simulating the Flavigny tests by giving only a measure of the dilatancy and the friction angle of perfect plasticity.

5. Model Error And Measures Error

5.1. MODEL ERRORS

The mathematical model described in this article still remains to be fine or simplified; it has been defined on the basis of reference tests, i.a. of perfect tests in the actual state of the art or with incompressible uncertainties. This mathematical equation is not a behavior model but a test representation model. But, since it gives the opportunity to adjust the models parameters, it will enable us to define different errors including model and measures errors.

Those errors are elements of two basic errors composing the total error: the code error D1 and the site error D2 (Favre, Touati, 1995), (Favre, 1996), we like to remained that the model error D11, is a code error element and corresponds to a systematic bias due to the fact that the model is not able to represent perfectly the reality and to a purely hazardous error D12, caused by the fact that perfect tests are not perfect. A third error D13, is linked with imprecision during the model integration in order to take into account the boundary problem (foundation sinking,...). Errors D11, D12 and D13 are not explicitly separable but here, D11 can be represented through the difference between the mathematical equation and the test curve simulated with the help of the behavior model, and D12 through the difference between the "perfect" behavior and the mathematical equation. Only D12 can be evaluated on the basis of the actual research. A quadratic distance measure minimizing the deviation-weighted sum between two curves has been proposed, and done (Touati 1994, Favre and Touati, 1885). It is a real variance δ_0^2. We have used here a similar notion, i.a. the correlation coefficient, but the adjustment of the mathematical model is not perfect.

5.2. MEASURES ERRORS

The measure errors D21 are not out of the two D2 site error components in addition to the uncertainty due to the natural variability D22. On one hand, the material is not automatically representative for the area (bad sampling) or for the local behavior (disturbance), and on the other hand the measures errors are due to the fact that the standard tests used for the rheological model parameters selection and for the typical behavior (mathematical equation) are showing a series of error factors in conjunction with the tests or the reference hypothesis (see 3.2.). One or more factors can be sufficiently significant to make the test absolute or to correct it following a pertinent procedure; it is in any case one of the research object. The standard test shows then systematic bias (in part of the curves) compared to the reference test or to the typical behavior. But the factors can play in average in a hazardous way (in a given contest), and result in a variance higher than D12 reference tests. Those two elements constitute the measure error. We have now the necessary means to separate them consequently to eliminate the systematic part either through the test cancellation or trough a curve correction on the basis of the reference behavior. We anticipate estimating with the help of Modelisol database (Favre and al.; 1993), the hazardous part of the measure errors and the correction rules according to the different error factor.

5.3. ERRORS ESTIMATION

Since we have only at our disposal the standard test B of the reference behavior A', and the simulation M of the adjusted model A now on A', after B deduction, we can only estimate the model (M-A) and measures (B-A) errors. They will be estimate on the typical behavior A', and the A-A' distance will be previously estimate using the reference tests data.

6. Conclusion

The typical behavior notion is very important since it provides at last an objective reference point on the soil behavior in unifying it under the grain assembly concept with or without glue. It allows appreciating a test quality, to decide to correct it or even to cancel it. It replaces the "perfect" test, highly costly for the industry and difficult to define. The perfect test uncertainty compared to the typical behavior is calculated once for all (and updated), it is then possible to estimate directly the model error and parts of the measures errors (without taking into account the representatively). This notion gives us the opportunity to lift a major obstacle in order to establish the total error starting from the different components. Up till now, it was necessary to start from the total error and define its components in order to estimate them and to select parameters in order to find the total error (Favre, 1996). Those finding are opening new perspective in the area of work reliability calculation and rheological model performance estimation.

References

J. Bairez & P. Y. Hicher. Elementary Mechanics of Soil Behavior, Saturated, Remoulded soils. A. A. Balkema / Rotterdam / Brookfield ; 1994

D. Bouvard. Rhéologie des Milieux pulvérulents : étude expérimentale et identification d'une loi de comportement. Thèse de Docteur ingénieur, Université Scientifique et Médicale et Institut Polytechnique de Grenoble, 19982

E. Flavigny, H. Bousquet, M. Djeddid, J. Lanier. Compilation des essais triaxiaux de révolution sur le sable d'Hostun RF. Rapport interne de recherche, mai 1993.

E. Frossard. Caractérisation pétrographique et propriétés mécaniques du sable. Thèse de Docteur ingénieur, Université Paris VI ; 1978

K. H. Roscoe, A. N. Schofield & C. P. Wroth. On the yielding of soils. Géotechnique, N°. 8, pp.22-53.

L. Lancelot, I. Shahrour, M. Al Mahmoud. Comportement du sable d'Hostun sous faibles contraintes. Revue Française de Géotechnique N° 74, pp.63-74, 1996.

B. Al Rechane. Palliatifs pour les difficultés expérimentales de l'appareil triaxial-Comportement type pour les sables et argiles remaniées. Thèse de docteur ingénieur, Ecole Centrale Paris, 1995

C. Zervoyannis. Etude synthétique des propriétés mécaniques des argiles et des sables sur chemins oedomètrique et triaxial de révolution. Thèse de Docteur ingénieur, Ecole Centrale Paris, 19982

F. Ziani. Contribution à l'étude du comportement des sols, cas particulier du comportement des sables très peu denses. Thèse de Docteur ingénieur, Université de Gembloux, 1987.

A COMPARISON BETWEEN "MEASURED" AND "CALCULATED" VALUES IN GEOTECHNICS

An Application to Settlements

CLAUDIO CHERUBINI,
Technical University of Bari, 70126 Bari, Italy
VENANZIO RAFFAELE GRECO
University of Calabria, 87030 Roges di Rende (Cs), Italy

1   Abstract

Computation models used in geotechnics are usually affected by error. Estimations of settlements of shallow foundations in cohesionless soils on the basis of the results of in situ tests, particularly the Standard Penetration Test, can be made by different methods. Each of the methods has been developed and/or calibrated according to specific considerations, to which can be added the specific typology of soils covered in the analysis. Only recently, efforts spent in attempts made to evaluate successfully, on the basis of data from observed settlements, the precision, accuracy and reliability of the many available computational models, have met some success in terms of acceptance by professional circles. This paper reports on statistical/probabilistic elaborations and discusses the correlations between measured and calculated settlements (or viceversa) and thereby presents circumstantial judgements about the positive aspects of each method as compared to the other methods considered in the analysis.

2   Introduction

It is often recognised that the allowable bearing capacity of shallow foundations is nearly always controlled by considerations about settlement admissibility. In cohesionless soils, however, settlement forecasting is complicated by the difficulty, and often the impossibility, to have undisturbed samples for laboratory tests. This has determined the need to use results of in situ tests, chiefly standard penetration tests (SPT) and cone penetration tests (CPT)..

Terzaghi and Peck (1948) developed the first method based on SPT results to provide a conservative empirical prediction from data of foundation settlement in sand. Modifications, aimed to improve the prediction accuracy of this method, were proposed by Gibbs & Holtz (1957), Meyerhof (1965), Bazaraa (1967), Peck et al. (1974), Meigh & Hobbs (1985).Other empirical methods based on SPT results were presented by Parry (1978), Arnold (1980), Burland & Burbidge (1985), Anagnostopoulos et al.(1991). Methods based on the elasticity theory were suggested by D'Appolonia et al. (1970), Schultze & Sherif (1973), Berardi & Lancellotta (1991), Papadopoulos (1992). Methods based on CPT results were also developed by De Beer (1965), Meyerhof (1965), Schmertmann (1970), Schmertmann et al. (1978). However, methods for predicting settlements are often used without regard to type of available data using correlation between CPT and SPT results.

All these methods assess soil compressibility through driving resistance. The fundamental shortcoming of this approach is the evident difficulty to evaluate

481

G. N. Frantziskonis (ed.), PROBAMAT – 21st Century: Probabilities and Materials, 481–498.
© *1998 Kluwer Academic Publishers.*

compressibility, a parameter connected to low level of deformation, by means of failure tests, where deformations are large. It does not surprise, therefore, that settlements measured by checking the real behaviour of structures are often considerably different from the predicted settlements. According to D'Appolonia et al. (1968, 1970), the empirical methods of Terzaghi & Peck (1948), Meyerhof (1965) & Bazaraa (1967) greatly overestimate the predicted settlements; but when corrected for overburden pressure the method of Meyerhof predicts settlements accurately. However, the method of Terzaghi & Peck (1948) was devised as a temporary expedient to provide a conservative basis for design. Therefore, it is suggested to reduce the calculated settlement by one third (Sutherland, 1974), or two third (Jorden, 1977).

Simons et al. (1974), by comparing the observed settlements of six structures with those calculated by means of eight methods, found that the method of Alpan (1964) yields the best prediction. From a comparative study, made by Talbot (1981) on a large number of foundations, it was found that the methods of D'Appolonia et al. (1970) and Parry (1971) produce the best forecasts. Another study performed by Tan & Duncan (1991), using 76 cases of measured settlements among the most popular prediction methods, showed that the methods of Alpan (1964), Schultze & Sherif (1973), Parry (1971) e D'Appolonia et al.(1970) are the most accurate ones, while those of Terzaghi & Peck (1948) and Peck et al. (1974) are largely conservative. More recently Berardi and Lancellotta (1994) made a comparative analysis, using 125 of the cases reported by Burland & Burbidge (1985), of which the methods of Burland & Burbidge (1985), D'Appolonia (1970) and Berardi & Lancellotta (1991) proved to be the most accurate while those Alpan (1964) and Terzaghi & Peck appeared to be the most conservative ones.

This paper aims at giving a further contribution in this field by comparing observed and calculated settlements for a large number of cases extracted from data collected by Burbidge (1982).

3   Methods used to predict settlements

In the present comparative analysis, the following methods have been used:
1. Method of Terzaghi & Peck (1948)
2. Method of Meyerhof (1965)
3. Method of Meigh & Hobbs (1975)
4. Method of Arnold (1980)
5. Method of Burland & Burbidge (1985)
6. Method of Anagnostopoulos, Papadopoulos & Kavvadas (1991)
7. Method of Schultze and Sherif (1973)
8. Method of Berardi & Lancellotta (1991).

Method of Meyerhof (1965) and Meigh & Hobbs (1975) are really modifications of the method of Terzaghi & Peck (1948). In all the methods, which will be recalled summarily further down, notation is used:

B (m) least width of a rectangular foundation, or diameter of a circular foundation;
B_o (m) width of one-foot-square loading plate (B_o = 0.3048 m);
D (m) depth of footing embedment below ground surface;
D_w (m) depth of ground water level below base of foundation;
E (kPa) modulus of soil stiffness;
H (m) depth of the incompressible soil;
H_s (m) thickness of the compressible layer below the foundation;

I (-) influence factor for computing settlement from elasticity theory;
L (m) length of a rectangular foundation;
N_{SPT} (-) number of blows in the SPT (corrected for grain size of soil, as suggested by Terzaghi & Peck (1948));
N_{AV} (-) N_{SPT} value averaged over the depth of influence;
N_1 N_{AV} value corrected for overburden pressure;
q (kPa) net increase of the effective pressure at foundation level;
S_c (mm) calculated settlement;
S_m (mm) measured settlement;
$\sigma'_{vo,z}$ (kPa) overburden effective pressure at depth z;
$\Delta\sigma'_{v,z}$ (kPa) increase of effective vertical pressure at depth z.

3.1   METHOD OF TERZAGHI & PECK (1948)

The predicted settlement is calculated using the following equation:

$$S_c = 3.18 \frac{q}{N_{SPT}} \left(\frac{B}{B+B_o} \right)^2 C_d C_w \qquad (1)$$

where depth coefficient, C_d, is assumed to be 1 for D = 0, 0.75 for D ≥ B (Jorden, 1977), and is linearly interpolated among these values for 0 < D < B (Pasqualini, 1983), i.e.:

$$C_d = \max \begin{cases} 1 - 0.25 \dfrac{D}{B} \\ 0.75 \end{cases} \qquad (2)$$

Water table coefficient, C_w, is assumed equal to 1 for D_w ≥ 2B, to 2 for D_w ≤ 0, and is linearly interpolated among these values for 0 < D_w < 2B, i.e.:

$$C_w = \begin{cases} 2 & \text{if } D_w \leq 0 \\ 1 & \text{if } D_w \geq 2B \\ 2 - 0.5 \dfrac{D_w}{B} & \text{otherwise} \end{cases} \qquad (3)$$

3.2   METHOD OF MEYERHOF (1965)

Settlement prediction is calculated using the following equation:

$$S_c = 2.12 \frac{q}{N_{SPT}} \left(\frac{B}{B+B_o} \right)^2 C_d \qquad (4)$$

where C_d, is calculated as in the method of Terzaghi & Peck (1948).

484

3.3   METHOD OF MEIGH & HOBBS (1975)

Settlement is calculated by following equation:

$$S_c = 3.18 \frac{q}{N_{SPT}} \left(\frac{B}{B + B_o}\right)^2 C_d \frac{4}{\dfrac{q_c}{N_{SPT}}} \tag{5}$$

where C_d, is calculated as in the method of Terzaghi & Peck (1948), and the ratio q_c/N_{SPT} depends on grain size distribution of soil.

3.4   METHOD OF ARNOLD (1980)

This method permits settlement calculation for incremental layers. If referred to a single value of N_{SPT}, the predicted settlements can be calculated by the following equation:

$$S_c = \frac{43.065 B \alpha}{\left[1 + (3.281 B)^m\right]^2} H_s \ln\left(\frac{Q}{Q - 0.5q}\right) \tag{6}$$

H_s being the depth of the compressible layer below the foundation (in m), i.e.:

$$H_s = \min(2B, H - D) \tag{7}$$

and:

$$\alpha = 0.032766 - 0.0002134 D_R \tag{8}$$
$$Q = 19.63 D_R - 263.3 \tag{9}$$
$$m = 0.788 + 0.0025 D_R \tag{10}$$

where D_R is the relative density, given by:

$$D_R(\%) = 25.6 + 20.37 \sqrt{\frac{1.26(N_{SPT} - 2.4)}{0.0208\sigma'_{vo,B+D/2} + 1.36} - 1} \tag{11}$$

Because eq.(4f) is valid for $N_{SPT} \geq 6$ and $45\% \leq D_R \leq 100\%$, the out of in this range are not utilised for the comparative analysis.

3.5   METHOD OF BURLAND & BURBIDGE (1985)

The immediate settlement is calculated as:

$$S_c = 1.71\left(q_{gross} - \frac{2}{3}\sigma'_{vo,D}\right)\frac{B^{0.7}}{N_{AV}^{1.4}}\left(\frac{1.25L}{L + 0.25B}\right)^2 \frac{H_s}{Z}\left(2 - \frac{H_s}{Z}\right) \tag{12}$$

where q_{gross} is the gross bearing pressure at foundation level (kPa), and Z is the depth of influence. If the N_{SPT} values decrease with depth, Z is equal to 2B, otherwise it has been calculated by the following interpolating equation:

$$Z = 0.933B^{0.779}$$
(13)

3.6   METHOD OF ANAGNOSTOPOULOS ET AL (1991)

With reference to the paper of these Authors settlements have been calculated using the following two equations (formula 3 and 4 of the original paper):

formula 3:

$$s_c = 2.37 \frac{q^{0.87} B^{0.70}}{N_{SPT}^{1.20}}$$
(14)

formula 4:

$$S_c \begin{cases} 0.57q^{0.94} B^{0.90} N_{SPT}^{-0.87} & \text{for} \quad 0 < N_{SPT} \leq 10 \\ 0.35q^{1.01} B^{0.69} N_{SPT}^{-0.94} & \text{for} \quad 10 < N_{SPT} \leq 30 \\ 604q^{0.90} B^{0.76} N_{SPT}^{-0.82} & \text{for} \quad N_{SPT} > 30 \end{cases}$$
(15)

3.7   METHOD OF SCHULTZE & SHERIF (1973)

The predicted settlement is calculated using the general formula:

$$s_c = \frac{qB}{E} I$$
(16)

where the factor of influence, I, has been calculated using method of Steinbrenner (1934) for rectangular footing and has been obtained by numerical integration of Boussinesq solution for circular footing. Modulus of stiffness, E (in kPa), is given by:

$$E = 1678 \left(N_{SPT}\right)^{0.87} \sqrt{B} \left(1 + 0.4 \frac{D}{B}\right)$$
(17)

3.8   METHOD OF BERARDI & LANCELLOTTA (1991)

The predicted settlement is calculated using the elasticity equation:

$$s_c = \frac{qB}{E} I$$
(18)

where the factor of influence, I, has been calculated by numerical integration of Boussinesq solution considering rigid footing. The stiffness modulus, E, is given by:

$$E = K_E P_a \sqrt{\frac{\sigma'_{vo,D+\frac{B}{2}} + \Delta\sigma'_{v,D+\frac{B}{2}}}{P_a}} \qquad (19)$$

where p_a is the reference pressure, and K_E is initially evaluated as:

$$K_E = 100 + 900 D_R \qquad (20)$$

$$D_R = \min\left\{1; \sqrt{\frac{N_1}{60}}\right\} \qquad (21)$$

$$N_1 = \frac{2}{1 + \dfrac{\sigma'_{vo,D+B/2}}{98.1}} N_{SPT} \qquad (22)$$

Once the settlement has been calculated, K_E is corrected through the equation:

$$K_{E,corr} = K_E\, 0.1912\left(\frac{S_c}{B}\right)^{-0.6248} \qquad (23)$$

and the predicted settlement is recalculated using $K_{E,corr}$ instead of K_E.

4   The probabilistic approach

Probability theory is a powerful tool for the analysis of problems presenting non-negligible sources of uncertainty, Uncertainties encountered in geotechnics are partly due to imprecise knowledge about soil properties, stress tension states, water pressures and applied loads and, partly, to deficiencies in the physical and mathematical models used in the analysis. As has often been pointed out, uncertainties about soil properties arise from three different sources:
- the natural heterogeneity of natural deposits resulting from the deposition processes;
- the limited amount of information available about soil properties;
- experimental errors, sampling disturbances, different conditions under which measurements are taken as against in situ measurements, the use of correlations for indirect evaluation (estimation) of certain parameters.

Uncertainties related to the spatial variability of geotechnical factors are an important source of inaccuracy and risk. Consequently, many researchers have felt encouraged to develop models for a statistical interpretation of such variability (Alonso & Krizek, 1975; Campanella et al., 1987). Generally, these models are constructed on the assumption that the soil is statistically homogeneous. With respect to the investigated properties, it is characterised by random fluctuations of values about a mean value, which is possibly a function of the position of the analysed point. Uncertainties about the second point can be reduced by adequately increasing the number of tests. From this point of view, statistical methods can afford a useful tool for the best choice of an exploration program, hence for avoiding redundancy of data (Veneziano & Faccioli, 1975; Kulatilake & Miller, 1987) or for the identification of areas with statistically homogenous soil in terms of geotechnical properties. A certain amount of caution has been suggested when data collected by penetration tests are interpreted using geostatistic tools such as variograms (Azzouz et al.,1987). The third group of causes for uncertainty has aroused less interest in statistical interpretations. With reference to possible experimental errors, which have been discussed in detail by Pasqualini (1983), for SPTs, and by Lancellotta (1983), for CPTs, researchers have been chiefly oriented so far to identify and correct the different sources of error rather than to analyse statically whether and how they could have influenced the calculation of settlements. Just in the same way, often the correlation of magnitudes is not worked out in a statistical form, and the presentation of envelopes is preferred instead although, at times, they may lend themselves to overconservative design. Particularly the latter two points have added more difficulties to a probabilistic analysis of settlements that would take into account all of, or at least the main, intervening uncertainties.

Recently, however, statistical methods based on experimental observation have been proposed; these methods combine various levels of probability with different values of predictable settlement. For example, the method of Burland & Burbidge (1984) provides an estimate developed according to the probability theory but based on the observation of the behaviour of 204 foundations of varying dimensions. With this approach, settlement follows a lognormal probabilistic distribution with a 0.14 coefficient of variation (C.V.). Russell & Byrne too (1987) propose a probabilistic model that is based on experimental observations and on a finite-element analysis in which the distribution of settlement seems to be characterised, in practice, by a C.V. of about 0.60. No doubt these methods benefit from the fact that they offers analysis models that are both fast and simple to use and that cover globally all the uncertainties contained in the analysis. Since they were developed starting from experimental observations, they also guarantee sufficiently good agreement with actual conditions. Still, in different cases they do not allow different weights to be attributed to the various intervening uncertainties.

5   Data used for comparison

Data for comparison have been drawn from Burbidge (1982). The data used in the analysis have been tabulated, in Appendix I together with the settlement calculated by the above mentioned methods. Because some methods require the stress state to be evaluated, the following parameters were arbitrarily assumed:

unit weight of soil above water table: $\gamma = 18.2 \text{ kN m}^{-3}$
effective unit weight of soil below water table $\gamma' = 9 \text{ kN m}^{-3}$

For cases where the position of the water table was unknown, predicted settlements were not calculated for methods requiring the affective stress state to be known. Consequently the number of data used for comparison is less of those reported by Burbidge (1982).

In the computations, the values of N_{AV} were used in the formulas of the previous section in place of N_{SPT}. An attempt was also made to examine the influence of the overburden pressure, using N_1.

TABLE I. Mean values and CV values of the S_c/S_m ratio.

Forecasting method	using N_{SPT}		using N_1	
	$\overline{R}$	CV_R	$\overline{R}$	CV_R
Terzaghi & Peck (1948)	3.488	0.747	2.942	0.649
Meyerhof (1965)	1.769	0.828	1.352	0.673
Meigh & Hobbs (1975)	1.670	0.772	1.344	0.655
Schultze & Sherif (1973)	1.204	0.599		
Arnold (1980)	1.054	0.639	0.965	0.717
Burland & Burbidge (1985)	1.689	0.663	2.603	0.892
Berardi & Lancellotta (1991)			0.866	0.614
Anagnostopoulos (formula 3)	1.812	0.609	1.659	0.810
Anagnostopoulos (formula 4)	1.282	0.782	1.308	1.190

TABLE II Mean values and CV values of the S_m/S_c ratio.

Forecasting method	using N_{SPT}		using N_1	
	$\overline{r}$	CV_r	$\overline{r}$	CV_r
Terzaghi & Peck (1948)	0.570	1.183	0.548	0.928
Meyerhof (1965)	1.465	1.380	1.379	1.130
Meigh & Hobbs (1975)	1.466	1.246	1.332	0.984
Schultze & Sherif (1973)	1.537	1.414		
Arnold (1980)	1.505	0.850	1.616	0.773
Burland & Burbidge (1985)	0.838	0.666	0.636	0.852
Berardi & Lancellotta (1991)			2.101	1.504
Anagnostopoulos (formula 3)	0.791	0.769	0.921	0.747
Anagnostopoulos (formula 4)	1.129	0.795	1.422	1.003

6   A comparison between observed and calculated settlements

Based on available data, the following method was devised to compare measured versus calculated data, and viceversa.
The ratio

$$R = \frac{S_c}{S_m} \tag{24}$$

was evaluated and a mean $\overline{R}$ and a variation coefficient CV_R were obtained, and also for the ratio:

$$r = \frac{1}{R} = \frac{S_m}{S_c} \tag{25}$$

On this basis, it appears possible to estimate which one of the methods is best according to some defined criteria.
"A perfectly accurate method of calculating settlements would be one that resulted in calculated settlements equal to the measured settlements in every case (Tan & Duncan 1991)". In a statistical sense an accurate method would be one that yielded, for a set of n values of the S_m/S_c values (or viceversa), a mean close, or equal to, one.

Similarly, a statistically precise method (Chapra and Canale 1988) would be one that, for the same set of n values of the S_m/S_c ratio, resulted in very low scatter (which would be estimated, for instance, by means of the coefficient of variation).

A perfectly reliable method of calculating settlements would be one that never resulted in estimated values of settlement that were smaller than the actual settlement. In a statistical sense, one method is more reliable than another if the frequency of the ratios, $Sm/Sc < 1$, is higher (or if the frequency of the ratios, $S_c/S_m < 1$, is lower). The term reliable involves complex concepts. We propose, for this criterion to use the term "conservative" instead of reliable.

The results given in Tables I and II and in Fig.1a,b where are plotted the cumulative frequencies of S_c/S_m and S_m/Sc show that:

- The $\overline{R}$ values closest to one (greater accuracy) relates to the method of Arnold, and the corresponding coefficient of variation is not very high either (Fig.2a,b);
- Concerning the ratio, S_c/S_m, the method of Terzaghi & Peck show a very high mean thereby indicating that when accuracy is (rather) low, "reliability" is remarkable high, in the sense indicated by Tan and Duncan.
- The same considerations as above apply whether the N_{SPT} or the N_1 values are used.
- Concerning the ratio, (Sc/Sm), the best mean obtained by using N_{SPT} is given by Anagnostoupolos et al. (expression 4), and by N_1, again by Anagnostoupolos et al. (expression 3).
- The remarkable reliability and simultaneous lack of accuracy of the method of Terzaghi & Peck are obviously confirmed.
- Concerning precision (coefficient of variation), while with the ratio , S_c/S_m, CV values were rather similar to one another, (0.6-0.828 when N_{SPT} was used), with the ratio, S_m/S_c, the CVr values are not quite similar to one another, and more precise method seems to be that Burland and Burbidge. For N_{SPT}, the highest value is the one obtained with the method of Schultze & Sherif (1973), while the lowest value is given by the method of Burland & Burbidge (1985). In Fig.2a,b are plotted, for the method of Arnold, the histogram of S_m/S_c and the scatter diagram of S_c on S_m. It can be observed that the method, together with perhaps all the methods examined in this paper, does not fit sufficiently the highest values of measured settlements.

7   Conclusions

The calculations methods discussed above are some of the most widely accepted ones. When they are used, the deviation generally occurring between observed and calculated settlements is fairly large. As a matter of fact, while the various methods succeed in estimating settlements, they do not take into due account some of the several factors acting upon the settlements and, obviously, the variability of the soil mass..

Among these factors, the following deserve attention:
- The soil's initial relative density. As this rises, the ratio between compressibility modulus and penetration resistance decreases.
- The surface roughness of soil particles that hampers their mutual displacement but makes them more apt to form more open, and thus more readily yielding, structures.
- Particle size distribution, particularly the presence of small amounts of silt or clay partly filling the voids so that compressibility is greatly reduced
- Capillary forces tending to reduce or to delay compressibility.
- The stress state, the history of strains undergone by the material, the degree of overconsolidation, the presence of the water.

490

These factors can determine important differences between settlements predicted by calculation and those actually observed. A large number of data were processed statistically: in the process, the peculiarities of the different calculation methods being compared were highlighted using the criteria of accuracy, precision and reliability. The comparison was made on the basis of the S_m/S_c, as well as the reverse, ratio.

The results obtained appear to be significant, though certainly not exhaustive. Obviously, a still larger data base is required.

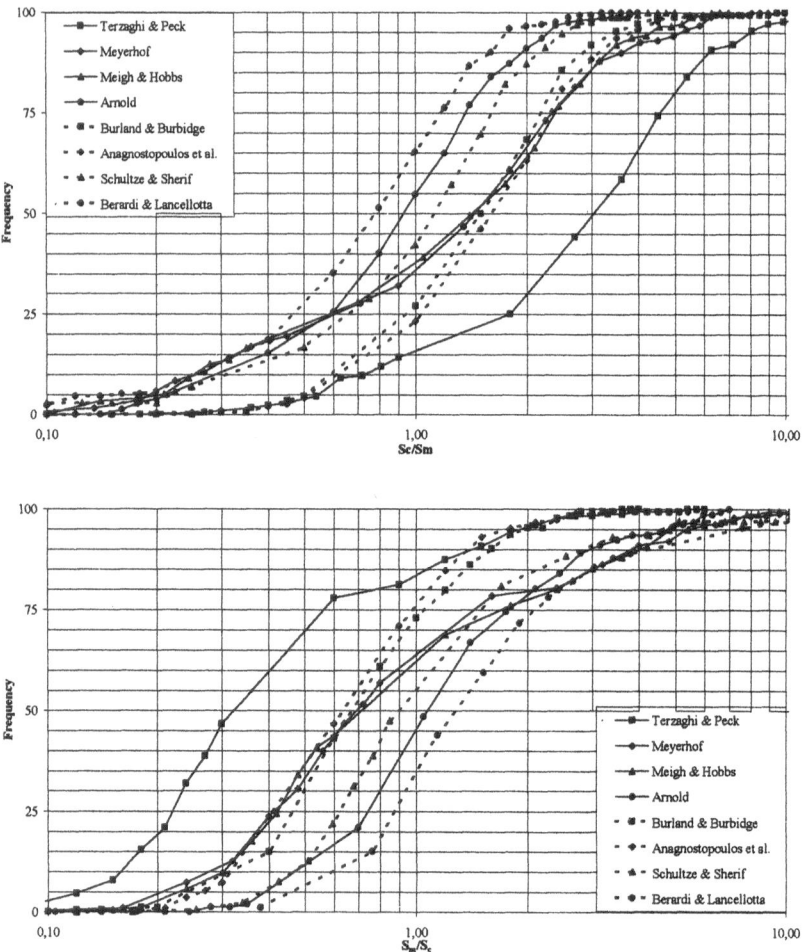

Figure 1a,b Cumulated frequencies of the S_c/S_m and S_m/S_c ratio for the methods reviewed in this paper.

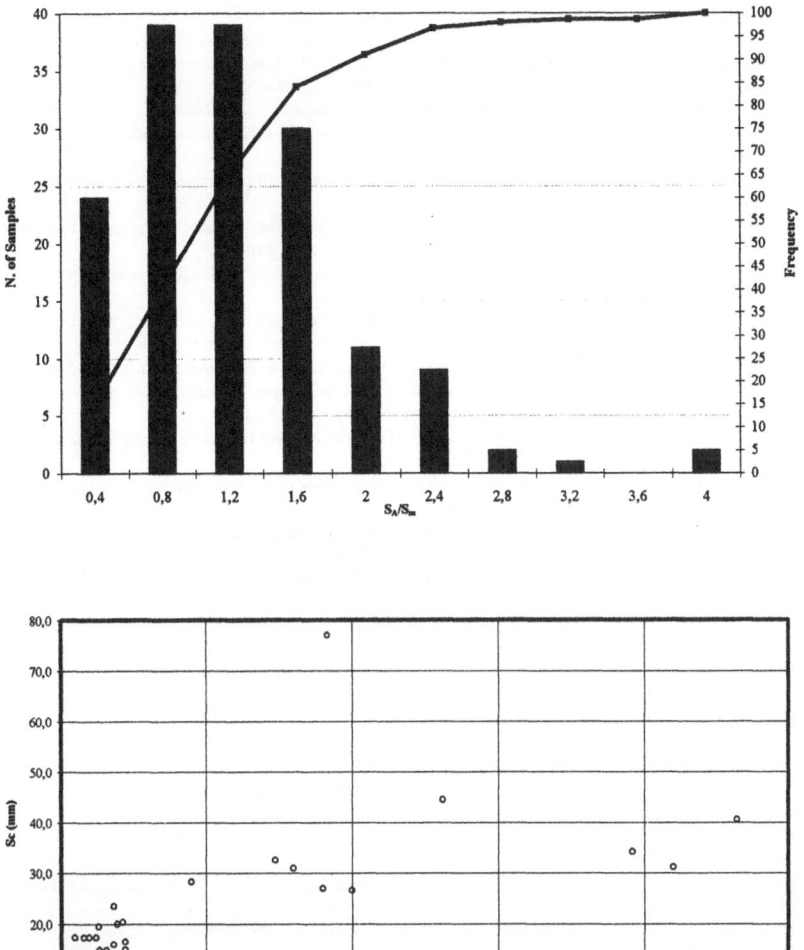

Figure 2a,b Histogram and scatter diagram relative to the method of Arnold

References

- Alpan, I. (1964). Estimating the settlements of foundations on sands. *Civil Eng. & Pub. Works Review.*
- Anagnostopoulos, A.G., Papadopoulos B.P., Kavvadas, M.J.(1991). Direct estimation of settlement on sand, based on SPT results. *Proceedings of the 10th ECSMFE. Florence. 293-296.*
- Arnold, M.(1980). Prediction of footing settlements on sand. *Ground Eng., n.2, 40-49.*
- Bazaraa, A.R.S. (1967). Use of standard penetration test for estimating settlements of shallow foundations on sand. *PhD thesis, University of Illinois, Urbana.*
- Berardi, B., Lancellotta, R. (1991). Stiffness of granular soils from field performance. *Geotecnique 41, n.1, 149-157.*
- Berardi, B., Lancellotta, R. (1994). Prediction of settlements for footings on sand: accuracy and reliability. *Proc. of Settlement '94. College Station, Texas. ASCE Geot Eng. Div., 640-651.*
- Burbidge, M.C. (1982). A case study review of settlements on granular soil. *MSc/Dissertation, Imperial College, University of London.*
- Burland, J.B., Burbidge, M.C. (1985). Settlement of foundations on sand and gravel. *Proc. Inst. Civil Eng.., part 1, 78, 1325-1381.*
- Chapra, G.C., Canale, R.P. (1988). Metodi numerici per l'ingegneria. *McGraw Hill. Libri Italia. Milano.*
- Cherubini, C., Cucchiararo, L., Orr, T.L.L. (1995). Criteria to compare calculated and observed bearing capacity of piles. *Proceedings of the ICASP 7th Conference Paris, pp.9-14.*
- D'Appolonia, D.J., D'Appolonia, E., Brissette, R.F. (1968). Settlement of spread footings on sand. *Journ. Soil Mech. Fou. Div., ASCE, 94, n.3, 735-760.*
- D'Appolonia, D.J., D'Appolonia, E., Brissette, R.F. (1970). *Closure of* Settlement of spread footings on sand. *Journ. Soil Mech. Fou. Div., ASCE, 94, n.3, 754-7601.*
- De Beer, E. (1965).Bearing capacity and settlement of shallow foundations on sand. *Proc. Symp. Bearing capacity and settlement of foundations, Duke University, 15-33.*
- Gibbs, H.J., Holtz W.G. (1967). Research on determining the density of sands by spoon penetration testing. *Proc. IV Int. Conf. S.M.F.E., London.*
- Jorden, E.E. (1980). Settlement in sand - methods of calculating and factors affecting. *Ground Eng., n.1,30-37.*
- Meigh, A.C., Hobbs, N.B. (1975). Soil Mechanics. *Section 8, Civil Engineer's Reference Book, 3rd Ed. Newnes-Butterworth, London.*
- Meyerhof, G.G. (1965). Shallow foundations. *Jour. Soil Mech. Found. Div. ASCE, 91, n.SM2.*
- Papadopoulos, B.P. (1992). Settlements of shallow foundations on cohesionless soils. *Jour. Geot. Eng. ASCE, vol.118, n.2, 377-393.*
- Parry, R.H.G. (1971). A direct method of estimating settlements in sand from SPT values. *Proc. Symp. Interaction of structures and foundations. Midlands SMFE Soc. Birmingham, UK.*
- Pasqualini, (1983). Standard penetration test. *Atti dell'Istituto di Scienza delle Costruzioni del Politecnico di Torino. Pubbl.n.608.*
- Peck, R.B., Hanson, W.E., Thorburn, T.H. (1974). Foundation Engineering. *J.Wiles & Sons, New York, 2nd Ed.*
- Schultze, F., Sherif, G. (1973). Prediction of settlements from evaluated settlement observation for sand. *Proc.8th ICSMFE, Moscow, URSS, 3, 225-230.*
- Schmertmann, J.H. (1970). Static cone to compute static settlement over sand. *Jour Soil Mech. Found. Div., ASCE, 96, n.SM3, 1011-1042.*
- Schmertmann, J.H., Hartman, J.P., Brown, P.R. (1978). Improved strain influence factor diagrams. *Jour. Geot. Eng. Div., ASCE, 104, n.GT8, 1131-1135.*
- Tan, C.K. & Duncan, J.M. (1991). Settlement of footings on sands. *Accuracy and reliability. Geot. Eng. Congress 1191. Geotech.Publ.n.27, ASCE, New York, 446-455.*
- Terzaghi, K. & Peck, R.B. (1948). Soil mechanics in engineering practice. *J.Wiles & Sons, New York, N.Y.*

8   Appendix I - Data tables

Case	N_{AV}	B	L	D	D_w	H	q_{gem}	q_{test}	S_{min}	TP	M	MH	A	BB	APK	SS	BL
													S_{calc}				
1	28	36.60	36.60	0.00	1.50	>12.20		193	18.0	42.7	14.4	12.3	15.9	36.9	52.6	10.1	16.4
2	17	12.20	12.20	1.20	7.30	>7.80	150	130	22.0	38.4	15.1	22.6	14.9	25.1	31.5	9.5	8.2
3/a	8	3.30	14.00	1.80	1.60	>25.00		52	20.0	26.3	10.0	9.2	7.4	18.8	14.0	12.3	3.1
3/b	8	3.30	14.00	1.80	1.60	>25.00		52	36.0	26.3	10.0	9.2	7.4	18.8	14.0	12.3	3.1
6/p	30	6.00	16.00	2.80	-1.50	>15.00		162	10.5	27.5	9.2	8.5	4.0	11.6	11.7	11.9	9.7
6/r	30	6.00	16.00	3.60	-2.30	>15.00		162	11.0	26.5	8.8	8.1	4.4	11.8	11.7	11.5	9.6
7/a	35	5.50	16.00	2.85	-1.55	>15.40		93	6.5	13.2	4.4	4.1	1.8	5.4	5.7	6.5	4.0
7/p	38	3.00	14.25	2.85	-1.55	>15.40		140	3.0	14.7	4.9	4.5	2.7	4.9	4.8	5.6	3.8
8/b	10	4.50	24.00	2.50	0.50	>18.00		93	8.0	43.5	14.9	25.5	11.5	30.0	22.1	19.6	8.0
8/p	10	2.60	22.00	2.00	0.00	>18.00		147	12.0	60.5	20.2	34.6	13.0	31.1	22.4	19.0	9.2
9/n	60	2.50	9.50	3.00	?	>13.60		284	1.0						4.5		
9/s	60	2.50	9.50	3.00	?	>13.60		284	3.0	6.0	6.0				4.5		
13/a	15	19.00	19.00	0.00	1.00			80	52.0	32.4	11.0	18.8	8.9		32.7		
13/b	15	150.00	200.00	0.00	1.00			60	80.0	25.3	8.5	14.5			108.1		
13/c	15	0.80	0.80	0.00	1.00			78	7.0	11.9	5.8	9.9	1.6	2.6	3.5		
14	7	29.40	29.40	0.00	0.00	25.00		164	143.0	146.0	48.7	73.0		195.7	206.7	47.4	73.9
15(3)	6	14.50	64.00	1.00	1.70	>23.00		74	74.0	71.8	24.7	22.8		101.1	75.9	37.3	23.1
15(4)	6	14.50	64.00	1.00	1.70	>23.00		74	75.0	71.8	24.7	22.8		101.1	75.9	37.3	23.1
15(8)	6	25.00	25.00	1.00	0.60	>23.00		70	121.0	71.3	23.9	22.1		100.5	105.9	28.4	21.9
15(9)	6	25.00	25.00	1.90	0.80	>23.00		63	84.0	63.5	21.3	19.7		98.5	96.6	30.2	17.2
15(9a)	6	25.00	25.00	2.30	-0.80	>23.00		75	87.0	75.9	25.3	23.3		114.3	112.4	33.4	22.4
15(9b)	6	25.00	25.00	2.10	0.90	>23.00		76	85.0	76.3	25.7	23.7		117.3	113.7	35.1	22.5
15(11)	6	25.00	25.00	2.80	-0.60	>23.00		75	87.0	75.5	25.2	23.2		119.1	112.4	35.9	21.4
15(17)	6	22.40	84.00	1.00	1.60	>23.00		75	92.0	75.2	25.5	23.6		136.3	104.1	36.2	31.9
15(18)	6	25.00	25.00	2.60	0.80	>23.00		86	120.0	86.0	28.9	26.7		134.5	126.6	40.2	26.1
27	47	60.00	60.00	5.20	-3.70	60.00	417		45.0	46.7	15.6	7.2	28.3	51.5	68.1	32.9	120.8
30(1)	20	3.00	4.80	1.50	4.00	8.00		231	6.6	35.3	17.7	21.2	9.2	15.6	16.0	13.7	9.0
30(2)	20	3.00	4.80	1.50	4.00	8.00		231	7.9	35.3	17.7	21.2	9.2	15.6	16.0	13.7	9.0
30(3)	20	3.00	4.80	1.50	4.00	8.00		231	8.6	35.3	17.7	21.2	9.2	15.6	16.0	13.7	9.0
30(4)	20	3.00	4.80	1.50	4.00	8.00		231	7.9	35.3	17.7	21.2	9.2	15.6	16.0	13.7	9.0
30(5)	20	3.00	4.80	1.50	4.00	8.00		231	7.1	35.3	17.7	21.2	9.2	15.6	16.0	13.7	9.0

Case	N_{AV}	B	L	D	D_w	H	q_{gen}	q_{net}	S_{min}	TP	M	MH	A	BB	APK	SS	BL
30(6)	20	3.00	4.80	1.50	4.00	8.00		231	11.2	35.3	17.7	21.2	9.2	15.6	16.0	13.7	9.0
30(7)	20	3.00	4.80	1.50	4.00	8.00		231	7.1	35.3	17.7	21.2	9.2	15.6	16.0	13.7	9.0
30(8)	20	3.40	5.40	1.70	4.00	8.00		247	12.2	40.9	19.3	23.2	10.8	18.2	18.5	15.3	11.0
30(9)	20	3.70	5.90	1.80	4.00	8.00		139	7.4	24.2	11.1	13.3	6.3	11.3	11.9	9.5	6.0
30(10)	20	3.70	5.90	1.80	4.00	8.00		215	15.0	37.4	17.1	20.5	9.9	17.0	17.4	13.8	10.1
30(11)	20	3.70	5.90	1.80	4.00	8.00		215	6.4	37.4	17.1	20.5	9.9	17.7	18.1	13.8	10.7
30(12)	20	3.70	5.90	1.80	4.00	8.00		225	7.4	39.2	17.9	21.5	10.4	19.7	20.0	14.3	12.2
30(13)	20	3.70	5.90	1.80	4.00	8.00		252	16.5	43.9	20.0	24.0	11.7	21.8	21.8	15.8	13.7
30(14)	20	3.70	5.90	1.80	4.00	8.00		279	8.6	48.6	22.2	26.6	13.1	21.8	21.8	17.3	13.7
30(15)	20	3.70	5.90	1.80	4.00	8.00		290	11.2	50.5	23.1	27.7	13.6	22.6	22.6	17.9	14.3
30(16)	20	4.00	6.40	2.00	4.00	8.00		97	6.1	17.5	7.8	9.3	4.7	8.7	9.2	7.3	4.0
30(17)	20	4.00	6.40	2.00	4.00	8.00		145	7.4	26.1	11.6	13.9	7.1	12.5	13.0	10.0	6.7
30(18)	20	4.00	6.40	2.10	4.00	8.00		225	9.1	40.6	18.0	21.6	11.2	18.8	19.1	14.4	11.5
30(19)	20	4.30	6.90	2.10	4.00	8.00		102	7.1	19.1	8.3	9.9	5.2	9.6	10.1	7.7	4.5
30(20)	20	4.30	6.90	2.10	4.00	8.00		134	10.2	25.0	10.9	13.1	6.9	12.3	12.8	9.5	6.4
30(21)	20	4.30	6.90	2.10	4.00	8.00		139	7.1	26.0	11.3	13.5	7.2	12.7	13.2	9.7	6.8
30(22)	20	4.30	6.90	2.10	4.00	8.00		145	6.6	27.1	11.8	14.1	7.5	13.2	13.7	10.1	7.1
30(23)	20	4.30	6.90	2.10	4.00	8.00		145	11.2	27.1	11.8	14.1	7.5	13.2	13.7	10.1	7.1
30(24)	20	4.30	6.90	2.10	4.00	8.00		145	4.1	27.1	11.8	14.1	7.5	13.6	14.1	10.3	7.4
30(25)	20	4.30	6.90	2.10	4.00	8.00		150	6.4	28.0	12.2	14.6	7.8	13.6	14.1	10.3	7.4
30(26)	20	4.30	6.90	2.10	4.00	8.00		150	9.4	28.0	12.2	14.6	7.8	13.6	14.1	10.3	7.4
30(27)	20	4.30	6.90	2.10	4.00	8.00		161	4.6	30.1	13.1	15.7	8.3	14.5	15.0	10.9	8.1
30(28)	20	4.30	6.90	2.10	4.00	8.00		161	3.6	30.1	13.1	15.7	8.3	14.5	15.0	10.9	8.1
30(29)	20	4.30	6.90	2.10	4.00	8.00		177	6.4	33.1	14.4	17.2	9.2	15.9	16.3	11.8	9.2
30(30)	20	4.30	6.90	2.30	4.00	8.00		113	8.1	21.7	9.2	11.1	6.2	11.1	11.6	8.4	5.4
30(31)	20	4.60	7.40	2.30	4.00	8.00		166	5.1	31.8	13.5	16.3	9.2	15.8	16.2	11.2	9.0
30(32)	20	4.60	7.40	2.50	4.00	8.00		97	8.1	19.0	8.0	9.5	5.6	10.2	10.6	7.5	4.6
30(33)	20	4.90	7.80	2.50	4.00	8.00		97	4.3	20.8	8.0	9.5	5.6	10.2	10.6	7.5	4.6
30(34)	20	4.90	7.80	2.50	2.50	8.00		102	6.9	21.9	8.4	10.0	5.9	10.7	11.1	7.8	4.9
30(35)	20	4.90	7.80	2.50	2.50	8.00		107	3.6	23.0	8.8	10.5	6.2	11.2	11.5	8.0	5.3
30(36)	20	4.90	7.80	2.50	2.50	8.00		113	8.9	24.3	9.3	11.1	6.6	11.7	12.1	8.4	5.7

Grouping headers: the column S_{min} is a single column; the columns TP, M, MH, A, BB, APK, SS, BL fall under the group heading S_{calc}.

Case	N_{AV}	B	L	D	D_w	H	q_{gmax}	q_{gut}	S_{min}	TP	M	MH	A	BB	APK	SS	BL
30(38)	20	4.90	7.80	2.50	2.50	8.00		123	5.8	26.4	10.1	12.1	7.2	12.6	13.0	8.9	6.4
30(39)	20	4.90	7.80	2.50	2.50	8.00		123	7.4	26.4	10.1	12.1	7.2	12.6	13.0	8.9	6.4
30(40)	20	4.90	7.80	2.50	2.50	8.00		182	8.4	39.1	14.9	17.9	10.8	18.0	18.3	12.0	10.6
30(41)	20	4.90	7.80	2.50	2.50	8.00		182	19.1	39.1	14.9	17.9	10.8	18.0	18.3	12.0	10.6
30(42)	20	4.90	7.80	2.50	2.50	8.00		188	15.0	40.4	15.4	18.5	11.1	18.6	18.8	12.3	11.1
30(43)	20	4.90	7.80	2.50	4.00	8.00		199	11.7	42.7	16.3	19.6	11.8	19.6	19.8	12.9	11.9
30(44)	20	5.50	8.80	2.60	4.00	8.00		139	9.4	28.6	11.7	14.0	8.8	15.4	15.7	9.8	8.3
30(45)	20	6.10	9.80	3.00	4.00	8.00		161	10.2	34.1	13.6	16.3	11.4	19.1	19.2	10.7	10.9
30(46)	20	6.40	10.20	3.20	4.00	8.00		150	14.5	32.1	12.7	15.2	11.2	18.7	18.7	10.0	10.4
30(47)	21	6.70	10.70	3.40	4.00	8.00		113	5.8	23.3	9.1	10.9	8.2	14.2	14.2	7.7	7.0
30(48)	21	6.70	10.70	3.40	4.00	8.00		113	4.1	23.3	9.1	10.9	8.2	14.2	14.2	7.7	7.0
30(49)	22	7.00	11.20	3.50	4.00	8.00		177	7.6	35.3	13.7	16.5	12.7	20.4	20.5	10.0	13.1
30(50)	22	7.00	11.20	3.50	6.40	8.00		177	8.9	35.3	13.7	16.5	12.7	20.4	20.5	10.0	13.1
31/a	21	42.70	42.70	0.00	6.40	27.40		166	80.0	47.7	16.5	19.8	31.0	55.2	72.6	19.0	37.0
31/b	19	33.50	33.50	0.00	6.40	27.40		156	90.0	48.9	17.1	20.5	27.0	50.4	65.4	19.8	36.7
31/c	17	27.40	27.40	0.00	6.40	27.40		154	100.0	53.1	18.8	22.6	26.6	50.5	64.2	21.5	38.2
31/d	20	38.10	38.10	0.00	-7.20	27.40		241	131.0	72.3	25.2	30.2	44.5	79.3	98.3	29.1	66.1
32	60	55.00	101.00	9.70	0.60	21.30	289		65.0	17.9	6.0	9.0	9.9	23.8		6.5	6.8
35/a	11	24.40	24.40	0.00	0.61	9.60		182	232.0	102.1	34.2	31.6	40.6	99.1	115.5	19.9	24.9
35/b	11	23.80	23.80	0.00	3.70	9.60		158	196.0	88.5	29.7	27.4	34.2	84.9	100.3	17.4	20.8
36	25	22.90	88.40	0.30	10.00	7.30	245	180	39.4	42.7	14.8	13.7	13.5	38.0	40.9	8.7	10.3
39°af1	16	1.00	1.00	0.00	10.00	>50	245		8.0	28.6	19.1	17.6	5.3	8.6	10.2	8.7	3.5
39°af4	16	1.00	1.00	0.00	10.00	>50	245		10.4	28.6	19.1	17.6	5.3	8.6	10.2	8.7	3.5
39°cg4	16	1.00	1.00	0.00	10.00	>50	245		7.0	28.6	19.1	17.6	5.3	8.6	10.2	8.7	3.5
39°e4	16	1.00	1.00	0.00	10.00	>50	245		14.0	28.6	19.1	17.6	5.3	8.6	10.2	8.7	3.5
39°af6	16	2.50	2.50	0.00	10.00	>50	245		20.5	38.7	25.8	23.8	8.8	16.4	19.4	13.8	8.5
39°af7	16	2.50	2.50	0.00	10.00	>50	254		10.8	40.1	26.7	24.7	9.1	17.0	20.0	14.3	8.9
39°af8	16	2.50	2.50	0.00	10.00	>50	254		6.6	40.1	26.7	24.7	9.1	17.0	20.0	14.3	8.9
39°cg6	16	2.50	2.50	0.00	10.00	>50	254		17.8	40.1	26.7	24.7	9.1	17.0	20.0	14.3	8.9
39°tm1	16	2.50	2.50	0.00	10.00	>50	245		-4.4	38.7	25.8	23.8	8.8	16.4	19.4	13.8	8.5
39°tm2	16	2.50	2.50	0.00	10.00	>50	254		9.4	40.1	26.7	24.7	9.1	17.0	20.0	14.3	8.9
39°tm3	16	2.50	2.50	0.00	10.00	>50	254		8.4	40.1	26.7	24.7	9.1	17.0	20.0	14.3	8.9

Case	N_{AV}	B	L	D	D_w	H	q_{gross}	q_{net}	s_{mis}	s_{calc} TP	M	MH	A	BB	APK	SS	BL
39/ca1	16	2.50	2.50	0.00	10.00	>50	254		12.5	40.1	26.7	24.7	9.1	17.0	20.0	14.3	8.9
39/a5	16	2.50	2.50	0.00	10.00	>50	254		10.4	40.1	26.7	24.7	9.1	17.0	20.0	14.3	8.9
39/a6	16	2.50	2.50	0.00	10.00	>50	245		7.6	38.7	25.8	23.8	8.8	16.4	19.4	13.8	8.5
39/c3	16	2.50	2.50	0.00	10.00	>50	245		18.4	38.7	25.8	23.8	8.8	16.4	19.4	13.8	8.5
39/c4	16	2.50	2.50	0.00	10.00	>50	245		6.8	38.7	25.8	23.8	8.8	16.4	19.4	13.8	8.5
40	12	16.00	20.50	1.50	0.00	26.70	220		210.0	96.1	32.0	48.0	31.2	80.9	81.4	42.1	49.0
44/p1	35	1.52	1.52	0.60	3.00	>3.00		150	2.1	8.6	5.7	5.2	2.6	2.4	3.5	3.1	1.9
44/p2	50	1.52	1.52	0.60	3.00	>3.00		150	1.0	6.0	4.0	3.7	2.6	1.5	2.3	2.3	1.8
44/m1	28	1.22	1.22	0.60	3.00	>3.00		150	1.3	9.6	6.4	5.9	2.4	2.8	3.9	3.5	1.8
44/m3	45	1.22	1.22	0.60	3.00	>3.00		150	0.6	6.0	4.0	3.7	2.4	1.5	2.2	2.3	1.5
44/H2	50	3.05	3.05	1.50	-0.90	>4.60		150	3.8	13.8	4.6	4.3	2.9	2.4	3.7	2.7	3.3
44/H3	45	3.05	3.05	1.50	-0.90	>4.60		150	6.2	15.4	5.1	4.7	2.9	2.8	4.2	3.0	3.3
44/H5	30	3.05	3.05	1.50	-0.90	>4.60		150	7.2	23.1	7.7	7.1	2.9	5.0	6.8	4.3	4.0
45/a	18	13.00	31.50	2.10	-0.10	19.20	193		21.0	50.4	16.8	14.4	12.8	38.8	35.9	23.9	26.5
45/b	18	13.00	27.40	2.10	-0.10	19.20	193		18.0	50.4	16.8	14.4	12.8	37.7	35.9	23.4	25.9
45/c	18	13.00	22.50	2.10	-0.10	19.20	193		14.0	50.4	16.8	14.4	12.8	36.0	35.9	22.6	25.0
47/a	29	1.20	1.20	2.60	2.50	>20.00		215	2.5	11.2	7.5	3.2	4.1	4.0	5.1	3.5	2.3
47/b	26	1.20	1.20	2.60	2.50	>20.00		215	1.5	12.5	8.4	3.6	4.7	4.7	5.8	3.9	2.5
47/c	18	1.20	1.20	2.60	2.50	>20.00		215	8.6	18.1	12.1	5.2	7.6	7.8	9.0	5.4	3.2
48	30	34.00	57.00	7.90	-0.90	>27.00	270		22.0	26.4	8.8	8.1	16.5	36.6	33.6	18.9	16.7
49	6	18.00	18.00	0.00	1.20	8.70		75	44.5	75.6	25.6	23.7		78.7	89.3	14.3	10.9
50/a	20	18.30	18.30	0.30	1.80	>8.90		41	4.8	12.3	4.2	7.2	3.1	8.4	12.6	2.9	1.8
50/b	20	15.20	15.20	0.30	1.80	>8.90		33	2.8	9.7	3.3	5.7	2.2	6.0	9.2	2.5	1.4
51(A)	37	4.00	7.00	5.00	6.40	7.00		518	7.6	34.6	19.2	6.4	17.3	18.8	18.9	8.2	9.4
51(B)	37	4.00	7.00	5.00	6.40	7.00		518	7.6	34.6	19.2	6.4	17.3	18.8	18.9	8.2	9.4
51(C)	37	4.00	7.00	5.00	6.40	7.00		518	11.9	34.6	19.2	6.4	17.3	18.8	18.9	8.2	9.4
51(F)	37	4.00	7.00	5.00	6.40	7.00		518	9.5	34.6	19.2	6.4	17.3	18.8	18.9	8.2	9.4
51(G)	37	4.00	7.00	5.00	6.40	7.00		518	9.5	34.6	19.2	6.4	17.3	18.8	18.9	8.2	9.4
51(H)	37	4.00	7.00	5.00	6.40	7.00		518	4.6	34.6	19.2	6.4	17.3	18.8	18.9	8.2	9.4
52/c	50	1.22	1.22	0.46	3.00	4.10		300	4.5	11.1	7.4	3.2	5.0	2.5	3.6	4.2	3.0
52/a3	30	0.91	0.91	1.22	3.66	6.10		300	4.0	13.4	8.9	3.8	4.5	4.2	5.4	4.5	2.5
52/d3	20	0.91	0.91	3.05	0.91	6.10		300	6.7	30.1	13.4	5.7	8.7	7.7	8.7	4.6	3.3

52/j	20	0.91	0.91	1.22	1.83	3.40		300	2.7	20.1	13.4	5.7	6.3	7.4	8.7	6.4	3.3
										s_{calc}							
Case	N_{SV}	B	L	D	D_w	H	q_{gros}	q_{net}	s_{min}	TP	M	MH	A	BB	APK	SS	BL
53/1-2	≤2	4.50	30.50	2.70	-1.10	7.10		70	7.0	27.7	9.2	15.8	6.2	18.2	13.9	8.8	5.0
53/3-4	≤2	4.50	30.50	2.70	-1.10	7.10		70	3.0	27.7	9.2	15.8	6.2	18.2	13.9	8.8	5.0
58/a	≤3	1.10	1.10	1.20	1.50	2.70		78	2.0	11.6	5.9	2.3	3.0	4.3	5.2	3.0	1.2
58/b	≤3	1.50	1.50	1.20	1.50	2.70		77	2.1	15.6	6.9	2.8	3.5	5.3	6.3	3.3	1.6
58/c	≤3	1.50	1.50	1.20	1.50	2.70		77	1.3	15.6	6.9	2.8	3.5	5.3	6.3	3.3	1.6
59/a	35	23.60	26.90	3.00	1.50	>25.00	167		15.4	19.0	6.4	7.7	6.0	14.8	18.5	11.5	13.7
59/b	25	1.80	1.80	3.00	1.50	>25.00		230	3.4	25.4	10.7	12.8	6.9	7.1	8.5	6.0	4.1
59/c	25	1.40	1.40	3.00	1.50	>25.00		230	3.9	21.7	9.9	11.8	6.2	5.9	7.1	4.7	3.2
59/d	25	2.20	2.20	3.00	1.50	>25.00		284	10.5	34.7	13.9	16.7	9.4	9.9	11.8	8.5	6.4
59/e	35	4.50	5.70	3.00	1.50	>25.00		195	3.9	23.8	8.6	10.4	4.8	7.8	9.4	8.8	9.4
59/f	35	15.00	72.90	3.00	1.50	>25.00	81		5.4	4.3	1.5	1.8	1.0	4.9	3.8	6.8	1.4
59/g	25	1.40	140.00	0.40	2.60	>25.00		226	7.7	19.3	12.0	14.4	3.9	8.5	7.0	9.9	4.6
59/h	25	1.60	12.60	0.40	2.60	>25.00		250	9.3	25.0	14.0	16.8	4.4	9.7	8.4	11.2	5.5
59/i	25	1.20	12.70	0.30	2.70	>25.00		250	10.0	19.0	12.6	15.2	4.1	8.0	6.9	9.8	4.2
59/j	25	0.80	18.40	0.30	2.70	>25.00		294	5.8	17.8	11.8	14.2	4.1	7.3	6.0	9.1	3.3
59/k	25	1.80	24.10	0.30	2.70	>25.00		206	16.9	23.0	12.2	14.7	3.7	8.9	7.7	10.3	5.3
59/m	40	1.00	1.00	0.00	3.00	>25.00		294	5.0	13.7	9.2	11.0	4.5	2.9	4.0	5.1	2.6
59/n	40	3.30	5.70	3.00	1.50	>25.00		304	11.0	27.8	10.4	12.5	6.1	8.6	9.4	9.8	8.3
59/o	40	3.30	5.70	3.00	1.50	>25.00		304	12.2	27.8	10.4	12.5	6.1	8.6	9.4	9.8	8.3
59/p	40	3.60	6.30	3.00	1.50	>25.00		304	12.7	29.2	10.8	13.0	6.1	9.2	10.0	10.5	9.1
59/q	40	3.60	6.30	3.00	1.50	>25.00		304	13.6	29.2	10.8	13.0	6.1	9.2	10.0	10.5	9.1
59/r	40	4.50	6.80	3.00	1.50	>25.00		304	18.3	32.4	11.8	14.1	6.2	10.4	11.7	11.9	11.0
60/a	30	22.90	32.60	3.00	1.50	>25.00	165		20.4	21.7	7.4	8.8	7.4	19.0	21.5	13.6	15.7
60/b	30	21.70	22.20	3.00	1.50	>25.00	148		19.8	18.3	6.2	7.4	6.0	14.2	17.9	11.4	11.0
60/c	25	1.00	1.00	3.00	1.50	>25.00		196	6.0	13.7	7.3	8.8	4.4	4.0	4.9	2.7	1.9
61/a	34	1.00	1.00	0.00	2.00	2.00		220	3.6	12.1	8.1	12.1	3.4	2.7	3.8	4.4	2.0
61/b	45	1.00	1.00	0.50	2.00	2.00		564	4.4	20.5	13.7	12.6	9.1	4.7	6.1	6.8	4.3
61/c1	≤1	1.00	1.00	0.50	0.00	2.00		339	6.0	100.8	33.6	31.0	13.1	20.3	21.2	14.2	6.3
61/c2	≤1	1.00	1.00	0.50	0.00	2.00		284	4.7	84.4	28.1	26.0	10.9	17.1	18.2	11.9	4.9
65	25	1.20	1.20	0.00	?	?		320	2.8		17.3				8.6		
66/a	≤2	2.40	121.90	2.00	5.30	5.30		168	15.7	27.8	18.5	27.8	12.6	27.1	19.1	14.4	8.9

The following table includes a scale/reference row (Case "66/b"). The label S_{calc} spans the columns TP, M, MH, A, BB, APK, SS, BL.

Case	N_{AV}	B	L	D	D_w	H	q_{gem}	q_{ber}	S_{mld}	TP	M	MH	A	BB	APK	SS	BL
66/b	12	4.60	121.90	2.00	5.30	5.30		188	19.1	55.6	26.1	39.1	20.0	15.7	35.3	15.8	12.7
76	20	22.50	65.00	10.00	-2.50	>30.00	270		21.0	30.6	10.2	8.7	20.4	49.4	3.6	29.3	19.9
77	60	10.00	10.00	1.50	9.80	12.00		240	7.0	17.4	7.7	7.1	4.9	6.9	10.3	7.2	11.5
78/a	5	20.00	20.00	3.00	-1.00	32.00	85		116.0	47.1	15.7	26.9		79.9	68.7	34.4	9.1
78/b	5	20.00	20.00	3.00	-1.00	45.00	85		81.0	47.1	15.7	26.9		79.9	68.7	38.0	9.1
79/a	5	27.50	27.50	0.00	0.00	40.00		130	286.0	161.8	53.9	107.9		237.1	241.4	65.4	75.6
79/b	5	27.50	27.50	0.00	0.00	40.00		176	1124.0	219.1	73.0	146.1		321.0	311.1	88.6	-14.9
81/c	5	0.90	0.90	0.30	10.00	10.00		133	7.6	43.3	28.9	43.3	9.4	22.4	22.5	12.3	3.9
81/d	6	0.90	0.90	0.90	10.00	10.00		113	6.4	25.1	16.7	25.1		15.3	15.7	7.9	2.7
81/e	7	1.20	1.20	0.20	10.00	10.00		199	13.0	55.1	36.7	55.1	14.8	25.5	26.1	16.5	6.2
81/f	8	1.20	1.20	0.90	10.00	10.00		268	12.7	55.1	36.7	55.1	19.5	28.8	28.8	16.9	7.4
83	20	17.60	84.00	10.70	-2.20	>37.00	240		21.2	17.1	5.7	2.4	11.1	33.4	18.4	31.0	12.3
84	14	16.00	43.00	7.30	-1.80	>23.00	228		17.9	43.3	14.4	6.2	23.5	58.1	42.1	30.0	21.2
85	10	20.50	20.50	3.50	2.50	>26.00	173		8.0	62.7	21.5	9.2		73.4	75.5	32.3	24.2
86	26	14.50	14.50	3.50	7.50	21.50	225		15.5	31.0	11.9	5.1	14.8	21.2	25.7	15.7	17.3
87	34	33.00	33.00	5.30	-2.50	8.20	216		43.8	26.4	8.8	3.8	8.8	19.9	31.0	2.6	1.3
89/a	37	2.60	10.70	1.00	?	5.10		293	10.9		12.2				8.5		
89/b	43	1.70	170.00	1.60	?	3.40		180	19.2		4.9				3.5		
91	27	24.40	24.40	0.00	?	?		120	14.3		9.2				27.4		
92/a	50	2.10	2.40	2.40	?	?		584	4.4		14.2				9.3		
92/b	50	2.10	2.10	1.50	?	?		697	2.3		18.5				10.8		
92/c	50	1.80	2.80	1.50	?	?		575	2.7		14.1				8.2		
92/d	50	2.10	2.40	3.00	?	?		584	4.6		14.2				9.3		
92/e	50	2.10	4.10	3.00	?	?		347	1.8		8.4				5.9		
93/A	5	8.20	61.00	0.00	?	?		35	13.0		13.8				35.0		
94/A	18	30.20	30.80	2.70	6.50	22.30	386		91.6	108.0	38.0	32.6	77.1	115.4	126.9	43.3	85.0
94/B	50	3.80	380.00	7.00	6.00	17.00	383		4.8	12.7	7.0	3.0	6.4	8.4	6.9	8.5	8.1
97/cmb	7	20.00	6.00	0.00	0.90	18.00		145	120.0	126.4	42.6	63.9		132.1	141.8	63.0	50.8
97/plate	7	6.00	6.00	0.00	0.90	18.00		190	74.0	150.6	52.1	78.2	32.6	74.5	77.2	43.7	27.6
98/A	4	2.80	14.00	1.00	?	10.00		142	97.0		55.8				68.8		
98/B	4	3.30	14.50	1.00	?	10.00		99	37.0		40.7				56.4		

INFORMATIONAL METHODS IN OPTIMIZATION
OF TOOLS

L. L. MISHNAEVSKY JR

Staatliche Materialprüfungsanstalt (MPA), University of Stuttgart,
Pfaffenwaldring 32, D-70569 Stuttgart and
Max-Planck-Institut für Metallforschung,
D-70174 Stuttgart, Germany

AND

S. SCHMAUDER

Staatliche Materialprüfungsanstalt (MPA), University of Stuttgart,
Pfaffenwaldring 32, D-70569 Stuttgart, Germany

1. Introduction

This paper seeks to develop a mathematical model of contact interaction, which is applicable to contacting bodies of complex shapes and made from disordered materials; this model should be able to serve also as a theoretical basis for the improvement of efficiency of technological processes which includes the contact interaction between tool and workpiece (among them are drilling, milling, machining, etc). In the first part of the paper, a method of description of the stress state in disordered materials based on the entropy maximization is presented. In the second part, a general characteristics of shapes of contacting bodies is developed on the basis of the information theory methods. Interrelations between these two models, possibilities of their practical application and further development are discussed as well. The main area of application of the developed model should be an optimization of destructing tools; therefore the stronger and weaker contacting bodies will be called in the text below "tool" and "workpiece", respectively, what does not influence the generality of results as applied to the modelling of contact interaction of heterogeneous and complex bodies. Table 1 and Figure 1 are presented here with kind permission from Elsevier Science Ltd.

G. N. Frantziskonis (ed.), PROBAMAT – 21st Century: Probabilities and Materials, 499–510.
© 1998 *Kluwer Academic Publishers.*

2. Stress distribution in a disordered material

2.1. APPLICATION OF THE MAXIMUM ENTROPY METHOD

The stress distribution in disordered materials depends on a number of random factors: spatial distribution of inclusions, distributions of their sizes, shapes, properties and orientations, superposition of stress field caused by each stress concentrators in materials, variability of properties and types of local behaviour, cooperative effects, etc. These effects cause high variability of the local stresses from the values which are calculated with the use of the averaged constants and elastic or plastic solutions. Lippmann [1] noted that the stress in disordered materials presents a sum of an averaged stress and microstress caused by the local microstructure and heterogeneities. This microstress can be considered as a local fluctuation of stress field. Mishnaevsky Jr and Schmauder [2] have shown, that the materials become more heterogeneous with increasing damage parameter and approaching to failure. It leads also to the increase of stochastic component in the local stress in the materials.

The information about the microstructure of heterogeneous material, distribution of inclusions, their sizes and orientation is lacking usually. So, the problem of description of stress state in disordered materials is a problem of determination of parameters under conditions of the lack of required information. In order to solve such problems, one can use the maximum entropy method [3]. Consider the stress distribution in a loaded material. A stress in a point is determined by interaction between stress fields of all nearest stress concentrators (inclusions, hard particles, microcracks, pileups of dislocations). A stress distribution which is determined by superposition of stress fields of a wealth of randomly distributed stress concentrators can be taken as random and may be described by a probability function, for instance, by a probability function of the von Mises equivalent stress (for simplicity, we shall use here this scalar value, but this approach can be applied to describe a distribution of vector components as well). In the case of homogeneous material, this probability function is reduced to the Dirac delta function. This function is determined from a condition of maximal likelihood of a given stress state, which corresponds to a maximal entropy of the system. In order to determine the probability function which describes the stress state, all available data about the stress distribution are introduced into the corresponding Lagrangian, and the probability function is determined on the basis of the condition of the maximum of this Lagrangian.

For example, consider the case when the material is extremely heterogeneous and almost no information, except for the energy of loading A and strained volume V is given. One should note here that this model may be applicable to thermic loading, but not to mechanical loading at which conditions and direction of loading are known always; the other case in which the model can be applied is that when there is two levels process (for example, micro- and macrofracture) in a problem to be considered and one may neglect geometrical effects on one of the levels. The restrictions on possible probability distributions of stress are minimal and look as follows:

$$\sum_a p_a(a_e) = 1 \tag{1}$$

$$\sum_a p_a(a_e)a_e = \frac{A}{V} \tag{2}$$

where a_e - the specific elastic strain energy per unit volume, $p_a(a_e)$ - probability function of the specific elastic energy, A - energy of loading, V - strained volume. In order to simplify the equations (1), (2), we wrote them for the distribution of specific elastic energy $a_e = k_p \sigma_{eq}^2$, and not stress. The condition of maximum entropy is written in such a way:

$$-\sum_a p(a_e) \ln p_a(a_e) \to max \tag{3}$$

The Langangian of the system looks as follows:

$$Z_{(\lambda_1, \lambda_2, \lambda_3)} = \lambda_1[1 - \sum_{a_e} p_a(a_e)] + \lambda_2[A/V - \sum_{a_e} p_a(a_e)a_e] - $$
$$- \lambda_3 \sum_{a_e} p_a(a_e) \ln p_a(a_e) \tag{4}$$

Minimizing the Lagrangian $(dZ/dp_a(a_e) = 0)$, one can obtain:

$$p(\sigma) = \lambda_1 \sigma \sqrt{2k_p} \exp(-0.5 k_p \lambda_1 \sigma^2) \tag{5}$$

where $\lambda_1 = \lambda_2 = \lambda_3 = V/A$, σ -von Mises equivalent stress in a point, $p(\sigma)$ - probability function of σ. If one states the condition of local failure (the critical level of equivalent stress, for instance), one can determine the probability of microcrack formation (or damage density) as well:

$$R = \int_{\sigma_{cr}}^{\infty} p(\sigma)d\sigma = \exp(-\lambda_1 k_p \sigma_{cr}^2) \tag{6}$$

where σ_{cr} - critical level of the equivalent stress. This formula relates the microcrack density and the probability function of stress distribution (which depends on the heterogeneity of the material).

One can introduce into the Lagrangian (4) also other boundary conditions than the simplest ones used above, what may allow us to take into account the structure of the material. For example , one can exert into the Lagrangian the correlation function of stress in different points of strained volume (that can be considered as an analogue of strain compatibility equations as applied to the brittle disordered materials), or the average deviation of stress in strained volume points from the stress being calculated by elastic problem solution (with averaged elastic constants). For instance, if a heterogeneous material consists of inclusions (grains, filler particles or quartz) of elliptical shape and relatively soft matrix, and the distribution of inclusions is known, the stress in any point of matrix is proportional to the applied stress which includes the stochastic components and to the value $\sqrt{d/r_i}$, where d - length of an elliptic inclusion, r_i - radius of the tip of inclusion. With the use of the formulas of function of random values and having the probability distributions of d and r_i (for sintered materials or cast alloys, they can be determined by consideration of kinetics of material formation), one can find the distribution of σ, which should be then substituted into the Lagrangian eq.(4) for the function $p_a(a_e)$. Minimizing the Lagrangian, one finds the probability functions of stress and strain energy distribution in the material with given distribution of inclusion sizes. Thus, the developed method makes it possible to obtain the probabilistic description of stress distribution and damage in disordered material, taking into account the lack of information about the material structure and the avaiable (insufficient) information about it.

Practically, it means that the problem of stress state modelling is solved from the "other end": usually, one supposes that an ideally elastic material (or a material with other fully known properties) is considered, and then one begins to introduce corrections, which should allow for heterogeneity and other deviations of the material from the ideal case; here, we accept initially, that the information about the material structure and behaviour is lacking, and then begin to introduce any available information into the Lagrangian.

2.2. SUPERPOSITION OF STRESS FIELDS

In order to illustrate possibilities of practical using of developed model, consider the influence of superposition of stress fields and the effect of indexing on the damage in a brittle material (for example, rock). The indexing effect is caused by interaction between the stress fields or crack systems from neighbouring teeth or cutters on a drilling bit. Here, only the effect of superposition of stress fields is considered. Practically, the simultaneous loading of rock by several indenters takes place when the teeth on the drilling bit are placed in pairs or in groups, and their stress fields are superimposed. Consider simultaneous and successive loadings of a rock by several (in this case, two) indenters. In first case (i.e. the simultaneous indentation), there are the superposition of stress fields from the indenters (it is supposed that the indentations are made on rather small distance, which is about 1... 10 indenter radii). In the second case (the successive loading), the rock is unloaded during the time interval between indentations and the stress fields do not interact (the influence of as-formed damage on the formation of new damage is neglected).

Let the function $p(\sigma)$ for single indentation be denoted by $p_0(\sigma)$. The damage parameter corresponding to the single loading is designated by R_0. Consider the values of damage R for simultaneous and successive loading by two equal loads. In line with the damage accumulation hypothesis, one can write for the case of successive loading: $R_{suc} = 2R_0$. In the simultaneous loading (when the stress fields are superimposed), the damage is equal to the sum of damage parameters from each indenter plus some value ΔR which is determined by stress field superposition: $R_{sim} = 2R_0 + \Delta R$. By its meaning, the value ΔR is a probability of local failure (i.e. the microcrack formation) due to the stress fields superposition in a point in which the rock does not fail at the successive loading. This probability can be determined as a probability of coincidence of two events : an event A that $\sigma < \sigma_{cr}$ in the point at successive loading, and an event B that $\sigma > \sigma_{cr}$ in the same point at simultaneous loading:

$$\Delta R = Prob_A \ Prob_B \qquad (7)$$

where $Prob_{A,B}$ is the probability of the event A or B. Determine the value $Prob_A$ and $Prob_B$. It is clear that $Prob_A = 1 - R_{suc}$ and

$$Prob_B = \int\limits_{\sigma_c}^{+\infty} p_{sim}(\sigma)d\sigma \qquad (8)$$

where $p_{sim}(\sigma)$ is the probability function of stress distribution for simultaneously applied loads. Then, the distribution of stress for the case of superimposed fields can be found as a convolution of two distributions $p(\sigma)$:

$$p_{sim}(\sigma) = \int\limits_{-\infty}^{+\infty} p_0(\sigma - z)p_0(z)dz \qquad (9)$$

If one defines a coefficient of indexing as a ratio between the damage in rock in simultaneous and successive loading by the same load $\eta_i = R_{sim}/R_{suc}$, one can calculate this value with the use of eqs. (6) - (9) as follows:

$$\eta_i = 1 + 0.5 \int\limits_{\sigma_c}^{+\infty} p_{sim}(\sigma) d\sigma \tag{10}$$

To calculate the value of η_i one can use the approximate formula (6). Substituting eq. (6) to eqs. (9) and (10), one can obtain after some rearrangements: $Prob_B = (1 + \lambda a_{cr}) R_0$. Taking into account eqs. (6), (9) and (10), we obtain:

$$\eta_i = 1 + 0.5(1 - 2R_0)(1 - \ln R_0) \tag{11}$$

Let us take $R_0 = 0.3$ (i.e. the cracked part of loaded volume of rock is considered; the value 0.3 presents the critical damage density at which a large crack is formed [4]) Then, we have from eq.(11) : $\eta_i = 1.44$.

It means that the intensity of material destruction increases sufficiently (almost 1.5 times) in simultaneous indentation of several indenters as comparing with the successive indentation of the same indenters.

3. Informational description of shapes of contacting bodies

3.1. HOW TO CHARACTERIZE SHAPES OF CONTACTING BODIES?

The intensity of material destruction in indentation is determined by the shape of indentors. Let us consider three simplest forms of indentors: spherical, conical and cylindrical ones. Although the shapes of indenters differ evidently and the peculiarities of destruction for each of the indenters have been well investigated, there is no quantitative parameter ("input" or apriopi characteristic) which can characterize the form and serve as a criterium for their comparison and which may be generalized for more complex cases.

The experiments on the indentation of differently shaped indentors, described in [5], have shown that the volume of craters of spalled rock is maximal for conical, minimal for spherical and medium for cylindrical indentors. If one compares the result with the contact stress distributions for these cases [6], one can see that the maximal volume of crater corresponds to the most sharp curve of contact stress distribution, whereas the minimal volume corresponds to the most homogeneous contact stress distribution.

One can suppose that the "sharpness" (i.e. non-homogeneity) of contact stress distribution is a parameter which determines the intensity of material destruction (in this case, the volume of crater). To characterize this "sharpness" of distributions quantitatively for arbitrary tool shape (including, for example, non-axisymmetric bit with many teeth), one can use an informational entropy of contact stress distribution [7].

Let us suppose that a contact stress distribution function is given in following general form:

$$\sigma_c = F(x, y, z) \tag{12}$$

where σ_c is the contact stress in a point, x, y, z - coordinates of a contact point. The function (12) is determined by the tool shape and the stress-strain relationship for a given material. Peaks of this function correspond to stress concentrators on the tool surface. Quantifying the range of contact stress variation, one can obtain from eq. (12) the probability distribution of contact stress over the contact surface:

$$p(\sigma_c) = \frac{1}{N_L} \sum_j Y[F(x, y, z); \sigma(c)] \tag{13}$$

where $p(\sigma_c)$ is the probability that the contact stress in a point is equal to the value σ_c, N_L - the amount of quantization levels of σ_c, j - the number of a contact point, Y[] - step function, $Y[x_1; x_2] = 1$, when $x_1 = x_2$ and is equal to 0, otherwise.

The informational entropy of contact stress distribution can be calculated by the formula:

$$H_c = - \int_{\sigma_c} p(\sigma_c) \ln p(\sigma_c) d\sigma_c \tag{14}$$

The greater the parameter H_c, the more non-homogeneous the contact stress distribution. This value characterizes the "sharpness" (non-homogeneity) of contact stress distribution for arbitrary function F, and thus, for arbitrary drilling bit shape. This parameter does not depend on any kind of symmetry of tool, and may be used for tool shapes of any complexity.

3.2. EFFECT OF SHAPES OF CONTACTING BODIES ON DAMAGE GROWTH

In order to investigate the influence of the parameter H_c on the intensity of material destruction, one takes a series of contact stress distributions, which differ by the parameter H_c and each of them corresponds to some shape of indenter, and calculates the damage parameter in the material for each distribution. The damage evolution in contact interaction proceeds as follows: first, the surface damage is formed in the vicinity of the contact surface, and then the damage density begins to grow. The damage growth rate is the more the greater the damage parameter [4]; therefore, the initial damage (in this case, surface damage) determines the damage in the material at later stage of destruction. That is why we use here the surface damage as a characteristic of the damaged state of the material. A contact stress distribution (i.e. the function (12)) can be presented as a power function in the rather general case:

$$\sigma_c(x) = q_1(x/a_c)^{q_2} \qquad (15)$$

where $2a_c$ is the width of contact area, q_2 -a power coefficient which determines the appearance of contact stress distribution (when $q_2 > 1$, tool is extremely sharp; when $0 < q_2 < 1$ it corresponds to the more realistic case when the tool has a convex surface), q_1 - a coefficient which depends on the applied load, x - a distance between a point and the axis of tool. The applied force is supposed to be constant. The multiplier q_1 is determined by integrating equation (15): $q_1 = (q_2 + 1)P_a/(a_c^{q_2+1})$, where P_a is the applied load. The values of initial damage R_o and the contact stress entropy were calculated by formulas (14) and (13). The damage parameter (density of microcracks) was determined by the formula (6). The coefficient q_2 was varied from 0.25 to 1.8. The following input data were used: $2a_c = 10$; $P_a = 25$, the number of quantifying levels for stress $N_L = 800$, the step of discretization of contact stress was 0.1, the contact surface was discretized for 1000 elements, the average local strength of the material σ_{cr} is equal to 170. A plot of surface damage R_o versus the contact stress entropy obtained as a result of the calculations is presented in Figure 1. From Figure 1 one can see that the surface damage increases monotonically with increasing the entropy of contact stresses (at constant load). Therefore one can conclude, that the damage in material is the more, the greater is the informational entropy of contact stress distribution. If one applies this result to a destructing tool, it follows herefrom that the destruction ability of tool can be increased if one increases the contact stress entropy. This conclusion is verified below by comparison with technical solutions in this area.

3.3. COMPARISON WITH EXPERIMENTS AND PRACTICAL RECOMMENDATIONS

In this section, the possibility of practical application of developed approaches in design and optimization of tools are discussed, and the theoretical results are compared with experiments and technical solutions. To test the results about the indexing effect, one can use the experimental data from [8]. In this work, the maximal distance between indenters which ensures the spalling of barriers between craters in successiv and simultaneous indentations of two and three indenters in marble blocks was determined. It was shown that this distance for simultaneous indentation of two indenters is equal aproximately to 8 radii of indenters and for the case of the successive indentation presents 6 radii. So, the simultaneity of indentations leads to the increase by 33 % in the linear size of as-formed (supposedly, Hertzian cone) cracks. The area of new surface may be taken as being proportional to

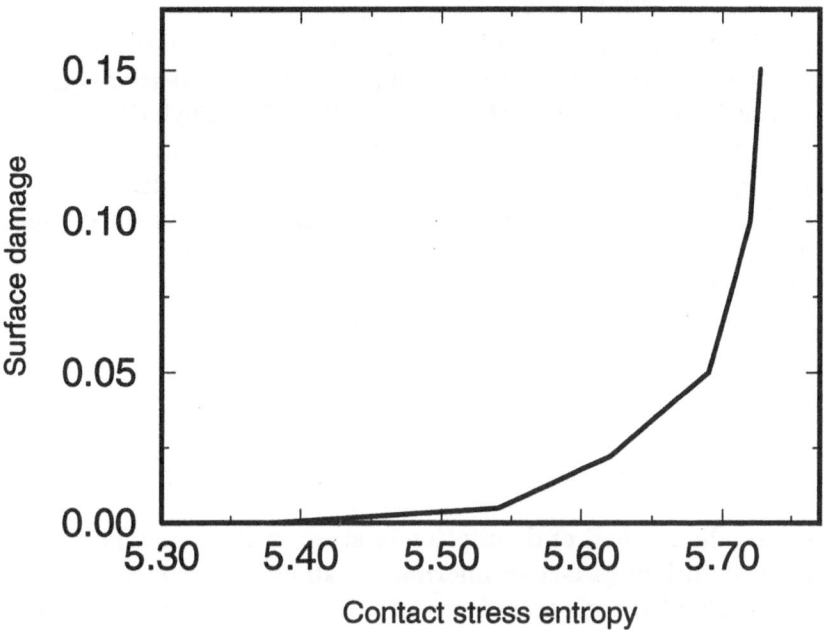

Figure 1. Surface (initial) damage plotted versus entropy of contact stresses [7]

the damage parameter to the power 2.42 [2]. So, the linear size of as-formed cracks should increase at the expense of the simultaneity of indentation by the value $\eta_i^{1.21}$, in accordance with the present model. If we use the value $\eta_i = 1.44$, it corresponds to the increase by 1.55 in the linear crack size. The deviation is about 16 %. One can see that the developed model gives the results which are rather close to the experimental data despite the rough assumptions used in the model. The conclusion that an increase in the informational entropy of contact stress distribution leads to an increase of damage in work material and to the improvement of destructing ability of tool was obtained on the basis of the consideration of the elementary forms of indenters and then confirmed numerically. Now let us compare this result with technical solutions in the area of the tool improvement. To do this, we use the Table 1 from [7], which presents a result of an analysis of about 250 patents of different countries in the area of tool improvement. The main ideas of all patents and technical solutions were studied and compared with ideas of other patents.

Correlating the ideas from all of the groups, one can see that all considered methods of tool improvement consist in introducing some heterogeneity (in other words, information) in tool constructions. The groups 1 and 2 from this Table correspond fully to the above conclusion: the increase in the destructing ability of tool is achieved by increasing the informational entropy of contact stress distribution. If one compares this observation with the numerical conclusion above and the results obtained on the basis of investigations of tool wear [7,9], and generalizes them, one can state the following general principle of drilling tool design: the efficiency of the drilling tool increases with increasing the heterogeneity of distribution of local parameters of tool. In terms of the information theory it is formulated as follows: the greater are the informational entropies of distributed parameters of drilling tool the higher is the tool efficiency. Therefore, the comparison of the informational approach to the tool improvement with the technical solutions allows not only to confirm but also to generalize this approach.

4. Discussion and directions of further investigation

In this paper, the method of description of stressed state in deformed material under conditions of lack of information about the material structure and properties, and the general characteristic of contacting bodies, which determines the destruction in them and can be used as a criterium of optimization of complex tool, are developed on the basis of information-entropic approach. One can see that this approach allows to find solutions of problems which are difficult to solve with the use of the traditional mechanical methods.

Two extreme cases are considered above: in first case, the effect of contact stress distribution (as well as other geometrical effects) is neglected and the damage growth is supposed to be determined only (or mainly) by local fluctuations of stress field, and the second case, when the contact stress distribution is supposed to determine fully the damage evolution in loaded bodies. Actually, both local fluctuations of stresses and the distribution of stresses averaged over some representative volume determine the damage evolution in materials. In order to take both these effects into account, one may proceed as follows. The contact interaction of two bodies can be considered as a transmission of information from a tool to a work material. As a clear example of such model, one may take the indentation into a plastic material: the information contained in the indentor shape is transmitted into the shape of hollow; the correspondance between input and output is evident. For the case of the contact interaction of disordered materials, the information contained in tool (the shape of tool, for instance) does not determine the results of interaction fully, like in above example, but influences the probability distribution of possible states of loaded material after interaction. It means that the entropy term in the Lagrangian (4) decreases by some value related with the information contained in the

TABLE 1. Technical solutions in the area of tool design [7]

Main ideas	Some examples
1. Unevenness of the tool work surface: making a cutting face convex or concave, or prismatic or cylindrical lugs on cutting face; stepped working surface; cavities, bevels, slopes on the tool working surface	No 1044765A, 1023062A, No. 1323706A1, 623958 (USSR); No. 1284539 (UK) No 57- 35357 (Japan)
2. Asymmetry of tool working surface about direction of tool movement: a cutting face or its parts are inclined to the cutting vector	723123 and 1046465A (USSR)
3. Using teeth of dissimilar shapes or orientations on the same bit: combination of radial and tandential cutters; different cutting and wedge angles on teeth from one bit; using different materials of inserts; the strength of inserts changes from axis of auger to periphery.	No. 395559, 153680A1, 1366627A1 (USSR)
4. Irregular arrangement of teeth on a bit: teeth or cutters are placed in pairs or in groups; various distance between teeth	No. 3726350, 3158216 (USA) No. 1472623A1 (USSR)
5. Elements with different mechanisms of loading on a bit are combined: combination of mobile and fixed elements,or rotating and progressively moving elements, or cutting and impact elements	No. 52-48082 (Japan) 697711 (USSR)
6. Different wear-resistances of different points of tool working surface: layers with different strengths in a cutter; diamond coatings and graded materials; cavities of required shape in tool	No. 714003, 281349. 145496, 693000, 609884 (USSR)
7. Self-sharpening and self-organization of tool	No.4230193 (USA) No.717327, 719192 (USSR)

510

tool. It can be described by the following way: the interrelation between stresses in neighbouring points is introduced into the Lagrangian (as an autocorrelation function, for instance); the stresses in some points (namely, in the points near the contact surface) are prescribed; the function (5) is dependent on the coordinates of the point; the probability distribution of stress in the loaded material is determined on the basis of the entropy maximization as well. Such procedure allows to investigate the effect of tool shape on the stress distribution and damage in disordered material. The effect of the term with the contact stress entropy in the Lagrangian (4) on the damage density in material can be estimated qualitatively in the simplest case when the effect of correlation of stress levels in neighbouring points is neglected: the introduction of new constraints in the Lagrangian leads to the more sharp distribution $p(\sigma)$ and, consequently, to an increase of damage density in the material.

In conclusion one may formulate the following ways of complex tool improvement which result from the above considerations. The destructing ability of a tool can be increased by increasing the informational entropy of contact stress distribution over the contact surface between the tool and work material. Destructing elements (teeth, cutters) on the complex (multiteeth) tool should be arranged in pairs or in groups; that allows to use the indexing effect and to increase the intensity of the destruction of the work material. Generally, the efficiency of a complex tool can be improved if the informational entropies of distributions of local parameters of the tool are increased.

References

1. Lippmann, H. (1996) Sense and nonsense of averaging stress and strain, *Crystal plasticity modelling: abstracts of workshop*, MPA, Stuttgart
2. Mishnaevsky Jr, L.L. and Schmauder, S. (1997) Damage evolution and localization in heterogeneous materials under dynamical loading: stochastic modelling, in *Proceedings IUTAM Symposium on Innovative Computational Methods for Fracture and Damage, June 30 - July 5, 1996, Dublin, Ireland* (to be published)
3. Wilson, A.G. (1970) *Entropy in urban and regional modelling*. Pion Ltd, London
4. Mishnaevsky Jr, L.L. (1996) Determination for the time to fracture of solids, *Int. J. Fracture* **79** (4), 341-350
5. Zhlobinsky, B.A. (1970) *Dynamic fracture of rocks under indentation*, Nedra, Moscow
6. Galin, L.A. (1961) *Contact problems in the theory of elasticity*, North Carolina State College
7. Mishnaevsky Jr, L.L. (1996) A new approach to the design of drilling tools, *Int. J. Rock Mech. Min. Sci. and Geomech. Abstr.* **33**(1), 97 - 102
8. Eigheles, R.M., Strekalova, R.M. and Mustafina. N.N. (1975) Choice of optimal sizes of rock-destructing elements and their arrangement on the drilling bit, in R.M. Eigheles (ed) *Rock fragmentation*, VNIIBT, Moscow, pp. 136-150
9. Mishnaevsky Jr, L.L. (1995) Mathematical modelling of wear of cemented carbide tools in cutting brittle materials, *Int.J. Machine Tools and Manufacture*, **35** (5), 717-724

STOCHASTIC MODELING OF FATIGUE CRACK GROWTH IN METALS

K. DOLIŃSKI
Centre of Mechanics, Institute of Fundamental Technological Research
ul. Świętokrzyska 21, PL-00-049 Warsaw, Poland

1.Introduction

In fatigue damage problems there are two sources of stochastic uncertainty that significantly influence the fatigue damage process and, eventually, the assessments of the structural lifetime. Firstly, the fatigue damage process, in particular the fatigue crack growth, is observed to be of stochastic character even in very well-controlled fatigue experiments under deterministic constant amplitude loading, e.g. Virkler et al. [1], Ghonem and Dore [2]. In these cases the random nature of material non-homogeneity is the only reason to produce a scatter of results. Some material parameters in fatigue damage evolution equations are usually admitted to be random variables or fields. Such a randomization in the fatigue damage model reflects the effect of the first source of stochastic uncertainty originated by material randomness, cf. [3, 4].

The second source of stochastic scatter in fatigue damage problems may result from loading features. Many kinds of fatigue loading are of stochastic nature. The problem appears, however, more complex due to the load cycle sequence effect influencing the fatigue damage process. For fatigue crack growth it is experimentally well documented, e.g. a paper by Schijve [5], a review by Kumar [6], that load cycles with single or multiple peak tensile overloads result in retardation of fatigue crack growth or even in crack arrest. The transient diminution of the crack propagation rate, the duration of the retardation phase and the magnitude of the retardation effect depend on many factors including specimen geometry, environmental effects, material properties, the magnitude of the overload and of subsequent extremes. The physical nature of this phenomenon has not been completely explained, yet. Various mechanisms have been suggested to rationalize it: crack tip blunting [7], crack tip strain hardening and the generation of a favorable residual stress field ahead of the crack tip [8, 9], plasticity-induced fatigue crack closure [10], crack branching and micro-roughness of fracture surfaces [11, 12]. All of these mechanisms are present and observed in fatigue experiments to affect the post-overload fatigue crack growth. The plasticity-induced fatigue crack closure is, however, generally considered as a dominant cause of the retardation in Mode I of fatigue crack growth, cf. [13].

Accounting for effects of previous cycles makes the fatigue crack growth models to be load history dependent. Hence, the fatigue crack growth cannot be anymore considered as a memoryless process. Most of the models proposed in the literature to

G. N. Frantziskonis (ed.), PROBAMAT – 21st Century: Probabilities and Materials, 511–530.
© 1998 *Kluwer Academic Publishers.*

predict the fatigue crack growth with regard to the load sequence effects refer to the overload-induced plastic zone and a diminution of the effective stress intensity factor range after an overload, see e.g. [14]. Such an approach was also applied in modeling of fatigue crack growth under stochastic loading by Ditlevsen and Sobczyk [15], Doliński [16], Veers [17], Veers *et al.* [18]. Mathematical tools and solution methods differ, however, substantially in the quoted papers. Birth process, averaged load characteristics, diffusion Markov process, numerical simulation were there, respectively, used in derivation of statistical characteristics of the structural lifetime when a critical fatigue macro crack length defines the structural failure due to stochastic loading. The complexity of the fatigue phenomena is very difficult to be satisfactory taken into account by pure analytical approach. On the other hand the numerical simulation alone appears often to be ineffective. A very great number of load cycles to failure requires many very long samples of stochastic loading to be simulated and used in fatigue calculation. It is also inefficient in eventual reliability analysis of a structure.

A mixed, partially numerical, partially analytical approach has been originally suggested by Colombi and Doliński in [19] and modified and improved in [20] and [21]. The approach leads eventually to analytical form of probability distribution of the fatigue lifetime while the retardation effects due to stochastic loading as well as the random material properties are taken into account. In the present contribution this approach is presented and shown to yield some quite accurate lifetime assessments for stochastically loaded specimens investigated by Sarkani *et al.* [22] in a series of fatigue experiments. Another example illustrates the effect of the loading power spectrum bandwidth on fatigue lifetimes of a initially cracked specimen under stationary Gaussian stochastic excitation.

2. Retarded Fatigue Crack Growth

The fatigue crack length increment, Δa_i, due to the i-th load cycle can be written in the following general form

$$\Delta a_i = F\left(a_i, S_i^+, S_i^- | \mathbf{x}\right) \tag{1}$$

where a_i denotes the current crack length at the moment of application of the i-th load cycle. The quantities S_i^+ and S_i^- denote, respectively, the stress maximum and minimum in the i-th cycle of the far-field stress applied to a cracked element. The vector $\mathbf{x} = [x_1, x_2, ..., x_K]$ represent the material parameters. In general, the parameters may be assumed to be some random variables. In that case the vector $\mathbf{x}$ denotes a sample of a random vector $\mathbf{X} = [X_1, X_2, ..., X_K]$.

Among a great number of fatigue crack growth laws proposed in the literature, see e.g. [23], there is a very wide class of equations for which the function $F\left(a_i, S_i^+, S_i^- | \mathbf{x}\right)$ can be written as a product of two functions, i.e.

$$\Delta a_i = g(a_i | \mathbf{x}) \cdot \Xi\left(S_i^+, S_i^-\right) \tag{2}$$

Such a form has the well-known fatigue crack growth equation proposed by Paris and Erdogan in [24]

$$\Delta a_i = C \cdot \Delta K_i^m \tag{3a}$$

Since the stress intensity factor range corresponding to the i-th stress cycle is given by $\Delta K_i = Y(a_i) \cdot \Delta S_i \cdot \sqrt{\pi \cdot a_i}$ the functions $g(\cdot)$ and $\Xi(\cdot, \cdot)$ in the Paris-Erdogan equation take the following forms (subscript "i" omitted)

$$g(a) = C \cdot Y^m(a) \cdot \left(\sqrt{\pi \cdot a}\right)^m \quad \text{and} \quad \Xi(S^+, S^-) = (S^+ - S^-)^m = \Delta S^m \tag{3b}$$

with the coefficient C and the exponent m being some parameters that generally depend on material and load conditions. The dimensionless function Y(a) depends on the crack and specimen geometry.

In 1971 Elber [10] noticed that the fatigue crack may close before reaching the minimum, S^-, by the stress process and remains closed during the stress increase when $S^- \leq S \leq S_{op}$, where S_{op} denotes the crack opening stress. Since the fatigue crack growth is affected only by the part of a stress cycle above S_{op} the effective stress cycle amplitude, $\Delta S_{eff} = S^+ - S_{op}$, should be considered in fatigue crack equations instead of ΔS if $S_{op} > S^-$. The crack opening stress, S_{op}, remains constant under constant amplitude loading but depends on the load condition. In the literature there is no universal formula describing such a relation. Most of the proposals are based on experimental data, see e.g. [25, 26], and introduce a function, $q(S^-, S^+)$, depending on the current extremes or on the stress asymmetry ratio alone, $R = S^-/S^+$, so that $S_{op} = q(S^-, S^+) \cdot S^+$.

The crack closure and opening effects result from crack tip plasticity and are usually modeled by referring to the plastic zone that develops at the crack tip due to the stress cycle maximum. The range of the plastic zone can be estimated [27] as

$$r_Y(a, S) = \frac{\gamma}{\pi} \cdot \frac{K^2}{\sigma_Y^2} = \frac{\gamma \cdot Y^2(a)}{\sigma_Y^2} \cdot S^2 \cdot a \tag{4}$$

The coefficient $\gamma = 1$ for plane stress and $\gamma = (1 - 2\nu)^2$ for plane strain condition with ν as the Poisson coefficient. It is observed in fatigue experiments under constant amplitude loading with a single overload, S_{ol}, that the crack opening stress, S_{op}, increases transitorily after the overload application and then returns to its pre-overload value, e.g. [28]. As a consequence it lessens the effective stress amplitude, ΔS_{eff}, and the fatigue crack growth rate, eventually. In some retardation models, e.g. [29, 30], the retarded growth of fatigue crack after an overload is assumed to continue so long as the current plastic zones, $r_Y(a, S^+)$, due to the maxima, S^+, following an overload, S_{ol}, are contained in the plastic zone, $r_Y(a_{ol}, S_{ol})$, created by the overload at $a = a_{ol}$, cf. Fig. 1.

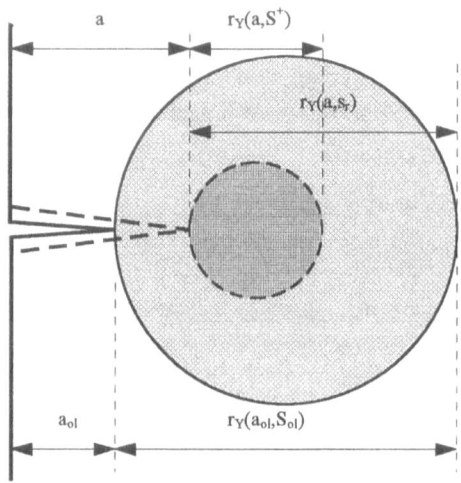

Fig. 1. The overload-induced plastic zone range, $rY(a_{ol}, S_{ol})$, the current one, $r_Y(a, S^+)$, and the one corresponding to the reset stress, $r_Y(a_{ol}, S_r)$.

In the Willenborg model the reduction of the fatigue crack growth rate after an overload is associated with the stress, s_r, called the reset stress, necessary to create a plastic zone, $r_Y(a, S_r)$, that would reach the boundary of the overload-induced plastic zone, $r_Y(a_{ol}, S_{ol})$. The reset stress is calculated from the equality $a_{ol} + r_Y(a_{ol}, S_{ol}) = a + r_Y(a, S_r)$, where a is the current crack length, and is given as follows

$$S_r = S_r(a_{ol}, a; S_{ol}) = \frac{\sigma_Y}{\gamma \cdot Y(a)} \sqrt{\frac{a_{ol}}{a} \cdot \left(1 + \frac{\gamma \cdot Y^2(a_{ol}) \cdot S_{ol}^2}{\sigma_Y^2}\right) - 1} \qquad (5)$$

Large compressive stresses around the fatigue crack tip after unloading associated with the large overload-induced plastic zone hinders the crack opening due to the subsequent stress cycles. The opening stresses corresponding with the post-overload maxima, $S_{op,r}$, transitorily increase while the current plastic zones move within the overload-induced one. In order to specify the retardation intensity Veers [17] assumed the augmented opening stress, $S_{op,r}$, to be equal to $S_{op,r} = (s_r/S^+) \cdot S_{op}$, where $S_{op} = q(R) \cdot S^+$ denotes the opening stress corresponding with the current stress cycle (S^-, S^+) (without retardation effects). Considering the effective amplitude, ΔS_{eff}, relevant to the fatigue process as a difference between the stress maximum and the effective minimum, $\Delta S_{eff} = S^+ - S_{eff}^-$, the latter is explicitly defined as follows

$$S_{eff}^- = S_{eff}^- \left(a, S^-, S^+ \middle| a_{ol}, S_{ol};\right) = \begin{cases} q(R) \cdot S^+ & \text{if } S_r < S^+ \text{ and } S^- < S^+ \\ q(R) \cdot S_r(a \middle| a_{ol}, S_{ol}) & \text{if } q(R) \cdot S_r < S^+ < S_r \\ S^+ & \text{if } q(R) \cdot S_r > S^+ \text{ or } S^- > S^+ \end{cases} \qquad (6)$$

Introducing the reset stress, $S_r(a \middle| a_{ol}, S_{ol})$, as a variable governing the retardation effects and depending on the current crack length, a, on the crack length, a_{ol}, at the time of application of the last overload, S_{ol}, and on the overload value itself we loose a very convenient product form property of the fatigue crack growth equation as given in (2). The crack length increment, Δa_i, due to a single stress cycle application can be written as follows

$$\Delta a_i = g(a_i \middle| \mathbf{x}) \cdot \Xi \left(S_i^+, S_i^-; S_r\left(a_i \middle| a_{ol}, S_{ol}\right)\right) \qquad (7)$$

The phenomenon of fatigue crack growth retardation introduces a memory effect. It excludes a separation of variables allowing for definition of any damage parameter, Γ, depending on the load process alone and satisfying the Palgrem-Miner hypothesis on linear accumulation of damage. Thus, an involved cycle-by-cycle incremental analysis has to be always performed to determine the development of damage and, eventually, the fatigue crack length due to variable-amplitude loading. Such a time consuming procedure is apparently impractical for stochastic loading for which any load path is only a sample of a stochastic process. The entirely incremental approach to complete statistical information about stochastic behavior of the fatigue process adequate for reliability analysis would require extensive numerical simulation. In the method described herein some features of the fatigue crack growth retardation phenomenon under stochastic loading and some statistical properties of extremes of stochastic processes are noticed and used to significantly simplify and restrict the numerical simulation that furthermore, provides some initial and sufficient data to continue the probabilistic analysis analytically.

3. Block Structure of Fatigue Crack Growth Process

Considering the fatigue loading as a stochastic process it is easily seen that the retardation phase of the fatigue crack growth may be initialized by any random maximum provided that some conditions about the last previous maximum and the subsequent one are satisfied. The necessary and sufficient condition for a maximum, S_i^+, to be an overload, $S_i^+ = S_{ol,m} = S_{I_o(m)}^+$, initializing the m-th new retardation phase can be formulated as follows

$$S_i^+ = S_{ol,m} \quad \text{iff} \quad \left\{ S_i^+ \geq S_{i-1}^+ \text{ and } S_i^+ \geq S_{r,m-1}(a_i) \text{ and } S_{i+1}^+ \leq S_{r,m}(a_{i+1}) \right\} \qquad (8)$$

where $S_{r,m}(a) = S_r(a_{ol,m}, a; S_{ol,m})$ and $S_{r,m-1}(a) = S_r(a_{ol,m-1}, a; S_{ol,m-1})$ as defined in (5) with the crack lengths, $a_{ol,m} = a_{I_o(m)}$ and $a_{ol,m-1} = a_{I_o(m-1)}$, and the corresponding maxima, $S_{I_o(m)}^+$ and $S_{I_o(m-1)}^+$, supposed to be the m-th overload, $S_{ol,m}$, and the overload before, $S_{ol,m-1}$, respectively. The subscript $I_o(\cdot)$ numbers the maxima at which the retardation phases begin.

The maximum, $S_{I_o(m)}^+$, satisfying the conditions (8) becomes an overload and starts a new retardation phase which continues as long as (the subscript "m" indicating the number of the last actual overload having initialized the current retardation phase is omitted)

$$S_i^+ \leq S_r\left(a_{ol}, a_i; S_{ol}\right) \quad \text{for } i > I_o(m) \qquad (9)$$

If the condition (9) for continuation of the retardation phase is not satisfied for a maximum, $S_{I_r(m)}^+$ say, where $I_r(\cdot)$ numbers the first maxima after the ends of retardation phases, this maximum can be the next overload, $S_{I_r(m)}^+ = S_{I_o(m+1)}^+ = S_{ol,m+1}$, or it can start a post-retardation phase which will continue as long as

$$S_{i-1}^+ \leq S_i^+ \leq S_{i+1}^+ \quad \text{for } i > I_r(m) \qquad (10)$$

If the condition for continuation of the post-retardation phase is not satisfied for a maximum S_i^+, this maximum is assumed to start the next retardation phase, i.e. $S_i^+ = S_{ol,m+1} = S_{I_o(m+1)}^+$.

This scheme can be extended on the whole fatigue crack propagation process which appears to alternately consist of retardation and post-retardation phases. The couples of these successive phases are considered as *retardation blocks* starting and terminating with overloads. It will be shown that the block structure appears a decisive feature in stochastic approach to fatigue damage problems.

3.1. FATIGUE DAMAGE WITHIN A RETARDATION BLOCK

Equation (7) can be transformed into the form

$$\frac{\Delta a_j}{g(a_j|\mathbf{x})} = \Xi\left(S_j^+, S_{j,eff}^-\left(S_j^+, S_j^-, a_j|a_{ol}, s_{ol}\right)\right) = \Delta\Gamma_j\left(S_j^+, S_j^-; a_j, |a_{ol}, s_{ol}\right) \qquad (11)$$

where the subscript "j", $j = 1,2,..$, numbers the subsequent crack length increment, $\Delta a_j = a_{j+1} - a_j$, due to the j-th subsequent load cycle. The left-hand side of (11) denotes an increment of the fatigue damage parameter due to the crack length increment Δa_j. The right-hand side of (11) can be considered as an elementary increment of the load fatigue indicator corresponding with an effective single random stress cycle, $\Delta S_{eff,j} = S_j^+ - S_{eff,j}^-$, given the last overloading $S_{ol} = s_{ol}$ has occurred at $a = a_{ol}$. Considering only one retardation block with crack length increment, $B(s_{ol}, a_{ol}, \mathbf{x})$, and with a number of stress cycles, $N_B(s_{ol}, a_{ol}, \mathbf{x})$, the increment of the fatigue damage parameter, $\Gamma_B(s_{ol}, a_{ol}, \mathbf{x})$, within a retardation block can be alternatively written as follows

$$\Gamma_B(s_{ol}, a_{ol}, \mathbf{x}) = \int_{a_{ol}}^{a_{ol} + B(s_{ol}, a_{ol}, \mathbf{x})} \frac{da}{g(a|\mathbf{x})} = \sum_{n=1}^{N_B(s_{ol}, a_{ol}, \mathbf{x})} \Delta\Gamma(S_n^-, S_n^+; a_n|s_{ol}, a_{ol}) \qquad (12)$$

where the subscript "n", $n = 1,2,..,N_B(s_{ol}, a_{ol})$, runs over stress extremes within the block.

Looking at the reset stress expression given in (5) and accounting for the strong inequalities $B_m \ll a_{ol,m}$ appearing to be satisfied for any $a_{ol,m}$ from a crack length interval $[a_0, a_F]$ where a_0 and a_F denote the initial and admissible ultimate crack length, respectively, and the subscript "m" numbers the retardation blocks during the fatigue crack growth, it is easily seen that the effect of the reset stress depends on the ratio $a_{ol,m}/a$ with $a_{ol,m} \le a \le a_{ol,m} + B_m$, rather than on the location, $a_{ol,m}$, of the block within the interval $[a_0, a_F]$ alone. Thus, the increments of the load fatigue indicator over a block are almost independent of the block location, $a_{ol,m}$. Moreover, the reset stress does not depend on the material parameter vector, $\mathbf{x}$. Hence, the number of cycles within a block can be considered as independent both of $a_{ol,m}$ and of $\mathbf{x}$, i.e.

$$N_{B,m}(s_{ol}, a_{ol,m}, \mathbf{x}) \approx N_B(s_{ol}) \qquad (13)$$

All of these makes also the fatigue damage parameters over blocks, B_m, approximately independent of $a_{ol,m}$ and $\mathbf{x}$, i.e.

$$\Gamma_{B,m}\left(s_{ol}, a_{ol,m}, \mathbf{x}\right) \approx \Gamma_B(s_{ol}) \qquad (14)$$

4. Simulation Procedure

Under stochastic loading the damage parameter, $\Gamma_B(s_{ol})$, and the number of cycles, $N_B(s_{ol})$, within a retardation block, B, are random variables. According to the arguments given above they do not depend on the stage of the fatigue process. Thus, they have got the same probability characteristics for every retardation block. In this section some properties of extremes of stationary stochastic processes are briefly reported and their usefulness in computation of the joint probability distribution of (Γ_B, N_B) is presented.

4.1. EXTREMES OF STATIONARY STOCHASTIC LOADING

Extremes of the stress process are the only load parameters involved in fatigue crack growth equations. Therefore, just their probabilistic characteristics are desired to predict the structural lifetime due to the macrocrack propagation. Unfortunately, a full stochastic description of sequence of extremes can be obtained in very specific cases of stochastic processes only. Recently, Frendahl & Rychlik [31] has shown on a very wide numerical simulation basis that a homogeneous Markov chain is a very good approximation of the random sequence of extremes of stationary Gaussian and non-Gaussian processes with various spectral characteristics. Any homogeneous Markovian sequence, $[S_k] = [S_1, S_2, ...]$, is fully described by a transition probability density function $p(s_k|s_{k-1})$ of any current term of the sequence, e.g. S_k, given the last previous term $S_{k-1} = s_{k-1}$. Since analytical derivation of the transition probability density function of extremes of stochastic process is, in general, hardly possible a numerical procedure proposed in [31] can be applied.

The range of possible values of extremes is discretized into a finite number of levels, $u_1 < u_2 < \cdots < u_{N_u}$. Samples of stochastic load process obtained from numerical simulation or from actual observations are statistically analyzed to estimate two transition probability matrices. The first matrix, $\mathbf{P}$, contains the probabilities of transitions of the process from its local maximum to the subsequent minimum. Elements of the second matrix, $\hat{\mathbf{P}}$, denote the probabilities of transitions of the process from its local minimum to the subsequent maximum. The distributions of maxima, π, and minima, $\hat{\pi}$, can be determined analytically, if possible, directly from statistical analysis or calculated from $\pi = \pi \cdot \mathbf{P} \cdot \hat{\mathbf{P}}$ and $\hat{\pi} = \hat{\pi} \cdot \hat{\mathbf{P}} \cdot \mathbf{P}$ with normalizing conditions $\sum_{i=1}^{N_u} \pi_i = 1$ and $\sum_{i=1}^{N_u} \hat{\pi}_i = 1$, respectively.

The assumption of Markovian character and the transition probability matrices, $\mathbf{P}$ and $\hat{\mathbf{P}}$, completely define the random sequence of extremes. They suffice to propose a simple numerical procedure that allows us to follow all probabilistic characteristics associated with fatigue crack propagation.

4.2. JOINT PROBABILITY DISTRIBUTION OF (Γ_B, N_B)

The Markov property of stress extremes allows us to apply a simple numerical simulation scheme to estimate the probability distribution of (Γ_B, N_B), cf. (14) and (13). Simulation begins with assumption of a crack length, $a = a_{ol}$, and an overload, $S_{ol} = u_k$. A discrete crack length increment, δa, is assumed so that the current crack lengths $a_i = a_{ol} + i \cdot \delta a = a_{ol} + b_i$, $i = 1, 2, \dots$ Successive calculation (cycle after cycle) is performed according to a random walk scheme from the initial state, a_{ol}, to any state, $a = a_i$, with the transition probabilities, $\mathbf{P}$ and $\hat{\mathbf{P}}$, while. the inequality conditions (8), (9), (10) and equation (7) for the crack length increment due to a current stress cycle, $\left(S_j^-, S_j^+\right)$ are taken into account. The calculation directly provides the joint probability distribution

$$P_{B,N_B}(b_i, n | u_k, a_{ol}, \mathbf{x}) = P\left[B(u_k, a_{ol}, \mathbf{x}) = b_i \wedge N_B(u_k, a_{ol}, \mathbf{x}) = n\right] \tag{15}$$

of the crack length increment, $B(u_k, a_{ol}, \mathbf{x})$, within the retardation block which has started with an overload $S_{ol} = u_k$ at $a = a_{ol}$ and of the number of stress cycles, $N_B(u_k, a_{ol}, \mathbf{x})$, within this block. Repeating the calculation for all values of maxima as overloads, $u_1 < u_2 < \dots < u_{N_u}$, we can release the overload condition as follows

$$P_{B,N_B}(b_i, n | a_{ol}, \mathbf{x}) = P\left[B(a_{ol}, \mathbf{x}) = b_i \wedge N_B(a_{ol}, \mathbf{x}) = n\right] =$$
$$= \sum_{k=1}^{N_u} \pi_k \cdot P_{B,N_B}(b_i, n | u_k, a_{ol}, \mathbf{x}) = \sum_{k=1}^{N_u} \pi_k \cdot P\left[B(u_k, a_{ol}, \mathbf{x}) = b_i \wedge N_B(u_k, a_{ol}, \mathbf{x}) = n\right] \tag{16}$$

Applying the relation (12) we transform the probability distribution of (B, N_B), into the probability distribution of (Γ_B, N_B) as follows

$$P_{\Gamma_B, N_B}(\gamma_i, n) = P_{B, N_B}(b_i, n | a_{ol}, \mathbf{x}) \tag{17}$$

where the substitution $\gamma_i = \int_{a_{ol}}^{a_{ol}+b_i} da/g(a|\mathbf{x})$ allows us to release the condition $(a_{ol}, \mathbf{x})$ as explained at the end of section 3.

5. Number of Stress Cycles to Failure

At failure when the crack, initially of the length a_0, reaches its critical size, a_F, the fatigue damage parameter takes a value $\gamma_F(\mathbf{x}) = \gamma(a_0, a_F | \mathbf{x})$. i.e.

$$\gamma_F(\mathbf{x}) = \gamma(a_0, a_F | \mathbf{x}) = \int_{a_0}^{a_F} \frac{da}{g(a|\mathbf{x})} \approx$$

$$\approx \sum_{m=1}^{M_F(\mathbf{x})} \Gamma_{B,m} = \sum_{m=1}^{M_F(\mathbf{x})} \sum_{n=1}^{N_{B,m}} \Delta\Gamma(A_{m,n}, S_{m,n}^+, S_{m,n}^- | A_{ol,m}) \qquad (18)$$

where the subscript "m", m = 1,2,...,M(x), numbers the retardation blocks, $B_m = B(A_{ol,m}, \mathbf{x})$, that start at the random crack lengths $A_{ol,m}$ while $A_{ol,m} = A_{ol,1} + \sum_{\mu=1}^{m-1} B_\mu$, subsequently for m = 2,3,...,M. The approximate equality sign in (18) points out an error resulting from an inconsistency of the initial crack length, a_0, with the beginning of the first block, $A_{ol,1}$, and of the critical crack length, a_F, with the crack length at the end of the last M-th block. The strong inequality $B_m \ll (a_F - a_0)$ assures this error to be negligible.

The fatigue damage parameter, $\Gamma_F(M|\mathbf{x})$, is the random variable depending on the number, M(x), of the retardation blocks, B_m. Until the fatigue failure the fatigue damage parameter should be less than a critical value, i.e. $\Gamma_F(M_F|\mathbf{x}) < \gamma_F(\mathbf{x})$, corresponding with the critical crack length, a_F, given the material parameter vector $\mathbf{X} = \mathbf{x}$. Thus, the probability of failure given $\mathbf{X} = \mathbf{x}$ is defined as

$$P_F(\mathbf{x}) = P[\Gamma_F(M_F|\mathbf{x}) > \gamma_F(\mathbf{x})] = P\left[\sum_{m=1}^{M_F(\mathbf{x})} \Gamma_{B,m} > \gamma_F(\mathbf{x})\right] \qquad (19)$$

where $M_F(\mathbf{x})$ denotes the random number of blocks to failure given $\mathbf{X} = \mathbf{x}$.

There is another very important observation that enables us to continue the calculation of the number of stress cycles to failure, N_F, analytically. All examples show that the following strong inequalities $n_{corr} \ll N_B \ll N_F$ are satisfied both for narrow- and wide-band spectra of stochastic load processes where n_{corr} denotes the correlation length (in number of cycles) of the random sequence of stress maxima. It allows us to consider the fatigue damage parameter, $\Gamma_F(M_F|\mathbf{x})$, after M_F blocks as a sum of M_F independent random variables having the same probability distributions given in (17). Statistical moments of the k-th order of the damage parameter, Γ_B, and of the number of cycles, N_B, within a retardation block, B, are easily calculated as follows

$$\overline{\Gamma_B^k} = \sum_{i=1}^{\infty} \gamma_i^k \cdot \sum_{n=1}^{\infty} P_{\Gamma_B, N_B}(\gamma_i, n)$$

$$\overline{N_B^k} = \sum_{n=1}^{\infty} n^k \sum_{i=1}^{\infty} P_{\Gamma_B, N_B}(\gamma_i, n)$$

It is well-known that the mean, $\overline{M_F}(\mathbf{x})$, of the random number, $M_F(\mathbf{x})$, of random variables, $\Gamma_{B,m}$, in the sum in (19) is equal to

$$\overline{M_F}(x) = \gamma_F(x)/\overline{\Gamma}_B \qquad (20)$$

For a great $\overline{M_F}$ value the probability distribution of the random number of independent random variables in such a sum can be approximated by the inverse Gaussian probability distribution, see [32], with the variance

$$\sigma^2_{M_F}(x) = \sigma^2_{\Gamma_B} \cdot \gamma_F(x) \big/ \overline{\Gamma}_B^3 = v^2_{\Gamma_B} \cdot \overline{M_F}(x) \qquad (21)$$

where $\overline{\Gamma}_B$ is the mean of Γ_B and $\sigma^2_{\Gamma_B} = \overline{\Gamma^2_B} - \overline{\Gamma}^2_B$ denotes its variance.

The total number of cycles to failure, $N_F(x)$, given $X = x$ is then given as the sum

$$N_F(x) = \sum_{m=1}^{M_F(x)} N_{B,m} \qquad (22)$$

with $N_{B,m}$ denoting the random number of cycles within the m-th retardation block. The numbers $N_{B,m}$ are statistically independent random variables. The number of blocks to failure, $M_F(x)$, is usually sufficiently great to apply the central limit theorem modified for a sum of a random number of random variables, cf. [33]. Thus, the probability distribution of the number of cycles to failure, $N_F(x)$, can be approximated by the Gaussian probability distribution

$$F_{N_F}(n;x) = P\big[N_F(x) \le n\big] \approx \Phi\left[\frac{n - \overline{M_F}(x) \cdot \overline{N}_B}{\sigma_{N_F}(x)}\right] \qquad (23)$$

where the variance of $N_F(x)$ is given as follows

$$\sigma^2_{N_F}(x) = \overline{M_F}(x) \cdot \sigma^2_{N_B} + \sigma^2_{M_F}(x) \cdot \overline{N}^2_B = \frac{\gamma_F(x)}{\overline{\Gamma}_B} \cdot \left(\sigma^2_{N_B} + v^2_{\Gamma_B}(x) \cdot \overline{N}^2_B\right) \qquad (24)$$

with $\overline{N}_B$ as the mean of N_B and $\sigma^2_{N_B} = \overline{N^2_B} - \overline{N}^2_B$ denotes its variance.

6. Time to Failure

A general relation between the random number, N_t, of cycles of a stochastic process and the corresponding time interval, T_n, results from the renewal theory

$$F_{N_t}(n|t) = P[N_t \le n|t] = 1 - P[T_n \le t|n] = 1 - F_{T_n}(t|n) \qquad (25)$$

and involves the probability distribution, $F_{N_t}(n|t)$, of the random number, N_t, of cycles occurring within a time interval $[0,t]$ and the probability distribution, $F_{T_n}(t|n)$, of length of the random time interval $[0,T_n]$ containing a given number of cycles, n. For long time intervals and some usually satisfied assumptions concerning the rate of decay of correlation function Malevich [34] proofed the limit theorem assuring the probability distribution of the number of zeros of a zero mean stationary Gaussian stochastic process to be approximately normal as $t \to \infty$, i.e.

$$F_{N_t}(n|t) = P[N_t \le n|t] \approx \Phi\left[\frac{n - \overline{N_t}(t)}{\sigma_{N_t}(t)}\right] \qquad (26)$$

where $\overline{N_t}(t)$ and $\sigma_{N_t}^2(t)$ denote the mean and variance of the number of zeros being some functions of the time interval range. In the fatigue analysis we have to count the maxima of the process. The number of maxima, $N_t^+(t)$, is equivalent to the number of zero upcrossings of the first derivative of the process. Thus, substituting $\overline{N_t}(t)$ and $\sigma_{N_t}^2(t)$ in (26) with the mean of maxima

$$\overline{N_t^+}(t) = v^+ \cdot t \qquad (27)$$

and variance of maxima which for $t \gg t_{corr}$ becomes also linearly dependent on time

$$\sigma_{N_t^+}^2(t) = S_N^+ \cdot t + \overline{S}_N^+ \qquad (28)$$

the probability distribution, $F_{N_t^+}(n|t)$, of the number of maxima, $N_t^+(t)$, within the time interval $[0,t]$ is obtained whereas v^+ in (27) denotes the mean rate of maxima, $v^+ = \sqrt{\lambda_4/\lambda_2}/(2\pi)$, while λ_i's are the spectral moments of the i-th order given by $\lambda_i = \int_{-\infty}^{\infty} \omega^i \cdot g_S(\omega)\, d\omega$ with $g_S(\omega)$ denoting the power density function of the stress process S(t). The coefficients S_N^+ and $\overline{S}_N^+$ in (28) depend on the correlation function of the stress process and their analytical derivation is given in [35], say, and applied to fatigue analysis in [36]. The coefficients can also be estimated from the stress path sample used in calculation of the transition probability matrices in the Markov approximation of the sequence of extremes.

Equation (25) and the probability distributions (23) and (26) allow us to derive the conditional probability distribution of the fatigue lifetime T_F given $\mathbf{X} = \mathbf{x}$ and write it down as follows

$$F_{T_F}(t|x) = P[T_F \le t|x] = \int_{-\infty}^{\infty} F_{T_n}(t|n) \cdot f_{N_F}(n|x) \, dn = 1 - \int_{-\infty}^{\infty} F_{N_t}(n|t) \cdot f_{N_F}(n|x) \, dn \approx$$

$$\approx 1 - \int_{-\infty}^{\infty} \Phi\left[\frac{n - \overline{N_t}(t)}{\sigma_{N_t}(t)}\right] \cdot \frac{1}{\sigma_{N_F}(x)} \cdot \varphi\left[\frac{n - \overline{M_F}(x) \cdot \overline{N_B}}{\sigma_{N_F}(x)}\right] dn = \Phi\left[\frac{\overline{N_t}(t) - \overline{M_F}(x) \cdot \overline{N_B}}{\sqrt{\sigma_{N_t}^2(t) + \sigma_{N_F}^2(x)}}\right] \tag{29}$$

The probability (29) has got the form similar to the so-called Birnbaum-Saunders probability distribution [37] with the mean and variance of the lifetime $T_F(x)$ as follows

$$\overline{T}_F(x) = \frac{1}{v^+} \cdot \left[\overline{N}_F(x) + \frac{S_N^+}{2 \cdot v^+}\right]$$

$$\sigma_{T_F}^2(x) = \frac{S_N^+}{v^{+2}} \cdot \left[\overline{T}_F(x) + \frac{\overline{S}_N^+ + \sigma_{N_F}^2(x)}{S_N^+} + \frac{3}{4} \cdot \frac{S_N^+}{v^{+2}}\right]$$

Depending on the dimension, K, of the material parameter vector $X = [X_1, X_2, \ldots, X_K]$ the unconditional probability distribution of the lifetime can be calculated by direct integration

$$F_{T_F}(t) = \int_{-\infty}^{\infty} F_{T_F}(t|x) \cdot f_X(x) \, dx \tag{30}$$

where $f_X(x)$ denotes the probability density function of the parameter vector X or by employing some approximate methods of reliability analysis involving a search for design points, first and second order reliability method, importance sampling, say.

7. Examples

Thought a lot of fatigue experiments were performed under pseudo-stochastic or even really stochastic loading conditions showing the load cycle sequence effects the published information on the experiment conditions is usually too scanty to be used in exhaustive verification of an approach proposed elsewhere. In the first example below some results of fatigue experiments published by Sarkani *et al.* in a series of papers [38, 39, 22] are compared with predictions based on the approach described above. The second example shows some possible effects of the retardation phenomenon on the statistical lifetime parameters for various bandwidths of the loading spectra.

7.1. COMPARISON WITH EXPERIMENTS

In experiments reported in [38, 39, 22] the welded cruciform specimens of steel plate were subjected to ideally zero-mean narrow-band Gaussian and non-Gaussian fatigue loading of various standard deviations. An auto-regressive procedure described in [40] was used to simulate the loading samples with correlation coefficient between the neighboring maxima kept to be equal to 0.95. The tests were conducted until failure which was defined in terms of specimen compliance. Due to development of cracks from the toe of welds at the center of the specimen towards the edges the compliance of the specimen increased and the tests were terminated when the compliance exceeded two times that at the initiation of the test. Thus, only one value: the number of cycles to failure, was obtained from one specimen subjected to a sample of stochastic load process.

In order to calibrate a damage model the constant amplitude fatigue tests were performed before with fully reversed stress cycles at various stress amplitude levels. In the original papers a Wöhler (S-N) curve, $N_F = K \cdot (S^+)^{-m}$, and Palgrem-Miner rule of damage accumulation were assumed for lifetime calculation. The parameters of the model was determined from the constant amplitude test results yielding $K = 10^{9.998}$ and $m = 3.564$. The approach proposed in the present paper requires a model of damage evolution. Unfortunately, the data set in the hand does not allow us to perform any rigorous analysis of an evolution model. Nevertheless, we can try to adopt a fatigue crack growth equation and fit its parameters so that the lifetime estimates calculated from the model could be accepted comparing with the experimental results from the tests under constant amplitude loading. The simplest Paris-Erdogan equation (3a) with effective stress amplitudes, cf. (6), accounting for the crack opening factor, $q(S^-, S^+)$, was considered as a candidate of the fatigue crack evolution model. Data fitting procedure resulted in the following relations

$$\Delta a = C \cdot \left(S^+ - S_{eff}^- \right)^2 \cdot a \qquad (31)$$

where the effective minimum definition, cf. (6), involves the crack opening factor, $q(S^-, S^+)$, in the following form

$$q(S^-, S^+) = q(S^+) = 1 - \left(\sqrt{\sigma_u / S^+} - 1 \right)^{-5/8} \qquad (32)$$

The coefficient $C = 1.91 \cdot 10^{-9}$. The form of $q(S^+)$ assures $\Delta a \rightarrow \infty$ as a stress maximum, S^+, tends to the ultimate stress, $\sigma_u = 683$ MPa. It is also seen that for $S^+ > 0.25 \cdot \sigma_u$ the parameter $q(S^+) < 0$ so that $S_{eff}^- < 0$ and a negative part of a cycle participates in the damage increment diminishing retardation effects. It seems that this feature may be specific for fatigue processes in welds. The initial and critical crack lengths in (18) are assumed to be $a_0 = 0.02$ mm and $a_F = 15$ mm. It should be pointed out that the forms of the equations, (31) and (32), and the values of the parameters result from data fitting rather than from reasoning based on mechanics. But it is also a similar situation with searching for a S-N curve. In Fig. 2 the S-N curve, experimental results and the predictions from the current model are plotted in logarithm coordinates $\log N_F$-$\log S^+$.

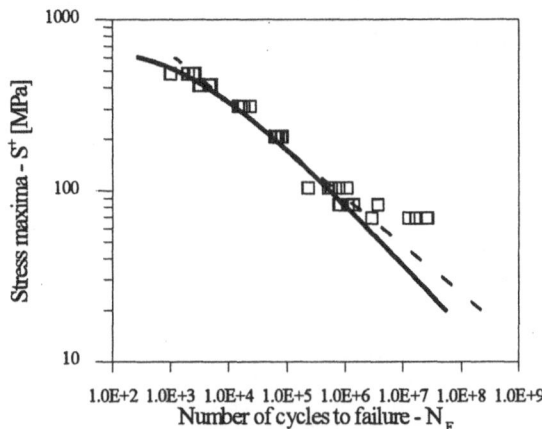

Fig. 2. Lifetimes under constant amplitude loadings: current model (solid line), Wöhler curve (dashed line), experimental results (□).

It is seen that the experiment results for the lowest amplitude values show the lifetime values to increase much faster than the linear relation in the log-log coordinates. This tendency would suggest an endurance limit to exist as it is often assumed in fatigue problems. As written also in [22] the hypothesis that the endurance limit observed under constant amplitude loadings does not exists (or at least is suppressed to a lower level) under variable amplitude loadings seems to be true looking at the experimental and analytical results. The zero endurance limit is also admitted in the paper. Any other value of it introduced here a serious error in predictions for stochastic loadings.

Due to the strict narrow-band property of the variable amplitude loadings preserved in experiments, the Rayleigh probability distribution of maxima could be assumed for Gaussian loading, X. Moreover, a non-linear transformation

$$Z = G(X) = \frac{X + \beta \cdot \left[\operatorname{sgn}(X)\right] \cdot |X|^n}{c}$$

was used by the authors to model non-Gaussian loadings, $Z = G(X)$, where the parameters β and n specify the intensity of the non-linearity and the coefficient c preserves a desired unit variance of the standard non-Gaussian process, Z. Two sets of values of the parameters (β, n, c) where assumed to generate non-Gaussian loading samples used in fatigue tests and calculations, namely (0.342, 2, 1.563) and (1.735, 0.5, 2.527) resulting, respectively, in two kurtosis values, $\kappa = 5$ and $\kappa = 2$, of the probability distribution of the non-Gaussian loading process. For the Gaussian process the kurtosis is equal to 3.

Experimental results and mean lifetime predictions published in [22] are compared with the predictions obtained from the approach presented in this paper. The lifetimes versus standard deviations of loading are shown in Fig. 3. The predictions with neglected retardation effects are also plotted.

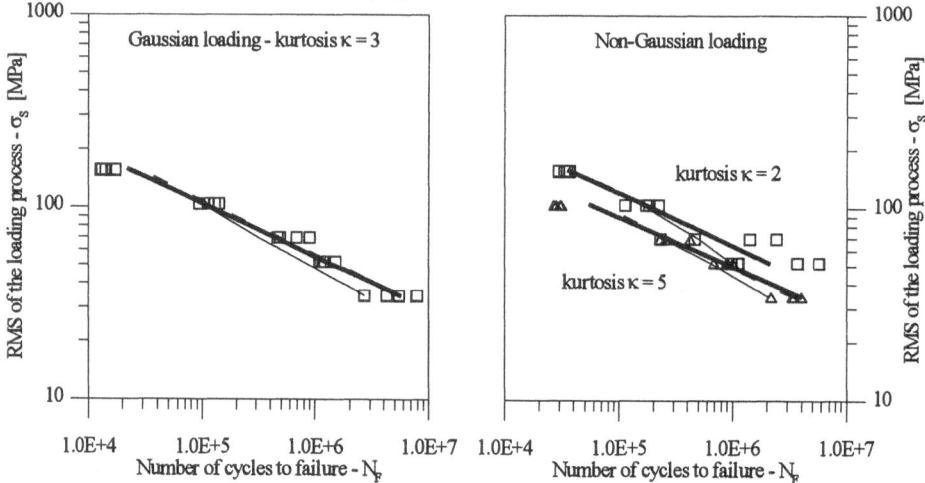

Fig. 3. Lifetimes under variable amplitude loadings: current model with retardation (thick line), current model without retardation (thin line), Wöhler curve (dashed line), experimental results (□ and .Δ).

7.2. EFFECT OF LOADING SPECTRUM BANDWIDTH ON FATIGUE LIFETIME

The fatigue crack propagation in a metal sheet in a region of stress concentration is considered. The yield stress of the material $\sigma_Y = 350$ MPa. The Paris-Erdogan equation (3) with $C = 4.1 \cdot 10^{-13}$, $m = 3.5$ and $Y(a) = (K_t - 1) \cdot \exp(-\beta \cdot a^\delta) + 1$ is assumed to describe the crack growth, where $K_t = 3.3$, $\beta = 1.0648$ and $\delta = 0.44$.

The basic external loading is modeled as a stationary zero mean Gaussian stochastic process with the following bimodal one side power spectrum density often used as stress spectrum in joints of offshore structures [41]

$$G(\omega) = \sigma^2 \cdot \frac{\kappa \cdot \exp[-1050/(\omega \cdot T_D)^4]}{\omega^5 \cdot T_D^4 \cdot \left\{[1 - (\omega/1.797)^2]^2 + (0.041 \cdot \omega/1.797)^2\right\}} \qquad (33)$$

where $\omega = 2 \cdot \pi \cdot f$ is the frequency in radians/sec with f in 1/sec, κ is the normalizing coefficient, T_D is the dominant wave period and σ denotes the standard variation of the stress process admitted to be equal to 75 Mpa in the example. The nine values of T_D, namely T_D = 46.15, 27.91, 21.33, 17.50, 14.77, 12.54, 10.48, 8.18, were considered resulting, respectively, in nine values of the bandwidth parameter $\alpha = \lambda_2/\sqrt{\lambda_0 \cdot \lambda_4}$ = 0.2, 0.3, 0.4, 0.5, 0.6, 0.7, 0.8, 0.9. For α = 0 the process is the ideally broad-band one and for α = 1 it is the ideally narrow-band one. The load spectra and sample stress paths for α = 0.2 and 0.9 are plotted in Figs. 4.

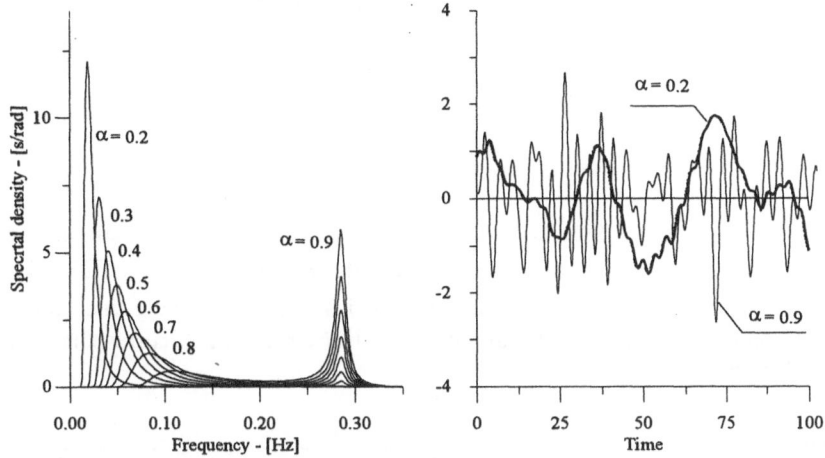

Fig. 4. Spectral power densities for various bandwidth parameters α and segments of the standard stochastic process samples for α = 0.2 and α = 0.9.

The stress amplitude, ΔS, in the stress intensity factor range in (3) is assumed to be the effective one, $\Delta S_{eff} = S^+ - S_{eff}^-$, with S_{eff}^- defined in (6). The crack opening factor, $q(S^+, S^-)$, is assumed to depend on the stress ratio, $R = S^-/S^+$, alone. In the present example any negative values of $q(R)$ are not allowed. Also negative parts of cycle ranges are assumed to not affect the crack growth. These limitations and the bilinear form $q_V(R) = \max\{q_0 \cdot (1+R/|R_0|)R\}$ proposed by Veers [17] and admitted here with the parameters q_0 = 0.496, R_0 = -5 lead to the following relation, cf. Fig. 5

$$q(R) = \max\{0, \min\{1, q_V(R)\}\}$$

528

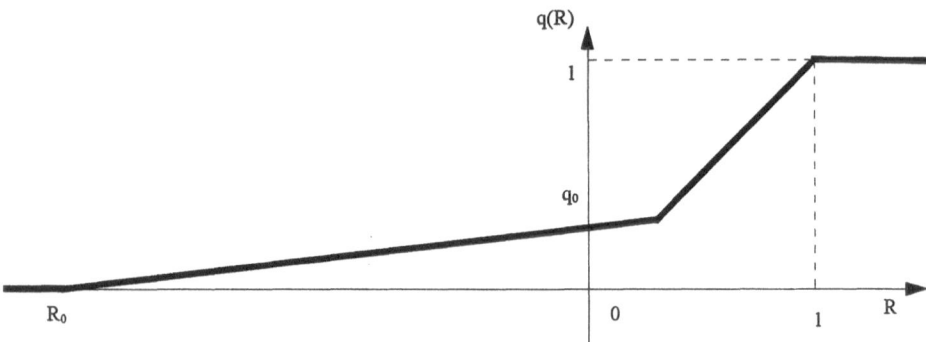

Fig. 5. Plot of the crack opening factor - $q(R) = S_{op}/S^+$

For the initial crack length, $a_0 = 0.15$ mm, and critical crack length, $a_F = 4$ mm, the means of lifetimes, $\overline{T}_F$ in hours, and coefficients of variations, COV, resulting from the current crack growth model are plotted in Fig.6. The solid lines correspond with the retarded crack growth solutions and the dashed lines correspond with the unretarded ones.

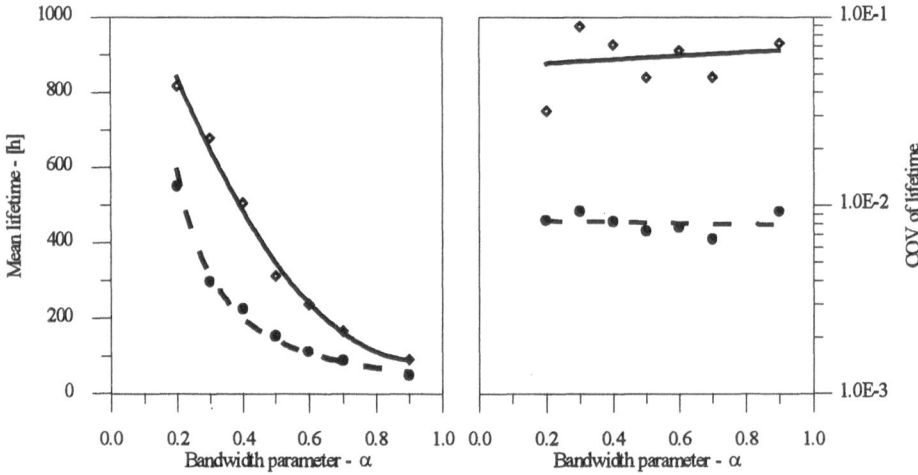

Fig. 6. Estimates of mean lifetimes, T_F, and coefficients of variation, COV, for *retarded (solid lines)* and *unretarded (dashed lines)* crack growth models.

It is seen that neglecting the retardation effects results in the fatigue lifetimes two times shorter than in the case when the retardation is considered in the model. It was also observed in the previous example for lower RMS values. This benefit is, however, reduced by an tenfold increase of the coefficient of lifetime variation. It must be especially taken into consideration in structural reliability analysis where any random scatter significantly affects the assessment of the structural failure probability.

8. Concluding Remarks

The analysis of fatigue damage process under stochastic loading proposed in the paper is based on two features:
- Markov property of extremes of stationary stochastic processes;
- block structure of the fatigue damage process reflecting the discrete memory marked by successive overloads.

The first one allows us to calculate the joint probability density function of the retardation block parameters by a simply numerical random walk procedure. The second one allows us to apply an analytical approach to derive the probability distribution of the fatigue lifetime. It is seen that the lifetime probability distribution function (29) depends directly and analytically on the following parameters:
- mean and variance of the number of stress cycles, N_B, within a retardation block, B, of the fatigue damage process;
- mean and variance of the fatigue damage parameter increment, Γ_B, over that block;
- coefficients, S_N^+ and $\overline{S}_N^+$, of the linear approximation of the variance,

$$\sigma_{N_t^+}^2(t) = S_N^+ \cdot t + \overline{S}_N^+ ;$$

- damage parameter value, $\gamma_F(\mathbf{x})$, corresponding with the damage increment from the initial to the critical damage state.

The division of the fatigue crack growth process into the blocks and Markov assumption about the sequence of extremes of the stress process shortens the simulation procedure providing all necessary parameters involved in the lifetime probability distribution. Many retardation models and most of the fatigue crack growth equations can be easily and well implemented into the numerical code.

9. References

1. Virkler, D.A., Hillberry, B.M., Goel, P.K. (1979) The statistical nature of fatigue crack propagation, *J. Enging Materials and Technology* **101**, 148-152.
2. Ghonem H., Dore S. (1987) Experimental study of the constant-probability crack growth curves under constant amplityde loading, *Enging Fract. Mechanics* **27**, 1-25.
3. Doliński, K. (1993) Formulation of a stochastic model of fatigue crack growth, *Fatigue Fract. Engng Mater. Struct* **16** (9), 1007-1019.
4. Doliński, K. (1993) Comparison of a stochastic model of fatigue crack growth with experiments, *Fatigue Fract. Engng Mater. Struct.* **16** (10), 1021-1034.
5. Schijve, J. (1973) Effect of load sequences on crack propagation under random and program loading, *Engng Fract.Mech.* **5**, 269-280.
6. Kumar, R. (1992), A review on crack closure for single overload, programmed and block loadings, *Enging Fracture Mechanics* **42** (1), 151-158.
7. Christensen, R.H. (1959) *Metal Fatigue*, McGrow-Hill, New York.

530

8. Jones, R.E. (1973), Fatigue crack growth retardation after single-cycle overload in Ti-6Al-4V titanium alloy, *Engng Fract.Mech* **5**, 585-604.

9. Schijve, J., Broek, D. (1962) Crack propagation. The results of a program based on a gust spectrum with variable amplitude loading, *Aircraft Engineer* **34**, 314.

10. Elber, W. (1971) The significance of fatigue crack closure, in Damage Tolerance in Aircraft Structures, *ASTM STP 486*, 230-242.

11. Vecchio, R.S., Herzberg, R.W., Jaccatd, R. (1984) On the overload induced fatigue crack propagation behaviour in aluminum and steel alloys, *Fatigue Engng Mater. Struct* **7** (3), 181-194.

12. Suresh, S. (1983) Micromechanisms of fatigue crack growth retardation following overloads, *Engng Fract.Mech.* **18** (3), 577-593.

13. Shin, C.S., Fleck, N.A. (1987) Overload retardation in a structural steel, *Fatigue Fract. Engng Mater.Struct* **9** (5), 379-393.

14. Wanhill, R.J.H., Schijve, J. (1988) Fatigue Crack Growth Under Variable Amplitude Loading, in J. Petit et al. (eds.), Elsevier, London.

15. Ditlevsen, O., Sobczyk, K. (1986) Random fatigue crack growth with retardation, *Enging Fract. Mech.* **24**, 861-878.

16. Doliński, K. (1987) Fatigue crack growth with retardation under stationary stochastic loading, *Enging Fract.Mech..* **27** (3), 279-290.

17. Veers, P.J. (1987) Fatigue crack growth due to random loading, *SAND87-2037, Sandia National Laboratories.*

18. Veers, P.S., Winterstein, S.R., Nelson, D.V., Cornell, C.A. (1989) Development of Fatigue Loading Spectra, in J.M. Potter & R.T. Watanabe (eds.), *ASTM STP 1006.*

19. Colombi, P., Doliński, K. (1994) Markov approach to fatigue crack growth under stochastic load, in M.H. Aliabadi et al. (eds.), *Localized Damage III, Computer-Aided Assessment and Control*, Comp. Mech. Publications, Southhampton Boston, 89-96.

20. Doliński, K., Colombi, P. (1996) Fatigue crack growth under stochastic loading, in A. Naess and S. Krenk (eds.), *Advances in Nonlinear Stochastic Mechanics*, Kluwer Acad.Publs, Dordrecht, 143-152.

21. Doliński, K. (1996) Stochastic modeling of fatigue processes in metals, in J. Petit *et al.* (eds.), *ECF-11 Mechanisms and Mechanics of Damage and Failure*, EMAS, Warley, U.K., 43-52

22. Sarkani, S., Kihl, D.P., Beach, J.E. (1994) Fatigue of welded joints under narrowband non-Gaussian loadings, *Probabilistic Engineering Mechanics* **9**,179-190.

23. Kocańda, S. (1978) *Fatigue Failure of Metals*, Sijthoff & Noordhoff Int. Publ., Alphen aan den Rijn.

24. Paris, P., Erdogan, F. (1963) A critical analysis of crack propagation laws, *J. of Basic Enging, Trans. ASME*, **85**, 528-534.

25. Bulloch, J.H. (1991) The influence of mean stress or R-ratio on the fatigue crack threshold characteristics of steels - A review, *Int.J.Pres. Ves. & Piping* **47** (3), 263-292.

26. Finney, J.M., Deirmendjan, G. (1992) Delta-K-effective: which formula?, *Fatigue Fract.Engng Mater.Struct.* **15** (2), 151-158.

27. Irwin, G.R. (1960) Fracture mode transition for a crack traversing a plate, *J.Basic Engineering, Trans. ASME* **82.**

28. Reynolds, A.P. (1992) Constant amplitude and post-overload fatigue crack growth behaviour in PM aluminium alloy AA 8009, *Fatigue Fract. Engng Mater.Struct.* **15** (6), 551-562.

29. Wheeler, O.E. (1972), Spectrum loading and crack growth, *J.Basic Engineering* **94**, 181-186.

30. Willenborg, G J., Engle, R.M., Wood, H.A. (1971) A crack growth retardation model using an effective stress concept, *Report TM-71-1-FBR*, Wright-Patterson Air Force Base, Ohio.

31. Frendahl, M., Rychlik, I. (1992) Rainflow analysis - Markov method, *Res.Report, Univ. of Lund, Dept. of Mathematical Statistics*, No. 3.

32. Johnson, N.L., Kotz, S. (1970) *Distributions in Statistics: Continuous Univariate Distributions -1*, Houghton Mifflin Comp. Boston.

33. Renyi, A. (1962) *Wahrscheinlichkeitsrechnung mit einem Anhang über Informationstheorie*, VEB Deutscher Verlag der Wissenschaften, Berlin.

TWO PLASTICITY MODELS CONSIDERING
MICROMECHANISMS OF OBSERVED PHENOMENA

Nonproportional cyclic plasticity, transformation plasticity

P.V.TRUSOV†, I.E.KELLER‡ AND A.V.KLUEV‡
†*Professor,* ‡*Post-graduate students*
PSTU, Komsomolsky ave., 29a, 614600, Perm, Russia

1. Introduction

The damage accumulation of many plastically deformable metals, alloys and ceramics under complex thermomechanical loading depends essentially on plasticity micromechanisms. Therefore, for fracture simulation in these conditions it is important to have adequate notions of the preceeding plastic deformation process. The attempts to model some plasticity effects connected with complex loading, transformation plasticity as a rule describe experimental results not quite accurately. In the present paper two plasticity models are proposed and, in order to formulate them, corresponding physical mechanisms on the crystal lattice scale were investigated. The first model is intended to describe the additional hardening effect under nonproportional cyclic loading, the second is to describe transformation plasticity in ceramics. Since the physical mechanisms of both phenomena take place in grain scale, the corresponding models use orientation and statistic averaging.

2. A model of nonproportional cyclic plasticity describing additional hardening effect

2.1. THE EFFECT

Nonproportional cyclic loading is a kind of mechanical loading presented by closed origin-symmetric curves in two-dimensional deviatoric strain subspace, usually of the following shapes: segmental (proportional loading), cruciform, stellate, butterfly-shaped, elliptic, square and circular [1]. All the strain paths in the experimental investigations usually had the maximal equivalent strain value of the von Mises type up to 1%; tests were carried out

G. N. Frantziskonis (ed.), PROBAMAT – 21st Century: Probabilities and Materials, 531–537.

under strain rate of 10^{-5}-10^{-3} c^{-1} at room temperature and the material of specimens was isotropic. Material responses were stabilized during 5-20 cycles. In the stable state such materials as nickel, copper, some alloys and chrome-nickel austenitic stainless steels showed maximal equivalent stress of the von Mises type depends on the strain path shape. The given magnitude under nonproportional cyclic loading along the path of any shape was found to be significantly higher than the same parameter under proportional cyclic loading. Maximal additional hardening, as high as 80%, appears under cyclic loading along the circular path. True prediction of cyclic hardening in dependence on strain path shape presents certain difficulties so that many cyclic hardening measures (including well-known McDowell's parameter ϕ^* [2]) do not describe material responses well.

2.2. THE PHYSICAL MECHANISM OF THE PHENOMENON

We have found that all metals and solid solutions sensitive to complexity of the cyclic strain path had FCC crystal lattice and the crystallographic slipping as the basic mechanism of inelastic deformation. Comparison of materials disposition to additional hardening at various homological temperatures (the ratio of testing temperature to melting one in Kelvin scale) permits us to establish temperature independent of the effect. Correlation of measures of dislocation structure sensitivity to nonproportionality of cyclic loading and additional hardening with stacking fault energy was experimentally discovered by the authors of [3]. With the aid of the cited paper data we have found both measures correlate closely with the dimensionless ratio of stacking fault energy to averaged shear modulus and the absolute value of the Burgers vector. When this complex decreases, the ratio of the mean sizes of dislocation structures cell under nonproportional and proportional cycling decreases while additional hardening measure increases. The complex is inversely proportional to the stacking fault width in dissociated dislocation. Its relatively low magnitude results in formation of widely dissociated dislocation and makes processes which require contraction of dislocation, such as destruction of dislocation barriers, more difficult. For this reason metals with low stacking fault energy are effectively hardened by formation of Lomer-Cottrell barriers under interaction of dislocations in the certain pairs of slip systems.

During the nonproportional loading cycle, especially of such a shape when strain and strain rate tensors are not proportional at each moment in time (square, circular paths), cases of simultaneous activity of certain pairs of slip systems appeared to be more probable which forms strong dislocation barriers during their interaction. Thus, a probable physical mechanism of additional hardening effect is the likely formation of strong dislocation barriers such as Lomer-Cottrell ones.

2.3. THE MODEL

Constitutive equations of single crystal elastoplasticity were proposed taking into account the slipping mechanism through the FCC-crystal slip systems and the possibility of their different hardening (critical shear stresses τ_*^k, $k=1..12$ may be not equal):

$$\dot{E} = \dot{E}^r + \dot{E}^p, \tag{1}$$

$$\dot{E}^r = pq(q-1)\tilde{A} : \dot{S}, \tag{2}$$

$$\dot{E}^p = j\lambda q\tilde{A} : S, \tag{3}$$

where

$$j = 1 \text{ under } F(S) = \nu \text{ and } dF(S) = 0,$$

$$j = 0 \text{ under } F(S) < \nu \text{ or } F(S) = \nu, \ dF(S) < 0,$$

$$F(S) \equiv \sum_{k=1}^{12} \left| \frac{S : M_k}{\tau_*^k} \right|^q, \quad \tilde{A} \equiv \frac{1}{q(q-1)} \frac{\partial^2 F(S)}{\partial S^2} = \sum_{k=1}^{12} \frac{M_k}{\tau_*^k} \left| \frac{S : M_k}{\tau_*^k} \right|^{q-2} \frac{M_k}{\tau_*^k},$$

M_k is symmetrized dyad setting k-th slip system, $k=1..12$, $q \geq 2$ is a parameter, defining yield surface shape, $\nu = \Phi(S_{*j})$, $S_{*j} = \tau_* M_j$ — this condition guarantees the presence of 24 common points of the yield surface $F(S) = \nu$ at any q and the Bishop-Hill polyhedron, corresponding to the single slippings. The criterion formulated takes into account contributions of all resolved shear stresses $S : M_k$, $k=1..12$ at the yielding, the unlike the Bishop-Hill criterion. The given criterion predicts yielding condition at the moment when energy of reversible strains (2) reaches the limiting value. Constants τ_* and q are determined by size and shape of the yield surface obtained for the virgin single crystal, p is determined by the slope of the stress-strain curve obtained from single crystal extension along the direction $< 001 >$ at the moment preceeding yielding.

Plastic shear rates (slip system hardening measure rates) are determinated from the following equations:

$$\dot{\gamma}^k = \lambda q \frac{S : M_k}{\tau_*^k} \left| \frac{S : M_k}{\tau_*^k} \right|^{q-2} \frac{1}{\tau_*^k}, \quad k=1..12, \ \nexists k. \tag{4}$$

Evolution equations were formulated for intrinsic variables τ_*^k, $k=1..12$

$$\tau_*^k = \left[H \left\{ \sum_{i=1}^{12} \gamma_M^i + a \sum_{j(k)} h(\gamma^j \dot{\gamma}^k) \gamma_M^i \right\} - S\{\tau_*^k - \tau_*\} \right] |\dot{\gamma}^k|, \quad \nexists k. \tag{5}$$

Here $\gamma_M^i = \max\limits_{\tau \in [0,t]} |\gamma^i(\tau)|$ (τ is parameter of deformation process), H, S, a are positive constants, τ_* is the initial value of critical shear stresses. Study of the Lomer-Cottrell reaction geometry shows that the FCC single crystal slip systems are distributed into four nonoverlapping groups so that a barrier is formed when any two slip systems from the same group are interacting. Relevant reaction takes place with certain combinations of the Burgers vector signs and the movement directions of interacted dislocations (in other words, it is under certain configuration of simple shears on slip systems). It is possible to orient the symmetrized dyads M_k, $k = 1..12$ of the slip systems in such a way that a triple of these dyads of the same group could form the closed path in the symmetric deviator space. With that in mind the criterion for the barrier forming was written with the aid of cut-off function $h(x) \equiv \{1, \ x > 0; \ 0, \ x \leq 0\}$ in the evolution equations (5); $j(k)$ in (5) is "the Lomer-Cottrell group" including k-th slip system.

The component $H\{..\}$ in (5) considers work-hardening by self-hardening and latent hardening of the slip systems (the first term) and by the Lomer-Cottrell barriers formation (the second term). The component $S\{..\}$ considers strain softening. Under cyclic straining the values γ_M^i, $i=1..12$ are bounded together with the expression in curly brackets at H in (5). Then under the cyclic stabilization $d\tau_*^k/|d\gamma^k| = 0$ and from (5) $\tau_*^k = \tau_* + \frac{H}{S}\{..\}$, whence $\tau_*^k \geq \tau_*$ and $\frac{H}{S}\{..\}$ is the magnitude of cyclic hardening of the given slip system. The constant a in (5) defines relative strength of the the Lomer-Cottrell reaction. In the computer experiments it has been found that at $a = 0$ the values of the cyclic hardening were practically identical to those under the cyclic straining along the segment and the circle, at $a \neq 0$ the latter value was almost equal to the previous and the former was controlled by the parameter a in the sufficiently wide limits. These properties of the proposed model (1)-(5) agree with the studied physical mechanism of the additional hardening effect and permits its description.

The ratio H/S is found from the experiment on the cyclic proportional straining, the constant a is found from the experiment on the cyclic nonproportional straining along the circle, the constant S is defined by the cyclic stabilizing rate of the plastic responses in the second experiment. It should be noted that the intensity of the slip systems interaction in the model depends on the material parameter q controlling its relative activity.

2.4. THE RESULTS

The relations formulated have been used in combination with Lin's polycrystal hypotheses. The model has given a careful description of the stress paths and both elasto-plastic hysteresis loops. The most remarkable point is that this model describes the dependence of cyclic hardening magnitude on cyclic strain path shape accurately. This magnitude increases accordingly in the following sequence of shapes: segmental, cruciform, stellate, butterfly-shaped, square and circular ones, as it was in the experiments. Cyclic hardening under cyclic straining along the square path was a little less then it was for the circular straining, but the cruciform and stellate paths were accompanied by cyclic hardening even less then in the previous two cases. Similar responses have been found experimentally. The model has described a complicated inflection curve of the dependence of the cyclic hardening from the ratio of ellipse axes during cyclic straining along the elliptic paths.

3. A transformation plasticity model

3.1. THE TRANSFORMATION PLASTICITY EFFECT

Responses of some metals, alloys and ceramics deformable by transformation plasticity are difficult to describe comprehensively by the macrophenomeno-logical approach. On the other hand, the physical mechanism of transformation plasticity is well known and it is polymorphic transformation (phase transition of the first kind) activated by thermomechanical loading. On the macroscopic scale it is impossible to regard such process as an equilibrial one as far as the transformation leads to energy dissipation and, consequently, to irreversible deformations. Nevertheless, on the mesoscopic scale such process can be described in the framework of equilibtium thermodynamics. This assumption is accepted in the manner of the local balance hypothesis used in nonequilibrium thermodynamics.

3.2. A STRUCTURAL MODEL USING THERMOSTATICAL TRANSFORMATION CRITERION

Let us suppose that the homogeneous structural unit (lamella) is in the condition of thermodynamical equilibrium before and after the tratsforma-tion occurs. With the help of the first and the second laws of thermodyna-mics and a series of hypotheses, the following isothermal transformation criterion for the structural unit has been formulated

$$S : N = \Delta u(1 - \frac{T}{T_m}), \qquad (6)$$

where Δu is the transformation heat, N the tensor of crystal lattice distortion under the transformation, S the stress tensor and T_m the transformation beginning temperature. The right term of the criterion is the difference of structural unit Helmholtz energy and the left term is the work of applied forces on the distortion. If the left term of (6) is less then the right, the structural unit is straining elastically with the modules of the initial lattice. Under condition (6) the structural unit is strained by transforming on the value N; next it continues to strain elastically with the modules of the transformed lattice.

Each structural unit had one epitaxial plane and the direction of transforming. For connection of meso- and macrolevels, the Voigt and Reuss hypotheses in combination with orientation averaging were used. The constants of the model, initial and transformed lattice modules and the values of N, Δu and T_m were taken for circonia stabilized ceramics from [4]. The computer experiments on pressure loading of the sample have shown that the model proposed describes the corresponding behavior well. The stress-strain curve has expressive three-stage character observed in the corresponding experiments [4]. The initial stage was induced by the elastic behavior of the cubic phase, the intermediate one was stipulated by the transformation of the cubic lattice into the tetragonal one, and the final stage was induced by the secondary elastic behavior of the transformed material having the tetragonal lattice.

The materials investigated in [4] cannot endure other homogeneous mechanical loadings, except the compression. However, manufacturing and working of ceramic details usually accompanied with more rich evolution of stress state in each material particle. Hence, the model formulated can be used for the prediction (not quite accutate, of course) of the behavior of the similar hard deforming materials under complex loading. In this work the model has been used for the identification of the thermodynamical constitutive equations which are stated by the macrophenomenological way below.

3.3. THE THERMODYNAMICAL APPROACH

The framework of nonequilibrium thermodynamics was used to formulate macrophenomenological constitutive equations. The linear dependence of thermodynamical flows and forces has accepted, that it is valid under infinitesimal deviations from the equilibrium state. As a result the following expression for the entropy producing χ_s was stated

$$\chi_s T = -\rho \boldsymbol{S} : \dot{\boldsymbol{E}}_t - \Delta\mu\chi_c - \frac{1}{T}\boldsymbol{I}_q \cdot \nabla T, \tag{7}$$

where $\dot{\boldsymbol{E}}_t$ is the transformation strain rate tensor, $\Delta\mu$ the chemical potential difference of the phases, χ_c the concentration of the transformed phase, $\boldsymbol{I}_q$ the heat flow vector. According to the local form of the least energy dissipation principle [5] in terms of thermodynamical forces it yields

$$\delta[\chi_s - \Psi]_I = \sum_k (I_k - \frac{\partial\Psi}{\partial X_k})\delta X_k = 0, \tag{8}$$

where Ψ is local dissipation potential in terms of forces

$$\Psi(X,X) = \frac{1}{2}\sum_k L_{ik}X_iX_k, \tag{9}$$

where L_{ik} are kinetic coefficients.

3.4. IDENTIFICATION OF THE THERMODYNAMICAL CONSTITUTIVE EQUATIONS

Homogeneous compression loading of ceramics stabilized with circonia was considered as an example. The constitutive equations (7)-(9) were reduced to the following form

$$\dot{\varepsilon}_t = \ell_1\sigma + \ell_2\dot{c}, \tag{10}$$

where $\ell_1 = (\frac{L_{vc}^2}{L_{cc}} - L_{vv})$, $\ell_2 = \frac{L_{vc}}{L_{cc}}$. It has been accepted that the sample was loaded at the constant strain rate $\dot{\varepsilon} = \dot{\varepsilon}_t + \dot{\sigma}/E$, where E is Young modulus. The computer experiments carried out with the structural model have shown that the dependence of the concentration parameter on the stress and the temperature can be approximated by the following function

$$c = A(T)\ln(\sigma) + B(T). \tag{11}$$

Considering this result as well as the constancy of strain rate the last two equations have been resolved relative to stress

$$\frac{\sigma}{\sigma_0} = \frac{\exp(\frac{E\varepsilon}{A\ell_2})}{1 + \frac{\sigma_0\ell_1}{\varepsilon}(\exp(\frac{E\varepsilon}{A\ell_2} - 1)}, \tag{12}$$

where σ_0 is the threshould value of the transformation stress. Equation (12) has described the experimentally obtained stress-strain curve at room temperature well at the following values of kinetic coefficients: $\ell_1 = -3.2 \cdot 10^{-8}$ and $\ell_2 = 2.9 \cdot 10^9$.

References

1. Tanaka, E., Murakami, S. and Ooka, M. (1985) Effect of strain paths shapes on nonproportional cyclic plasticity, *J. Mech. Phys. Solids* **33**, 559-575.
2. McDowell, D.L. (1987) An evaluation of recent developments in hardening and flow rules for rate-independent, nonproportional cyclic plasticity, *Trans. ASME. J. Appl. Mech.* **54**, 323-334.
3. Ishikawa, H. and Sasaki, K. (1992) Application of the hybrid constitutive model for cyclic plasticity to sinusoidal loading, *Trans. ASME. J. Eng. Mater. Technol.* **114**, 172-179.
4. I-Wei Chen and Reyes-Morel, P.E. (1987) Transformation plasticity and transformation toughening in Mg-PSZ and Ce-TZP, *Mat. Res. Soc. Symp. Proc.* **78**, 75-87.
5. Gyarmati, I. (1970) *Non-equilibrum thermodynamics: field theory and variational principles*, Springer-Verlag, Berlin.

STATISTICAL KINETICS OF FRACTURE
OF HETEROGENEOUS SOLIDS

V.S. KUKSENKO, N.G. TOMILIN, E.E. DAMASKINSKAYA
Ioffe Physico-Technical Institute, Russian Academy of Sciences
Polytekhnicheskaya 26, St.Petersburg, 194021 Russia

1. Physical Notions on the Fracture of Solids

Up to the middle of the 50's, in the physics of strength the agreed-upon notion was that the fracture of solids occurs when stress or deformation reaches their limiting values. The kinetic concept of the strength of solids opened up an essentially new approach to the problems of the physics of fracture [1]. From basing on extensive experimental material it was proved that the macroscopic fracture takes place not only when the stress limit is reached, but also at lower loading in case of the duration of their action is long. For a uniaxial tension, an empirical expression for lifetime (τ) of a specimen was obtained:

$$\tau = \tau_0 \exp \left(\frac{U_0 - \gamma \times \sigma}{kT} \right) \tag{1}$$

where U_0 is the activation energy of fracture whose value is close to the energy of sublimation; σ is the applied stress; T is the absolute temperature; k is the Boltzmann constant; γ is the structure-sensitive parameter which can be equal to 10–10^3 of atomic volumes. The pre-exponential factor τ_0 coincides with the period of thermal vibrations of atoms in a solid. Relation (1) is valid for a wide range of materials: metals and alloys, halide and semiconductor crystals, glasses, polymers, composites, and rocks [2]. It was established that exponential relations similar to (1) describe different processes: chemical reactions, diffusion, phase transitions. As the rate of these processes increase with temperature, they are called as thermoactivation processes. The basis of fracture is the transition over the barrier U_0 which is decreased by the value ($\gamma \times \sigma$) by the applied stress and overcome by thermal fluctuations. The kinetic nature of fracture is unique not only

G. N. Frantziskonis (ed.), PROBAMAT – 21st Century: Probabilities and Materials, 539-556.

for a process in the field of mechanical forces, but also for a beam fracture of an optically transparent environment [3]. Experimental data on a time-temperature dependence of the electrical strength of polymers [4] and ceramics [5] have been obtained, which show that the electrical fracture under the conditions of the suppression of partial discharges at a constant stress is also a kinetic process.

Within the framework of the kinetic concept of strength a question arises as to the nature of an elementary act of fracture, though at present there is no final answer. Two approaches to the given problem may be stated: structural and thermofluctuational. The structural approach is based on the experimental fact that the origination of cracks always comes before the microplastic deformation, and the fracture is the critical act of the evolution of defect structure. These representations are widely used in a number of models, [6] being most known for crystal solids. The kinetic approach [7] which was formed later assumes the determining role of thermal fluctuations in the formation of nucleus cracks. Their sizes vary from tens up to thousands nanometers for different materials, depending on the parameters of fluctuations, which, in their turn, are related in an ordered environment to the free path length of phonons. For polymers [8], metals [9] and crystals [10] there was established the experimental relationship for the rate (N') of the accumulation of nucleus cracks:

$$N' = N_0 \exp(-U_c/kT), \tag{2}$$

where for polymers U_c coincides with the value of the activation energy of the formation of exited bonds while for metals it is $\sim U_0$. It was shown in [11] that relations (1) and (2) are interconnected and are caused by the statistics of fracturing thermofluctuations [12].

An important fundamental consequence following from the kinetic notions on strength is the essentially stochastic character of fracture which is determined by the statistics of thermal fluctuations. The physically caused randomness of the process of fracture at a microlevel and the statistical spread of local values of physical-mechanical properties of heterogeneous materials justifies the development and use of statistical models of the phenomenon when studying the fracture.

Thus, for a wide range of materials there are two generalizing conditions: a unified kinetic nature of fracture at a microlevel and the fact that the major lifetime of a body under load falls at the process of the accumulation of damages. It is well known that the character of further fracture, i.e. from nucleus cracks up to macroscopic ones depends on a type of a material and for many cases it is defined by its structure and the type of a stressed state. Therefore we shall restrict the further analysis of the kinetics of fracture and the construction of the model of the process by considering the case

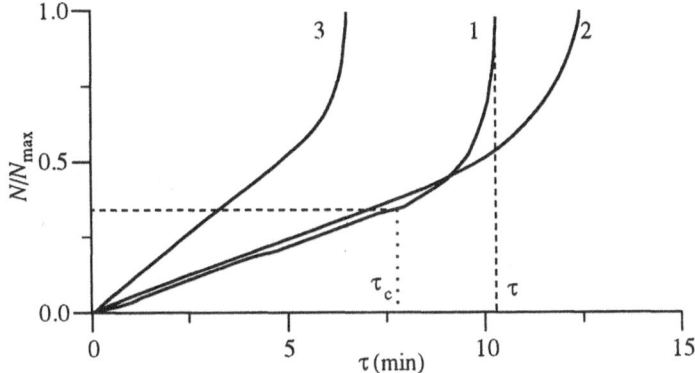

Figure 1. Accumulation of the cracks in time recorded by acoustic emission in samples with a cut under a constant load: under the bending of a steel plate (curve 1); under compression of a granite box (curve 2); under tension of a polymer fibre (curve 3) (from [13]).

of crystal bodies with an expressed heterogeneous structure, e.g., metals, alloys, rocks.

2. Kinetics of the Fracture of Heterogeneous Solids

The detection of cracks originating under load and the study of the regularities of the accumulation of defects can be carried out by applying a number of experimental methods one of which is the registration of the elastic energy release of signals originating at the generation of cracks. The method of acoustic emission allows us to determine the rate of the generation of cracks N' and to estimate their sizes from the value of the signal amplitude and duration [14].

Fig. 1 presents the results of the measurements of the accumulation of cracks for various materials in a specimen with a cut. If the temperature and stress are constant, the accumulation of cracks in a close to the cut occurs with a constant rate N', which agrees with relation (2). However, the stationary accumulation proceeds only up to a certain value N_c, after which occurs the stage of accelerated accumulation followed by the rupture of a specimen.

It was experimentally established that by the moment of fracture the ratio of a mean distance between cracks R to their size L is practically constant: $R/L = $ const. Proceeding from this experimental fact, the condition has been formulated for the transition from a disperse origination of cracks to their enlargement [12]. According to this criterion, the concentration of ensembles consisting of S coalescing cracks is determined by the following relation:

$$C_s = \frac{C}{S^{(3/2)}} \left(\frac{e}{K}\right)^s, \tag{3}$$

where S is the number of coalescing cracks of size L_j, C is the concentration of initial cracks, e is the basis of a natural logarithm. The value $K = C^{(-1/3)}/L_j$ which is called as "concentration parameter", describes the mean distance between the cracks in terms of their sizes. It follows from relation (2) that the concentration of large ensembles of cracks ($S \gg 1$) is low for $K > e$ and increases quickly at $K = e$. This result confirms the existence of the critical value of the concentration parameter K, at which the disperse formation (generation) of defects is replaced by the second stage of development of fracture characterized by the enlargement of defects.

On this basis it is possible to explain the cause of the acceleration of generation of cracks in an interval $\tau - \tau_c$ (Fig. 1): the presence of enlarged cracks increases local stresses σ in their vicinity and thus in accordance with (2) it leads to an increase of the rate of the crack formation.

Various studies [13] have proved the feasibility of the concentration criterion at the fracture of materials of different types: polymers, composite materials, rocks for a wide range of sizes of cracks being formed which vary from micron cracks in specimens up to ruptures extended to hundreds kilometers which are formed under rock bursts and earthquakes [15, 16]. This is demonstrated by comparing the threshold concentration C_* and sizes L for different scale levels [13] (Fig. 2). In Fig. 2 the continuous line corresponds to the theoretical value $K_c = e$ while the points describe the experimental values $K_c = (C_*)^{(-1/3)}$ and L that are calculated by the moment of macrofracture. It can be seen that the experimental values agree well with the theoretical relationship. It should be noted that the value $K_c = e$ has been received in [12] from the statistical consideration only, without taking into account the fields of stresses upon the interaction of cracks.

3. Two-Stage Model of Fracture of Heterogeneous Bodies

The kinetic approach to the strength of solids and the extensive experimental material on the origination of cracks allow constructing the model of the development of fracture of heterogeneous materials. Its basic statements are as follows (Fig. 3).

The structural heterogeneity of a material results in a non-uniform distribution of applied external loading. As a result, there originate some elements whose probability of fracture is higher than the average one for the body as a whole due to the decreased thermoactivation barrier. The cracks arising upon the fracture of these volumes that are randomly distributed in space are stabilized at heterogeneity boundaries. Thus, the first stage of

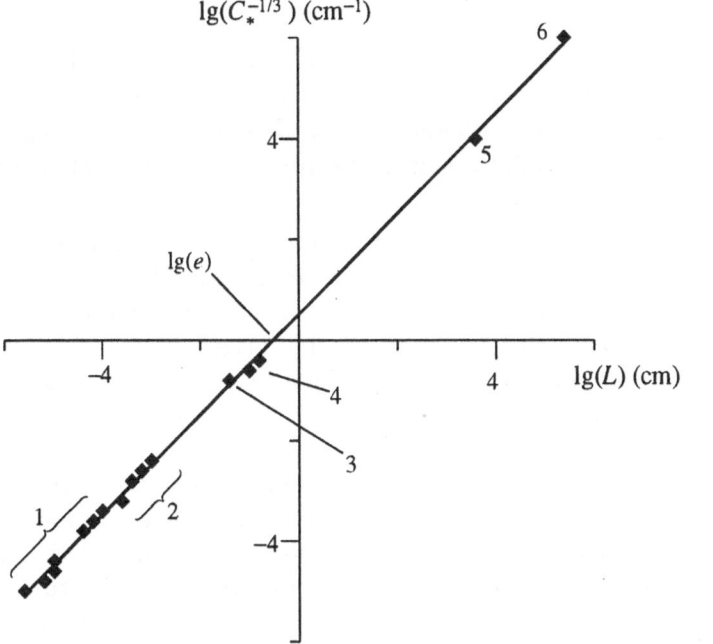

Figure 2. Mean distance $C_*^{-1/3}$ between cracks of size L immediately before fracture: polymers (curve 1); metals (curve 2); composites (curve 3); rocks (curve 4); rock bursts(curve 5); earthquakes (curve 6) (from [13]).

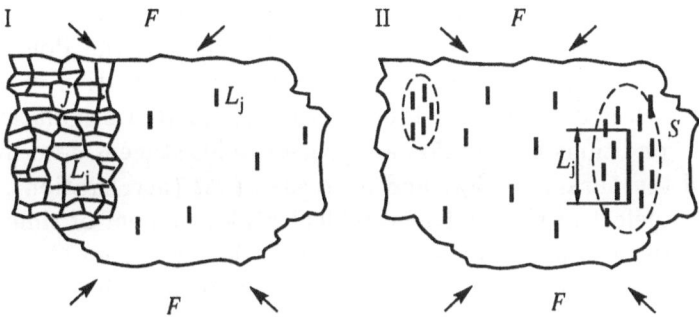

Figure 3. Two stages during accumulation and development of cracks.

the process consists of the multiple disperse accumulation of non-interacting cracks whose size is determined by a given level of structural heterogeneity J. During the process of accumulation of defects their concentration in a certain area exceeds accidentally an average value for the body as a whole. Such an area is called a fracture nucleation site. When the threshold concentration is achieved, the interaction between cracks arises, which in its turn stimulates further formation of defects. The conditions are formed for which the nucleation zone loses stability, and a defect of the $(J + 1)$ rank is formed which corresponds to the next characteristic heterogeneity size. If there is the hierarchy of heterogeneity sizes in a material, e.g., as it is in rocks [6], then the model assumes the similar development of the process of fracture at all the present scale levels.

The random character of fracture of heterogeneous materials determines the choice of statistical methods for its description. The defects originating upon the loading can be represented as a flow of discrete events, each of them being characterized by the moment of origination, coordinate, and the size of a corresponding defect. It is possible to correspond an experimentally recorded seismic or acoustic emission (AE) sequence to such a flow, in which the energetic parameters of a signal serve as a specific feature of the size of a defect. By reformulating the two-stage model in terms of the statistics of a flow of discrete events, it is possible to consider the first stage as satisfying to the conditions a quasi-stationary Poisson process so that the violation of these conditions are the criterion of the formation of a fracture nucleation site.

We can choose average values of temporal $(\bar{\Delta}t)$ and spatial $(\bar{\Delta}r)$ intervals between chronologically sequential events and corresponding coefficients of the variation of intervals $(V_{\Delta t}, V_{\Delta r})$ as the parameters characterizing the spatial-temporary peculiarities of fracture. The changes of these characteristics are schematically shown in Fig. 4. At the first stage $(t < T_1)$ of the noncorrelated formation of defects their flow is the Poisson one of intensity $\bar{\Delta}t$. By the moment of time T_1 the concentration of defects is achieved which is sufficient for their further stimulated origination, which breaks the conditions for the Poisson process. This stage is described by an increase of the parameter $V_{\Delta t}$ and decrease of $\bar{\Delta}t$ (acceleration). The formation of a defect of the next hierarchical rank (moment of time T_d) does not necessarily complete the second, non-stationary, stage. As a result of the relaxation of a nucleation zone, from the moment of time T_3, there may arise the reverse tendencies (T_3-T_4), i.e. the increase of $\bar{\Delta}t$ and decrease of $V_{\Delta t}$. For brevity, further we shall call the above qualitative change in the characteristics under study in the range T_1-T_4 as an image of a fracture nucleation site. It has been found out and checked as to the statistical significance in the computer simulation of the process.

Generally, the fracture is realized simultaneously at various scale levels and at each of them as two stages following each other. Therefore the detection of the considered image of the nucleation site is possible if three conditions are fulfilled. These are as follows: only the events falling within the spatial region of the preparation of fracture nucleation site should enter the sampling under analysis; secondly, their energy (amplitude) should correspond the hierarchical level of the process, and finally, the temporary

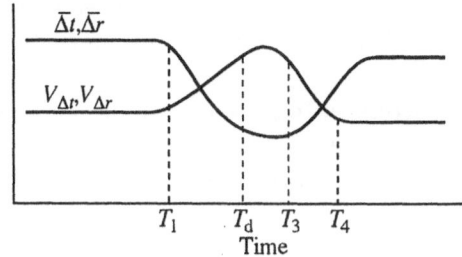

Figure 4. Statistical criterion of formation of fracture nucleation site.

step-type behavior of the analysis should guarantee revealing all the peculiarities of the process. The first two conditions include the double space-energy selection of the flow of events, where the criterion is the reception of trends in the described characteristic change of statistical parameters.

4. Computer Simulation of Fracture of Heterogeneous Materials

Computer simulation is widely used together with the experimental methods of studying the development of fracture. Computer experiments allow differentiating the effect of both the properties of material itself and loading conditions on the regularities of fracture.

The present study suggests a simulation model based on the kinetic concept of the strength of solids. Its advantage is the possibility of studying the development of fracture (accumulation of defects) both in space and in time.

The simulated material is a system of identical structural elements. The system is under the action of external load so that its variation can be controlled. To simplify the calculations, we have used a one-dimensional model, because both the analysis of publications and specially spent researches have shown comparability results received in models of various dimensionalities. The lifetime of each element is determined from the assumption that the most probable time before fracture depends exponentially on stress:

$$\tau_i = A \exp(-\gamma_i \times \sigma_i), \qquad (4)$$

where τ_i is the lifetime of the i-th element, γ_i is a structure-sensitive factor, σ_i is the stress on the element, $A = $ const. In order to take into account the random character of fracture, it is set the probability of fracture of the i-th element:

$$P_i(\tau_i) = 1 - \exp(-t_i/\tau_i) \qquad (5)$$

where t_i is the actual time before the fracture of the i-th element which differs from the most probable time (τ_i) and reflects the thermofluctuation nature of strength. The values of the probabilities of the fractures of elements, P_i, are set by generator of random numbers distributed uniformly in the interval $(0;1)$.

In the model, the stress on each element of the system changes with changing external load and as a result of redistribution of stresses. Then the random time before the fracture of an element ti is given by the relation:

$$\int_0^{t_i} \frac{\mathrm{d}t}{A \exp[-\gamma_i \times \sigma_i(t)]} = \ln\left(\frac{1}{1 - P_i}\right). \qquad (6)$$

The heterogeneity of the system is modeled by assigning a parameter γ_i from a given distribution to each element.

During the computer experiment the lifetimes of all intact elements are calculated. An element with the minimum lifetime is considered to be destroyed, and the stress from this element is redistributed on the R nearest intact elements (the stress decreases with increasing distance).

The computer experiment comes to an end at the moment $T_{\max}$ of the fracture the last element remaining in the system, and as a result, there is obtained a succession of defects each of which is characterized by the time of formation, coordinate and size.

Let us use for the analysis of the development of fracture in the simulation experiment the statistical parameters $\bar{\Delta}t$, $\bar{\Delta}r$, $V_{\Delta t}$, and $V_{\Delta r}$ whose characteristic change at the formation of a fracture nucleation site was considered in section 3. At the first stage there occurs the disperse quasi-stationary non-correlated formation of defects. This is reflected in low-changing values of $\bar{\Delta}t$, $\bar{\Delta}r$ and $V_{\Delta t}$, $V_{\Delta r}$ (Fig. 5). Fig. 5(a) shows the plot of the change of the concentration parameter at the fracture of the simulation system. It can be seen that by the moment of time T_1 the value K becomes equal to 3, i.e. K reaches the threshold value K_c. It indicates that the concentration of defects in a local area (in a fracture nucleation site) has reached the threshold value. At the moment T_1 the simultaneous decrease of $\bar{\Delta}t$, $\bar{\Delta}r$ and increase of $V_{\Delta t}$, $V_{\Delta r}$ takes place (Fig. 5(b,c)). Such a specific change of the parameters indicates the transition of the process of fracture into the second stage and the formation of a nucleation site. The loss of stability by the nucleation site leads to the formation of a defect of a larger size (moment T_d).

This criterion was verified when changing the conditions of the computer experiments. For this purpose, the parameters of the distribution of the structural factor γ and the law of the redistribution of stresses were varied. The development of fracture was studied at a constant and linearly increasing external loading. The limiting cases were considered: the brittle fracture from one site for which the image degenerates and the total dispersed fracture of system, for which there is no formation of site. This

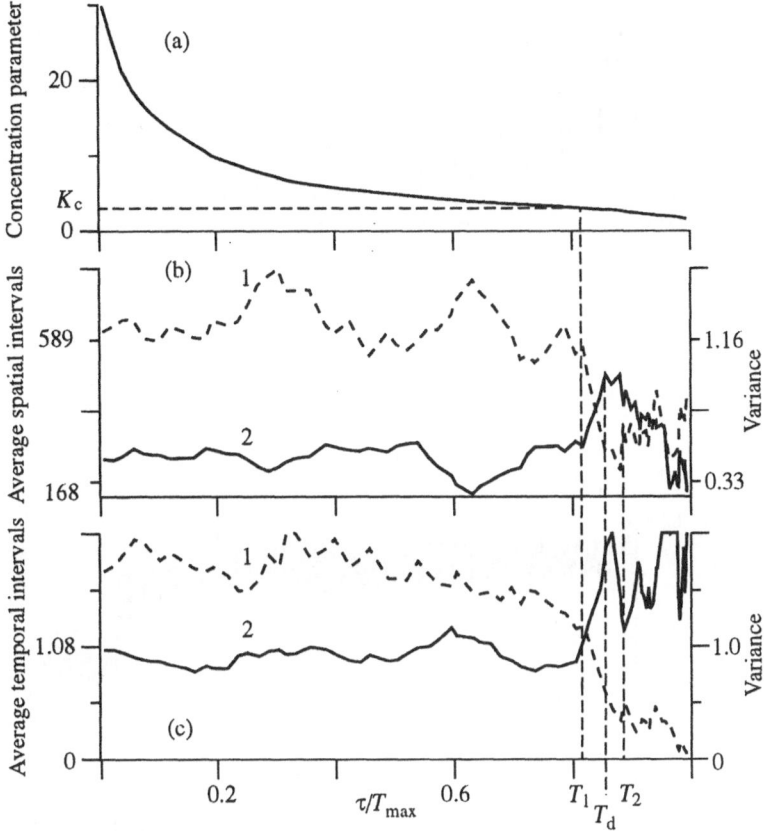

Figure 5. Variations of (a) concentration parameters, (b) average spatial interval (1) and variance of spatial intervals (2), (c) average temporal interval (1) and coefficient of variation of temporal intervals (2) during computer simulation of fracture.

allows a conclusion to be made that the criterion of the formation of a fracture nucleation site is fulfilled in a wide range of physical-mechanical properties of a material under study and loading conditions.

5. Practical Example of the Application of Statistical Kinetics Approaches to the Analysis of Fracture of Heterogeneous Solids

We shall demonstrate the applicability of the basic statements of the two-stage model of fracture of heterogeneous solids by considering the processes proceeding in rocks which are characterized by a clearly expressed block structure [17] as an example. The uniqueness of the given class of materials when solving the problems of the physics of fracture is possible when studying the processes in a wide scale range: from laboratory samples up

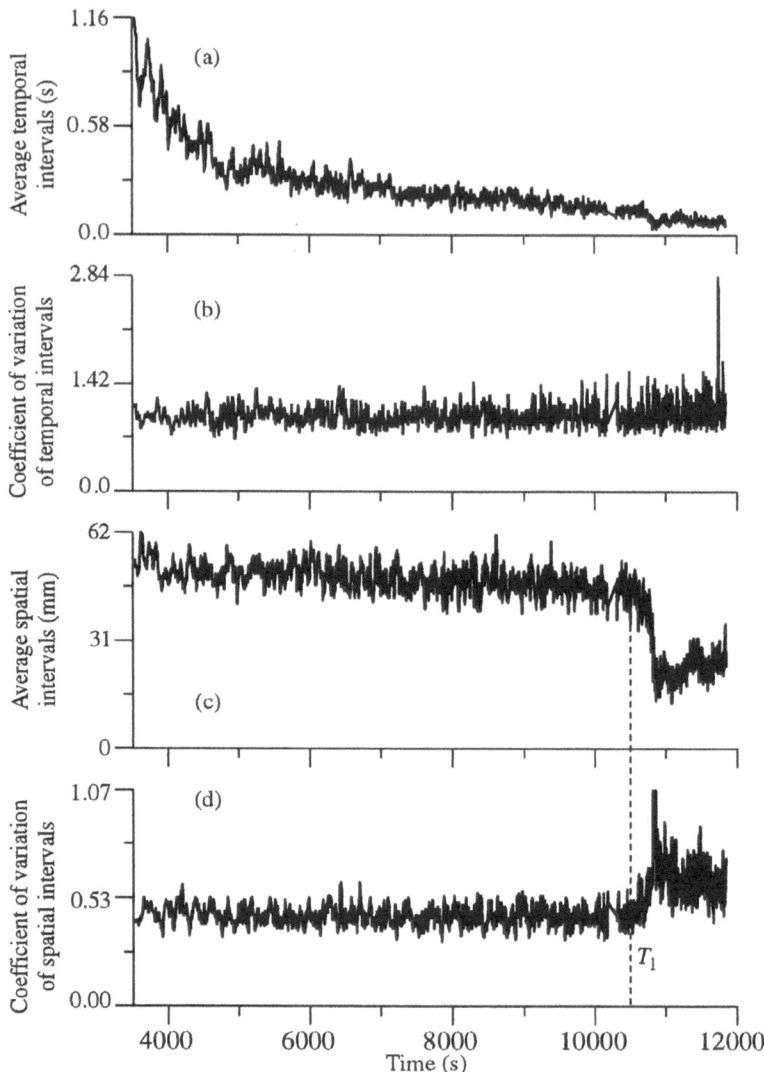

Figure 6. Variations of statistical parameters during deformation of sample of Westerly granite (all AE-signals): (a) average temporal intervals; (b) coefficient of variation of temporal intervals; (c) average spatial intervals; (d) coefficient of variation of spatial intervals.

to mountain rocks and regions in which fracture is accompanied by seismic events.

The present study analyses the experimental data that were obtained at the deformation of cylindrical specimens ($h = 190.5$ mm, $d = 76.2$ mm) of Westerly granite [18]. The experiments were carried out under the conditions of constant confining pressure and uniaxial loading. In the experiments were measured both the longitudinal and transverse deformations and parameters of signals of elastic energy release - acoustic emission (AE), whose main source in rocks are cracks being formed [19]. In order to carry out further fractographic analysis, the experiment was stopped before the final fracture of the specimen at the moment of time T_k. The most general regularities for a series of experiments will be presented. Hereinafter the AE signals will be called events, each of them being characterized by the time, three hypocentre coordinates and amplitude recalculated to a reference-sphere $R = 10$ mm.

When analyzing the experimental space - time succession of AE events we shall use the above statistical parameters: average temporal ($\bar{\Delta t}$) and average spatial ($\bar{\Delta r}$) intervals between chronologically successive events and the corresponding variances $V_{\Delta t}$, $V_{\Delta r}$. Fig. 6 shows temporary dependencies $\bar{\Delta t}$, $V_{\Delta t}$ and $\bar{\Delta r}$, $V_{\Delta r}$ describing the development of the process during the entire experiment. The complete file has been used for the analysis, i.e. all the signals recorded in the experiment. It can be seen that the simultaneous decrease of $\bar{\Delta r}$ and increase of $V_{\Delta r}$ indicating the formation of a fracture nucleation site occurs at the moment of time T_1. Here the change of temporary parameters does not show specific features which would allow the identification of the nucleation of fracture to be made. Essentially another picture is displayed in case the analysis of AE signals is carried out in a certain range of amplitudes.

Fig. 7 presents similar parameters for the analysis of signals with amplitudes $A > 40$ mV. The choice of such a threshold of discrimination is not accidental and will be discussed below.

It can be seen that during the period of time up to T_1 there is but a little change in average temporal intervals $\bar{\Delta t}$ and variances of temporal intervals $V_{\Delta t}$. This means that the process is quasistationary. There occurs a dispersed non-correlated accumulation of defects, which is characterized by values of mean space intervals $\bar{\Delta r}$ determined by sizes of a specimen at stable and low values of $V_{\Delta r}$. It is illustrated by the distribution of hypocentres of AE sources (Fig. 8(a)).

Thus, up to the moment of time T_1 the flow of defects may be considered as a quasistationary Poisson one that is distributed disperely throughout the entire specimen (Fig. 7, 8(a)).

From the moment T_1 (Fig. 7), according to the two-stage model, the formation of a fracture nucleation site starts, which is accompanied by the decrease of $\bar{\Delta t}$, $\bar{\Delta r}$ and simultaneous increase of $V_{\Delta t}$, $V_{\Delta r}$ (Fig. 7). The process of fracture passes to the second stage. To moment T_2 (Fig. 8(b)), defect formation is "focused" in a local zone and the nucleation site loses

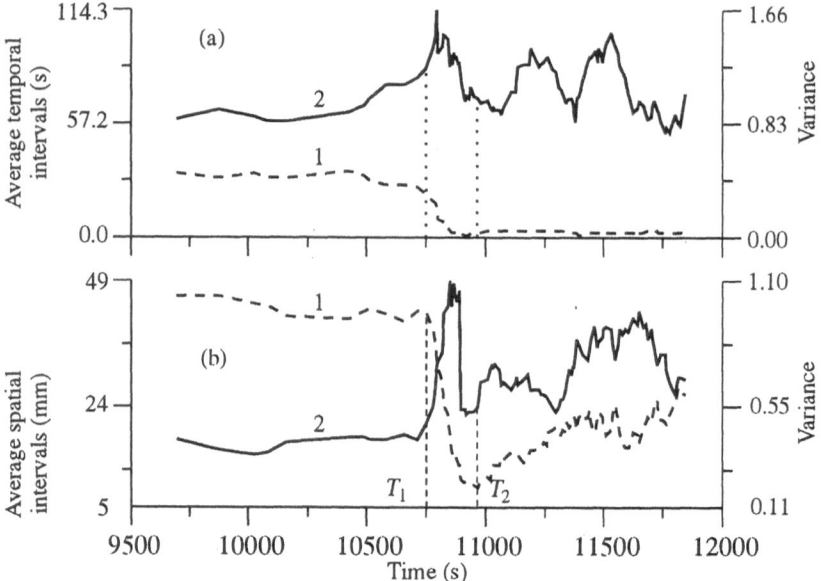

Figure 7. Temporal dependencies of average temporal $\bar{\Delta}t$ (a, curve 1) and spatial $\bar{\Delta}r$ (b, curve 1) intervals and corresponding variances $V_{\Delta t}$, $V_{\Delta r}$ (a, curve 2; b, curve 2) during deformation of sample of Westerly granite ($A > 40$ mV).

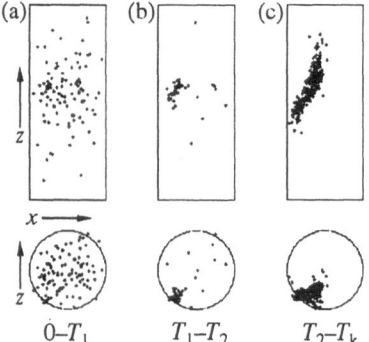

Figure 8. Spatial distribution of hypocentres of AE sources. $0-T_1$ — disperse accumulation of defects; T_1-T_2 — formation of fracture nucleation site; T_2-T_k — faulte propagation.

stability. This is proved by the propagation of a fault (Fig. 8(c)) during the time interval T_2-T_k.

We have succeeded in selecting the detected specific stages of the AE flow in a set of experiments, which confirms their regularity. As was noted above, the maximum contrast range of the image of a nucleation site is achieved when analyzing the AE signals with $A > 40$ mV. In this case the

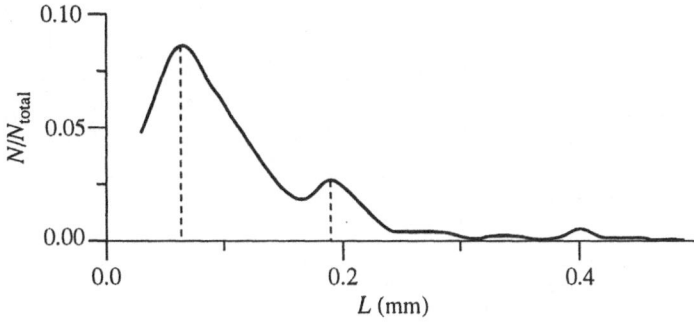

Figure 9. Distribution of the structural elements in the granite sample ($N_{\text{total}} = 553$, $L_{i+1}/L_i = 3$).

attributes of the corresponding stages remain also in a wider amplitude range of signals $A_1 = 20 \pm 5\,\text{mV} - A_2 = 65 \pm 5$ mV. The constancy of the considered amplitude range in the set of experiments under analysis can be explained as follows. Extensive experimental data [17] indicate that rocks have a multilevel structure which is characterized by a multimodal distribution of blocks structural sizes. The ratio of two adjacent predominant sizes of heterogeneity L_{i+1}/L_i is in the range 2−5 and does not depend on the scale of these sizes. Because the boundaries of blocks are the locks for the cracks being formed, one may assume that the sizes of the cracks correspond to those of the blocks and also have a multimodal distribution. This allows us to introduce the notion about the rank of the crack formation or the process of fracture of rocks. This can be confirmed by the distribution of sizes of structural elements (Fig. 9) formed in a granite specimen by moment T_k at which the loading is removed. By taking into account the correlation between the size of the defects being formed and the energy parameters of corresponding elastic pulses, let us assume that the selected amplitude range of signals in which the image of the fracture nucleation site is most distinctly displayed, corresponds to one J-th structural level and hence to the J-th rank of the process.

When analyzing the signals with amplitudes less than lower boundary A_1 of the range under study ($A_1 = 20 \pm 5\,\text{mV} - A_2 = 65 \pm 5$ mV), it is observed the alternation of the signs of stationary and nucleation stages. In our opinion, this is an indication of the fact that the defects being classified not only to the J-th level, but are also those related to the previous one, i.e. to the $(J-1)$-th level fell within the range under analysis. The fracture at the $(J-1)$-th scale level, as well as that at the J-th one, is realized as two successive stages. Therefore when analyzing the flow which includes the events of two ranks, no unique quasi-stationary stage is observed. From the beginning of the experiment, the AE sources of $A > 65$ mV are generated

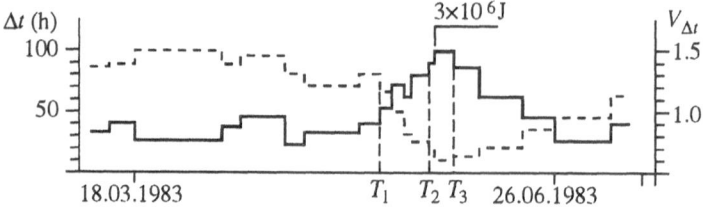

Figure 10. Variation of average temporal intervals (1) and coefficient of variation of temporal intervals (2) during development of fracture nucleation site of rock burst.

in a local nucleation site zone passing the stage of dispersed nucleation. This allows a conclusion to be made that with an increase of an amplitude threshold over a certain value, there are observed the exhibitions of the next $(J + 1)$ rank of fracture.

Thus, the above data on the fracture of a heterogeneous material presented in this section prove the basic statements of the two-stage model of fracture and the applicability of the statistical criteria of the formation of a fracture nucleation site that have been formulated on its basis.

6. Prediction of Fracture of Heterogeneous Solids

According to the notions on the limit of strength, a body under load can be in two states: initial and destroyed. The problem of the prediction of a service lifetime is reduced to calculating the probabilities of overlimiting loadings and the necessary safety assurance factor. The thermofluctuation nature of fracture leads to an essentially different formulation of the problem of prediction. Thermal fluctuations in a loaded body result in the formation of cracks, the fracture of a body being the process of the accumulation of defects which proceeds in time. The physical basis of the prediction is the estimation of a measure of the presence of defects and the rate of its approach to the critical value.

Practical problems of the prediction of fracture can be reduced in the first approximation to two types: estimation of residual service life of a mechanically loaded system and prediction of the origination of a local defect. One can refer to the second type such a vital problem as the prediction of seismic phenomena which are the release of elastic energy with the formation of a defect inside rocks. Below we shall describe the opportunity to solve in practice the problem of prediction by using the above statements of the statistical kinetics of fracture. Let us take as an example the results of the application of the prediction technique of rock bursts. It is based on the two-stage model of fracture and uses the formalized statistical criteria of the formation of a fracture nucleation site described above (Fig. 10).

For the first time the technique was tested at the Severoural'sk bauxite mine which turns refers to the most dangerous mines in Russia with respect to the manifestation of rock pressure. General characteristics of seismic conditions and systems of its registration have been presented in [20]. The future prediction of rock bursts has been made by the seismic services of the mine since 1986 within the limits of mine fields whose characteristic sizes are equal to $5 \times 3 \times 1$ km. In this case the following problems are solved: prediction of a place, time and energy of technogenous seismic phenomena; transition of a nucleation region into a nonburstdangerous state; checking of the efficiency of methods controlling the rock pressure and applied preventive maintenance.

The lower boundary of the dynamic range of recording the energies of rock bursts is equal to 10 J. The confidence value in the entire field of seismicity is equal to 10^2 J. It was found out in the retrospective analysis that the lower energy threshold of bursts being predicted exceeds the lower confidence level of registration by three orders of magnitude. During the period of observation in the regions under testing 57 rock bursts with energy $E > 10^5$ J took place, a part of them being presented in Table 1.

The result of the checking (C) included: successful realization of the prediction (Y), objective omission (N) and retrospective prediction (R), i.e. the case for which the objective omission occurred according to subjective reasons which did not depend on the scope of the techniques itself. Below are presented the size of a spatial zone (S) in which an exposed alarm is located (prediction by a place), predicted (E_{pr}) and actual (E) value of energy in logarithms, predicted (T_{pr}) and actual (T) time in days from the moment of exhibiting of alarm up to a seismic event (prediction by time).

In some cases there was a significant excess of the predicted value by the actual time of alarm (Table 1). This is related to the fact that the chosen techniques assumes the alarm removal by the appearance of the physical criteria for the transition of a nucleation region under checking into a nonburstdamage state, rather than by the fact of the predicted seismic phenomenon. It increases the alarm period but at the same time it allows predicting group bursts (23.01.90) that occur within small intervals practically in one and the same place. The last point has a paramount significance when making decision about the renewal of technological processes in the direct vicinity from a relaxing nucleation region. The efficiency of the prediction by place is not analyzed in the present study, though the alarm spatial localization in a volume with the characteristic size equal to $200-250$ m (for the events with $E = 10^6$ J) with the total area of checking equal to thousands meters is the convincing proof of the opportunities of the present approach. The accuracy of the prediction of elastic energy release lies within the limits of the accuracy of its measurement.

TABLE 1. Parameters of the prediction of rock bursts in the mines "Sevural-bauxiteruda" Co. in 1988–1990.

Date	C	S (m)	E_{pr} (J)	E (J)	T_{pr} (d)	T (d)
25.03.88	Y	200×200×200	6.1	5.5	13	85
12.04.88	Y	200×200×200	6.1	6.2	13	46
15.09.88	Y	300×250×300	6.7	6.7	20	68
27.12.88	Y	250×250×250	6.4	5.8	17	8
01.03.89	R	100×200×300	6.1	6.2	13	55
21.03.89	Y	225×175×200	6.3	6.4	15	14
14.05.89	Y	400×400×250	7.2	5.3	26	26
21.05.89	Y	250×250×250	6.4	6.3	16	50
24.05.89	Y	200×325×375	6.8	6.2	22	65
06.07.89	R	50×225×225	6.3	5.8	15	19
29.07.89	N		5.7			
21.08.89	Y	200×200×250	6.1	6.5	13	6
23.10.89	R	200×225×300	6.3	5.8	15	33
07.01.90	Y	225×150×300	6.3	6.3	15	10
13.01.90	Y	200×150×150	6.1	6.7	13	8
23.01.90	Y		5.6			
		300×200×400	6.7		20	25
23.01.90	Y		5.8			
06.03.90	R	200×320×395	6.8	6.7	21	16
23.04.90	R	235×175×150	6.3	6.1	10	1
03.08.90	R	200×200×200	5.9	5.7	12	51
29.08.90	Y	210×245×245	6.4	6.4	16	30
31.08.90	Y	185×175×275	6.2	6.2	12	6
10.09.90	N		6.2			
27.11.90	Y	200×300×175	6.7	6.6	20	33

The high efficiency of the prediction of rock bursts in the mines "SUBR" is provided on the one hand, by a reliable and sufficiently precise registration of seismic phenomena and, on the other hand, by the application of universal criteria of the formation of a local fracture nucleation site of rocks that have been found out within the framework of the two-stage model of the process. It should be noted that only two events could not be predicted by objective reasons (N) in the temporal interval given in Table 1. This indicates the high potential of the present technique of prediction.

7. Conclusion

The kinetic concept of strength of solids formulated on the basis of theoretical and experimental studies considers the fracture as a process which develops in time and space, which opens up an essential opportunity for its checking and predicting. One of the fundamental properties of fracture is the stochastic character of the accumulation of defects, which determines the choice of statistical methods of research.

Within the framework of the kinetic approach the two-stage model of fracture of heterogeneous bodies has been developed that is invariant as to the scale of a process. Basing on this model, the universal statistical criteria for the formation of a fracture nucleation site have been suggested, which make it possible to solve in practice the problem of prediction.

Acknowledgments

The work was supported by the Russian Foundation for Basic Research (project No. 96-05-64585).

References

[1] Zhurkov, S.N. (1965) Kinetic concept of the strength of solids, *Int. J. Fract. Mech.* **1**, 311–323.
[2] Regel, V.R., Slutcker, A.I., and Tomashevskii, E.E. (1974) *Kinetic nature of solids strength*, Nauka, Moscow.
[3] Kusov, A., Kondyrev A., and Chmel A. (1990) Common approach to the problem of defect nucleation in solids under "pre-threshold" laser irradiation, *J. Phys.: Condens. Matter* **2**, 4067–4080.
[4] Berezhanskii, V.B., Bykov, V.M., Gorodov, V.V., Zakrevskii, V.A., and Slutsker, A.I. (1985) Electrical strength of polymers in the absence of partial discharges, *Sov. Phys. Tech. Phys.* **30**, 969–971.
[5] Dakhiya, M.S., Zakrevskii, V.A., and Slutsker, A.I. (1986) Accumulation processes in the mechanism of electrical damage to ceramics, *Sov. Phys. Solid State* **28**, 1513–1516.
[6] Cottrell A.H. (1958) Theory of brittle fracture in steel and similar metals, *Trans Met. Soc. AIME* **212**, N2, 192–198.
[7] Bronnikov, S.V., Vettegren, V.I., and Frenkel, S.Y. (1996) Kinetics of deformation and relaxation in highly oriented polymers, *Adv. Polymer Science* **125**, 103–146.
[8] Tamuzh, V.P. and Kuksenko, V.S. (1981) *Fracture micromechanics of polymer materials*, Martinus Nighoff Pub., The Hague, Boston, London.

[9] Betekhtin, V.I., Kadomtsev, A.G., Petrov, A.I., and Vladimirov, V.I. (1976) Reversibility of the first stage of fracture in metals, *Phys. Stat. Sol.(a)* **34**, 73–78.

[10] Zhurkov, S.N., Novak, I.I., Poretskii S.A., and Yakimenko I.Yu. (1987) Light-scattering study of microcrack nucleation kinetics in alkali halide crystals, *Sov. Phys. Solid State* **29**, 87–91.

[11] Gezalov, M.A., Kuksenko, V.S., and Slutsker, A.I. (1972) Kinetic of the formation of submicroskopic cracks in polymers under load, *Sov. Phys. Solid State* **14**, 344–348.

[12] Petrov, V.A. (1979), Mechanisms and Kinetics of Macrofracture, *Sov. Phys. Solid State* **21**, 2123–2126.

[13] Zhurkov, S.N., Kuksenko, V.S., and Petrov, V.A. (1984) Principles of the kinetic approach of fracture prediction, *Theoretical and Applied Fracture Mechanics* **1**, 271–274.

[14] Kuksenko, V.S., Lyashkov, A.I., Mirzoev, K.M., Negmatulliev, S.H., Stanchits, S.A., and Frolov, D.I. (1982) Correlation between sizes of cracks occurring under load and duration of release of elastic energy, *Sov. Phys. Dokl.* **264**, 846–848.

[15] Gor, A.Yu., Kuksenko, V.S., Tomilin, N.G., and Frolov, D.I. (1989) Applicability of the concentration criterion to prediction of rock shocks, *Phys. Tech. Problems of Mining* **3**, 54–60, (in Russian).

[16] Sobolev, G.A., and Zavialov, A.D. (1981) A concentration criterion for seismically active faults, in Amer. Geophys. Union (ed.), *Earthquake Prediction – An International Review*, pp. 377–380.

[17] Sadovskiy, M. A., Golubeva, T.V., Pisarenko, V.F., and Shnirman, M.G. (1984) Characteristic dimensions of rock and hierarchical properties of seismicity, *Izv. Acad. Sci. USSR, Phys. Solid Earth* **20**, 87–96.

[18] Lockner, D.A., Byerlee, J.D., Kuksenko, V., Ponomarev, A., and Sidorin, A. (1992) Observations of quasistatic fault growth from acoustic emissions, in B. Evans and T.-F. Wong (eds.), *Fault Mechanics and Transport Properties of Rocks*, Academic Press, London, pp. 3–31.

[19] Myachkin, V.I., Kostrov, B.V., Sobolev, G.A., and Shamina, O.G. (1974) Laboratory and theoretical investigations of process of earthquake preparation, *Izv. Akad. Nauk SSSR, Fiz. Zemli* **10**, 2526–2530 (in Russian).

[20] Voinov, K.A., Krakov, A.S., Lomakin, V.S., and Halivin, N.I. (1987) Seismological investigations of rock bursts at the Severoural bauxite mine, *Izv. Akad. Nauk SSSR, Fiz. Zemli* **10**, 98–104.

DETERMINISTIC AND STOCHASTIC ASPECTS OF STRAIN LOCALIZATION IN THE FATIGUE OF METALS

M. ZAISER*, M. AVLONITIS AND E. C. AIFANTIS**
*Laboratory of Mechanics and Materials, Polytechnic School,
Aristotle University of Thessaloniki, 54006 Thessaloniki, Greece*
AND

* *Max-Planck-Institut für Metallforschung, Institut für Physik
P. O. Box 800 665, D-70506 Stuttgart, Germany*
** *Center for Mechanics of Materials and Instabilities,
Michigan Technological University, Houghton, MI 49931, USA*

1. Introduction

In recent years, a long-standing issue is the understanding of the localization and postlocalization behaviour of plastic deformation in materials that exhibit non-monotonic stress-strain relationships (hardening, softening and possible rehardening). Such type of non-convex stress-strain graph implies that there is an unstable region where the flow stress required for further plastic flow decreases with increasing plastic strain. Several reasons for such type of behaviour may be envisaged: (i) *multiplication softening* arising from a rapid increase in the density of mobile dislocations after yield [1]; (ii) *structural softening* resulting either from a dislocation glide resistance that decreases in the course of straining due to dislocations destroying or sweeping obstacles to their motion (for an overview, see [2]) or from a transition between different types of dislocation arrangements; (iii) *geometrical softening* commonly caused by lattice rotation [3].

As strain fluctuations become undamped when the stress-strain graph enters the region of negative slope, localization of plastic deformation at various length scales is a generic feature of plasticity in this unstable regime. The phenomena include travelling waves (Lüders bands, for an overview see [4] and references therein), as well as the formation of localized shear bands [5]. It is now commonly accepted that a workable theoretical description of

G. N. Frantziskonis (ed.), PROBAMAT – 21st Century: Probabilities and Materials, 557–572.
© 1998 *Kluwer Academic Publishers.*

those spatial and spatio-temporal phenomena can be obtained by adopting the approach developed by Aifantis [6, 7], introducing gradient terms into the constitutive equations. Such terms provide a mechanism that stabilizes deformation after the material has entered the unstable regime and thus allows for well-defined localized solutions. The adaption of this gradient approach to the analysis of cyclic strain localization patterns in terms of a non-convex cyclic stres-strain graph and a second-order dependence of the cyclic flow stress on the cyclic shear strain was suggested by Aifantis [8] as early as in 1983.

Recently, another approach has been proposed by Zaiser and Hähner [9, 10] for describing strain localization phenomena that take place at a smaller length scale, e.g. at the range of one micrometer. Using the framework of stochastic dislocation dynamics [11], dislocation interactions in a deforming crystal were considered in terms of fluctuations of the internal stress field and the local shear strain rate. The approach, while not giving direct access to the spatial distribution of slip, has proved successful in deriving important statistical characteristics of microstructural slip localization patterns in strain softening materials [9, 10].

In the present work we use both approaches to study the formation of so-called persistent slip bands (PSB) in fatigued metals. We shall demonstrate that cyclic strain localization patterns can be treated within the general theoretical framework developed for materials exhibiting non-convex stress-strain graphs. The paper is organized as follows: an overview of the experimental findings and of previous theoretical work on PSB formation is given in Section 2. Sections 3 and 4 present an outline of the theoretical treatment using both deterministic (gradient) and stochastic approaches, respectively. The results are discussed in Section 5.

2. Matrix-PSB patterns in fatigued metals

In strain-amplitude controlled cyclic deformation of fcc metals and alloys, strain localization is indicated by the emergence of coarse slip markings on the specimen surface. They show up if the imposed shear strain amplitude $\hat{\gamma}_{ext}$ exceeds a critical amplitude $\hat{\gamma}_M$ which is typically of the order of 10^{-4}. The slip markings indicate the formation of the so-called persistent slip bands (PSB). Cyclic deformation is almost completely localized within these bands where, on the microstructural level, a peculiar dislocation arrangement evolves consisting of a ladder-like pattern of parallel dislocation walls arranged perpendicular to the primary Burgers vector. As can be seen from Figure 1, the PSB are embedded into a second phase commonly known as 'matrix' which consists of irregularly arranged dislocation-rich regions ('veins'). A comprehensive overview of experimental findings on PSB fea-

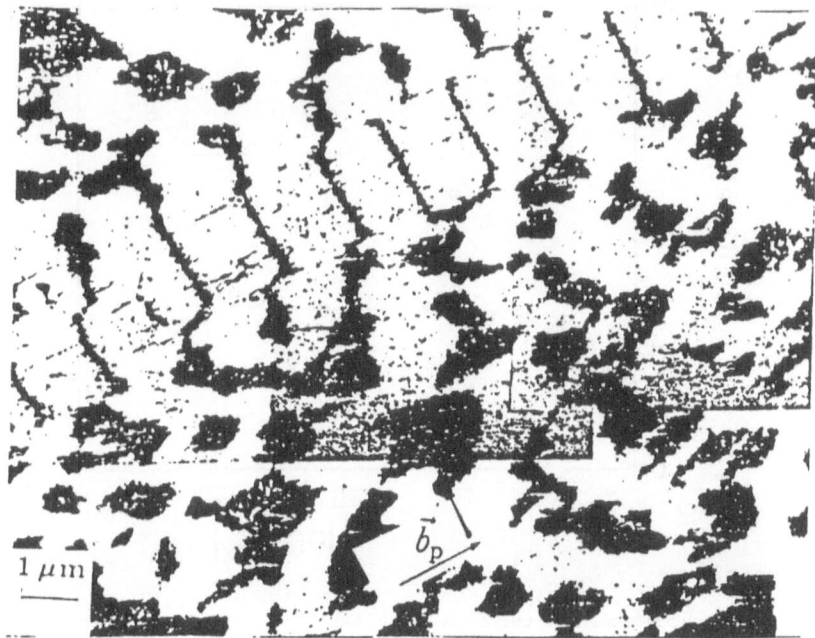

Figure 1. Electron micrograph from a Cu single crystal fatigued into saturation with an imposed plastic shear strain amplitude $\hat{\gamma}_{ext} = 2.2 \times 10^{-3}$ at a temperature $T = 300$ K, showing the ladder-like dislocation pattern of a PSB and the more irregular dislocation arrangement in the surrounding matrix, b_p indicates the primary Burgers vector; after [13].

tures may be found in the recent review by Basinski and Basinski [12] and references therein.

In strain-controlled experiments, the co-existence of matrix and PSB goes along with a plateau in the cyclic stress–strain curve (the plot of the saturation flow stress $\hat{\tau}_{ext}$ vs. the applied plastic shear strain amplitude $\hat{\gamma}_{ext}$). This is depicted in Figure 2: Between the two characteristic strain amplitudes $\hat{\gamma}_M$ and $\hat{\gamma}_{PSB}$, the saturation flow stress remains at an almost constant level. Throughout the plateau regime, the strain amplitude in the matrix remains close to $\hat{\gamma}_M$, while the local shear strain amplitude $\hat{\gamma}_{PSB}$ in the persistent slip bands exceeds this value by about two orders of magnitude. If the applied strain amplitude is raised, this is accomodated by an increase in the volume fraction f_{PSB} occupied by the PSBs [14, 15]. It has been pointed out by Winter [14] that throughout the plateau regime, the volume fractions of matrix and PSB fulfill the law of mixtures

$$\{\hat{\gamma}\} = f_{PSB}\hat{\gamma}_{PSB} + (1 - f_{PSB})\hat{\gamma}_M \quad , \tag{1}$$

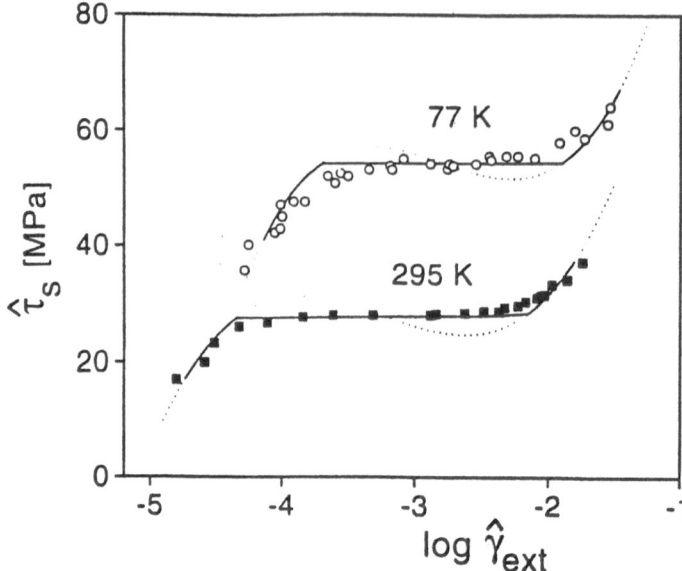

Figure 2. Cyclic stress-strain curves at 295 K and 77 K. Data points after Mughrabi [15]. Dotted lines: $\hat{\tau}_s(\hat{\gamma})$ graph [Eq. (2)] fitted to the data outside the plateau. Full lines: calculated cyclic stress-strain curves $\hat{\tau}_{ext}(\{\hat{\gamma}\})$.

where the average strain amplitude $\{\hat{\gamma}\} = \hat{\gamma}_{ext}$. The spatial distribution of slip is governed by two characteristic lengths: i.e., the mean width Δx and spacing λ of the PSB. As Δx is found to be almost independent of $\{\hat{\gamma}\}$ and $f_{PSB} = \Delta x/\lambda$, an increase in the average strain amplitude implies a decrease in PSB spacing.

Theoretical work on persistent slip bands has mainly focused on two questions: (i) Several semi-phenomenological models have been developed for describing the particular features of dislocation glide in the fully developed slip bands [16, 17]. (ii) The formation of the charateristic dislocation patterns (Figure 1) associated with the matrix and the PSB has been treated theoretically in several works. In the Walgraef-Aifantis model, a reaction-diffusion approach was used for describing the coupled dynamics of mobile and immobile dislocation populations in order to adress the problem of wavelength selection and stability of the ladder structure of PSBs [18]. [For more detailed results see a recent revisit of this model by Glazov et al. [19].] Finally, we mention the work of Hähner [20] who used stochastic methods to derive statistical characteristics of the dislocation arrangements in the matrix and the PSBs.

While significant work has been done on the dynamics of dislocations within the PSBs and the formation of ordered dislocation arrangements.

little attention has been given to the simultaneously occurring localization and patterning of plastic deformation. This is remarkable in view of the fact that the characteristic time scale of deformation localization (which is of the order of a few cycles only) is actually much shorter than the characteristic time of dislocation microstructure evolution, which proceeds on a time scale of several hundreds of cycles [13]. The present investigation outlines a phenomenological model of the aforementioned strain localization which *precedes* dislocation patterning within the PSBs and is thus crucial for the complete understanding of PSB formation; especially in relation to the width and spacing of these deformation bands.

3. Deterministic Modelling

Several authors have pointed out that the matrix-PSB transition with its characteristic stress plateau and the co-existence of two 'phases' has many similarities with first-order phase transitions in equilibrium thermodynamics (see e.g. [14]). In the present work, we shall elaborate on this idea by considering the co-existence of matrix and PSBs to be a consequence of the existence of an unstable region (a regime of negative slope) in the graph that relates the saturation stress and the strain amplitude.

While various kinds of such relationships may be envisaged, it turns out that a third-order polynomial in [log $\hat{\gamma}$] is the simplest form to model a non-convex graph that is suited for the present problem[1]. Thus, we describe the $\hat{\tau}_s(\hat{\gamma})$ 'ideal' graph by

$$\hat{\tau}_s(\hat{\gamma}) = C_1 - C_2 \log\left[\frac{\hat{\gamma}}{\hat{\gamma}_0}\right] + C_3 \log^3\left[\frac{\hat{\gamma}}{\hat{\gamma}_0}\right] \quad . \tag{2}$$

In order to determine the constants C_1, C_2, C_3, and $\hat{\gamma}_0$, Eq. (2) is fitted to the experimental cyclic stress-strain curves in the region outside the plateau. The appropriate values are given in Table 1. The $\hat{\tau}_s(\hat{\gamma})$ graphs

T	C_1	C_2	C_3	$\hat{\gamma}_0$
293 K	28 MPa	8 MPa	7 MPa	5.6×10^{-4}
77 K	54.5 MPa	8 MPa	10 MPa	1.6×10^{-3}

TABLE 1. Parameters of the $\hat{\tau}_s(\hat{\gamma})$ 'ideal' graphs (dotted lines in Figure 2).

[1]In other situations different representations may be appropriate; e.g., in the analysis of PLC band propagation in [5] an exponential representation was used.

obtained in this manner are indicated by the dotted lines in Figure 1. The procedure implies that in the region outside the plateau – but not in the plateau regime – the $\hat{\tau}_s(\hat{\gamma})$ graphs coincide with the actual cyclic stress-strain curves. The validity of this procedure is demonstrated later.

As Eq. (2) applies to cyclic saturation only, it has to be supplemented by a dynamic equation specifying how the saturation state is reached in the course of time. To this end, we assume that the dynamics of $\hat{\gamma}$ is governed by a relaxation equation of the form

$$\partial_t \hat{\gamma} = K \frac{\hat{\gamma}}{t_{\text{cycl}}} \left[1 - \frac{\hat{\tau}_s(\hat{\gamma})}{\hat{\tau}_{\text{ext}}} \right] \quad . \tag{3}$$

This corresponds to the assumption that for a given external stress amplitude $\hat{\tau}_{\text{ext}}$, the strain amplitude relaxes to a value where $\hat{\tau}_{\text{ext}} = \hat{\tau}_s(\hat{\gamma})$ with a rate that is proportional to the distance from the $\hat{\tau}_s(\hat{\gamma})$ graph as well as to the strain rate $\sim \hat{\gamma}/t_{\text{cycl}}$ (where t_{cycl} is the duration of one deformation cycle).

According to Eq. (3), the system behaves in an unstable manner if the strain amplitude lies in the region where the $\hat{\tau}_s(\hat{\gamma})$ graph [Eq. (2)] has negative slope. To discuss the implications of this, we note that the $\hat{\tau}_s(\hat{\gamma})$ graph may be considered as the cyclic analogue of the constitutive relation $\tau_f = \tau_f(\gamma)$ relating the flow stress τ_f to the plastic strain γ in (rate-independent) unidirectional plastic deformation. As pointed out in [5], this constitutive relation provides a reasonable description as long as deformation remains homogeneous, but not in the unstable regime where $\tau_f(\gamma)$ has negative slope and deformation proceds in an unstable, localized manner (shear banding). To describe deformation in the postlocalization regime, Aifantis [7] and Zbib and Aifantis [5] modified the flow stress expression according to $\tau_f = \tau_f(\gamma) - c\Delta\gamma$, where $\Delta\gamma$ is the Laplacian of γ.

In view of the success of this approach in modelling shear banding, it appears straightforward to use an analogous generalization of the $\hat{\tau}_s(\hat{\gamma})$ graph in order to model the matrix-PSB patterns. However, it is important to keep certain differences in mind: for instance, the relaxation equation (3) allows for a decrease in plastic strain amplitude $\hat{\gamma}$, whereas the plastic strain γ cannot be reversed. As the strain amplitude $\hat{\gamma}$ (and not the cumulative strain) is considered as the relevant variable, we have to account for gradients of the local strain amplitude or, equivalently, the deformation rate.[2] To give a physical motivation for the corresponding stress contribution, we consider the spatial coupling that arises between mobile dislocations moving with different velocities (giving rise to different local strain rates)

---

[2]Since, in the following, gradients of the logarithm of strain rate are considered, any proportionality factors between local strain rate and amplitude do not affect a replacement of one quantity by the other.

on parallel slip planes. As the faster-moving dislocations 'push' the slower ones, this provides a stress contribution which enhances plastic flow in the regions of reduced plastic activity, while impeding it in the regions of maximum strain rate. Thus, one expects this spatial coupling to provide the required stabilizing properties. A detailed treatment of this mechanism has been given in a recent article [21]. It turns out that the coupling gives rise to a diffusion-like term which is proportional to the internal-stress contribution of the mobile dislocations, τ_m, and to the second-order gradient of the logarithm of the strain rate. Accordingly, we put

$$\hat{\tau}_s(\hat{\gamma}) \rightarrow \hat{\tau}_s(\hat{\gamma}) - (\xi^2 \tau_m)\frac{\partial^2 \ln \hat{\gamma}}{\partial x^2} \quad , \tag{4}$$

where x denotes the direction normal to the slip plane normal, and ξ is the characteristic length for the interaction of two dislocations [cf. Eq. (11)].

Since the rate-dependent contribution to the saturation stress amplitude (the 'effective stress') is small [22], we may identify $\hat{\tau}_s(\hat{\gamma})$ with the long-range internal stresses that develop for a given strain amplitude. Denoting by β the fraction of these stresses which are made up by mobile dislocations, we have $\tau_m = \beta \hat{\tau}_s(\hat{\gamma})$; i.e., the gradient modification of the relaxation equation (3) reads

$$\partial_t \hat{\gamma} = K \frac{\hat{\gamma}}{t_{\text{cycl}}} \left[1 - \frac{\hat{\tau}_s(\hat{\gamma})}{\hat{\tau}_{\text{ext}}} + \frac{\hat{\tau}_s(\hat{\gamma})}{\hat{\tau}_{\text{ext}}} \xi^2 \beta \frac{\partial^2 \ln \hat{\gamma}}{\partial x^2} \right] \quad . \tag{5}$$

The stationary strain amplitude patterns associated with the PSB are characterized by the steady-state solutions of this equation. Introducing non-dimensional variables $u = \ln(\hat{\gamma}/\hat{\gamma}_0)$, $x = \xi\sqrt{\beta}\,\tilde{x}$ and using Eq. (2), the steady-state version of Eq. (5) is

$$u_{\tilde{x}\tilde{x}} = 1 - \frac{\hat{\tau}_{\text{ext}}}{C_1 - C_2 u + C_3 u^3} \quad . \tag{6}$$

This belongs to a general class of equations the solutions of which have been studied in detail in [23]. In the present work, two kinds of solutions are of interest: (i) solutions describing a localized region of enhanced plastic activity (a single band), and (ii) solutions describing periodic arrangements of such bands. (In fact, solutions of the first type – called 'reversals' in [23] – may be considered the limiting case of a periodic solution with infinite wavelength.)

Integrating Eq. (6) twice gives

$$\tilde{x} - \tilde{x}_0 = \int_{u_1}^{u} \frac{1}{\sqrt{2\tilde{\Phi}(u')}} du' \,, \quad \Phi(u') = \int_{u_0}^{u'} 1 - \frac{\hat{\tau}_{\text{ext}}}{C_1 - C_2 u'' + C_3 u''^3} du'' \,. \tag{7}$$

564

Periodic solutions exist provided that $\hat{\tau}_{\text{ext}}$ is in the stress range where $\hat{\tau}_{\text{s}}(\hat{\gamma})$ has negative slope. In this case, the function $\Phi(u)$ exhibits two minima, Φ_1 at u_{I} and Φ_2 at u_{II}. We choose the integration constant u_0 such that $\Phi_1, \Phi_2 \leq 0$ and denote by u_1 and u_2 the zeros of $\Phi(u)$ where $u_{\text{I}} \leq u_1 \leq u_2 \leq u_{\text{II}}$. Then, for $u \in [u_1, u_2]$, Eq. (7) describes a half-period of a periodic solution with wavelength

$$\lambda = 2 \int_{u_1}^{u_2} \frac{1}{\sqrt{2\Phi(u')}} du' \quad . \tag{8}$$

For a given $\hat{\tau}_{\text{s}}(\hat{\gamma})$ 'ideal' graph (fixed values of $\hat{\gamma}_0, C_1, C_2, C_3$), the shape and wavelength of the solutions depend on the external stress amplitude $\hat{\tau}_{\text{ext}}$, as well as on the integration constant u_0.[3]

As the gradient model leads us to a two-parameter manifold $u_{\hat{\tau}_{\text{ext}}, u_0}(\tilde{x})$ of stationary periodic solutions, one may ask for additional criteria to determine the pattern that actually emerges. In strain-controlled fatigue, the average plastic strain amplitude $\{\hat{\gamma}\}$ is prescribed. For a periodic pattern, this is given by

$$
\begin{aligned}
\{\hat{\gamma}\} &= \frac{1}{\lambda(\hat{\tau}_{\text{ext}}, u_0)} \int_0^{\lambda(\hat{\tau}_{\text{ext}}, u_0)} \hat{\gamma}[u_{\hat{\tau}_{\text{ext}}, u_0}(\tilde{x})] d\tilde{x} \\
&= \frac{\hat{\gamma}_0}{\lambda(\hat{\tau}_{\text{ext}}, u_0)} \int_0^{\lambda(\hat{\tau}_{\text{ext}}, u_0)} \exp[u_{\hat{\tau}_{\text{ext}}, u_0}(\tilde{x})] d\tilde{x} \quad .
\end{aligned}
\tag{9}
$$

While this relation may be used to obtain, for given $\hat{\tau}_{\text{ext}}$ and $\{\hat{\gamma}\}$, the integration constant u_0, an additional relationship between $\hat{\tau}_{\text{ext}}$ and $\{\hat{\gamma}\}$ is required for uniquely determining the pattern. To provide this relation, which actually corresponds to the cyclic stress-strain curve, a complementary stochastic approach to cyclic strain localization and patterning is used in the next section.

4. Stochastic Modelling

As discussed in [11], an alternative manner of incorporating dynamic interactions between glide dislocations into a constitutive description consists of considering the concomitant spatio-temporal fluctuations of the effective stress or local strain rate. Due to dislocation interactions, the stress required for sustaining locally a strain amplitude $\hat{\gamma}$ flucutates in the course of time. Thus, we put

$$\hat{\tau}_{\text{s}}(\hat{\gamma}) \rightarrow \hat{\tau}_{\text{s}}(\hat{\gamma}) + \delta\tau_{\text{eff}}(t) \quad . \tag{10}$$

---

[3]Note that λ becomes infinite ('reversal'-type of solution) if either Φ_1 or Φ_2 become zero.

This may be considered as the stochastic counterpart of Eq. (4). Using dislocation dynamics arguments, Hähner [11] estimated the magnitude and correlation length ξ of the effective-stress fluctuations as

$$\langle \delta \tau_{\text{eff}}^2 \rangle \approx \langle \tau_{\text{int}} \rangle S \quad , \quad \xi \approx \frac{Gb}{4\pi \sqrt{\langle \tau_{\text{int}} \rangle S}} \quad . \tag{11}$$

Here $\langle \tau_{\text{int}} \rangle$ is the mean internal stress experienced by the ensemble of glide dislocations, and S is the strain-rate sensitivity of the flow stress. We note that, as shown in [21], the fluctuation correlation length ξ also defines the effective interaction distance [entering Eq. (4)] of two mobile dislocations [21]. This fact underlines the close relationship between the gradient and stochastic approaches pursued in the present work.

Substituting Eq. (11) in Eq. (3) and introducing scaled variables according to $t = [t_{\text{cycl}}(\ln 10)/K]\tilde{t}$ and $u = \log(\hat{\gamma}/\hat{\gamma}_0)$, one obtains

$$\partial_{\tilde{t}} u = \left[1 - \frac{\hat{\tau}_{\text{s}}(u)}{\hat{\tau}_{\text{ext}}} \right] + \frac{\delta \tau_{\text{eff}}}{\hat{\tau}_{\text{ext}}} \quad . \tag{12}$$

The correlation time of the effective-stress fluctuations is of the order of the cycle time t_{cycl} [20]. As we are interested in the behaviour at saturation (i.e. at large cycle numbers) we may idealize $\delta \tau_{\text{eff}}$ by a delta-correlated stochastic process,

$$\frac{\delta \tau_{\text{eff}}}{\hat{\tau}_{\text{ext}}} \approx Q\dot{w} \quad , \quad \langle \dot{w}(\tilde{t}) \dot{w}(\tilde{t}') \rangle = \delta(\tilde{t} - \tilde{t}') \quad . \tag{13}$$

Using $\langle \tau_{\text{int}} \rangle \approx \hat{\tau}_{\text{ext}}$ and putting $K \approx 1$, one finds from Eq. (11) that the 'noise amplitude' entering Eq. (13) is approximately $Q^2 \approx S/\hat{\tau}_{\text{ext}}$.

The Langevin-type equation (12) may then be written as

$$\partial_{\tilde{t}} u = -\frac{\partial \Psi(u)}{\partial u} + Q\dot{w} \quad , \quad \Psi(u) = -\int \left[1 - \frac{\hat{\tau}_{\text{s}}(u')}{\hat{\tau}_{\text{ext}}} \right] du' \quad . \tag{14}$$

In the following, we consider the distribution $P_{\hat{\tau}_{\text{ext}}}(u, \tilde{t})$ of local strain amplitudes: $P_{\hat{\tau}_{\text{ext}}}(u, \tilde{t})du$ gives the probability to find at time $\tilde{t}$ in a randomly chosen volume element a strain amplitude that lies in the interval corresponding to $(u, u + du)$. [Note that the dynamics given by Eq. (14) and thus also the resulting strain amplitude distributions depend on the externally applied stress amplitude $\hat{\tau}_{\text{ext}}$.]

Assuming the stochastic process $\dot{w}$ to be Gaussian, one obtains from Eq. (14) the Fokker-Planck equation that governs the temporal evolution of $P_{\hat{\tau}_{\text{ext}}}(u, \tilde{t})$, i.e.,

$$\partial_{\tilde{t}} P_{\hat{\tau}_{\text{ext}}} = \partial_u \left[\partial_u \Psi(u) \right] P_{\hat{\tau}_{\text{ext}}} + \frac{Q^2}{2} \partial_{uu} P_{\hat{\tau}_{\text{ext}}} \quad . \tag{15}$$

For details of the mathematical derivation of this equation, the reader is referred to standard works, e.g. [24]. In the following, we are mainly interested in the steady-state solutions $P_{\hat{\tau}_{ext},s}$ that characterize the strain localization patterns at saturation. These are given by the relation

$$P_{\hat{\tau}_{ext},s}(u) = \mathcal{N} \exp\left[-\frac{2\Psi(u)}{Q^2}\right] \quad , \tag{16}$$

where $\mathcal{N}$ is a normalization constant. The average strain amplitude corresponding, at saturation, to a given external-stress amplitude $\hat{\tau}_{ext}$ is readily calculated using the respective steady-state strain amplitude distribution, i.e.

$$\{\hat{\gamma}\}(\hat{\tau}_{ext}) = \int P_{s,\hat{\tau}_{ext}}(u)\hat{\gamma}(u)du = \hat{\gamma}_0 \int P_{s,\hat{\tau}_{ext}}(u)\exp(u)du \quad . \tag{17}$$

This relation may be inverted to yield, for the case of strain-controlled deformation, the external stress amplitude $\hat{\tau}_{ext}(\{\hat{\gamma}\})$ corresponding to a given or imposed strain amplitude, i.e. the cyclic stress-strain curve.

5. Results and Conclusions

5.1. CALCULATIONS USING THE STOCHASTIC APPROACH

Cyclic stress-strain curves have been calculated for $T = 77$ K and $T = 293$ K from the respective $\hat{\tau}_s(\hat{\gamma})$ graphs using the method outlined in the previous section. The fluctuation strength $Q^2 \approx S/\hat{\tau}_{ext}$ was calculated from the experimental values of the plateau stress and strain-rate sensitivity determined by Holzwarth [22]. The parameters used were $Q^2 = 4 \times 10^{-3}$ for $T = 77$ K and $Q^2 = 4.3 \times 10^{-3}$ for $T = 293$ K, while the $\hat{\tau}_s(\hat{\gamma})$ graphs were obtained by using the parameters compiled in Table 1. The results of the calculations are given in Figure 1 (full lines). It is seen that for the region where the ideal $\hat{\tau}_s(\hat{\gamma})$ graphs exhibit negative slope (unstable regime), the calculated $\hat{\tau}_{ext}(\hat{\gamma})$-curves exhibit a characteristic plateau. Outside this plateau, both types of curves almost coincide. If plotted as in Figure 1, the curves fulfil an equal-area condition: The $\hat{\tau}_s(\hat{\gamma})$ graphs (dotted lines) enclose the same areas above and below the respective $\hat{\tau}_{ext}(\{\hat{\gamma}\})$-curves.

We identify the upper and lower critical strain amplitude delimiting the stress plateau with the characteristic strain amplitudes in the PSB and the matrix, $\hat{\gamma}_{PSB}$ and $\hat{\gamma}_M$, respectively. This is motivated by the shape of the strain amplitude distributions in the plateau region. In Figure 2, distributions corresponding to $T = 77$ K and different average strain amplitudes are compiled. If an average strain amplitude is prescribed that lies in the plateau region, the 'potential' $\Psi(u)$ [Eq. (14)] exhibits two minima of almost equal depth at u values corresponding to $\hat{\gamma}_M$ and $\hat{\gamma}_{PSB}$. Accordingly, the

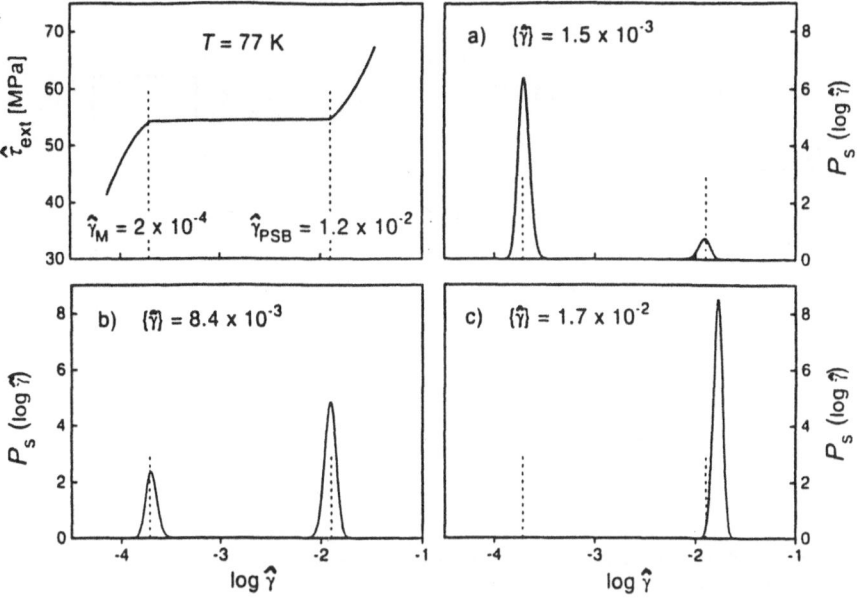

Figure 3. Distributions of local strain amplitudes calculated for $T = 77$K and different average strain amplitudes; for parameters see text.

strain amplitude distribution [Eq. (16)] exhibits two distinct peaks, i.e., part of the specimen deforms at the low amplitude $\hat{\gamma}_M$ corresponding to the left end of the plateau and the rest deforms at $\hat{\gamma}_{PSB}$, while intermediate strain amplitudes are extremely improbable. It is straightforward to identify those parts of the specimen deforming at $\hat{\gamma}_M$ with the matrix and those deforming at $\hat{\gamma}_{PSB}$ with the persistent slip bands. If the average strain amplitude is increased, the peak at $\hat{\gamma}_{PSB}$ grows at the expense of that at $\hat{\gamma}_M$. Since the area under each peak defines the volume fraction of the respective 'phase', this implies that the PSB volume fraction increases at the expense of the matrix (Figure 3 a,b). It reaches unity at the right end of the plateau where $\{\hat{\gamma}\} = \hat{\gamma}_{PSB}$. If $\{\hat{\gamma}\}$ is increased even further, this is accomodated by a shift of the remaining peak of the strain amplitude distribution to increasing values of $\hat{\gamma}$ (Figure 3 c). Below $\hat{\gamma}_M$ a similar behaviour is found: There, the matrix volume fraction is unity and the corresponding peak of

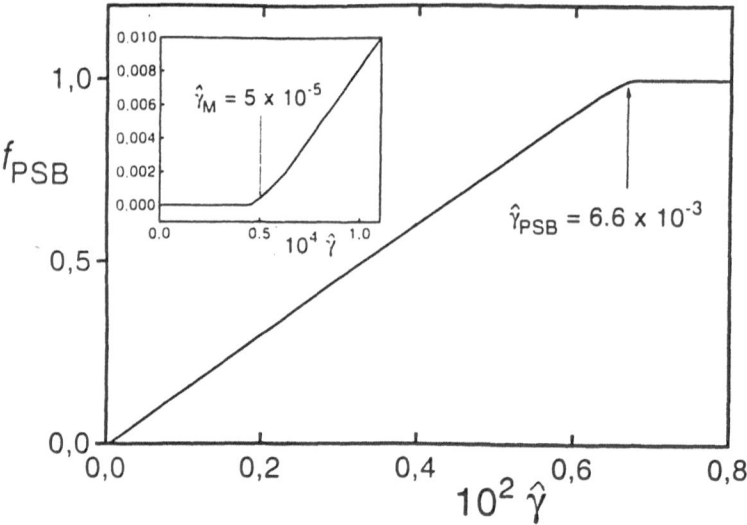

Figure 4. PSB volume fraction as a function of strain amplitude calculated for $T = 293$ K; for parameters see text.

the strain amplitude distribution shifts to decreasing $\hat{\gamma}$ if $\{\hat{\gamma}\}$ is decreased. In Figure 4, the volume fraction f_{PSB} of the persistent slip bands has been calculated as a function of the average strain amplitude for $T = 293$ K. It is seen that, in agreement with Winter's 'rule of mixtures' [Eq. (1)], f_{PSB} increases almost linearly throughout the plateau region.

We note that the results of our calculations indicate certain deviations from the idealized picture proposed by Winter [14], i.e. a strictly horizontal stress plateau throughout which f_{PSB} increases linearly. Owing to the finite amplitude of the effective-stress fluctuations, the plateau stress increases slightly with increasing strain amplitude. This is in agreement with the experimental findings (Figure 1). For the same reason, deviations from the linear increase of f_{PSB} with $\{\hat{\gamma}\}$ are found, in particular close to the upper end of the plateau. Furthermore, it is seen from Figure 3 that the two peaks of the strain amplitude distribution that correspond to the matrix and PSB phases have a finite width that remains constant throughout the plateau region. In physical terms, this finite width implies that the strain amplitudes in the PSBs and the matrix scatter by a factor of about two around the respective average values $\hat{\gamma}_M$ and $\hat{\gamma}_{PSB}$.

The Fokker-Planck Equation (15) may also be used to calculate the average rate of transitions between the minima of the 'potential' $\Psi(u)$. [For an overview of methods for such calculations, the reader is referred to the review by Hänggi et al. [25].] It turns out that the rate, corresponding to the rate of transitions between the matrix and PSB phases, is given by the relation

$$R_{\text{M,PSB}} = \frac{1}{2\pi}[\Psi''_{\min}\Psi''_{\max}]^{1/2} \exp\left[-\frac{2(\Psi_{\max} - \Psi_{\min})}{Q^2}\right] \quad . \qquad (18)$$

Here $\Psi_{\min}$ and $\Psi_{\max}$ denote respectively the values of the 'potential' Ψ at its minima and at the barrier in between, while $\Psi''_{\min}$ and $\Psi''_{\max}$ are the respective curvatures. Inserting typical values, it turns out that the rate of transitions between the two 'phases' at saturation is of the order of 10^{-14} ($10^{-14}K/(\ln 10)t_{\text{cycl}}$, in dimensional variables) only, i.e., the once established structure behaves in a truly 'persistent' manner.

5.2. CALCULATIONS USING THE GRADIENT APPROACH

While the stochastic approach gives access to important features of the matrix-PSB transition, it does not yield any direct information on the spatial distribution of slip (deformation patterning). To obtain such information, the gradient approach of Section 3 was used. By adopting the relationship between $\hat{\tau}_{\text{ext}}$ and $\{\hat{\gamma}\}$ that derives from the stochastic approach, it is possible to determine the spatial patters that correspond to a given strain amplitude in a unique manner.

Figure 5 shows slip patterns calculated from Eqs. (7) and (9) using the relationship between $\hat{\tau}_{\text{ext}}$ and $\{\hat{\gamma}\}$ that derives from Eq. (17). Calculations were performed with the parameter values given in Table 1 for $T = 77$ K for three different values of $\{\hat{\gamma}\}$ in the plateau region. While the 'reversal'-type solution in Figure 5a corresponds to a single PSB at the beginning of the plateau, the periodic patterns in Figure 5b and 5c correspond to PSB arrangements at higher average strain amplitude. To determine the length scale of the patterns, a value of $\beta=0.1$ was assumed [cf. Eq. (5)] and the characteristic interaction length ξ was calculated from Eq. (11) using experimental values of S given in [22]. It is seen from Figure 5 that the width of the PSB is almost unaffected by $\{\hat{\gamma}\}$. The calculated values $\Delta x = 0.8\mu$m at $T = 77$ K and $\Delta x = 1.8\mu$m at $T = 273$ K are in good agreement with observed average PSB widths of about $\Delta x = 0.9\mu$m and $\Delta x = 2.1\mu$m, respectively.

Figure 5 demonstrates that, while the strain amplitude in the PSBs is almost independent of the imposed strain amplitude, the PSB spacing λ decreases upon an increase in $\{\hat{\gamma}\}$. This decrease is found to be consistent with Winter's rule: Using $f_{\text{PSB}} = \Delta x/\lambda$, one readily deduces from Eq. (1) that the product $\lambda(\{\hat{\gamma}\} - \hat{\gamma}_M)$ is expected to be constant. This compares well with the results depicted in Fig.5: with $\hat{\gamma}_M = 2 \times 10^{-4}$ for $T = 77$ K

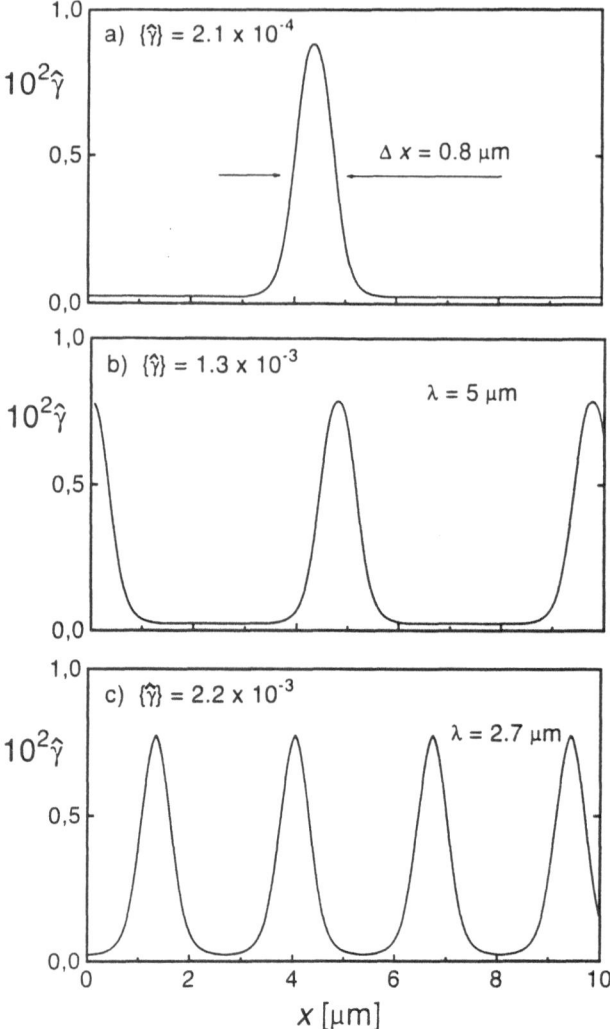

Figure 5. Strain amplitude patterns calculated for $T = 77$ K; a) 'reversal'-type solution (single persistent slip band) for $\{\hat{\gamma}\} \approx \hat{\gamma}_M$; b,c) periodic arrangements of PSB for $\hat{\gamma}_M < \{\hat{\gamma}\} < \hat{\gamma}_{PSB}$.

we find $\lambda(\{\hat{\gamma}\} - \hat{\gamma}_M) = 5.5 \times 10^{-3}\mu$m for Figure 4 b and $5.4 \times 10^{-3}\mu$m for Figure 4 c, which demonstrates the validity of the 'rule of mixtures' for the calculated patterns.

5.3. CONCLUSIONS

The present investigation of persistent slip band patterns in fatigued metals demonstrates the complementary character of deterministic (gradient) and stochastic approaches to strain localization. While stochastic models allow to determine important statistical characteristics of such patterns, they do not give access to spatial features such as the shape and width of persistent slip bands. On the other hand, gradient approaches are often – as in the present case – confronted with the problem that there are 'too many' nontrivial, space-dependent solutions, while a precise criterion for selecting the actually realized pattern is lacking.

In the present work, this problem was adressed by using additional information obtained from a stochastic approach which allows for the determination of the plateau regime in the stress-strain curve. In this manner, it was possible to determine uniquely the spatial patterns characterizing the strain amplitude distribution. The results of the stochastic and gradient treatment where found to be mutually consistent and in good agreement with the experimental findings. This gives hope that in the future it will be possible to give a comprehensive theoretical formulation that covers the various aspects of cyclic plasticity within a single model considering both stochastic and gradient terms.

Acknowledgement. The support of the European Commission under contract No. ERB-FMRX-CT 960062 is gratefully acknowledged. Partial support was also provided by AFOSR grant No. 49620-95-1-0208.

References

1. Estrin, Y and Kubin, L. P. (1986) Local strain hardening and nonuniformity of plastic deformation, *Acta Metall.* **34**, 2455–2464.
2. Luft, A. (1991) Microstructural Processes of Plastic Instabilities in Strengthened Metals, *Progr. Mater. Sci.* **35**, 97–204.
3. Deve, H. E. and Asaro, R. J. (1989) The developement of plastic failure modes in crystalline materials: shear bands in fcc polycrystals, *Metall. Trans. A* **20**, 579–593.
4. Estrin, Y., Kubin, L. P. and Aifantis, E. C. (1993) Viewpoint set on propagative plastic instabilities, *Scripta Metall.* **29**, 1147–1150.
5. Zbib, H. M. and Aifantis, E. C. (1988) On the Localization and Postlocalization Behaviour of Plastic Deformation I-III, *Res Mechanica* **23**, 261–305.
6. Aifantis, E. C. (1984) On the microstructural origin of certain inelastic models, *Trans ASME, J. Eng. Mat. Techn.* **106**, 326–330.
7. Aifantis, E. C. (1987) The physics of plastic deformation, *Int. J. Plasticity* **3**, 211–247.
8. Aifantis, E. C. (1983) Dislocation kinetics and the formation of deformation bands, in: *Defects, Fracture and Fatigue*, G. C. Sih and J. W. Provan (Eds.), Martinus-Nijhoff, 75–84.
9. Zaiser, M. and Hähner, P., (1996) A theory of the formation of slip channels in cold-worked bcc metals, *Phil. Mag. A* **74**, 287–298.

572

10. Zaiser, M. and Bay, K. (1997) Microstructural slip localization in strain softening materials, *phys. stat.sol (b)* **203** , 29–42.
11. Hähner, P. (1996) On the foundations of stochastic dislocation dynamics, *Appl. Phys. A* **62**, 473–481.
12. Basinski, Z. S. and Basinski, S. J. (1992) Fundamental aspects of low-amplitude cyclic deformation in face-centered cubic crystals, *Progr. Mater. Sci.* **36**, 89–148.
13. Holzwarth, U. and Essmann, U. (1993) Transformation of dislocation patterns in fatigued copper single crystals, *Mater. Sci. Engng. A* **164**, 206–210.
14. Winter, A. T. (1974) A model for the fatigue of copper at low plastic strain amplitudes, *Phil. Mag.* **30**, 719–738.
15. Mughrabi, H. (1978) The cyclic hardening and saturation behaviour of copper single crystals, *Mater. Sci. Engng.* **33**, 207–223.
16. Essmann, U. and Differt, K. (1996) Dynamic model of the wall structure in persistent slip bands of fatigued metals, *Mater. Sci. Engng. A* **208**, 56–68.
17. Pedersen, O. B. (1996) A static-dynamic model for the process of cyclic saturation in fatigue of metals, *Phil. Mag. A* **73**, 829–858.
18. Walgraef, D. and Aifantis, E. C. (1985) Dislocation patterning in fatigued metals as a result of dynamical instabilities, *J. Appl. Phys.* **58**, 688–691.
19. Glazov, M., Lanes, L. M., and Laird, C. (1995) Self-organized dislocation structures in fatigued metals, *phys. status solidi (b)* **149**, 295–321.
20. Hähner, P. (1996) Stochastic dislocation patterning during cyclic plastic deformation, *Appl. Phys. A* **63**, 45–55.
21. Zaiser, M. and Aifantis, E. C. (1997) On the dynamic interaction between moving dislocations, *Appl. Phys. A* , to be published.
22. Holzwarth, U. and Hähner, H. (1995) Temperature dependence of the strain-rate sensitivity of persistent slip bands in copper single crystals *Phil. Mag. A* **72**, 691–705.
23. Aifantis, E. C. and Serrin, J. B. (1983) The mechanical theory of fluid interfaces and Maxwell's rule, *J. Colloid Interface Sci* **96**, 517–529.
24. Haken, H. (1983) *Synergetics*, Springer, Heidelberg-New York-Tokyo.
25. Hänggi, P., Talkner, P. and Borkovec, M. (1990) Reaction-rate theory: fifty years after Kramers, *Rev. Mod. Physics* **62**, 251–341.

SCALES !

S. ROUX
Physique et Mécanique des Milieux Hétérogènes,
École Supérieure de Physique et Chimie Industrielles,
10 rue Vauquelin, 75231 Paris cedex 05, France.

The importance of the multiplicity of length scales appearing in a variety of problems pertaining to the field of solid mechanics or science of material in the past years has been recognized. Today, this concern has reached so many different subjects and development, that it is almost impossible to find any other common point in the following seven contributions. We will not try to provide a link between the texts. Rather, I appreciate the richness and diversity of these contributions, and I will simply try to underline some of the points which appear essential to me.

The first two texts which refer to synergetics, address very different questions. Ivanova's contribution is a philosophical essay on future materials. It is difficult not to agree on the fact that biomimetism will continue to be a fruitful source of inspiration for designing new materials. This apply at all levels of the conception. From basic chemistry — where the subtle associations of organic and inorganic parts which can be found in nature opens a basically unexplored route to new materials — to geometry and structure where optimal design is evidenced in a number of cases. In this domain of new material it is obvious that progress is expected through the mastering of smaller and smaller scales for the basic contituents, and evidently, their spatial arrangement. In this quest for a better and better micro-architecture, physico-chemistry is also essential, and amazing structure have been obtained recently with organized states of matter at supra-molecular scales (e.g. through liquid crystals formed by micelles where the elementary micelles play the role of micro-chemical reactors), and which preserve the geometrical structure at the final stage of elaboration. There, the first concern is often dictated by an ideal structure and the requirement of properties to be satisfied "on average", however, in most applications including non-linear behavior, the sensitivity to rare but severe defects also emphasizes the need for a control of the statistical properties of the structure. Hence the timelyness of this meeting. The status of synergetics in this

G. N. Frantziskonis (ed.), PROBAMAT – 21st Century: Probabilities and Materials, 573–577.
© 1998 *Kluwer Academic Publishers.*

574

context and its operational use for the purpose of creating new materials is a question which still appear unclear to me.

The second text by Shanyavskii is more specific, and attempts to account for fatigue patterning in a unified framework. This goal is quite appealing and ambitious. Our background in synergetics is however too limited to fully appreciate the status of the proposed general evolution equations issued from synergetics, and their relations with the underlying physics and mechanics.

In the third contribution, Borri-Brunetto, Carpinteri and Chiaia focus on a very different problem. In the past fifteen years, evidences have been accumulated showing that fracture surfaces for instance, but also other processed surfaces, display roughness over a wide spectrum of scales. Selecting *the* most relevant scale and studying an equivalent "monochromatic" rough surface, is a standard procedure which however has to be questioned for some applications. Contact between rough surfaces is typically such a case where there is no "equivalent monochromatic" rough surface. Therefore, the task of modelling the contact appear as extremely difficult. Fortunately, a rather ubiquituous symmetry has been experimentally observed for many such rough surfaces, an anisotropic scale-invariance called self-affinity. It turns out that this symmetry can be exploited for the contact problem to derive a number of important results, some of which are explicited in this contribution.

From a historical perspective, there has been an upsurge of activity in the past twenty years concerning the transport properties and behaviors of fractal (self-similar) structures. The interesting point was that a refined dimensional analysis could provide very robust informations on the behavior of the medium, and in particular shed some light of systematic scale effects. The observation of self-affinity for surfaces is much more recent. I feel that the exploration of physical instances where the topography of boundary conditions plays a key role, and controls the global response of the medium is yet to be done. The elastic contact problem is one obvious application. Other recently considered applications concern the electrical admittance of rough electrodes, anomalous diffusion from such boundaries, wave scattering (where more work is needed), permeability of cracks in geophysical applications, fracture induced by stress concentration along rough interfaces, ...

The fourth text by Chelidze, Gueguen and Le Ravalec is devoted to percolation. This theory is now one of the standard paradigms in statistical mechanics, and the important concepts are clearly presented.[1] I would like

---

[1] at the reservation of the problem of wave propagation which gives rise to localization ("fracton") for wavelength smaller than the correlation length, and no propagation. This subject has motivated a number of extensive studies and is used in practice to design

to stress two points in this connection. 1) The notion of correlation. For a two-phase (say A and B) medium close to the percolation threshold for phase A, one can define the pair correlation function $C_{AA}(r)$ (which gives the probability to encounter the A phase at a distance r from a point in the same phase. It is almost a matter of definition to state that this pair correlation function *does not* display any singular property around the percolation threshold, and past the size of the individual constituent, $C_{AA}(r)$ is simply a constant equal to the concentration of A phase. What is singular at the percolation threshold is the pair correlation function of the A-phase in *a connected cluster*. This simple remark is to underline the fact that *no specific organisation* of the A phase is present at the percolation threshold, but isolating clusters of connected elements singles out highly correlated (fractal) subdomains of vanishing density.

A point is worth being mentioned in relation with this contribution. In spite of the wealth of information which can be derived concerning percolative structures, their natural occurence can be suspected to be rare. The point is that there is a well defined concentration of one phase which corresponds to the percolation point, and there is a priori no reason why Nature would generate large structure precisely at this critical point. A fascinating paradigm has been proposed a few years ago by P. Bak and coworkers, which is known under the name of self-organized criticality. In essence, these authors observed that in some cases one can endowed a critical phenomenon with a dynamics such that the critical point becomes attractive for the dynamics. Some examples have been put forward, and they generally consists in having a retro action on the control parameter from a prescribed (infinitesimal) value of the order parameter. In this case, without any fine-tuning, the system naturally evolves towards its critical state and may display non-trivial scaling behaviors. This scenario cannot predict critical exponents, but justifies on extremely general grounds, the natural occurence of scaling characteristics from second-order critical phenomena. This may give a hint for producing material displaying some specific properties, (i.e. loss tangent angle independent of frequency, material with a localization gap in a broad range of frequencies, ...).

In the last three contributions, a convergence occurs on the analysis of acoustic emissions in damage and fracture.

The next contribution by Kuksenko, Tomilin and Damaskinskaya deals with a model of damage based on activated processes with a coupling to the local elastic energy. This model is extremely interesting, and due to the thermal fluctuations responsible for damage, a number of difficulties arising in the zero-temperature limit of this model vanish. In particular, the

"phantom" materials.

slow build up of correlations in the damage field can be neglected because they are smeared out in the thermal noise. Thus a simple local interaction criterion for brittle failure based on the density/size of micro-cracks is sufficient to account for the transition from diffuse damage to localized fracture. In view of this, a complete theory can be elaborated which yields a time-dependent damage evolution plus a criterion for "brittle" failure. I would have liked to find a discussion of the relevance of such a description for standard solids (nature of the solid, range of temperature where such a description is believed to apply, ...). Indeed one expects that a strong reduction in the activation barrier due to the elastic energy is necessary in order to use such an approach. Anyhow, the application of the model to mine earthquakes is extremely challenging (I must confess I missed a few steps from the initial model to the practical use in earthquake prediction), and the results presented in this paper are quite remarkable, and do constitute a major breakthrough, if confirmed on other systems.

The sixth contribution by Hansen and Hemmer concerns the analytical solution of an apparently simple problem which has far-reaching consequences. Considering the one-dimensional fiber bundle model with an equal load sharing rule, the question which is addressed in this text is the statistical distribution of "bursts" or avalanches of fiber breaking, i.e. the number of fibers which break simultaneously under constant applied load. It is shown that the latters are distributed as power-laws up to a maximum burst size which diverges at peak force for the average force-displacement relation. Such a form is reminiscent of the approach to a second-order critical point. The key concept which underlies this study is that in the thermodynamic limit of an infinite number of fibers in the system the force displacement characteristic is *not* differentiable. Fluctuations around the average force-displacement law can be shown to be a kind of random walk, for which derivatives of order larger than $1/2$ do not exist. The traditional way of tackling with this problem generally ignores fluctuations at the level of individual fibers. Considering only the mean behavior, the notion of burst becomes trivial (being either 0 or infinity). The lack of differentiability implies a sensitivity to the microscopic scales, and using a homogeneous average law is not legitimate for notions such as bursts which are based on local maxima of the characteristic. The natural physical manifestation of these bursts is to be found in acoustic emissions.

This observation is not as innocent as it might appear at first sight. Indeed, for larger dimensionalities, the divergence of avalanches corresponds to a point of localization. Hence it is of crucial importance to be able to quantify the statistical law of acoustic emissions as one approaches a localization point. It is also fundamental to stress that under different loading rules, the particular form of the burst size distribution may change completely. For local load sharing models, rupture appears to be due to a nucleation phenomenon of a critical defect whose propagation cannot be stopped. This gives rise to a first-order critical point. In this case, taking the thermodynamic is never justified. In particular, the peak force displays a strong system size dependency. This also underlines the clear difference between the model of Kuksenko et al and the present approach which can be seen as a "zero temperature" limit of the former. It would thus be of extreme interest to discuss the limit of applicability of the basic hypothesis of both modelling.

Finally, it appears unfair to us to comment on the final text of this section by Tanguy et al. We simply stress here the parallel with the previous approach in terms of non-smoothness of the fluctuation component of the toughness in a steady state situation. This arises naturally from the pinning of a crack front on local heterogeneities. Again the rapid fluctuation of the toughness are exected to be revealed through acoustic emissions. The emphasis put on the latter experimental technique to reveal the microscopic aspect of damage and fracture from three quite different perspectives is a strong motivation for performing accurate statistical analysis of acoustic emission and identification of local sources.

INTRODUCTORY REMARKS TO SESSION C: STRUCTURES AND STRUCTURAL COMPONENTS

DOMINIQUE JEULIN
Centre de Morphologie Mathématique
Ecole des Mines de Paris
35 rue Saint Honoré, F77300 Fontainebleau, FRANCE

Abstract. The reliability of parts in service depends on the scatter of mechanical properties, which operates at different levels: microstructural and structural properties, loading conditions. Probabilistic models are available for the description of these aspects and for the prediction of reliability. The types of models and the problems involved in their simulation for Structures and Structural Components are discussed.

1. Introduction

The reliability of parts in service, which has very important consequences on the cost and design of engineering systems, depends on the scatter operating at different levels, which are presented in the five papers of the session:

- on the scale of the microstructure of the materials involved in their fabrication [5, 8]
- on the assemblage of components, as in the case of glued laminated timber [16] or in general systems [4]
- on the geometry of parts [10]
- on the environmental and loading conditions [4, 5, 10]

All these levels involve a probabilistic approach coupled to a mechanical approach, for which specific problems are solved by appropriate types of models.

We briefly review here the main themes introduced in these presentations, details being found in the references. These introductory remarks are devoted to the choice of probabilistic models and to their implementation in simulations.

G. N. Frantziskonis (ed.), PROBAMAT – 21st Century: Probabilities and Materials, 579–583.

2. Representation of the variability by probabilistic models

In the context of the reliability of components submitted to mechanical loading, it is necessary to use probabilistic models. These enable us to represent phenomena with inherent fluctuations by descriptive models, and to generate random simulations.

2.1. USE OF RANDOM VARIABLES

When dealing with specimens presenting some statistical fluctuations, such as beams of timber [16], composite [1, 8] or metallic materials [3, 8, 5, 10], the overall properties of parts can be considered as random variables described by appropriate statistics: the strength is often modelled by a Weibull probability law [3, 8] , by some derived distribution [16], or by lognormal distributions [5], like other parameters. In reliability calculations it is common to transform initial variables into uncorrelated Gaussian variables [10], from which the response surface can be studied in a convenient way. In [16], the correlation between the MOE (modulus of elasticity) and the MOR (modulus of rupture) is used to generate simulations of the strength from simulations of laminae for which the effective MOE is easily derived analytically.

In any type of application, experimental data must be collected and statistically analyzed, in order to properly model the fluctuations inherent to the studied material. In [8] the statistical observed properties changes with the size of the specimen due to size effects that can be predicted from fracture statistics models involving random functions.

2.2. USE OF RANDOM FUNCTIONS

When considering the fluctuations of loads with time or the local variations of mechanical properties in space, we have to work with stochastic processes (for time varying processes like loading conditions or like fatigue life) [4], and with random functions, also called random fields. In this last case, the parameters to model may be [8]: some fracture criterion (like the fracture stress), the stiffness tensor for elastic media, an initial damage, or any parameter in the constitutive law.

The stochastic models proposed for loads are Gaussian processes and Markov chains [4]. Usual models are stationary, but non stationary situations have to be considered. In [4] various ways to consider non stationary cases are proposed, as for instance second order Markov processes (which are themselves stationary). Other classes of models, with stationary increments of order k, could be used for stochastic processes and more generally for spatial random functions [14].

Models derived from the theory of random sets [11, 15, 7, 8] provide a broad variety of morphological microstructures, largely used in materials, among which the Boolean, dead leaves, and mosaic models. These models are useful to describe and to simulate the morphology. They can also be introduced in change of scale models: prediction of the effective properties of random media [2, 6, 8, 9, 11, 12] from their microstructure, models of fracture statistics [8] where the occurrence of rare events is crucial. These aspects illustrate the predictive power of probabilistic models of random functions, when there description of the microstructure is correct.

For applications, the parameters of models must be estimated by statistical inference procedures.

3. Assessing structures properties by simulations

For many problems connected to the reliability of structures, no analytical solution can be obtained and simulations have to be performed [4, 5, 8, 10, 16].

3.1. SIMULATION OF RANDOM VARIABLES AND OF RANDOM FUNCTIONS

The simulations of random variables and of random functions makes use of various techniques, which are not detailed in the presentations: generation of random numbers, simulation of Gaussian variables and of Gaussian random functions, simulation of spatial point processes. New developments concern the non stationary simulations.

3.2. DISCRETIZATION OF RANDOM FUNCTIONS

When simulating the local evolution of a microstructure under loads, particular care must be taken for the discretization of the data, replacing a continuous random process by a discrete one. The main difficulty lays in the derivation of the statistics of the discretized medium to avoid distortions of the behavior, and in the construction of the optimal mesh.

Stochastic Finite Elements SFE procedures, were derived from the standard one for random media [5, 10]. They are largely applied in the evaluation of the reliability index of structures.

In [4] a system approach, with a decomposition of a structure in subsystems, is proposed. Simulations then operate on elements.

In [5], the size of the finite element (FE) mesh is suggested to be given by the correlation length of the microstructure.

Once the choice of a FE mesh is made, the statistical properties of random functions to simulate have to be known: the reduction of variance with the size of the mesh is known when locally averaging the random function [5]. This is a common practice in mining geostatistics where data with very different supports are available [13]: it can be shown for stationary models that the decrease of variance is asymptotically proportional to the integral range (calculated by integration of the correlation function over space) for large size fields. The distribution function of the average is not known in general, except for the obvious Gaussian case. Nevertheless, the use of averaging is legitime for additive variables only. This is not the case for effective properties (like the macroscopic stiffness), which do not show any simple composition law when increasing the size of the mesh [2, 9, 11, 12]; similarly, the ultimate strength of small volumes do not combine additively: weakest link models or more sophisticated models require other changes of support [8]. In [1, 8], a full procedure of discretization is proposed for the case of fracture with damage: a statistical mesh, named statistical volume element SVE, at a scale defined as a material property, is used for the simulation of the random fracture parameters, independently on the finite elements mesh.

3.3. EXPERIMENTAL DATA

The implementation of probabilistic models requires a considerable amount of data, the variability involving repetitions of experiments, as opposed to deterministic situations. This is the case for mechanical properties measured on specimens and for a correct estimation of probability distributions and of correlations between properties [10, 16]. In reliability calculations, the construction of the response surface requires the use of a careful design of experiments DOE [10], with new measurements to make during the process. For fatigue lifetime, there is a need of data implying multiaxial loads [4]. When modeling the morphological aspects of the structure, data can be obtained by means of image analysis [8].

In any situation, experimental data are needed for the validation of models, and for the inference of parameters.

4. Conclusion

The reliability of structures has to be modelled at different scales. There is presently a need of integration of the existing models for the different levels: microstructure, composite systems, parts, loading conditions. This has to be combined with a careful collection and statistical analysis of data.

References

1. Baxevanakis C., Jeulin D., Renard J. (1995) Fracture Statistics of a Unidirectional Composite, *International Journal of Fracture*, **73**, 149-181.
2. M. J. Beran (1968) *Statistical Continuum Theories*, J. Wiley, New York.
3. F.M. Beremin (1983) A local criterion for cleavage fracture of a nuclear pressure vessel steel, *Metall. Trans. A*. 14A, 2277-2287.
4. M. Bily (1997) Probabilistic approach to dimensioning of structures exposed to stochastic operating loads, *these Proceedings*.
5. A. Haldar, J. Huh (1997) Modeling of material uncertainties using the nonlinear stochastic finite element method, *these Proceedings*.
6. Z. Hashin and S. Shtrikman (1962) A variational approach to the theory of the effective magnetic permeability of multiphase materials, *J. Appl. Phys.*, **33**, 3125-3131.
7. D. Jeulin (ed) (1997) Proceedings of the Symposium on the *Advances in the Theory and Applications of Random Sets* (Fontainebleau, 9-11 October 1996), World Scientific Publishing Company.
8. D. Jeulin (1997) Probabilistic models of structures, *these Proceedings*.
9. E. Kröner (1971) *Statistical Continuum Mechanics*, Springer Verlag, Berlin.
10. Lemaire M. (1997) Finite element and reliability: combined methods by response surface, *these Proceedings*.
11. G. Matheron (1967) *Eléments pour une théorie des milieux poreux*, Paris.
12. G. Matheron (1968) Composition des perméabilités en milieu poreux hétérogène: critique de la règle de pondération géométrique, *Rev. IFP*, **23**, 201-218.
13. G. Matheron (1970) *The theory of regionalized variables and its applications*, Paris School of Mines.
14. G. Matheron (1973) The intrinsic random functions and their applications, *Adv. Appl. Prob.*, **5**, 439-468.
15. G. Matheron (1975) *Random sets and integral geometry*, J. Wiley, New York.
16. H. Mihashi, N. Itagaki (1997) Probabilistic design of performance in glued laminated timber, *these Proceedings*.

ABOUT THE USE OF PROBABILITIES IN MATERIAL RESPONSE PREDICTION AND MATERIAL DESIGN.

D. BREYSSE
CDGA, Univ. Bordeaux I, Av. des facultés, 33405 Talence cedex, France

I must confess that, after a first reading of the papers, I did not saw how to review them and find a common theme between them. Finally, in trying to replace them in a perspective, I realized that all of them pursue the same objective which is to identify the effects of material heterogeneity on the response at macro-scale. It is this question that I will discuss in the following, referring to the different papers when required.

1. Introduction

PROBAMAT subtitle is "tests, models and applications". Depending on one's school of thought, one can focus on each of these three words. However, since the final purpose of material science is to build structures. the question one has to answer in the field of probabilities is "when do we need statistics and probabilities in material studies and what do they add to our results ?". The answer is not straightforward, since for many years this field has been predominantly that of specialists of mathematics (statistics, probabilities, reliability...) who had found a new field game, than that of engineers and material scientist who tried to find new answers to old questions.
However, a number of applications have now been identified which intimately mix material science and statistical/probabilistic tools, and for which these tools bring answers which had not been brought before.

A natural tendency, after historical works by Weibull and few others, has been to consider probability theory as the only way for material studies. Of course, scattering of test results often appears as the first manifestation of randomness - it is also for us, as teachers. the first opportunity to see the birth of our students'critical thinking -. Scatter makes us - and them - ask questions : what is the reason behind two identical specimens not behaving identically ? The micro-scale, that is the material scale, is then natural, and material heterogeneity and probabilities are linked in a logical way. However, we cannot forget that a material is never used as it is, rather it is some matter used in a given environment (e.g. structure), to fulfill certain requirements. It follows that what would be in theory of primary importance (heterogeneity, scatter) may have, in many cases, very poor consequences on the structural response. that is of interest to the engineer, its user and the society in general.

G. N. Frantziskonis (ed.), PROBAMAT – 21st Century: Probabilities and Materials, 585–593.

The best point for us is to begin to recognize the problems in which material heterogeneity must be accounted for. In fact, as we will detail now, handling this problem will increase all costs (cost for identifying data, cost for numerical computations) and heterogeneity analyses have to be limited to the problems in which they improve the quality of prediction.

As Faravelli and Petrini say in their paper : "it is necessary to identify the parameters of the model to which the result of the structural analysis could show a significant sensitivity, the final goal being the capability of reliable numerical simulations. In this context, some parameters will always require the introduction of a probabilistic model, some of them will not need it at all and some others should be carefully checked before a decision is achieved".

From the first PROBAMAT Conference program, and from similar events (ICASP for instance), it is natural to propose three major sub-domains in the field of materials and probabilities :

(a) modelling of random media (soils, rocks...) or random microstructures, where the material heterogeneity will be described and its influence on higher scales looked for,

(b) probabilistic processes (failure, creep, fatigue), for which the macroscopic response is known to be related to material microscale evolutions. Thus it is appealing to try to link them in a same framework,

(c) structural reliability, for which a significant part of the structural performance comes from the structural properties. Therefore material heterogeneity must be analyzed with respect to its influence on the structural response.

2. Randomness of material structure

One often distinguishes between natural and artificial media. In natural media (soils, rocks...), the heterogeneity is large and exists at all scales. For this reason, specific tools have been developed quickly (geostatistics and random fields, now used in other fields, originate in mining science and geotechnics). In these media, the mechanician generally tries to quantify the risk of encountering a problem (foundation settlement, slope instability...), but the geometrical and material properties are fixed once for all (even if they are not fully known). In artificial, man-made, materials, the problem is doubled : in short-term, the mechanician wants to better understand how the microstructure will influence the macroscopic response. This will enable improving the material design by changing the microstructural characteristics in the longer term.

The conclusions of a NSF workshop (Beran 1993) devoted to probabilistic micromechanics defined the major challenges :

- identification of the experimentally accessible morphological information which is essential to the determination of the macroscopic response,
- multiscale statistical properties and response fields,
- modeling of stochastic evolution of microstructures

- damage-related properties such as strength, creep, toughness, and material instabilities,
- mechanics of granular and multi-phase materials.

We will focus now on the first of these items. Being able to build a valuable random microstructure means : (a) being able to identify the relevant micro-scale;the scale will be considered as relevant if by introducing a random phenomenology at this scale, one can derive correct macroscopic responses, (b) being able to identify the material parameters at that scale (as far as randomness is concerned, this means average values, pdf and correlations), (c) being able to input these properties in a model, the model being analytical or numerical.

The choice of a relevant scale (which may depend on the studied process) as well as the support will be discussed lower. Let us see in more details the point (b), i.e. the identification of all required material micro-parameters. Techniques for identification are many (from visual inspection to image analysis, with various tools of investigation). The key-question is *"until what level of detail must we go ? Do we need to look for the most detailed model or is it preferable to concentrate our efforts on some specific parameters ?"*. Such information is costful, and the cost increases significantly as soon as random fields are considered (a field for a given parameter means : probability density function, spatial correlation function and correlations with all other parameters). The researcher must, according to his knowledge of the process he wants to describe and to his knowledge of the material (and to the ability/cost for capturing more information), validate his choices. Examples of this strategy are described in the papers by Rackwitz (see § 4) and Van Mier (see § 3) respectively.

In his paper, Breysse tries to identify what are the more relevant parameters which influence the tensile strength of fiber reinforce concrete. Once it has been statistically verified that the effective number of fibers is a major parameter, efforts are made to quantify the spatial distribution of fiber density. This spatial distribution is quantified through direct observation (visual countings) and it is shown that microscopic disorder can explain an important percentage of macroscopic variability. It is also shown how a very phenomenological/empirical model can be built at very low cost, with predictions that fit the experimental observations. In the second part of this paper, a methodology which has been employed to identify the microstuctural disorder in timber wood is described. It was important to quantify the influence of this disorder on structural response, with engineering applications in the field of timber-grading and of structural failure. Breysse insists on the fact that, depending on the available amount of material data and on the ultimate purpose, one can build an empirical-statistical model (or semi empirical, enriched by a simultaneous mechanical analysis) or a complete numerical micro-macro model.

Studying shape-memory alloys, Faravelli and Petrini, sort out the parameters which have to be considered as random in the constitutive law (multisurface plasticity). Nine empirical (without direct relation with the material microstructure) parameters have been defined to allow the experimental data to be correctly reproduced while two sets of correlated vectors are required to make the model complete. The first set of parameters has to minimize the error from an experimental data set while the second set of parameters is identified by using a maximum likelihood estimation. The process for estimating the latter ones is detailed. It is also shown how local measurements reveal the need for a more detailed modelling: these parameters must be modelled as a random field, in order to incorporate an obvious spatial variability. It will be intersting to verify if the spatial variability effects are not averaged when the specimen (or structural element) is considered. Another question is to know until what point the macroscopic variability (experimental scatter due to specimen variability only) is related to the local spatial variability: would it be possible to draw information about spatial micro-variability from the quality of the likelihood estimation ?

3. Random processes

Many processes are considered to be intrinsiquely driven by microstructural heterogeneity, e.g. fatigue, creep, failure... When dealing with such processes, one always has several ways of handling the problem. Breysse (1993) defined several classes of models, these models being first defined according to the material scale at which the model is randomized (micro/meso-/macro) and the fact the process itself is -or not - randomized. In the case of micro-macro random models, one generally speaks about models into which a deterministic realization of a random microstructure is generated and for which strictly deterministic computations are following. This kind of models is the more frequent. However, one can consider models for which the microstructure does not really exist, and the process is probabilized (the phenomenological model proposed by Rackwitz in his paper is a good example) or wholly probabilistic models (stochastic finite elements).

Let us remark that the macroscopic response may seem deterministic, when some filtering effects through the range of scales are predominant. This is often the case in soils mechanics as was told by Gilbert and Magnan: "the effects that spatial variability in soil properties have on performance are minimised in many cases through spatial averaging and soil-structure interaction". This is also the reason why some mechanical properties in heterogeneous materials can be approached through the "magic filter" of homogenization theory: in certain cases, local effects are due to the micro-structure, but if one considers a variety of scales, averaging effects make this influence negligible. In summary, the averaging (or variation reduction through the wide range of scales from micro- to macro-) will depend both on the studied process, and on the material structure. A wholly disordered media like concrete cannot be compared with a ceramic where some local defects are put into a rather homogeneous matrix.

According to Sab, "a micro-mechanics model is a combination of three kinds of elements [among which...] an averaging procedure which defines overall variables from local ones [and] a localization modelling which refers to the derivation of local variables from overall ones...". Homogenization can be used in granular media, even knowing the discrete nature of the microscopic scale. A triangulation is mapped on a dense particles assembly and randomness is described through the branch orientation of tetrahedrons. Some fictitious branches are added to account for defects in the particle packing. The probability of having this kind of branches is taken as a function of average void ratio and it does not depend on local characteristic of packing. The micro-mechanics model requires 5 parameters describing normal and tangent stiffness, cohesion, void ratio and a scalar parameter α related to the localization field assumptions. It is shown that the model is capable of reproducing the stress/strain relation in compression even if the predictive ability is, in our opinion, not fully demonstrated since no indication is given on how to identify α and how the numerical result is sensitive to variations in the other parameters. The interest of such a model lays also in the fact that it can enable us to sort out what characteristics come from the constitutive/micromechanical model itself from what come from the microstructural randomness.

More generally, material models can be very rich, without practical limits, since one chooses to discretize the continuum. Therefore problems are not in solving the numerical equations (when problems occur with algorithms, they are a priori comparable to those encountered in homogeneous continuum and they do not deserve specific attention here) but in the way one identifies the many parameters one has to input in the numerical model. A second problem arises from the choice of the representative volume element (RVE): it is possible that the RVE size changes with the process development, as is probably the case when fracture/damage development is analyzed. Therefore, the unique purpose is to make the numerical experiments robust. It is clear now that none of the numerical micro-macro model developed for failure until recently could be considered as robust. Problems related to mesh sensitivity, algorithm dependence, or, more specifically parameter uniqueness had not been treated with sufficient care. We think that the major recent progress has been the fact that researchers are now working carefully on this problem. Two examples can be found in the papers by Bontempi and Casciati and by Van Mier and Vervuurt.

The paper by <u>Bontempi and Casciati</u> focuses on size-effect predictions in concrete. Since standard continuum mechanics model have proved to be inconsistent with the experimental response (softening, localization, size-effect) and since regularization techniques focuses attention on a given structural size level, the purpose is to build a model capable of reproducing the localization patterns and size effect whatever the scale being considered (particles, specimen or structural element). A Voronoi triangulation is mapped on the aggregate-matrix continuum, randomness affecting the spatial distribution of phases, the local stiffnesses and strengths. However, there is no spatial correlation of random properties within a phase. This comes to consider that the basic numerical element (bar) is much larger than the material correlation length in a given phase, which is probably not true for all scales. In reality, a double Voronoi triangulation is used, since it will better reproduce the 3D aspects of crack development and make the overall behaviour more ductile. Scale effects can be studied by two ways: (a) by considering a fixed scale of modelling the material microstructure and by predicting the influence of the structure/specimen size on the overall response, (b) by considering a fixed specimen size and verifying that the overall response remains unchanged when the mesh size is modified. In reality, the first way directly analyses size-effect while the second one ensures mesh-objectivity (all is OK if the result does not change). Numerical analysis show a certain ability to insure mesh-objectivity (which was not the case in previous models) and to reproduce - at less qualitatively - size effects. To ascertain that such a model can directly be used in engineering computations will also require that quantitative predictions (i.e. macroscopic strengths in tension and compression) are compatible with experimental data, which iwas not the purpose of this paper.

Van Mier and Vervuurt have developed a model with very similar basis and tools. Coming from the fact that the lack of quantitative predictions has, until these days, forbidden the use of such models by engineers, they insist on the fact that "combined experimental and numerical research [in fracture...] in different geometries and under a wide variety of loading conditions is equally important for developing a quantitatively sound model". This means that experiments should be designed such as to capture some essential micro-parameters if these parameters have proven to deeply influence the structural response in simulations. An additional problem is encountered when the required parameters are difficult to link to any measurable material property, which is the case for instance when one is looking for the strength of discrete bars modelling a continuous medium. In this case, the only method is to identify such parameters through reverse analysis, i.e., to consider the influence of these parameters on some well chosen structural responses. Two situations can be encountered: (a) the overall response is not sensitive to the parameter; one should then assume that it will be the case for different configurations (geometry + loading) and one can chose any value (or best, if possible, delete the parameter in the model), (b) the response varies with the parameter and its value can be directly identified. Such an inverse technique is developed in the paper for estimating the properties of the interfacial transition zone between aggregates and matrix in concrete (which causes many problems to researchers studying microstructure of concrete). The basis for comparison is a wedge splitting test for which the required parameter is assumed to be a multilinear function of many other parameters; e.g. matrix and aggregate stiffnesses and strengths. Numerical computations and experimental results are compared for crack development patterns, strength and work of fracture. Exhaustive analysis is however difficult due to the high cost of computations which also reveal the complexity of micro-mechanisms. Until now, it may be concluded that the interface stiffness is probably not a parameter of highest importance, but a quantitative identification of the interface strength has not been possible for any combination of micro-material parameters.

These two examples clearly show the care that must be (and is) taken in identifying material parameters which are used as inputs in the numerical model. It is only when this will be completed that we will be able to ensure the quantitative (better than descriptive) power of probabilistic models. When this will be reached, and we really think that no model can really claim to be today quantitatively predictive, we, as researchers and material scientists, will have to extend our field of applications. If it seems interesting to predict size-effect, or scatter in fatique service life, empirical models currently exist which do that. To be optimally efficient, robust numerical model will have to be used for specific applications in material design. We think that it will be possible, for instance, to optimize the way lams are packed in a glulam beam directly from their non destructive assessment, so as to optimize the beam strength, or , in another field, it will be possible to optimize the material thermal treatment, to change the microstructural properties and, consequently, the material response.

4. Reliability

The field of reliability theories has lead to many developments in engineering of offshore structures, bridges, and other applications. A consistent set of theorems and algorithms is available which can be directly used at another scale, in the material field. The ultimate purpose is the reliability assessment of complex systems, but a careless approach can yield results different from those expected. The use of complex models is not sufficient and the geotechnical field is characteristic of such a failure, as it has been told by Magnan (1995) : "Monte-Carlo simulations combined with the FEM cannot provide a practical solution to consolidation problems, nor can probabilistic methods significantly improve consolidation analyses".

A great danger must be recognized, and randomness has to be accounted for in three steps: tests on representative samples, model for the material and calculations using the model and the load. When randomness is used only in the third step (as it is often made in statistical geotechnics), it would lead to disappointing results. Developing a detailed reliability analysis with the best algorithms when the material data is raw and the mechanical model remains crude, is only a dead-end. Magnan concluded that "probabilistic analyses have come to the point where improvements should be sought in improved quality and increased number of input data than in the stochastic modelling itself".

The paper by Rackwitz and Gollwitzer is the only one of this session which analyses problems of reliability. They use the framework of structural reliability to analyse the relation between the values which are taken by several material parameters (and assumptions on their correlations) and the material constitutive law. However, through the problem of reliability analyses, the real problem is, once more, that of the identification of micro-material parameters (mean values, variation and correlations). The example is fiber reinforced plastic in which a plastic Coulomb-kind three dimensional criterion is assumed to describe the inter-fiber failure. Average values of the empirical material parameters are identified from a least square equation. Thus these parameters are studied for the sensitivity of a reliability index β with respect to changes in their value. The study focusses on three micro-strength parameters and one scalar parameter p. For the three first of them a Weibull pdf is assumed, p being assumed to be uniform. The reliability analyses also requires mean and c.o.v. of the stress components (σ_{22}, τ_{21}), but also correlations between τ_{21} and σ_{22} on one hand, and between the three strength parameters on the other hand. Here the assumptions are poorly validated : "the strength parameters must be stochastically dependent on each other because all three strength parameters are affected by the same material characteristics, but to a different degree. Unfortunately, such correlations are extremely difficult to verify experimentally". Of course, once the specimen has failed in a given failure mode, it would be necessary to make it fail also in other failure modes, such as to obtain the three parameters on the same specimen ! Since this is not possible, correlation degrees must be, to a large part, postulated. Consequently, the conclusions that one would reach on reliability cannot be considered from a quantitative point of view. Rackwitz prefers to insist on the sensitivity of the reliability index with respect to variations in these local correlations: it is shown that the safety indices do not depend to a large degree on the local correlations.

Michnaevsky, in a thought-provoking paper upon the determination of the stress field in an heterogeneous medium, reverses the point of view of the analysis. Telling first that, in an heterogeneous medium, it is very difficult to know the stress field at any point, he prefers to develop a global analysis, relying on entropy properties to handle the problem. Instead of assuming that the stress field $\sigma(x,y)$ is obtained through successive corrections which describe the effects of heterogeneities (local fluctuations), he first assumes that information is totally lacking and begins to introduce it. He writes that the pdf of the specific elastic strain energy must fulfill two conditions (total probability, total energy) from which he draws an equation on the maximization of entropy. The framework of the analysis seems to be appealing, even if it is surprising that a data like the critical value of the equivalent stress is assumed to be a constant in such a medium (it has no reason to be such). Other assumptions, like the linear damage accumulation concept are probably relevant to a given scale, but probably not at all scales (threshold effects f.i.). Coupling such an approach with micro-macro model as those discussed above would probably yield interesting results. However, some conclusions are yet reached about the more efficient shape of drilling tools, which appear to be in good agreement with experimental information.

Conclusion

We have tried to build a general overview of the use of probabilistic models of materials. Three steps have been identified, at each of which efforts have been done and also remains to be done: definition of relevant micro-scale parameters and identification of the field values, validation of numerical supports, reliability assessment. We think that the more difficult question to answer is "what amount of information must we get to build a predictive model". To answer this question, the first field of competence is that of a material scientist (what does this parameter means ?, what mechanisms can it be related to ? how can we measure it and for what cost ?).

It is only once probabilistic models will become reliable than they will be used. We all know that in all chains, the weakest link drives the overall response: the final quality is driven by the lower quality of the three steps defined above (identification of parameters through well-defined tests, development of numerical models, reliability computations). One good example of problem mixing all difficulties is that of assessment of differential settlement resulting from soil-structure interaction. In this case, we know that the soil is spatially varying, and, even if the mechanics is assumed to be known, we are not able to describe the process of regularization-averaging of this spatial variability which results in a given distribution of settlements. The day we will be able to tell the geotechnical engineer what must be the optimal grid for geotechnical investigation, probabilistic models will be easy to sell.

Acknowledgments

I must thank Pr. S. Ghosh, invited Professor at CDGA, who helped me in reviewing this paper.

References

Papers discussed in this session, written by F. Bontempi and F. Casciati, D. Breysse, L. Faravelli and L. Petrini, L.L. Mishnaevsky Jr and S. Schmauder, R. Rackwitz and S. Gollwitzer, K. Sab, J.G.M. Van Mier and A. Vervuurt

and

M.J. Beran, J.J. Mc Coy, final report of NSF Workshop, The statistical characterization of material muicrostructure and its relation to material performance, Catholic UniV. Amer., Washington, 1993.

D. Breysse, Failure and probabilities at various scales. A synthetic paper, pp. 29-38, Proc. PROBAMAT Conf., Cachan, Fr., 23-25/11/93, Ed. Breysse, Kluwer publ.

R.B. Gilbert, J.P. Magnan, Synthesis of Session 1a : soil mechanics and geotechnical engineering : foundations, pavements and tunnels, pp. 1403-1404, Proc. ICASP 7 Conf., Paris, 10-13/7/1995, Ed. Lemaire, Favre, Mébarki, Balkema publ.

J.P. Magnan, A. Bouheroua, Stochastic analysis of soil consolidation in theory and practice, pp. 77-84, Proc. ICASP 7 Conf., Paris, 10-13/7/1995, Ed. Lemaire, Favre, Mébarki, Balkema publ.

NATURAL MATERIALS AND COMPOSITES

L. L. MISHNAEVSKY JR
Staatliche Materialprüfungsanstalt (MPA), University of Stuttgart,
Pfaffenwaldring 32, D-70569 Stuttgart and
Max-Planck-Institut für Metallforschung
Seestr. 92, D-70174 Stuttgart, Germany

1. Introduction

Natural materials, concretes and composites are usually highly heterogeneous materials with random distribution of properties; many of them present granular and/or porous materials.

In recent years, the analysis of behaviour of granular and porous materials under different conditions became an important area of investigations in civil and mechanical engineering. It may be attributed to the following factors: the expansion of application of the powder materials in many areas of industry (sintered steels and composites, etc), the evidences of high efficiency of material models which takes into account local rotations (like Cosserat continuum) [1] or other mesoscopic models, and the evidences that just interaction between grains determines the behaviour of many natural materials [2, 3]. The bibliographic information about the granular and porous materials as well as the list of researchers working in these areas can be found in [4, 5].

Consider some of works in this area presented on the NATO Advanced Research Workshop "PROBAMAT - 21st CENTURY: Probabilities and Materials".

The presented papers can be divided conditionally into three groups: mathematical modelling of the structure of heterogeneous natural media, analysis of correspondance between theoretical and experimental results as applied to the natural materials, and analysis of possibilities of practical application of the studies of heterogeneous materials. The papers [6-8] may be assigned to the first group, the paper [10] and papers [11-12] to the second and third ones, respectively.

G. N. Frantziskonis (ed.), PROBAMAT – 21st Century: Probabilities and Materials, 595–599.
© 1998 *Kluwer Academic Publishers.*

2. Structure and properties of randomly heterogeneous media

Di Federico and Neuman [6], Auvinet [7] and Bolle [8] presented in their papers methods of description of structure and properties (anisotropy, spatial variability of local mechanical properties, flow and transport) of randomly heterogeneous media.

Di Federico and Neuman [6] have presented an hierarchical theory which allows to relate flow and transport parameters in random heterogeneous media and associated statistics, determined on one scale of measurement, to those determined on another scale. The results described in the paper present an extension of the results by Neuman [9], who has shown that any random field with homogeneous increments can be viewed as an infinite hierarchy of mutually uncorrelated homogeneous fields, characterised by exponential autocovariance functions and variances which increase as a power of scale.

The theory developed by Di Federico and Neuman [6] allows to bridge across scales by predicting the effect of viewing a multiscale random field, measured on the given support scale, through a larger widow defined by the domain under investigation. Expressions for effective hydraulic conductivities in multiscale random log permeability field under uniform mean flow, as well as for dispersion parameters have been obtained on the basis of the hierarchical theory.

Auvinet [7] simulated the random assemblies of both spherical and non-spherical granules numerically. A structural anisotropy with axial symmetry was observed for spherical particles in the case when the preferred axis of grain movement (like the gravity direction) exists.

In order to describe the structure from non-spherical granules, it was suggested to apply a linear geometrical transformation to the assemblies of spheres (they were "compressed" and "stretched" along several directions).

Although such structure can be considered as an extreme case of anisotropy, the author argued that the structure is similar to many natural granular materials. It was suggested to use the probability distribution of the orientations of vectors normal to the tangential planes in contact points to describe the random anisotropic fabric of granular media. This function characterises the geometrical anisotropy and can be determined from equations describing the transformation of the initial spherical grains.

It was suggested to describe the mechanical behaviour of granular materials through successive transformation of the probability function of normal vectors in contact points as well as other parameters due to progressive external loadings. The evolution of this function in loaded granular media was described as the "survival" of better-oriented contacts. On the basis of the described model, the stress-strain curve which allows for the stress hardening and Baushinger effect was obtained.

Bolle [8] discussed methods to describe the spatial variability of mechanical properties of rocks or soils. He reviewed the methods based on the autocorrelation functions and compared their behaviour with field measurements in natural soils.

A realistic description of the spatial variability, based on the following form of the power spectrum was developed:

$$g(k) = S \exp(-\frac{k}{\lambda}) \qquad (1)$$

where λ - a damping parameter, expressing the continuity of properties throughout remote sub-layers, L - wave length, related to the alternation of two types of sub-layers. By using the least squares method, the parameters of the autocorrelation function derived from the above spectrum were fitted to the experimental correlogram.

This approach was developed initially for linear autocorrelation and then generalised to the cases of two- and three-dimensional autocorrelation. For two- or three-dimensional cases, the parameters of the function (1) present two- or three-dimensional elliptic functions, not just numbers. The methods to obtain the elliptic functions are also discussed.

3. Applications in civil engineering, materials science and mining science

Cherubini and Raffaele Greco [10] reviewed methods used to predict settlements of shallow foundations in soils, and compared the calculated data with measured ones. In order to make such comparison, the authors determined the ratio between calculated and measured settlement, and compared this ratio and its scatter for different formulas.

Simultaneously, the authors determined the reliability for each method (a perfectly reliable method was supposed to be the method which secures always smaller calculated value as compared with measured one). The following conclusion has been drawn: when the accuracy of a method is low, its reliability may be rather high.

Brandt [11], and Mishnaevsky Jr and Schmauder [12] considered also practical aspects of the stochasticity inherent to the material behaviour and properties. Brandt [11] discussed technologies which allow to improve the reliability and performances of advanced materials. Mishnaevsky Jr and Schmauder [12] considered the methods of improvement of construction of drilling tools using the informational description of shapes of contacting bodies.

Brandt [11] has analysed possibility of use of several groups of advanced building materials and technologies of their production. The properties and principles of production technologies of high performance concretes, fibre reinforced concretes and fibre reinforced plastics were considered.

The high quality of high performance concretes, which includes both high performances and relatively low maintenance cost, may be achieved by increasing the packing and reduction of the thickness of interface around aggregate grains (by introduction of small silica fume particles into cement, high initial pressure into fresh mix, for instance). The effect of high damage resistance of fibre reinforced concretes (FRC) is caused by the fact that the fibres act as crack arrestors in FRC, they control the crack propagation and therefore influence the toughness and durability of FRC. The improvement of the strength of such materials may be also achieved by using the special technologies by slurry infiltration of fibres or mats.

The fibre reinforced plastics can become an important construction material although their costs are relatively high. Among main advantages of these materials there are their high tensile strength and high corrosion resistance.

Mishnaevsky Jr and Schmauder [12] discussed possibilities of the application of the information theory methods for the design and improvement of destructing tools. They have suggested to characterise the shape of contacting bodies by the informational entropy of the distribution of contact stresses over the contact surface. The relationship between this value and the intensity of rock fragmentation by an indenter of given form, which was obtained initially for the simple shapes of indenters, was generalised for complex shapes of tool. It was suggested to use the informational entropy of the distribution of contact stresses over the contact surface as a general criterion of the efficiency of destructing tool. The authors have analysed a number of patents in the area of the drilling tool improvement. It was concluded that a common feature of all technical solutions suggested in the patents is an introduction of some heterogeneity in the construction of tool (asymmetry, unevenness of working surface, combination of elements with different orientations, shapes or properties, etc). Based on the numerical investigations of the relationships between the contact stress entropy and the damage evolution in the material, and taking into account the results of the patent analysis, Mishnaevsky Jr and Schmauder suggested the following principle of tool improvement: the destruction ability of a tool can be increased by increasing the informational entropy of contact stress distribution over the contact surface between the tool and work material.

Another example of the application of informational methods in modelling of destruction of heterogeneous materials was the determination of stress distribution in heterogeneous materials with the use of the maximum entropy method. A probability function of equivalent stress distribution in a loaded heterogeneous material was determined by the minimisation of a

Lagrangian, which includes the given constraint and entropy. This model was applied to the description of the indexing effect in the loading of rock by several indenters. As a result, the authors gave the following recommendations for the drilling tool improvement: destructing elements (teeth, cutters) on the multiteeth tool should be arranged in pairs or in groups; that causes the indexing effect and increases the intensity of material destruction.

4. Conclusions

From the above consideration, one may draw several conclusions. The variability of properties of heterogeneous materials and interaction between their elements at mesolevel (grains, pores) influence the behaviour of heterogeneous materials to a large extent. The structure and properties of the heterogeneous materials can be efficiently modelled with the use of presented here methods. However, the applications-oriented works, presented here, are based mainly on the approaches, which use not so sophisticated mathematical techniques of simulation and data analysis. It is reasonable to assume that the application of advanced methods of modelling of structure and properties of natural (rocks) and artificial (composites, concretes) materials can be favourable for the development and improvement of technologies and materials in civil, mechanical and mining engineering.

References

1. Dai, C., Mühlhaus, H.-B., Duncan Fama, M. and Meek, J. (1996). Modelling of blocky rock masses using the Cosserat method. Int. J. Rock Mech. Mining and Geomech. Abstr. 4, 425-432.
2. Mühlhaus, H.-B. (1986). Shear band analysis in granular material by Cosserat-theory (in German). Ing.-Arch., 56, 383-388.
3. Herrmann, H.J. (1996) Granular media: some new results. *Simulationtechniken in der Materialwissenschaft.* Freiberg, pp.1-23
4. Powder Page. In: http://www.granular.com
5. Poroelasticity Resources Network. In: http://www.ce.udel.edu/faculty/cheng/poronet
6. Di Federico, V. and Neuman, S.P. (1997) Multiscale permeability and dispersion in randomly heterogeneous geologic media, *This volume*
7. Auvinet, G. (1997) Probabilistic modelling of granular media anisotropy, *This volume*
8. Bolle, A. (1997) Spatial variability of the mechanical properties in natural soil deposits observation and modelling, *This volume*
9. Neuman, S.P. (1990) Universal scaling of hydraulic conductivities and dispersivities in geologic media, Water Resour. Res. 26(8), 1749-1758
10. Cherubini, C. and Raffaele Greco, V. (1997) A comparison between "measured" and "calculated" values in geotechnics, *This volume*
11. Brandt, A.M. (1997) The role of new engineering materials in construction and rehabilitation of civil infrastructure, *This volume*
12. Mishnaevsky Jr, L.L. and Schmauder, S. (1997) Informational methods in optimization of tools, *This volume*

CRACK GROWTH AND DAMAGE ACCUMULATION

A review of five papers for PROBAMAT - 21st CENTURY

F. CASCIATI
Department of Structural Mechanics, Univesity of Pavia,
Via Ferrata 1 - 27100 Pavia - Italy

Abstract

Five papers among the contributions presented at the NATO workshop PROBAMAT-21st CENTURY cover *Fracture* and *Fatigue*, a very broad spectrum. The papers are not homogeneous but offer an update of problems, tools and applications in a branch of material science which represents a necessary link toward engineering.

Keywords: brittle behaviour, crack growth, damage accumulation, kinetic models, stochastic models.

1. Introduction

This paper reviews the contributions of five invited speakers to the NATO workshop PROBAMAT-21st CENTURY (under grant NATO.AWR-960993).
The reviewer tried to emphasize the work the author of the paper did, but also to check the easy access of the reader to such a scientific dissemination. A too deep immersion in the paper topic is avoided. One simply states the problem, the paper readability and the amount of original material the paper contains.

2. Crack growth

The paper by Dolinski [2] (*Stochastic Modelling of Fatigue Crack Growth in Metal*) follows the classical stochastic approach to fatigue crack growth. Early papers on the subject were authored by Y.K. Lin, F. Kozin and J. Bogdanov, fitting Virkler's experimental results. O. Ditlevsen and K. Sobczyk in 1986 put emphasis on the retardation effect. After that the author of the paper, Dolinski, began a deep investigation program leading him to the formulation of appropriate models in the general form:

$$\Delta a_i = F(a_i, S_i^+, S_i^- | x) \tag{1}$$

601

G. N. Frantziskonis (ed.), PROBAMAT – 21st Century: Probabilities and Materials, 601–604.
© 1998 *Kluwer Academic Publishers.*

where a_i denotes the current crack length when the i-th load cycle (of stress maximum S_i^+ and minimum S_i^-) starts and x represent the material parameters. The latter ones are generally assumed to be the realization of random variables.

The retardation effect accounts for the diminution of the actual stress intensity factor range after an overload. The analytical approach by Dolinski became an analytical-numerical approach when Colombi and Dolinski proposed to investigate the retardation effect by numerical simulation over a single cycle. Unfortunately, the idea was not adequately disseminated since there are not journal papers on it, but only two contributions to international conferences are available. In this sense the paper by Dolinski goes toward a better dissemination of its basic idea.

The comparison with experiment results of Figures 2 and 3 is an original contribution of this specific paper. The paper is readable.

The paper by Botvina [1] *Common Characteristics of Damage Accumulation and Fracture of Solids* moves its steps from the kinetic model, goes through the Paris scheme and conceive an extension to the ... earth crust fracture. The kinetic model equations for the lifetime t and the fracture rate $\dot{\epsilon}$ are:

$$t = t_0 \exp[\frac{U_0 - U(\sigma)}{RT}]$$
(2)

$$\dot{\epsilon} = \dot{\epsilon}_0 \exp[\frac{U_0 - U(\sigma)}{RT}]$$
(3)

where R is the gas constant, T the temperature, σ the applied stress, $U(\sigma)$ the activation energy and $\dot{\epsilon}$ the process rate. The Kachanov approach is used to define the damage measure: the relative area of cracks or pores cumulated during loading. The analogy of the model with the Paris-Erdogan law is discussed. Applications are pursued in metalic specimens as well as in the earth crust (Gutenberg-Richter relation).

The paper is readable but very synthetic.

A third paper, the one by Malkin [4], *Kinetic Models of Brittle Crack Growth: Crack Pattern Statistics and Longevity of Solids*, deals with the kinetic model (see Eqs. (2) and (3)). It is used to describe the brittle crack growth in mixed mode I and II. The geometry of the model is summarized in Figure 1. As the author says, "apparently the models considered are oversimplified to describe adequately the growth of real crack. However ..." there is "a qualitative agreement with the experimentally observed regularities".

The readability of this very long paper could be improved.

3. Two applications

The remaining two papers by Mishnaevsky and Schmauder [5], *Informational Methods in Optimization of Tools* and by Lamon and Lissart [3], *A Statistical-Probabilistic Approach to Microstructure-Structure Relations in Failure and Damage of Ceramic*

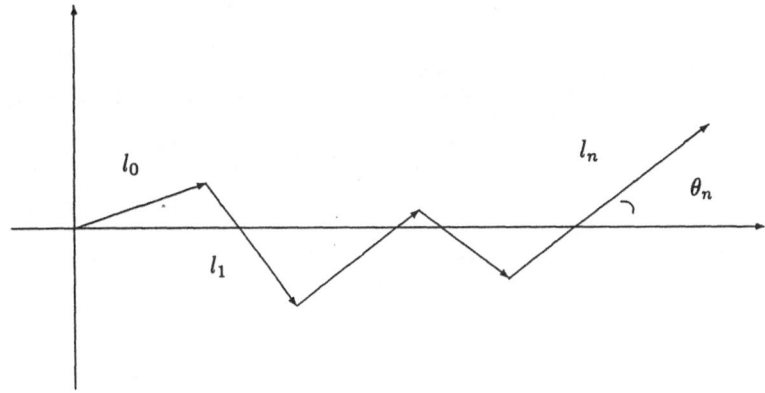

Figure 1. Geometry of the crack model in Malkin's paper

Composites, deal with two specific applications.

The first paper discussed the stress distribution in a disordered material toward an informational description of shapes of contacting bodies: a condition of maximum entropy is written. The ultimate goal is a sort of optimation of the woking condition of tools as the ones used in drilling, milling and machining.

The main conclusions are summarized by the authors: "The destructing ability of a tool can be increased by increasing the informational entropy of contact stress distribution over the contact surface between the tool and work material.". To arrange destructive elements in pairs or in groups "allows to use the indexing effect and to increase the intensity of the destruction of the work material."

The paper is readable.

The last paper approaches failure and damage of ceramic composites. The matrix cracking is first modelled, then the fracture of fibers is discussed and an equivalent fiber length is introduced. For SiC/SiC minicomposites, the force-strain behaviour were predicted in "excellent agreement with the experimental" data, thus validating the model. Therfore, it "provides quantitative guidelines for the selection of appropriate constituents".

The paper is very readable and the material is presented in a well organized framework.

Acknowledgement

The participation of the author in the NATO meeting was made possible by grants from the Italian National Research Council (CNR).

604

References

1. Botvina L.R. (1997), Common Characteristics of Damage Accumulation and Fracture of Solids, PROBAMAT - 21st CENTURY, NATO.ARW-960993.
2. Dolinski K. (1997), Stochastic Modeling on Fatigue Crack Growth in Metals, PROBAMAT - 21st CENTURY, NATO.ARW-960993.
3. Lamon K. and Lissart N. (1997), A Statistical-Probabilistic Approach to Microstructure-Structure Relations in Failure and Damage of Ceramic Composites, PROBAMAT - 21st CENTURY, NATO.ARW-960993.
4. Malkin A.I. (1997), Kinetic Models of Brittle Crack Growth: Crack Pattern Statistics and Longevity of Solids, PROBAMAT - 21st CENTURY, NATO.ARW-960993.
5. Mishnaevsky L.L. Jr. and Schmauder S. (1997), Informational Methods in Optimization of Tools PROBAMAT - 21st CENTURY, NATO.ARW-960993.

ROUND TABLE I. :

UNDERSTANDING MATERIALS AND TECHNOLOGICAL APPLICATIONS

D. BREYSSE and D. JEULIN

The round table is introduced by the following general questions emerging from the two sessions "Material structure, description and application" and "Scaling, fracture and applications" :

1. About universality of the material response :

- What are the reasons (lying in physics ?) and the limits of some universality of the behavior in fracture and what part is specific to a given material ?
- The response of materials depends on their nature. A large variety of materials has been reported in these two sessions. What properties and effects are material sensitive and what part remains universal ?

2. About micro-macro modelling

Global properties result, through a complex process, from microstructural information. Modelling tries to establish links between the different scales.
- How, from global measurements (of fracture energy) and from knowledge about material physics, can we feed micro-macro models ?
- What contribution can bring the models to the understanding of macro measurements ?
- How can models help us in designing materials ?

A last question mixes the two fields : in a micro-macro model, like that presented by B. Chiaia on fracture surfaces, what part of the model is material dependent ? If we would change the material, what would have we to modify in the model and with what macroscopic consequences ?

[J. Lamon] The determination of G_f is based on a calculation whose assumptions are valid for a continuum, not for heterogeneous media. In this situation, the conclusions drawn from macroscopic observation can be biased.
[D. Breysse] That is true. For G_f, it has been shown by Chudnovsky and Kunin (see Probamat first edition Proceedings) that if one assumes a stochastic field for fracture

G. N. Frantziskonis (ed.), PROBAMAT – 21st Century: Probabilities and Materials, 605–612.

energy at micro-scale - giving mean values and standard deviation-, what one obtains at macro-scale can not be simply derived from the microscale data.

[B. Chiaia] We have to distinguish between engineering approaches and mathematical models. It is not easy at all to eliminate "classical σ" or "classical G_f". We know, from a more sophisticated understanding, the reasons that produce scaling and size effects. We can keep the classical quantities for engineering and add scaling laws.

[A. Schanyavski] In the domain of synergetics, there is a very complex multiparametric influence on the fracture process, which develops simultaneously at different scales. The process changes with the material or with the fracture mode. The comparison of fractographic information and the measurements of fractal dimensions can help us to understand why different situations can be encountered.

[A. Haldar] We have not yet discussed about uncertainty on the parameters introduced in the models. What is the effect of uncertainty ? To predict the overall response, for instance in brittle (or ductile or semi-brittle) fracture, would it be preferable to build a simplified model in which uncertainty is introduced or a deterministic model as sophisticated as we can, describing the microstructure ?

[L.R. Botvina] There are interactions between the micro level and the macro level. Concerning the plastic deformation zone, what is the connection between its size and the fractal dimension ?

[A. Malkin] What is a fractal dimension for a non stationary surface ? How to measure it ?

[M. Bily] The problem of non stationary process is far beyond engineering applications up to now. No applicable mathematical model can be used for non stationary processes ; the correlation theory simply fails ; maybe Markov chains can help us. Once the process is non stationary, we know nothing of what we can do. But car, airplanes... must perform ! So we have to do something but we do not know practical rules. We should be aware of uncertainty but the problem is to identify data. If we consider the "worst case", making conservative assumptions about probability density functions, what is the quality of any prediction for engineers ? What if we say "the structure lifetime is between 20 and 2000 years" ? In theory, we can assume marginal distributions and so on... but practically, the best is probably, speaking about industrial products and uncertainty, to know that the latter exist : if you know the devil you can fight with him. A good quality control is probably the best solution.

[R. Rackwitz] Non stationary models are available in the fields of earthquakes and in reliability analysis.

[D. Jeulin] Some stationarity can be introduced to model non stationary data, like stationary increments of order k.

[R. Rackwitz] We can deal with strong non stationarities and non normal distributions. The problem is more here with non stationarity in space, i.e. ergodicity.

[D. Breysse] I would like to come back to the question of Prof. Haldar. The same macroscopic response can, in theory, be obtained by using sophisticated micro-macro model, or by introducing uncertainty at macro-scale. In a numerical micro-macro model for the non linear response of glulam that we have developed, we have identified the statistical information required at micro-level - main problem is to assume or to identify relevant cross-correlations between all data-. The macroscopic output (f.i. size-effect laws) is in terms of statistics but the work at another scale cannot be avoided. The question is to identify the cost of the model and to be sure that it is really predictive, i.e. useful for engineers.

[O. Naimark] There are two important sentences. Frenkel said that "*a model is a caricature of the phenomenon*". Landau said that "*if we have equations, the physical problem is closed*". Phenomenologically speaking, modelling starts with a thermodynamic potential. If we want to extend further the field of application, f.i. from elasticity to plasticity or failure, the physical problem arises: we have to list additional parameters. For instance, from the type of defect involved in fracture and from characteristic experiments, we can deduce the form of the thermodynamic potential. The free energy brings an additional term to the elastic term. This provides non linear equations depending on the type of involved statistics. We must have independent experiments for the determination of the thermodynamic and the kinetic parameters. The problem for engineers is to find more simple expressions for new statistics but the description must contain the typical non linearities. The critical parameters are critical stress or stress intensity factors which reflects the self-similarity at the crack tip for an elastic solid. We have to introduce new types of non linearity.

[A. Haldar] Between non linearity and two parameters or ten parameters, what do you choose, from an uncertainty point of view ?

[O. Naimark] It depends, there are different characteristic times for the different parameters. If I have ceramics and slow relaxation, I can have two independent deformations : elastic strain and strain due to the defects. Plastic strains can be excluded. That is my answer.

[A. Schanyavski] About the necessity to use probability, for aircraft crashes, the lifetime is variable, everything being equal, due to a varying critical crack length. In this situation, there is a very big influence of material variation. If we had first information, may be we should know what kind of probabilistic model can be used.

ROUND TABLE II. :

STOCHASTC MATERIAL STRUCTURE AND STRUCTURAL RELIABILITY

G. FRANTZISKONIS and O. NAIMARK

This round table addressed mainly two issues, (a) multiscale approach to material characterization and (b)structural reliability. Important questions such as what scale is important for reliability? came up. For example, imagine one can describe a material using a multiscale approach, what will then be the effect of this description on structural reliability?

[O. Naimark] I would like to discuss some aspects of the application of wavelet analysis in the study of non linear phenomena. Wavelet analysis has very good determination from my point of view as mathematical microscope for the study of nonlinear phenomena. Mathematically speaking, if we have a linear problem, the best variant of the approximation of the solution when we use some part of the Fourier expansion which corresponds to the eigenfunction spectrum of the linear problem, we have the type of differential equation and we have the opportunity to specify the least of the eigenfunctions according to boundary conditions for this equation. Practically the same problem we have for nonlinear problems but with the following important difference. According to group properties of nonlinear equations we have the opportunity to determine the spectrum of the eigenfunctions, but, the spectrum of the nonlinear problem distinguishes from the linear one in that its eigenfunction spectrum doesn't depend on the boundary conditions. And in fact, in my presentation, the spectrums of the eigenfunction in the blow up regions form a natural basis for the approximation of the nonlinear solution, and one notices the universality of wavelet basis because the least of the eigenfunctions for nonlinear problem in material science, for description of the transition from damage to fracture, from elastic deformation to plastic deformation. I think it's necessary to determine whether we have an opportunity to use some universal basis according to specific features of the eigenvalue problem for these phenomena in material science, due to the usage of universal basis like wavelet basis. It's a very powerful method but I think that for some phenomena it's necessary to be very careful in determining the approximation based on the wavelet analysis.

[M. Bily] I think we should be a bit careful as far as various terms are concerned. First of all the eigenfunction and eigenfrequency are terms applied to linear systems only.

[O. Naimark] I agree with you, but, this terminology was introduced first in Russia and is due to intensive investigations of very specific problems for parabolic equations and now some books are published in this area and probably it's some tradition in Russia for the group properties of linear equations.

[A. Haldar] I would like to know the effect of noise in the wavelet analysis. That part I think can create a lot of problems. For example, if I have energy at certain scale, and if there is noise at the same scale, I expect some energy present from it. How do you differentiate what is what, whether it is real (e.g. a defect in a material) or it is coming from some noise.

[G. Frantziskonis] Noise is a very general term. It could be part of the physical problem, it could be a result of something in the device you are using in the experiments, etc. You cannot denoise a system unless you understand where the noise is coming from. You cannot say, for example, I have a universal tool like wavelets and I have a system with unspecified noise and I use wavelets and thus resolve the noise problem. One has to understand where noise is coming from. As far as the energy, when studying a material for example, say you find out that scale 5 is an important one. The next step is to find out what kind of microstructure is present at that scale. For example, you might find out that the grain size scale is an important (or even a dominant) one. Then you study properties at that scale primarily and "ignore" information at other scales far from it, because they are not important according to this analysis.

[A. Malkin] I will try to pose the following question. Fourier transform is ill-posed, e.g. we must know the asymptotics with frequency going to infinity which we cannot do numerically. Is wavelet transform better from this point of view.

[G. Frantziskonis] Most of the problems that come with the Fourier transform are because the base of it, sine and cosines functions, never stop, I mean they go from minus infinity to infinity. The wavelets, and specifically some (orthogonal) wavelets, have a compact support and this allows them to look at things locally rather than globally.

[A. Malkin] Yes, though I think that we must, in order to form a convolution for wavelet analysis, we must know the window of frequencies to use in the integrant in the convolution. How we can find it. Before the experiment, before our investigation?

[G. Frantziskonis] You mean what scale to look at?

[A. Malkin] Yes.

[G. Frantziskonis] You don't know, it depends on the problem. It depends on what you're looking for. You cannot look at all scales from zero to infinity obviously. You have to understand the problem you're dealing with. A relevant problem is identifying a characteristic cell if it exists. If it exists, and since wavelets are compact in both real and frequency spaces and you are examining 5 or 6 orders of magnitude you should be able to identify the appropriate scales. The scales will probably be asymptotically identified by the wavelets.

[A. Hansen] The question of Nyquist frequency is still present in wavelets. However problems related to the periodicity, aliasing, etc., those are absent. When it comes to noise, noise should distinguish itself from the real signal by behaving in different way and being at a different scale. And we have tried to study this very carefully in connection with wavelets and we found it easy to recognize it and remove it. For example we produced many artificial signals, fractional Brownian motion ones, and added several types of noises on top of this and they have been very easy to pick out. So it seems to me that the wavelet analysis is very, very well suited for scaling analysis. Certainly it's the best kind of method we have ever found.

[A. Schanyavski] If we have damage, we can have fracture. If we have no damage we have no problem about describing the evolution of the system. If we analyze a physical phenomenon with an evolution body we can have many equations and introduce many points of view to describe evolution. But if we want to know how long we will have a stable system and what type of damage is present in this system we have to have scaling because, for instance, in low fatigue cycle loading damage maybe mesoscopic, yet damage may be microscopic during the first steps.

[R. Rackwitz] I would like to return to structural reliability and its relation to what you are doing here. I have observed two cases where there is a direct relation between say elemental strength and structural strength. If you know the distribution function or stochastic characteristics of the element then there are certain operations to give you the distribution function of a structural element or a structural unit which will go into its reliability analysis. It should be very clear that we cannot go into the micromechanical level in structural reliability in view of the computational capabilities we have. One I feel those two models or combinations thereof are the simple Weibull model and the Daniell system and that's all at the moment. This is a rather weak state of our science I would say. These are 2 nice cases but at the present I must state that there is not much more.

[D. Jeulin] I would like to add some information concerning your point. Of course you have the Weibull case which is one of weakest link but you have many distribution functions of weakest link type. This morning I have shown the Boolean random function. So we have many many distributions which are possible and what is important is the size effect that you obtain with this model. When you take the weakest link model you consider that you will not have fracture or you need to have zero defects in order to not break the specimen. But if you say I admit I can have one defect or two defects or three, up to n, n being given by some density, critical density, you will have all sorts of tools to calculate but the results will be very different from the Weibull model. For example the size effect will completely disappear. I cannot describe all the models but I think we have in our toolbox a full range of models one being the worst case, i.e. the Weibull or the weakest link model and intermediary situations also exist and can be applied.

[R Rackwitz] I think that the theoretical models may be there and you are maybe not the only one who has worked on the mathematical structure of these models but I have not seen in practice or practical applications even in the scientific community the transition from one to the other. One or two examples but it's still a bad state.

[S. Neuman] Just a very brief comment and that is that another aspect of scale effect is going to be discussed by me tomorrow morning. I'm not a material scientist so the focus is going to be very different but just to make clear that tomorrow I will again bring up some aspects of scaling which may or may not be relevant directly to the issues discussed here. But I think they may be, so I don't think this particular topic is going to be closed tonight.

ROUND TABLE III. :

STOCHASTC MATERIAL STRUCTURE AND STRUCTURAL RELIABILITY

P. CASTERA

During the closing session more general problems were discussed, especially concerning universal concepts, the use of multi scale approaches, and the prediction of the behavior of very complex systems through probabilistic approaches and small scale measurements.

[V. Moshev] I would like to touch one special point. There exists several scales of non uniformity that (in the simplest approximation) might be classified as macro- and micro non uniformity. In laboratory tests we obtain the probabilistic data that characterize macro non uniformity. These data, after mathematical treatment, are usually transformed into various quantities for practical use. The designers employ these data as they are. They cannot govern the non uniformity of materials. It is the duty of material science to work out methods of governing macro property distributions. Obviously it is the non uniformity at the structural level which gives rise to the observed macro stochastic behavior. In this connection, if we want to create an effective tool for controlling macro distribution of properties, first of all we must establish the interrelation between micro- and macro stochastic properties. However our knowledge of non uniformity at micro level seems to be poor. In my opinion the situation looks more or less satisfactory, only in the domain of local geometrical non uniformity estimations. However data on the chemical physical microscopic (structural) level (I mean polymeric composites before all) do not exist at all, as far as I know. Taking into account how complicate this problem is, I think more efforts should be applied in this direction. In future models the origination of damage and fatigue evolution ending in fracture should be regarded as one unified process. Models should be microscopically transparent through the whole life cycle of the material. Then the researcher will be able to make reasonable decisions concerning the criticality of the state and permissible safety levels.

The discussion arose then on the concept of universality of general laws with a comment of Prof. D. Breysse. He suggested to try to define the domain on which such laws apply. Due to the wide variety of materials we are treating, from soils and rocks to artificial and natural composites, concrete and steel, generality may be limited. If so then we have to treat two main questions: the first one is gathering data at the micro scale and the associated cost, and also the type of data we have to cover ; the second point is to build the models that make possible bridging across the scales. An remarkable example is provided in the paper of Prof. Vanmier about a model of numerical concrete, bridging across the scales with 6 or 7 parameters. One of these is the tensile or shear strength of the interface between the matrix and the aggregates. The author has no experimental possibility to identify this parameter, which may be very influent on the macro response of the material. Therefore he did a parametric study to show the influence of the parameter, and demonstrated that even with one model, one parameter, it is very difficult.

Prof. Neuman commented this problem that was raised by D. Breysse. He gave the example of permeability, which is a well defined parameter assuming Darcy's law holds. On the other hand when the phenomenology changes, as if when we go down to the level of the Navier-Stokes equation and try to upscale on to the Darcian level, then the same laws do not apply.

As pointed out by Prof. Frantziskonis this Probamat workshop showed that the scientific field is very wide, with many types of materials and many types of problems, fatigue or monotonic type of loading problems, and so on. On the other side one common thing of all these problems is that at one scale there must be some probability or stochasticity, whereas at other scales they can be deterministic. The purpose of this workshop, and also Probamat 1, has been to see if there is some common approach, through statistics and multi scale approaches, to treat this very diverse type of problems.

Upscaling means that properties are measured at a small scale, sometimes down to the atomic level, and this information is used to predict effective properties at a macroscale. Doing this we want to understand the behavior of very complex systems, materials or structural systems. But sometimes complex systems cannot be reconstructed from these observable data unless that information becomes so tremendous that we transform the system into a « Swiss cheese ».

According to Prof. Bily the reason why we cannot correlate micro scale phenomena to the macro scale behavior does not lie in the lack of knowledge of micro structural properties, which is huge. But it is difficult to find general correlation because the phenomena are multifactorial, and probability is in fact the measure of the multifactorial behavior of nature. Sometimes deterministic approaches may be better than probabilistic approaches.

At the end of this session Prof. Naimark gave an exhaustive presentation on these aspects in solids mechanics and non linear physics as well, such as turbulence. He pointed out the fact that in solids the picture of the evolution of the structure during and after loading exists and is frozen, whereas in turbulence situation it is more difficult. We need the physical background to deal with non linear systems, that is, thermodynamics of non equilibrium state.

It was pointed out by Prof. Frantziskonis that the Probamat workshop perhaps raised more questions than provided answers. There were several important presentations and very fruitful discussions. He concluded that this workshop provided the opportunity of exchanging knowledge on a unique field of research. The third Probamat meeting will be held in 2001.

INDEX

614